Using CADKEY

CADKEY® Keyboard Template

Text Editing/Direction Keys

Beginning of line	Home
Up one line	Up
Down one line	Down
Move one char. to right	Right
Move one char. to left	Left
End of current line	End
Deletes char. to left of cursor	BKSP
Deletes char. at current cursor position	Del
Toggles between insert and overwrite mode	Ins
(Insert mode) moves text from cursor location to new line	Enter
(Overwrite mode) moves cursor location to end of next line	Enter

Immediate Mode Commands
Remember to hold down the CTRL or ALT key when pressing the assigned letter key.

ARROWS	CTRL - A	DIGITIZE on/off	CTRL - Y	PEN #	ALT - Z
AUTO scale	ALT - A	DOUBLE scale	ALT - D	RECALL last	CTRL - U
BACK - 1	ALT - B	EXECUTE macro	ALT - E	RECORD macro on/off	CTRL - J
CALCULATOR	CTRL - I	GRID display	CTRL - G	REDRAW	CTRL - R
COLOR	ALT - X	HALF scale	ALT - H	SAVE file part	CTRL - F
CONST (2D/3D)	CTRL - W	LEVEL	CTRL - L	SCALE	ALT - S
CONST PLANE	ALT - K	LINE LIMITS	ALT - L	SLIDE file	ALT - F
CURSOR SNAP	CTRL - X	LINE TYPE	ALT - T	USER macro prompt	CTRL - O
CURSOR TRACK	CTRL - T	LINE WIDTH	ALT - Y	VIEW	ALT - V
DB SEARCH	ALT - R	MASKING level	ALT - N	VIEW/WORLD	CTRL - V
DELETE single	CTRL - Q	MASKING type	ALT - M	WINDOW	ALT - W
DEPTH	CTRL - D	PAN	ALT - P	WITNESS LINE	CTRL - B
		PAUSE macro on/off	CTRL - K		

CADKEY INC., 440 Oakland St., Manchester, CT 06040 (203) 647-0220/FAX: (203) 646-7120/TELEX: 510 600 7223

System Defined Views

⊙ Towards Viewer   ⊗ Away From Viewer

View 1: TOP
View 2: FRONT
View 3: BACK
View 4: BOTTOM
View 5: RIGHT
View 6: LEFT
View 7: ISOMETRIC
View 8: AXONOMETRIC

0030-0001

CADKEY®

VIEW DISPLAY CHART

⊙ TOWARD VIEWER
⊗ AWAY FROM VIEWER

View 1: TOP
View 2: FRONT
View 3: BACK
View 4: BOTTOM
View 5: RIGHT
View 6: LEFT
View 7: ISOMETRIC
View 8: AXONOMETRIC

0031 0014

Using CADKEY©

Version 5 and 6

Paul J. Resetarits, Ph.D.
Central Connecticut State University

Gary R. Bertoline, Ph.D.
Purdue University

Delmar Publishers Inc.®

I(T)P™

NOTICE TO THE READER

Cover photos: Courtesy of CADKEY, INC.

Delmar Staff

    Publisher: Michael McDermott
    Senior Project Editor: Laura Miller
    Production Supervisor: Larry Main
    Project Development Editor: Mary Beth Ray

For information, address Delmar Publishers Inc.
3 Columbia Circle, Box 15-015
Albany, New York 12212-5015

Printed in the United States of America.
Published simultaneously in Canada
by Nelson Canada
A division of the Thomson Corporation.

1 2 3 4 5 6 7 8 9 10 XXX 99 98 97 96 95 94 93

Library of Congress Cataloging-in-Publication Data

Resetarits, Paul J.
    Using CADKEY/Paul J. Resetarits, Gary R. Bertoline.—4th ed.
    p. cm.
    Includes index. ISBN 0-8273-5607-2
    1. CADKey 2. Computer-aided design. I. Bertoline, Gary R.
II. Title.
T385.R47 1993
620'.0042'02855369—dc20
                                 92-21484
                                     CIP

CONTENTS

Chapter 5 CADKEY 2D Sectional View Tutorial 74

Chapter 6 3D Dovetail Tutorial 102

Chapter 7 Deleting Entities 128

Chapter 8 The Creation of Entities 142

Chapter 9   The Immediate Mode   206

Chapter 10   The Position Menu   215

Chapter 11   Selecting Entities   226

Chapter 12   The Masking Menu   236

Chapter 13   File Types   244

Chapter 14   Display Options   275

Chapter 15   Editing Your Design   318

Chapter 16   Detailing Your Design   347

Chapter 17   X-FORM (Transform) Options   426

Chapter 18   3D Construction Techniques   450

Chapter 19   The CONTROL Options   461

Chapter 20   Plotting and Printing Your Design   491

This textbook is written as an aid to those who are learning how to use CADKEY software. CADKEY is a fully integrated 2D/3D drafting and design system that uses a menu structure to create part geometry. Every command from version 5 and 6 of CADKEY has been described and supplemented with figures and examples of their use. It can be used as a classroom textbook or as a supplement to *CADKEY's User Manual*. However, it is much more than a technical reference because each command is explained, and most commands are supplemented with "Possible Uses" and figures to further explain their use.

The chapters have been arranged so that those commands necessary to understand the use of CADKEY are presented first. The 2D and 3D tutorials are included to assist the beginner. To further assist the learner, a work disk called *The CADKEY Electronic Drafting Workbook* is available separately through Delmar Publishers Inc. (See order form in the back of the book.) The work disk offers additional drafting problems that can be used to supplement the text to learn both CADKEY and drafting principles and practices.

The fourth edition of Using CADKEY is a revision that updates the entire book to the current 6.0 version of CADKEY. The new features of CADKEY 5 followed by new features for CADKEY 6 are summarized here with specific reference to the chapters where they can be found.

CADKEY 5

■ **CHAPTER 1**
Introduction to CAD, CAM, and CIM

The terms of CAD, CAM, and CIM have been updated to reflect current trends in this field, including CADKEY's new CAM software called *The Cutting Edge* and their boundary elements analysis (BEM) software.

■ **CHAPTER 2**
The Components of a CAD System

The hardware used to run CADKEY has changed. This chapter has added coverage of Workstation technology for CADKEY's UNIX version, as well as the requirements for using the 386 version of CADKEY. The hardware lists have been updated to reflect current trends.

■ **CHAPTER 3**
The CADKEY Human Interface

The basic CADKEY human interface has not changed but some of the options have. The Status Window now includes options for QuickTrim, Undo, Layout Mode, and View Descriptors.

■ **CHAPTERS 4, 5, and 6**
The CADKEY Tutorials

The CADKEY tutorials have been revised to reflect the version 5.0 changes. The interaction hints have been moved to the front of Chapter 4 so that they may be reviewed prior to starting the tutorials.

■ **CHAPTER 9**
The Immediate Mode

New immediate mode commands include Alt-G Layout and Alt-I Undo.

■ **CHAPTER 13**
File Types

A new file type CDE (CADKEY Dynamic Extension), which allows the user to run third part programs from within CADKEY, is discussed.

■ **CHAPTER 14**
Display Option

New commands which are used in the drawing layout mode include VIEW-RENAME and View Descriptions.

■ **CHAPTER 15**
Editing Your Design

The Quick Trim option is discussed. The VERTRM option has been deleted because there is now an undo option with the Trim command.

■ **CHAPTER 16**
Detailing Your Design

Layout and all of its options have been added. Various detail options have been added or changed: DIMENSN-RADIUS options REGULAR, EDGE, and BENT; DIMENSN-ANGULAR options 2 LINES and 3 PTS; DIMENSN-DIAMETR options REGULAR and EDGE; DIMENSN CHAMFER; and X-HATCH.

■ **CHAPTER 17**
The X-FORM (Transform) Options

The UNDO command has been added to the X-FORM options, allowing the user to return the geometry to its original position.

■ **CHAPTER 20**
Plotting and Printing Your Design

The PLOT-OPTIMIZE option allows users to remove the back lines and plot only one copy of a line if there is more than one in the same location.

■ **CHAPTER 21**
Shading CADKEY Models

The CADKEY Solids 4.5 interface is discussed.

CADKEY 6

■ **CHAPTER 3**

The addition of dialog boxes to many commands.

■ **CHAPTER 11**
Selecting Entities

Selection Set, a new command, which allows users to use multiple selection methods in one command.

■ **CHAPTER 12**
The Masking Menu

The masking menu has been changed into a masking dialog box. All of the options can be selected in one screen rather than multiple menus.

■ **CHAPTER 13**
File Types

Dialog boxes have been added to all of the listing options. These dialog boxes allow file management such as sorting, deleting, copying, and moving to be in CADKEY. Two new file types have been added DWG and DXF. This allows drawings in these Autodesk formats to be read or written from within CADKEY.

■ **CHAPTER 14**
Display Options

Levels name list dialog box with the count option allows the user to list the number and type of entities on any level.

■ **CHAPTER 16**
Detailing Your Design

All detail drafting setting in the SET option are in a dialog box. Once you have your preferences set they can be saved as a Template file and used in future drawings. The X-HATCH command has been revised for easier boundary recognition and editing of existing hatched areas.

■ **CHAPTER 19**
The CONTROL Options

A dialog box has been added to the ATTRIBUTES option making it possible to change multiple attributes at one time.

■ **CHAPTER 20**
Plotting and Printing Your Design

All of the plotting and printing options have been placed in a dialog box, which allows you to setup your output preferences. You can then drag a box representative of the paper size over the design to locate the position of the output. It will automatically be centered if you press enter.

■ **CHAPTER 23**
Picture It

This easy-to-use rendering system replaces CADKEY Solids. It helps in the visualization of 3D wireframe models by removing hidden lines, changing the hidden lines to dashed lines or shading the model. It is also capable of producing suitable output for rapid prototyping systems.

We are grateful to the people who have assisted us in writing this text. We would like to thank Peter H. Smith from CADKEY Incorporated for his advice, encouragement, and help in planning and writing this textbook; CADKEY Incorporated for the use of their 3D Dovetail Tutorial; all those people at Delmar, especially Mike McDermott and Larry Main; John Larvick of DoDDS-Pacific Region; and a special thanks to Paul Mailhot, Steve Falusi, and the staff in the education department at CADKEY.

We would also like to thank our families for their assistance, and patience and understanding for the many hours that this project took us from them.

We would appreciate your input and feedback on *Using CADKEY*. If you have a possible use for a particular function that you feel would be of value to other CADKEY users, please send it to us.

Paul J. Resetarits
Central Connecticut State University
School of Technology
1615 Stanley Street
New Britain, CT 06050

Gary R. Bertoline
Purdue University
Technical Graphics Department
363 Knoy Hall
West Lafayette, IN 47907

DEDICATIONS

Dedicated to my wife Anni for her love and
support throughout this endeavor.
  Paul

To Gertrude Sullivan,
A most courageous woman.
Most people dream of having faith,
Some people think they have faith,
You have dedicated your life to your faith.
  Gary

CHAPTER 1

Introduction to CAD, CAM, and CIM

OBJECTIVES

After completing this chapter, you will be able to:

- Define CAD, CAM, CAD/CAM, CIM, and IGES.
- List three types of 3D models produced with CAD.
- List several applications of CAD.
- Describe the factory of the future.

The use of CAD (Computer-Aided Design or Drafting) has rapidly spread into the world of design/drafting. In a very short time, CAD has grown from only a handful of workstations to literally scores of different workstations, software, peripheral devices, and applications. CAD has been integrated into nearly all types of industries at some level. The recent explosion of CAD workstations into American industry can be explained by the proliferation of micro-based CAD systems and software. CADKEY is an example of a powerful, true 3D micro-based CAD software.

CAD DEFINED

CAD is an acronym for Computer-Aided Drafting or Computer-Aided Design, which can be defined as the use of computer hardware, software, and peripheral devices to produce graphic images. This generic term encompasses many different types of computer-generated images including engineering drawings, architectural drawings and renderings, electronics, mechanical, civil, business graphics, technical illustrations, and others. There are many applications of CAD; however, a CAD software package is normally limited to effectively producing only two or three different applications.

The advantages of using CAD over traditional methods are numerous. Many manual tasks are tedious and time consuming, such as lettering text, cross-hatching, and drawing symmetrical parts. CAD relieves some of the tedium associated with such tasks because it automates much of the work. For example, lettering becomes as easy as typing the words on the keyboard and selecting the type of font, height, and aspect ratio. Whole areas are filled very easily with the cross-hatching command. Parts are mirrored,

CADKEY 3 Helps Start-Up Car Manufacturer Get Into Production!

When the serial #001 VECTOR super-exotic sports car rolled out for delivery to its first customer, some of its high-tech features startled even aficionados of futuristic automotive design. Acceleration from 0 to 60 m.p.h. in four seconds. A 6-liter, twin turbocharged, all aluminum, 600hp V-8 engine with the highest horsepower and torque output of any automobile engine now in production, and a clutchless hydraulic transmission. Instead of the traditional array of dials and gauges, an advanced, electroluminescent, computer-programmed, multi-function, military-jet-fighter-type panel was designed that changes function automatically. This panel also incorporates actual aircraft circuit breakers and a heads-up windshield display. Among the first confirmed buyers were three Saudi Arabian princes. The base price of the car is $180,000 plus options, making it the most expensive automobile ever built for production in America.

The VECTOR is the first production model from Vector Aeromotive Corporation of Wilmington, California. As a start-up firm, Vector initially lacked the funds for a full-fledged, computer-assisted design system. Thus, although the VECTOR is technically advanced, the engineering staff drew up early plans with the most venerable of tools—pencil and paper. Only when the prototype began to draw some excited na-

tional attention and the car itself neared production was Vector finally able to computerize its design process.

This raised a formidable problem. How could the company's engineers learn a CAD system, enter hand-drawn plans into it, and master it thoroughly, yet in time to meet looming production deadlines? Since Vector Aeromotive was designing parts with some extremely exotic and complex geometric shapes, it needed a system capable of real—not simulated—3-D design. "Draftsmen think in two dimensions," said Gerald A. Wiegert, Vector's founder and president. "Real designers think in three dimensions." After careful consideration, Vector Aeromotive Corporation settled on software from Cadkey, Inc. CADKEY 3™ let the VECTOR's designers ease into CAD at their own pace: the learning curve was short without being steep. Designers found the transition from pen and paper into a computer environment surprisingly smooth.

This proved particularly useful as the car moved closer and closer to manufacture. For example, VECTOR's headers, differential housing, and its unique, jet V-shaped dashboard which resembles the cockpit of a fighter aircraft are extremely complicated, involving some highly sophisticated geometric shapes. All were edited and reworked repeatedly using CADKEY 3, even though Vector had been using the software for only ten weeks.

Despite their taste for cutting-edge technology in their cars, automotive engineers tend to be conservative when it comes to the process of design. They like the feel of pen and paper, and many have been reluctant to let computers into the process, as if afraid that mechanization would deprive it of its mystique. Thus, the plunge into CAD could prove traumatic. But, more advanced and more intuitive software is making it possible for designers to ease into computer-aided design at their own speed, and with a minimum of stress.

1990 VECTOR—the most powerful, production, exotic sports car ever built in America

Adapted with permission from 3-D World, v. 4, no. 1, Jan./Feb., 1990.

copied, scaled, moved, deleted, changed, and manipulated in a number of ways, as described in the following chapters. CAD is an extremely powerful tool that can enhance the design/drafting process significantly. However, it is important that the user of CAD learn the full capabilities of the hardware and software to receive the full benefits of CAD.

TYPICAL FEATURES OF A MICRO-BASED CAD SYSTEM

Most systems use a menu structure for selecting different commands. Basic entities that can be created include: lines, arcs, circles, splines, points, polygons, rectangles, fillets, and chamfers; detailed automatic dimensioning of lines, arcs, circles, and angles; display manipulation such as windows, zoom, grid, and pan; options for scaling, layering, copying, moving, rotating, extruding, and mirroring; geometric analysis of entities such as area, perimeter, moment of inertia, and centroid. All of these features and more are possible with CADKEY. One of the most important differences between CADKEY and other micro-based CAD software programs is that it can produce a true 3D model, which is very important for CAD/CAM operations and design visualization.

CAM DEFINED

CAD is not limited to simply creating graphics. Creating graphics is only one part of the total design and manufacturing of a product. It is possible to merge the two major phases of a manufactured product: design and manufacturing. The automation of product design is accomplished using CAD. The automation of the manufacturing process is accomplished with *CAM* (Computer-Aided Manufacturing).

CAM is the automation of the manufacturing process using computer-controlled machines and robots for fabrication, assembly, material handling, measuring, and inspecting. Some of these operations are created by using the geometric data base created with the CAD software. For example, the NC (Numerical Control) program for a milling machine can be produced from the geometric data base from the CAD system if the part geometry data is output in a format that can be interpreted by the NC machine. This merging of CAD and CAM is the key to the factory of the future which will automate the entire design and manufacturing process.

CAD/CAM (Computer-Aided Design/Computer-Aided Manufacturing)

The merging of the design and manufacturing process is called *CAD/CAM*. It is an umbrella process that combines the design and manufacture of a product into one integrated approach through the use of computers and the sharing of data. Alone, micro-based CAD systems are not capable of generating the data necessary for CAD/CAM. Most systems provide only some of the data but not all. However, a few micro-based CAD systems provide the link to other computers that can produce the data necessary for CAD/CAM. CADKEY can very effectively produce a 3D wireframe model of a part and it has the capabilities to share this data with other computers and software to produce the necessary data for CAD/CAM.

3D MODELING

Three-dimensional modeling of a part is very important for CAD/CAM. Basically, there are three different models that can be produced with CAD: a wireframe model, a surface model, and a solid model. CADKEY can create true 3D wireframe models of parts which can be translated and sent to minicomputer-based CAD/CAM systems. These systems can then produce a solid model from CADKEY's wireframe model. The model can then be analyzed before manufacturing. The part can be rotated, exploded, and viewed from any angle to check for interference and integrity of the design. A finite element model of the part is produced so that the part can be further analyzed under various loads, temperatures, and so forth. Tool pathing for NC machining can then be produced and the tool path is dynamically displayed on-screen for verification and editing. The part is now ready for manufacturing and the 3D model produced with CADKEY can be the initial step and the foundation for CAD/CAM.

CADKEY SURFACES

CADKEY Surfaces is a surface modeling program that works with CADKEY and permits the user to define any surface, including ruled, conical, surface of revolution, swept, and complex spline surfaces with three or four boundaries. Intersections between any surfaces can be defined and a constant or evolving radius fillet surface may be created between them (Figure 1-1).

The use of surfaces, intersections, fillets, and offsets is important in almost all product design applications. Many aircraft and automotive parts consist of complex spline surfaces. The designs created in CADKEY Surfaces can be directly transferred to any of the spline surface numerical control packages that are compatible with CADKEY.

CADKEY SOLIDS AND PICTUREIT

With the addition of CADKEY Solids and PictureIt, solid models are extracted from wireframe geometry. The solid model is used for object rendering, visualization, animation, and mass property calculations. CADKEY Solids and PictureIt produces filled and shaded models with hidden surfaces removed, wire frame models with hidden lines removed, or wireframe models with hidden lines replaced with dashes. The modeler also is capable of mass properties calculations, Boolean operations, and smooth shading (Figure 1-2).

REVERSE ENGINEERING

Reverse engineering uses intelligent input from a coordinate measuring machine interfaced with a CAD system to allow direct digitizing of 3D models and parts. Three-dimensional CAD

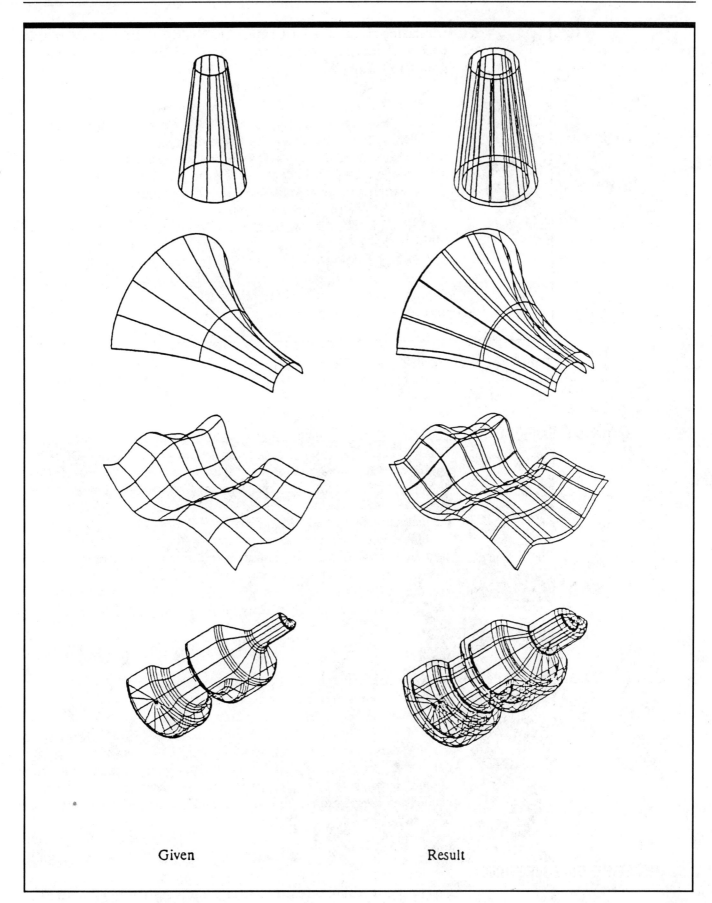

Given   Result

Figure 1-1 Offset surfaces (Courtesy of Cadkey, Inc.)

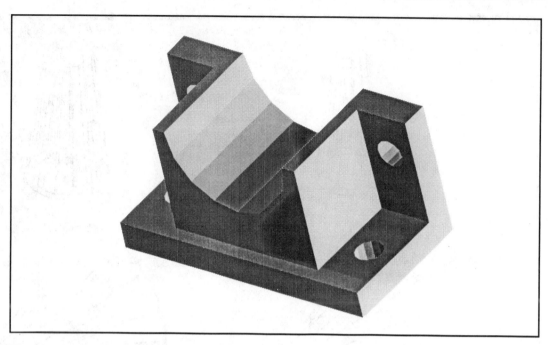

Figure 1-2 Smooth-shaded solid model

Figure 1-3 Cadkey's three-dimensional software, CADDINSPECTOR, is linked to a coordinate measuring machine to allow direct digitizing of 3D models. (Courtesy of Cadkey, Inc.)

software, such as CADKEY's CADDINSPECTOR, is linked to a co-ordinate measuring machine to allow direct digitizing of parts (Figure 1-3). This technology allows drawings to be created from a part and to compare manufactured parts with its CADKEY file.

CUTTING EDGE

Cutting Edge is a three-axis CAD/CAM software system that is integrated with CADKEY. It has all the capabilities of CADKEY combined with its own CAM machining features. The common user interface allows engineering, design, drafting, and manufacturing departments to communicate better with each other. Cutting Edge decreases CNC programming time and increases productive machine time. With Cutting Edge it is possible to edit a tool path graphically while creating it (Figure 1-4).

RAPID PROTOTYPING

Stereo lithography is a technology that allows designers to produce 3D plastic models without tooling. This process uses a laser beam to solidify layers of photosensitive liquid polymer. As each layer of the liquid hardens, it is lowered and another layer is exposed. This process is repeated until the 3D model is completed after which it must be cured (Figures 1-5 and 1-6).

CIM (Computer Integrated Manufacturing)

CIM is the total automation and computerization of the manufacturing process from receipt of the order to shipment of the

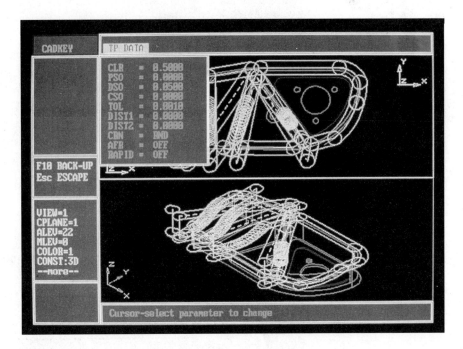

Figure 1-4   Three axis motion of a tool path (Photo courtesy of Cadkey, Inc.)

Figure 1-5 A stereo lithographic machine is used to produce 3D models without tooling.

product. For CIM to work effectively, a common data base pool must be available to all phases of the manufacturing operation. CIM is more theory than practice because it is difficult to get all the data produced with the design and manufacture of a product to be in a format that can be shared through out the manufacturing operation. Currently there are efforts to standardize the output of CAD software. One of these standards is called *IGES* (Initial Graphics Exchange Specification). This standard will allow the bidirectional exchange of data between different CAD systems. As this standard improves and as more software companies adopt IGES, the possibility of CIM becoming a reality will increase.

CADKEY: CAD, CAM, AND CIM

CADKEY is not limited to 2D and 3D wireframe modeling and production drafting. CADKEY can integrate its geometric data base with external programs used for numerical control, finite element analysis, bill of materials, and solids modeling. It is also possible to transfer 2D files from other systems with the CADKEY/DXF interface.

CADKEY offers an IGES translator module that provides 3D bidirectional data exchange between CADKEY and minicomputer-based CAD/CAM systems, such as Computervision, McAuto, CALMA, and GERBER. Figure 1-7 is a model of the IGES communications link used by CADKEY. It is also possible for CADKEY to link to third-party programs on microcomputers. These links make it possible for numerical control, finite element analysis, solid modeling, statistical control, and networking to be done on microcomputers. CADKEY could then be used as the central data base for manufacturing and design (Figure 1-8). CADL (CADKEY Advanced Design Language) is used to provide ASCII (American Standard Code for Information Interchange) input and output to CADKEY. Text and part geometry produced with CADKEY is used for other engineering applications: NC, bill of materials, finite

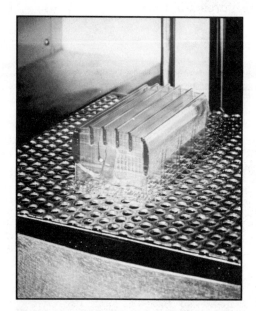

Figure 1-6 A 3D model produced by a stereo lithographic machine

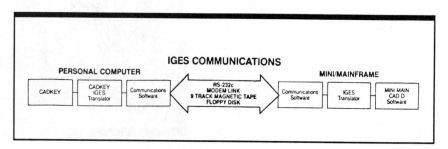

Figure 1-7 IGES communications model used by CADKEY (Courtesy of Cadkey, Inc.)

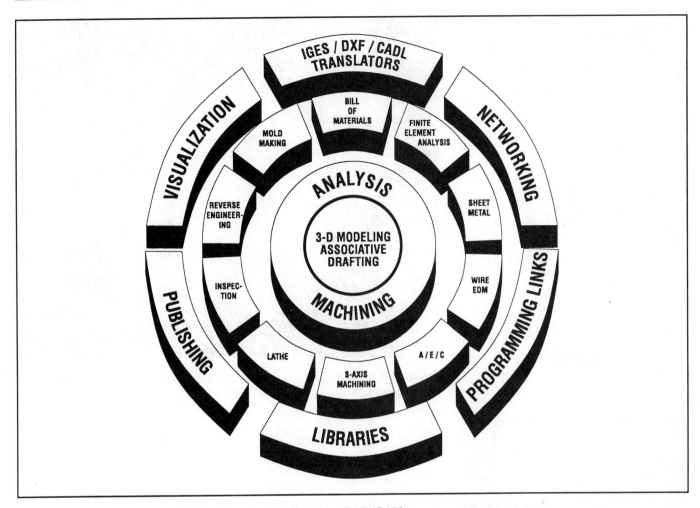

Figure 1-8   Model of CADKEY's shared data base for CAD/CAM (Courtesy of Cadkey, Inc.)

Figure 1-9   Numerical control output can be used for NC machining of the actual part. (Courtesy of Cincinnati Milacron)

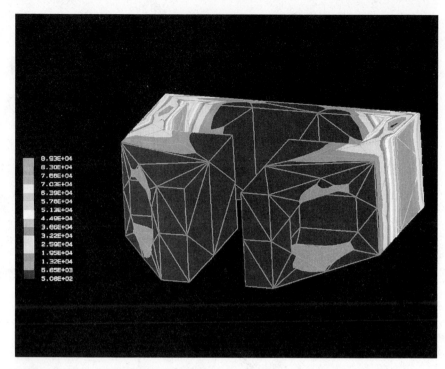

Figure 1-10 CADKEY analysis is used for finite element modeling and analysis. (Photo courtesy of Cadkey, Inc.)

element analysis, boundary element analysis, simulation, desktop publishing, and others (Figures 1-9 and 1-10).

SUMMARY   The goal of design and manufacturing is the factory of the future. This factory will integrate all phases of design and manufacturing through computers and the sharing of a common pool of data. The key to this CAD/CAM or CIM system is effective communications between computers of the geometric data base produced with a CAD system. CAD is the catalyst and the focal point for the generation of graphics used for engineering, design, manufacturing, documentation, technical illustration, biomedical engineering, desktop publishing, and any work requiring the display and documentation of graphics.

REVIEW QUESTIONS

1. Define CAD.
2. Define CAM.
3. Define CAD/CAM.
4. Define CIM.
5. List the three types of 3D models that can be produced with CAD and the type CADKEY can produce.
6. List some of the applications for CAD.
7. What is IGES?

The Components of a CAD System

OBJECTIVES

After completing this chapter, you will be able to:

- List the three types of CPUs.
- List the basic components of a CAD system.
- Name input devices used for CAD.
- Define the function of output devices.
- Explain graphics display cards.

The basic components of a CAD system are the same as any computer system. A CAD system is nothing more than a specialized computer system. It is a computer that has been developed and customized to meet the needs of computer-aided drafting and design by adding several specialized peripheral devices.

A CAD system, like other computer systems, is comprised of hardware and software. Software is a specific application program that has been written to run on a computer. It contains sets of instructions that tell the computer how to accomplish various functions or tasks. CADKEY is an example of a software package. Hardware is the combination of equipment such as input devices, central processing units (CPU), storage devices, display devices, and output devices. While many of these devices are standard for all computers, others have been developed specifically for use in computer graphics. A basic model of a CAD system is shown in Figure 2-1.

CADKEY, like most CAD systems, utilizes a digital computer. Digital computers operate using data that is in binary form or code. When information is input in the form of numbers, letters, or

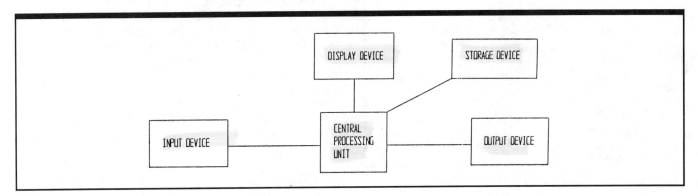

Figure 2-1 The components of a CAD system

11

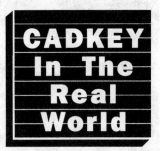

Vertical Machining Centers Using CADKEY!

All of the vertical machining centers at Fadal Engineering Company, Inc. of North Hollywood, California are designed and maintained using CADKEY. In 1989, Fadal Engineering Company bought its first CADKEY system. Now the company has seven CADKEY systems, all on stand-alone, 80386, IBM-compatible personal computers. "Settling on CADKEY was not easy for us," said Dean de Caussin, a Mechanical Engineer at Fadal. "We had tried other PC-based CAD software, and we had not been satisfied. After so many years in business, we had an enormous number of drawings to update. The system's dimensioning capabilities and ease of use were two critical factors. We bought CADKEY, and we're glad we did. It makes a wonderful replacement for drafting by hand. CADKEY and our draftsmen seem to think the same way. There was a very short learning curve."

Of its seven CADKEY systems, Fadal has dedicated one specifically for 3-D design—to see how a finished part will look and fit into a machine. "CADKEY gives you confidence that what you design will work," Dean said, "you can assemble the machine on the screen." Another CADKEY system is devoted totally to dimensioning part files. A third system serves for creating prototype part files and hardcopy prints for use in Fadal's machine shop. Fadal uses the fourth system for remanufacturing replacement parts for existing machines, to upgrade very old equipment into CNC machines. "We use these refurbished machines only for

our own internal use," Dean said. A fifth system serves the Engineering Department for developing new designs and special projects. The last two are systems that David de Caussin, Fadal's Chief Engineer, uses himself, one at work and one at home.

Safety Symbols Included in the Design

Even the safety symbols that appear on Fadal vertical machining centers are incorporated into the original part files using a combination of CorelDRAW!™ and CADKEY. The safety symbols are scanned into digital files in PCX format. An operator loads the PCX files into CorelDRAW!, traces over the symbols free hand, and creates output files in DXF format. Using CADKEY's DXF translator, the symbol files are converted form DXF format into CADL™ (CADKEY Advanced Design Language) format. The operator then reads the CADL file for the appropriate safety symbol into the original part file so that it will automatically appear in the hardcopy plot or blueprint of the part.

Fadal Engineering deliberately chooses to use American-made parts, as much as possible, in building its vertical machining centers. "We want to be loyal to our fellow manufacturers," Dean said, "and we want to help in keeping a base of machine-tool technology in the United States." According to articles published in **California Business** (February 1990), in **Business Week** (October 22, 1990), and in **American Machinist** (August 1991) magazines, Fadal Engineering Company appears to be succeeding in its effort.

Customer Enthusiasm

Fadal's customers also appear to appreciate what Fadal is doing. Fadal users speak passionately about the equipment that the company produces. (Editor's Note: "Passionately" is the most appropriate word to use here.) For example, Dean Alt of Alt's Tool and Machine, Inc., Santee, California, and Philip Durand of Connecticut Tool Company, Inc., Putnam, Connecticut, are both CADKEY users and Fadal users. Both extol Fadal's vertical machining centers.

"I have owned Fadals for six years," Dean Alt said, "they are very good machine tools. Dollar for dollar, they are the best you can get. I have done things with my Fadal machines that I could not do on other machines. They fit the job-shop market very well."

Exploded assembly drawing of rotary table.

Adapted with permission from 3-D WORLD, v. 6, no. 3, 1992.

special characters, it has to be converted to binary form or code for the computer to be able to understand it. The code is comprised of two digits, 1 and 0. The 1 is "on" and the 0 is "off." Communicating using binary codes is referred to as "machine language." Each 0 or 1 is considered a *bit*, short for binary digit. A sequential group of eight bits make up a byte by today's standards. Each number, letter, and special character input from a keyboard is comprised of eight bits or one byte. To store each character requires one byte of memory. The size of the byte varies with the type of computer. The sizes that are currently in use are 8, 16, 32, and 64 bits. Depending on the type of computer, it will have the capability to process 8 to 64 bits at a time. There are three categories of computers or central processing units. They are the mainframe computer, the workstation, and the microcomputer. The classifications of computers are becoming increasingly difficult to distinguish. The characteristics and capabilities of the different classifications continue to cross as the technology advances.

CENTRAL PROCESSING UNITS

MAINFRAME COMPUTER   The mainframe computer is the largest and most powerful category of central processing unit (CPU) or computer. Prior to the development of the microcomputer to its current level of sophistication, CAD software required the capabilities of a mainframe computer to run effectively. A mainframe system is generally very large in physical size and has multiple terminals or workstations connected to its CPU. Generally, a 64-bit or higher processor can be found in a mainframe computer. This type of system is used by large corporations, governmental agencies, and some educational institutions. Because of the high cost of a mainframe computer, its use is limited to those organizations with large budgets and many CAD operators.

WORKSTATION   A workstation is a class of computers that has been distinguished by its speed, multitasking, and networking abilities. The speed at which a workstation runs is 10 to 35 MIPS (Millions of Instructions Per Second) or faster, as compared to a microcomputer, which runs at 5 to 15 MIPS. The response time of a workstation for redrawing the screen and other options is noticeably faster. The operating system most commonly found on a workstation is UNIX. UNIX is a unified operating system with the capability of multitasking (the ability to run more than one program at a time through various windows). Another feature of workstations is that they are designed to run in a networked environment with multiple users. The **CADKEY UNIX** version runs on workstations by the following manufacturers:

> **SUN**—SPARC Station
> **SILICON GRAPHICS**—Personal Iris
> **SONY**—News
> **DIGITAL ELECTRONIC CORPORATION**—DEC Station

MICROCOMPUTER

The development of the microcomputer has radically changed the face of computing. The advancements in the amount of random access memory (RAM) available and the development of new classes of microprocessors have given microcomputers more power than was ever thought possible. The early microcomputers had 4 bit processors and 8KB (kilobyte equal to approximately one thousand bytes, specifically equal to 1,024 bytes) or 16KB of RAM. The current microcomputers have 8 bit, 16 bit, 32 bit, or 64 bit processors. Each time the number of bits processed doubles, the processing time is significantly reduced. The addition of an 8087, 80287, or 80387 math coprocessor chip to work with the 8088, 80286 or 80386 microprocessor will reduce the processing time by one-third. In September of 1986, the first microcomputer using the 80386 32-bit processing chip was announced by Compaq. The 32-bit machine ran at a clock speed of 16MHz as compared to the 8MHz of the 80286-based 16-bit machines. The development of the 80386-based PCs will further help the abilities of microcomputer-based CAD. Since that time, the 80486, 64-bit processing chip with a built-in math coprocessor has been introduced.

The RAM available in micros today range in size from 2MB to 16MB. (A megabyte equals approximately one million bytes, specifically 1,048,576 bytes.) CADKEY versions 5.0 and higher require 4MB. The differences between the classes of computers continue to become less with each new development. The 3D CAD work that can be done on a microcomputer today previously could only

Figure 2-2 A typical CADKEY workstation. New technical directions for CADKEY include a version which supports Sun Microsystem's OPEN LOOK interface. (Courtesy of Cadkey, Inc.)

be done on a mainframe CAD system. The microcomputer and microcomputer-based CAD have had and will continue to have a tremendous effect on the CAD marketplace. The cost of a microcomputer equipped with CADKEY software and the needed peripheral hardware is a small investment in comparison to a mainframe CAD system (Figure 2-2).

A COMPARISON OF IBM-COMPATIBLE MICROCOMPUTERS

Old Standard	New Standard	Microprocessor	Coprocessor	MHz
XT Model	25	8088 or 8086	8087	4.7-10
AT Model	30, 50, 60	80286	80287	8-20
	Model 70, 80	80386	80387	16-33
	75, 90	80486	BUILT-IN	25-66

INPUT DEVICES

An input device is the part of the computer system that is used to enter information, data, or commands to be processed by the CPU. The most commonly used input device with a computer is the keyboard. Commands and text can be entered and the cursor controlled through the use of the keyboard. There are several other input devices that have been developed to work in conjunction with the keyboard to make the input of graphic information easier. CADKEY can be operated using just the keyboard or a combination of the keyboard and a graphic input device such as a digitizer or mouse.

With CADKEY, the input device gives the user control of the screen cursor. The screen cursor is seen as a plus sign (+) on the screen. The size of the cursor is user definable in the **CONFIG** (configuration) program. The default size is 1 inch by 1 inch +, also called a crosshair cursor. It gives the user a visual reference on the screen of the current point or menu item that may be selected using the input device. The screen cursor can be moved through the use of the arrow keys found on the right of the keyboard or through the use of a graphic input device. The input device is used primarily for the selection of menu items or entities and for positioning points and entities. The data that is input comes in the form of X, Y, and Z Cartesian coordinates. The two major types of graphic input devices that are supported by CADKEY are the digitizing tablet and the mouse, often referred to as a pointing device.

DIGITIZING TABLET

The digitizing tablet is often referred to as a digitizer or tablet. In physical appearance, it is a square pad with a cursor control device attached to it. The tablet electronically sends to the computer the X and Y locations of a point selected from its surface. The cursor control

Figure 2-3　Digitizing tablets (Courtesty of Houston Instruments)

device that is moved on the tablet's surface can be a cursor or a stylus (Figure 2-3). A cursor, also known as a puck, has a clear area with crosshairs on it. The crosshairs are used for accurately lining up the point on the tablet surface to be selected or digitized. The cursor will have one to sixteen mouse buttons. The mouse button is pressed to make a selection or digitize a point. A stylus is a penlike object that is moved on the surface of the tablet to track the cursor on the screen in the same way that a cursor/puck does. To make a selection, you press down on the stylus or press a button located on the shaft of the stylus.

A digitizing tablet is an absolute input device. Every time that a specific point on the surface of the tablet is selected, that point will always have the same X and Y coordinates. With this accuracy a map or drawing can be taped to the surface of the tablet and traced or "digitized" accurately. A tablet menu can be placed over the tablet and used to select menu items. This method may be faster than paging through screen menu items (Figure 2-4).

Tablets can be as large as 42″ × 60″, but for most work the 11″ × 11″ size is adequate. Figure 2-5 lists some of the digitizing tablets supported by CADKEY. This list is always changing as drivers for new and different tablets and mice are added.

THE MOUSE

The mouse is also a common analog input device. A mouse is a small hand-held input device that looks like the cursor/puck used with the tablet. It generally has two or three pick buttons on the top that are used to select points or menu items. The mouse controls the screen

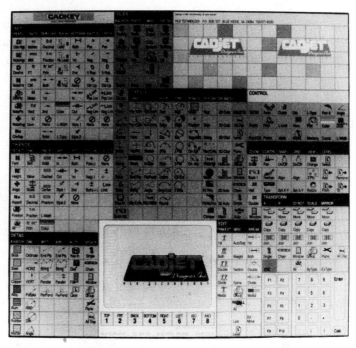

Figure 2-4   Tablet menu for CADKEY (Courtesy of HLB Technology)

cursor in the same way as the tablet. The screen cursor moves in direct relation to the movement of the mouse. If the mouse is moved to the left the screen cursor will also move left.

The advantage of a mouse is that it does not require a large tablet under it. A mouse can be used directly on a desktop or any

IBM Personal System/2 Mouse	Hewlett Packard Sketch Pro
Logitech Mouse	Hitachi Tiger Tablet
Microsoft Mouse	Houston Instruments HI-Pad DT-11A
Mouse Systems Mouse	
Summagraphic Summamouse 420	Houston Instruments True Grid 1011
Numonics Manager Mouse	Kurta Tablet, IS/Series 3
3Space Isotrak	Numonics 2200/Penpad 300 Tablet
Calcomp 2000, 2500, Drawing Board, WIZ Tablet	SAC GP-8 Sonic Digitizer
Cherry Tablet	Summagraphics Summa-sketch MM 1200, 1800
GTCO Digipad 5 series—low res	Summagraphics Summasketch MM 960
GTCO Digipad 5 series—high res	Summagraphics Microgrid Summagraphics MM 1100

Figure 2-5   Some of the mice and tablets supported by CADKEY

other flat surface. Mice can be constructed in three different ways. One style of mouse uses two wheels placed on the bottom side for tracking the screen cursor's X and Y movements. As the mouse is moved on the desktop the changes in the X and Y axes are displayed in the lower left corner of the CADKEY status window. When the screen cursor is at the desired point, one of the pick buttons is pressed to record the X and Y coordinates. Another style of mouse that works on the same friction principle is constructed using a small ball in place of the tracking wheels. The ball is inside the mouse's plastic case. When the mouse is moved across the desktop the X and Y coordinates change as a result of the ball rolling on the surface. The third style of mouse is an optical mouse (Figure 2-6). This mouse is constructed using optical technology to track movement. Unlike the other two styles of mice, this mouse cannot operate on any flat surface. The optical mouse must be used on top of a special reflective grid pad, which it uses for referencing its movement.

Another possible advantage of a mouse is its low cost. A mouse can be purchased in the $25 to $200 price range, about ten times less expensive than a tablet.

A mouse is a relative device. Unlike the tablet, which is accurate for tracing work, the mouse is not a tracing device. Each move of the mouse results in the screen cursor moving relative to its previous position. You may end up at the same point on the desktop, but it may not be the same point on the screen. Sometimes large moves require that the mouse be picked up and moved several times to track the cursor into position. This is referred to as skating.

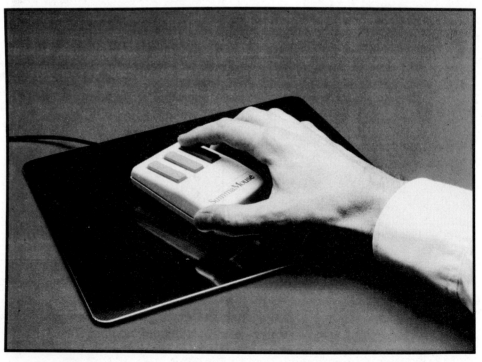

Figure 2-6 An optical mouse (Courtesy of Summagraphics Corp.)

Figure 2-5 lists some of the tablets and mice supported by CADKEY.

OUTPUT DEVICES

An output device is a piece of equipment that allows the user to view the data stored on the computer. The output that comes from this type of device is referred to as "hardcopy." Hardcopy is generally composed of an image/drawing and/or text which is output on paper, plastic film, or vellum. CADKEY supports two types of output devices: plotters and printers.

PLOTTERS

There are two major types of plotters: the electrostatic (Figure 2-7) and the pen plotter. The pen plotter is more commonly used with microcomputer-based CAD systems. A pen plotter utilizes a fiber tip pen, ball bearing tip, or a technical pen to produce a drawing or plot of the image generated on the screen. The number of pens that are available for use by the plotter at one time varies between models. The number of pens can range from one to fifteen. The pens can vary in color and thickness, thus producing lines that are distinguishable from one another. Lines of different colors and line weights (thicknesses) can be adapted to have different meanings and make reading the drawing easier. The overall appearance of a complicated drawing can be

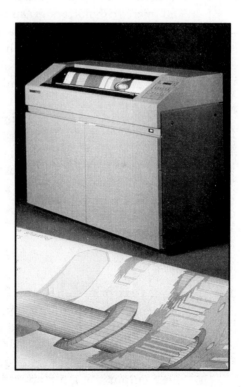

Figure 2-7   An electrostatic plotter (Photo courtesy of Hewlett Packard)

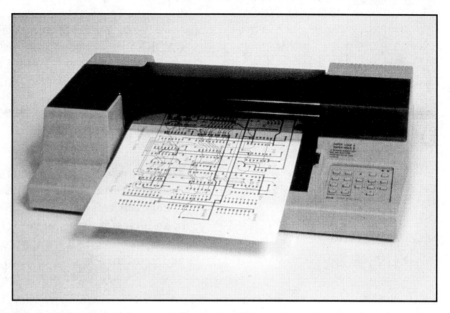

Figure 2-8 A desktop pen plotter (Photo courtesy of Hewlett-Packard)

American/Western
    Graphtec—GPGL

Calcomp Pen Plotter—PCI

Calcomp Electrostatic
    plotters—906/907

Hewlett Packard—HPGL

Houston Instruments—
    DMPL series

Ioline LP 400

Mutoh IP-230

Oce Graphics/Benson—BGL

Roland DXY-890

RS-274 Gerber Out

Figure 2-9 A partial list of plotters supported by CADKEY

greatly enhanced by the use of different pens. Plotters range in paper size from A size (8.5″ × 11″) to E size (34″ × 44″). There are also plotters that use rolls of paper, allowing the user to make large plots.

Most of the pen plotters are of the microgrip variety (Figure 2-8). The microgrip plotters grasp the paper by its edges and move the paper forward and back on one axis. While the paper is moving on one axis, the pen is moving on the other axis. This combination of movement enables the plotter to create on paper all of the geometric figures that have been created on the CAD system. The hardcopy output from a plotter is of excellent quality. The lines and circles are clear and accurate, as if drawn by hand-held instruments. Many of the illustrations in this book were generated using a pen plotter.

CADKEY supports plotters manufactured by a wide variety of vendors. Figure 2-9 shows some plotters supported by CADKEY.

PRINTERS   Dot matrix printers (Figure 2-10) have long been a way of producing inexpensive low-quality black and white hardcopy. In the past the dot matrix printer has been used for non-graphic alphanumeric data output. With the use of a new printer driver written for CADKEY, a dot matrix printer can output quality color graphic images that rival those output using a plotter. A dot matrix printer uses a head that is made up of a series of rows and columns of pins. The number of pins

Figure 2-10   A dot matrix printer (Courtesy of Epson America, Inc.)

in the rows and columns varies with printers. Two possible pin combinations are 5×7 and 9×9. The pins are controlled according to the data received from the computer. Those pins, given the signal to print, will strike through a carbon or ink ribbon which on impact leaves an image on the paper. By rasterizing the drawing data base, CADKEY is able to print clear images using a dot matrix printer.

In addition to the dot matrix printers, CADKEY also supports ink jet and laser printers. You can also print color with an ink jet or a laser jet printer. An ink jet printer uses small spray nozzles to deposit colored inks on the paper. A laser printer works with a laser beam to produce the image on paper. Figure 2-11 lists some printers supported by CADKEY.

Diablo 150	Hewlett Packard Paint Jet/XL	NEC Pinwriter P5, P6, P7, P9XL
Epson FX, JX, MX, RX	Hewlett Packard Think Jet	NEC PC-PR201
Epson LQ-2500	IBM Proprinter	Okidata 192, 193, 294
Fujitsu DL 2600	JDL 750+ (C size)	Postscript
Hermes Color	JDL 750-e/850 EWS	Toshiba P351SX
Hewlett Packard Laser Jet	MPI 350	
Hewlett Packard Paint Jet		

Figure 2-11   A partial list of printers supported by CADKEY

DISPLAY DEVICES AND GRAPHICS DISPLAY CARDS

The device that allows a CAD operator to observe input and eventual output is the cathode ray tube (CRT) display or monitor. The technology in this area of hardware is constantly changing. Vendors are continually developing new monitors and new graphics display controller cards to enhance and improve the image on the display screen. Most of the enhancements come in the areas of resolution and color. Resolution refers to the available number of horizontal and vertical pixels (picture elements, small lights inside the monitor that are lit to display the image). When resolution is being discussed, the larger the number of pixels, the greater the resolution. The greater the resolution, the clearer the image will appear on the screen. With low-resolution monitors, the image will appear distorted and may result in inaccuracies in the final drawing.

A graphics display controller card (Figure 2-12) works in conjunction with the monitor and controls the resolution and the color of the screen image. At the present time, several graphics display card standards are popular for use with CADKEY. The standards and the resolution available with each are:

Graphics Display Card	Resolution	Color
EGA (Enhanced Color Graphics)	640 × 350	16 colors
VGA	640 × 480	256 colors
Super VGA	1024 × 768	256 colors
High Resolution	1024 × 768 or 1280 × 1024	256 colors

Figure 2-12 A graphics display controller card

The EGA (Enhanced Color Graphics Adapter) is a standard that was developed by IBM. This standard has also been emulated and improved upon by third-party vendors. The standard offers the best combination of color and resolution. The EGA standard can support four different combinations of resolution and color. The preferred combination is 640 × 350 resolution with a palette of 16 colors. With a monitor supported by an EGA card, the CAD operator will be able to generate a colorful, precise image.

The VGA (Video Graphics Array) graphics card was introduced with IBM's PS/2 family of computers. This standard was quickly picked up by most graphics card vendors. This standard was an improvement over the widely accepted EGA standard in resolution (640 × 480) and color palette (256) and is affordably priced. Several of the Super VGA cards on the market will run at a resolution of 1024 × 768 with 256 colors.

The high resolution graphics cards are capable of producing resolution as high as 1280 × 1024 with a palette of 256 colors. These cards and the monitors that are required to support them have become very popular. They are more expensive, which will slow their rate of growth in the CAD marketplace. A good monitor and graphics card are necessary for the CAD operator who spends eight hours a day looking at the screen.

Specific monitors are designed to be compatible with each of the graphics display controller cards. Most monitors are compatible with multiple standards: the Hercules, VGA, and EGA graphics cards. There are also individual graphics cards on the market that are capable of running multiple standards.

The complete list of graphics display monitors and graphics cards supported by CADKEY is long and getting longer with new developments in this area. Figure 2-13 lists some of the monitors and display cards supported by CADKEY.

STORAGE DEVICES

The data that is input into the computer is resident in the RAM (random access memory) only while the computer is on. To save this data for retrieval at a later date, it must be stored on some type of magnetic medium. You may already be familiar with magnetic media for storage. If you have ever played or recorded a cassette tape, you have been using a form of magnetic media. The magnetic media used to store data on a microcomputer are most commonly found in three forms: a floppy disk, hard disk, or magnetic tape. There are two types of floppy disks: low-density, which is capable of storing 360KB of data, and high-density with a storage capability of 1.2MB. The storage capabilities of a hard disk range from 40MB to 150 or 300MB. A magnetic tape is generally used to back up, or make a copy of, all of the data on a hard disk. A single magnetic tape can store in the range of 100MB to 400MB of data. The hard disk is the fastest device for retrieving data.

Ahead Systems Wizard	NEC MVA 1024
AT&T Display Enhancement Board	Nth 3D Engine Bridge
AT&T Indigenous Board	Number Nine Pepper Pro
ATI EGA Wonder	Orchid ProDesigner VGA
ATI VGA Wonder	Orchid ProDesigner II VGA
ATI VIP VGA	Paradise 8514/A Plus
ATI VGA Wonder 1024 *Vesa	Paradise Autoswitch EGA 480
Compaq Advanced Graphics 1024	Paradise VGA Professional *Vesa
Compaq Portable Plasma Display	Paradise VGA 1024 *Vesa
Conographics 40	PCG Photon Mega
Control Systems Artist XJ/10	Pixelworks Clipper Graphics
Datapath Q-PC	Sigma Color 400
Everex View Point	Sigma EGA 480
DGIS-compatible cards	STB VGA/EM
Genoa Super EGA	STB VGA/EM-16 Plus *Vesa
Genoa Super VGA 5400	Tecmar Ega Master 480/800
Genoa Super VGA 6400	Tecmar Graphic Master
Hercules Graphics Station	Tecmar VGA/AD VGA
Hercules Monochrome	TIGA-compatible cards ver 1.1
Hewlett Packard Intelligent Graphics Controller 10	Toshiba T3100
	Tseng EVA/480
IBM Color Graphics Adapter	Vectrix Pepe
IBM Enhanced Graphics Adapter	Vectrix PRESTO!
IBM Personal System/2 8514/A	Vermont Microsystems Cobra/Plus
IBM Personal System/2 MCGA	Vermont Microsystems Image Manager Series
IBM Personal System/2 VGA	Vermont Microsystems X/Series
IMAgraph AGC/Image Plus	Verticom H, M series
IMAgraph TI-1210	Verticom CAD-480
Infotronic PGS-1289	VESA compatible cards
Matrox PG Series ver 1.00	Video-7 Deluxe
Metheus Premier VGA	Video-7 VEGA VGA
Metheus UGA series	Video-7 V-RAM VGA
Microfield Graphics T8	Video-7 V-RAM VGA 1024 *Vesa
Mylex AGA	Wyse-700

Figure 2-13   A partial list of the graphics display cards supported by CADKEY

The storage devices that place the data on the magnetic media are called drives. The three classifications of drives used with microcomputers are floppy disk drives, hard disk drives, and magnetic tape drives.

CADKEY HARDWARE CONFIGURATION

The minimum hardware requirements for CADKEY are:

- Intel-based 80386 or 80486 microcomputer
- Floppy disk drive
- Hard disk drive with 7MB to 13MB of free disk space
- Graphics display card
- Monitor
- 4MB of system memory
- MS DOS version 3.1 or higher
- Parallel port

The preferred hardware for CADKEY is:

- Intel-based 80486 DX microcomputer
- Floppy disk drive
- Hard disk drive with at least 20MB of free disk space
- Super VGA or higher resolution graphics display card
- 256 color monitor with a resolution of 1024×768 or better
- 8MB or more of system memory
- MS DOS version 5 or higher - SOR7U ALR
- Parallel (1 pt) port
- 2 serial (com) ports
- Mouse or digitizing tablet
- Plotter or printer

The preferred hardware for the CADKEY UNIX version is a worksta-tion by one of the following manufacturers:

SUN—SPARC Station
SONY—News
SILICON GRAPHICS—Personal Iris
DIGITAL—DEC Station

REVIEW QUESTIONS

1. What is the difference between hardware and software?
2. What five components make up a basic CAD system?
3. List an example of each of the components you listed in the previous question.
4. What are the standards for graphics display controller cards?
5. How many bits are there in a byte?
6. Identify the three classifications of computers.

CHAPTER 3

The CADKEY Human Interface

OBJECTIVES

After completing this chapter, you will be able to:

- Name the nine important areas of the CADKEY display screen.
- Manipulate the screen cursor.
- Remove temporary screen markers.
- List the special keys for CADKEY.
- Select menu items with the keyboard or the cursor.
- Select and change the Status Window.
- Control the display of the cursor tracking.
- List the nine 3D views.
- Define a viewport.
- Define a construction plane.
- Define the CADKEY standard angle convention.

CADKEY has been written and developed with special attention given to the details of the human interface. It was felt that in many cases CADKEY would be a person's first experience with CAD. To make this first experience a positive one the system had to be easy to work with. The first-time user must feel comfortable in the environment in which he or she will work. To facilitate the development of this type of environment, the R&D team at CADKEY reviewed the research on human interfaces with CAD. What they found was that some of the basic learning principles had not been employed by the other CAD systems. Using this information, they designed the CADKEY human interface to make learning and operating CADKEY easy.

The first step was to position the menus on the left of the display screen because, in normal reading, people's eyes scan from the left to the right. One of the first steps in operating a CAD system is to choose an option from the menu. The menu is the first item that your eyes see when they scan the display screen in CADKEY. This logical order of positioning the menus on the left side of the screen makes users feel more comfortable and less confused with the work environment.

When it was time to develop the options for the menus, again CADKEY looked to the research. This time they found that people can quickly recall two to ten items at a time. Menus that have more than ten options at one time become difficult to follow. After more than ten options, the human brain has difficulty discriminating between options. The menus in CADKEY have no more than nine options presented at one time.

THE DISPLAY SCREEN

Figure 3-1 illustrates the layout of the display screen while working in CADKEY. The display screen is divided into the following areas:

Menu Options Window
System Mode Indicator
Break Area
Status Window
History Line
Prompt Line
Pop-Up Menus
Cursor Tracking Window
Drawing Window (Graphic Viewports)

HISTORY LINE

The R&D staff at CADKEY also found that many CAD operators are intermittent users. That is, they leave their workstations from time to time to confer with their colleagues and then return to work on their projects. CADKEY displays a history line across the top of the display screen, so that when the user returns he/she knows exactly where he/she is in the CADKEY menu.

The history line is an on-screen record of the functions you have selected. The functions are listed in the sequence that they have been selected. This prevents you from losing your place when returning from a long break.

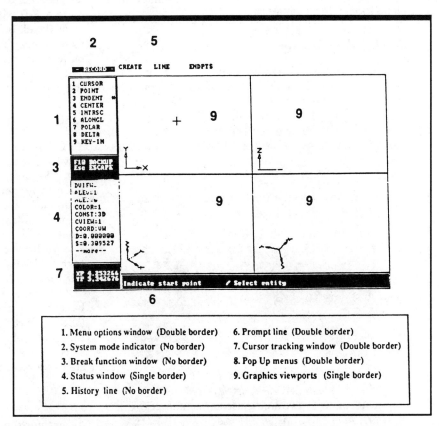

Figure 3-1   The interaction areas on a CADKEY screen (Courtesy of Cadkey, Inc.)

CADKEY Delivers Heavy Benefits for Ultralight Company

As a recreational activity, ultralight flying has a short history, starting in the early 1970s. By 1978, hang-gliding enthusiasts were seeking new thrills and better control over the flying conditions that limit unpowered aircraft. That's when engineer Dave Cronk attached a small motor to an existing hang glider and started the recreational aviation growth market of the 1980s: ultralight aircraft that weigh less than 350 pounds, yet can fly as high as 16,000 feet. This constituted an aeronautical revolution in which aircraft fly at safe, comfortable speeds between 35 and 50 miles per hour.

At the time, Dave's company, called Eipper Formance, was building a rigid-wing hang glider called Quicksilver. It was Quicksilver that Dave chose for his first experimental foray into powered flight.

Eipper Formance, now renamed Quicksilver Enterprises, Inc., upgraded facilities and tooling to stay competitive. One of the most important changes has been the introduction of CAD/CAM technology into Quicksilver.

CADKEY in Design

During the past several years, Dave has focused his use of CADKEY in the area of conceptual design. "I begin with a preliminary 3-D design in CADKEY of the basic aircraft," Dave continued. "CADKEY enables us to save lots of time by quickly finding areas, centroids, and perimeters. We also use CADKEY to determine the precise geometry of structural members, particularly moments of inertia of sections.

"After I am satisfied with the design, I pass it along to one of our draftsmen to convert the 3-D component drawings into a set of 2-D engineering drawings." Quicksilver currently operates three CADKEY systems and has six heavy users.

"After we have produced working drawings, we distribute them to the shop floor and to outside vendors for prototype parts." At this point, Quicksilver produces a prototype for more rigorous load and stability testing that involves instrumented hydraulic fixtures. After the prototype parts pass muster, the design is frozen; the part drawings are plotted and electronically stored.

CADKEY is also used to generate the many drawings that needed for the assembly manuals. According to Dave Cronk, Quicksilver has an interesting technique to make sure that the drawings in the owner's assembly manual always correspond to the most current versions of the parts included in each kit. "When we generate an engineering drawing of a part, we also store a 3-D wireframe model of the part on a hidden level (level 120) in the CADKEY part file. When the time comes for us to produce the owner's assembly manual, we make temporary pattern files of these hidden wireframe models of the parts, and insert them, in the appropriate sequence, into the manual.

CADKEY in Manufacturing

The precise way that Quicksilver produces aircraft, however, requires more than knowledgeable CAD users. "We design our own tooling," Dave said. "So, we have to do some custom programming to drive machine tools with CAD drawings. Our feeling is that we want to control as much of the process in-house as possible."

"We have a fairly complicated system in which we take CADKEY drawing data and translate it into input that our NC tools can understand.

"It's not a trivial task," Dave continued. "There can be up to 700 drawings that comply with DOD-STD-100 (U.S. Department of Defense, Standard 100) for even the relatively simple airplanes that we make. And since we want to punch out parts as efficiently as possible, we want to optimize the whole process."

QuickSilver Enterprises' Model GT 500, designed with CADKEY, displayed at CADKEY's CAD/CAM Solutions Fair at NDES '91, April 8-11, Chicago, Illinois.

—Continued, next page

CADKEY In The Real World—(Continued)

Optimization

At first Dave doubted whether CAD could optimize anything. He and others in the business have invested a lot of time in the *sail room*, the place where wings are designed and crafted. "We were afraid that with computers, we might lose some of the hands-on feel that is part of the thrill in the sport. People are attracted to ultralights because the craft aren't intimidating as 'real' airplanes are. It's really a feeling of being in direct control of everything, and also of reducing the experience to the bare essentials. I didn't want to give that up, at first. And, it seemed a lot of work to learn a CAD system."

As he began to use CADKEY, however, Dave became aware of the freedom to explore alternative designs quickly that is only possible with CAD. He also recognized that CADKEY would eliminate much of the drudgery of production drafting and save time for higher-level engineering.

I've never quantified it, but I'm sure we save immense amounts of time throughout the design and production cycle. The whole time to market has shrunk, and the products that we bring to market are of much higher quality because of the precision at every level of detail," Dave concluded. "Because we want our quality control to be second to none, CADKEY has become an integral part of our entire operation."

Adapted with permission from 3-D WORLD, v. 5, no. 2, Mar./Apr., 1991.

The history line is also particularly helpful in a learning situation. If a student is having a problem, the instructor can use the history line as a quick reference to find out what the student selected and what corrective actions need to be taken. If one of the functions on the history line is found to be wrong, with the use of the **BACK-UP (F10)** option you can move back through the command sequence one step at a time. With each step back, the appropriate menu will be displayed and you are given the opportunity to make a new selection to remedy the prior error.

PROMPT LINE The CAD interface studies also pointed out the need for on-line help. The prompt line, located at the bottom of the display screen, serves this purpose. It steps the user through a function with continuous prompts that tell the user when to choose an option, enter a value, locate a position, or select an entity. The prompt line also contains information about options such as position menu options that stay active until another option is chosen. The current prompt line has been designed to have the first word in each prompt describe the action being requested. The second part of the prompt line displays the positioning prompt when it is appropriate.

The prompt below is an example of a request that an endpoint of an entity be indicated using the **POLAR** option from the position menu:

Indicate endpoint / Indicate polar origin

The first part of the prompt line is requesting the endpoint for a line. The second part of the prompt line requests you to indicate the polar origin, which is an option from the universal position menu.

The following list includes commonly used first words on the prompt line and a description of each.

Indicate—requests a screen location using the cursor or an option from the position menu.

Choose, Accept, Save—requests a selection from the current menu displayed.

Enter—requests input from the keyboard and must be followed by a RETURN or ENTER.

Select—requests that entities be selected using a method displayed in the Selection Menu.

CR—carriage return, RETURN, or ENTER from the keyboard.

X, Y, Z—indicates to enter values in World Coordinate positions.

XV, YV, ZV—indicates to enter values in local View Coordinate position.

dX, dY, dZ—indicates to enter values in delta World Coordinate position.

dVX, dVY, dVZ—indicates to enter values in delta View Coordinate position.

XP, YP, ZP—indicates to enter values in construction plane position.

XT, YT—indicates to enter values for an on-screen position.

THE CURSOR

CADKEY uses a crosshair cursor (+) as its pointing symbol in the Drawing Window. The crosshair cursor appears on the screen as a cross or a horizontal and a vertical line; their intersection point is the selection point. The cursor is used to identify X, Y, and Z locations on the screen. The cursor can be moved on the screen through the use of a cursor control device, such as a mouse or a digitizer, or by using the arrow keys on the keyboard. The default size of the cursor is one inch by one inch. If you wish to make the cursor larger or smaller, the size can be changed using the **CONFIG** (configuration) program.

When the cursor is moved, using a digitizer or a mouse, from the drawing window to the menu options window, the status window or a pull-down menu, it changes from a crosshair to a reverse video display or arrow to highlight the option to be selected. When the cursor is moved into an icon menu it takes the shape of an arrow. The arrow is then used to select icons such as color, line type, level number and line width. By moving the cursor until it is on the menu option desired and pressing the cursor control button, the menu option will be selected. You will then be prompted what to do next. When the cursor is moved into an inactive area on the screen it appears as a small circle (o). This indicates that nothing is available to select in this area of the screen.

OTHER SCREEN SYMBOLS

Three additional screen symbols are found within the drawing window. All three of these symbols are temporary and can be removed from the screen by invoking the Immediate Mode command **CTRL-R** to redraw the screen without the temporary markers. In most cases, they indicate position or the selection of a specific entity.

Symbol	Description
X	used when identifying reference locations, such as those found with the Position Menu.
#	used to mark an entity selection that was canceled.
Highlight Style	(can be selected in the **CONFIG** program) used to mark an entity that has been selected to be removed using the **REDRAW** command.
Flashing Markers	reference or selection markers that are quickly displayed as circles and erased.
Drawn, highlighted markers	highlighted reference or selection markers that are displayed on the screen until they are removed with the **REDRAW** command.
Flashing, highlighted markers	highlighted reference or selection markers that are quickly displayed and erased.

DEFAULT ASSIGNMENTS

Many options that prompt you to enter a value or filename have a value or filename appearing in parentheses in the prompt line. This preassigned value is a default value. Specific default values have been set by the system so that every time CADKEY is initialized the defaults are active. In some cases the default value will be the value that you last entered. To accept the default value, press the **RETURN** key. If you wish to enter a value other than the default, type in the new value or filename, then press **RETURN.**

THE KEYBOARD

The keyboard that you are using may have some slight variations from the one described depending upon who manufactured your computer and what model you are using. The main section of a keyboard is similar to that of a typewriter except for several special keys that play a very important role in CADKEY. Below is a list of the special keys and a description of their functions.

> **ESC** (escape)—allows you to exit any function and return to the Main Menu. The **ESC** key is next to the "1" key on a PC or XT keyboard and above the numeric keypad on an AT.

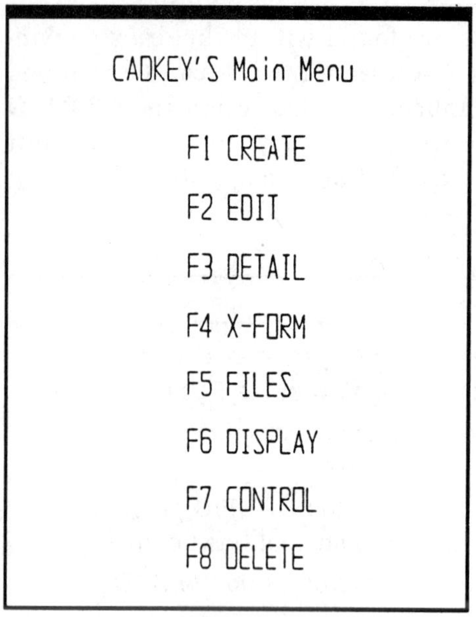

```
┌─────────────────────────────┐
│                             │
│   CADKEY'S Main Menu        │
│                             │
│     F1 CREATE               │
│                             │
│     F2 EDIT                 │
│                             │
│     F3 DETAIL               │
│                             │
│     F4 X-FORM               │
│                             │
│     F5 FILES                │
│                             │
│     F6 DISPLAY              │
│                             │
│     F7 CONTROL              │
│                             │
│     F8 DELETE               │
│                             │
└─────────────────────────────┘
```

Figure 3-2 CADKEY's Main Menu

CTRL (control)—used to invoke Immediate Mode commands. This key must be held down while the assigned mode key is pressed. The **CTRL** key is usually above the shift key on the left side of the keyboard.

ALT (alternate)—also used to invoke Immediate Mode commands. This key must be held down while the mode key is pressed. The **ALT** key is usually directly below the shift key on the left of the keyboard.

RETURN, ENTER, CR—accepts a default value and must be pressed every time after data is entered. The **ENTER** key is the large key next to the numeric keypad.

BKSP (backspace)—deletes values entered in the prompt line before **RETURN** is pressed. The backspace key is directly above the **RETURN** key.

Space Bar—used for designating position locations and other general data entry functions. Often, the space bar may be duplicated by a mouse button. The space bar is the long horizontal key at the bottom of the keyboard.

Function Keys

F1 to **F9**—select options from the menus displayed on the screen. The function keys are often located on the left of the keyboard in two columns. They may be located in a different position on your keyboard but they are easily recognized because each number has an F in front of it.

F10—returns you to the previous menu or cancels the last entry or selection made.

Numeric Keypad (with the NUM LOCK key off)

Arrow Keys—move the screen cursor in the Drawing Window when the keyboard is configured as the input device.

PgUp—doubles the current cursor snap value.

PgDn—divides the current cursor snap value in half. This is helpful when selecting a radius point.

All of the above keys are located on the keyboard in various locations depending on the style of your keyboard. The buttons on your mouse can also be configured to emulate many of these keys.

THE MENUS

CADKEY is a completely menu-driven system. It prompts you for all of the information that you need to enter to complete a specific function. With each prompt a new menu of options is displayed for you to select from. The Main Menu is the top of a many-tiered tree-structured menu system. Figures 3-2 and 3-3 show the Main Menu options and the second tier of options in the menu structure. See Appendix A for complete menus.

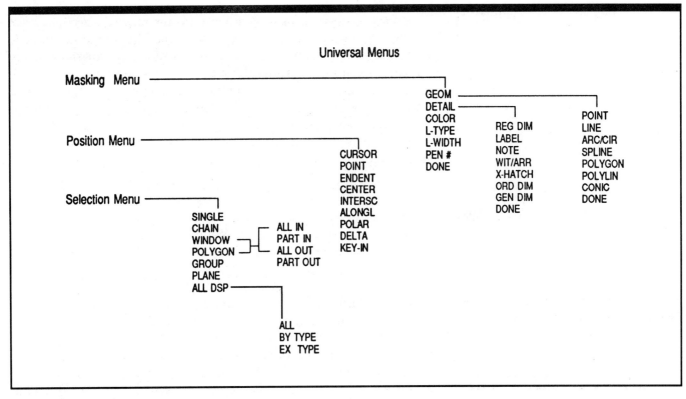

Universal Menus

Masking Menu

GEOM
DETAIL
COLOR
L-TYPE
L-WIDTH
PEN #
DONE

REG DIM
LABEL
NOTE
WIT/ARR
X-HATCH
ORD DIM
GEN DIM
DONE

POINT
LINE
ARC/CIR
SPLINE
POLYGON
POLYLIN
CONIC
DONE

Position Menu

CURSOR
POINT
ENDENT
CENTER
INTERSC
ALONGL
POLAR
DELTA
KEY-IN

Selection Menu

SINGLE
CHAIN
WINDOW
POLYGON
GROUP
PLANE
ALL DSP

ALL IN
PART IN
ALL OUT
PART OUT

ALL
BY TYPE
EX TYPE

Figure 3-3 CADKEY's universal menus

When the system is first initialized, the first menu that appears in the Menu Options Window is the Main Menu. Each time an option is selected, a new menu is displayed in the Menu Options Window.

Options can be selected from the menus in one of three ways. The cursor control device (digitizer or mouse), the keyboard, or a combination of both can be used to select options from the menus.

SELECTING FROM THE KEYBOARD

On the left side of the keyboard are two columns of keys numbered **F1** to **F10**. The number on each of these *function keys* corresponds to the number in front of the options found on the menu. The menu options 1 to 9 are constantly changing with each new option selected.

An **ESCAPE Code** is the sequence of function keys that are pressed following the **ESC** key. Every function has a specific sequence of key strokes that will bring you through the menu structure to a specific function. These ESCAPE codes can save you much time in making menu selections. If the ESCAPE Code **ESC-F1-F3-F2** was pressed from anywhere in the menu structure, you would be in the function **CREATE** a **CIRCLE** by the **CENTER** and **DIAMETER** method. The next prompt you would see would ask you to enter the diameter. For a function that you use often, you should either memorize its ESCAPE Codes or create a macro of the ESCAPE code to increase your efficiency in using CADKEY. Many of the ESCAPE Codes change between versions of CADKEY as more functionality is added to the system.

SELECTING USING A DIGITIZER OR A MOUSE

To make a selection using a mouse or digitizer, move the screen cursor to the left from the Drawing Window to the Menu Option Window. When the screen cursor enters the Menu Window, the crosshair cursor changes shape into a triangular pointer. This pointer or marker is then moved until it is in front of the option that you wish to select. Once it is in position, the mouse button is pressed to select the option. This method of selection will only work with the aid of a digitizer or mouse. You cannot use the arrow keys on the keyboard to make a menu selection.

THE BREAK AREA

The break area is found between the Menu Options Window and the Status Window. When the Main Menu is displayed the EXIT option **ALT-E** is displayed. Once you get past the Main Menu options the **BACKUP** and **ESCAPE** options can be found in the Break Area. The (**BACKUP**) option is assigned to function key **F10**. This option is used to cancel the last selection, prompt, menu, or option at any point in the menu structure. It will cancel only the last option, not the whole function. This gives you the ability to cancel a function one step at a time or, if one wrong selection has been made, it can be corrected without having to cancel the whole function.

To cancel or exit a function, press the **ESC** (ESCAPE) key. This will return you to the beginning, the Main Menu, to make a fresh start. The most common use of the ESCAPE key is to exit a function when you are finished with it. Note that if the **ESC** key is pressed while you are in an Immediate Mode command, only the Immediate Mode command will be canceled.

UNDO (Alt-I)

The **UNDO** options appears in the Break Option Window for a few commands. When it is selected the entities that were just created or moved will be removed. The part will be restored to the way it was before using the command. The commands that have the UNDO option available are **DETAIL-DIMENSN**, **LABEL**, **EDIT-TRM/EXT**, **BREAK**, **CREATE-FILLET**, and most of the **X-FORM** options.

THE STATUS WINDOW

The Status Window is a revolving set of two to four menu pages that contains information about the current part that is displayed: note text height, grid and snap status, current working scale, depth, current view, and so on. Choose the —**more**— option to display the next page of the menu. The —**more**— option can only be selected using a digitizer or a mouse. To change any of the options, select with the cursor the option you wish to change and check the prompt line to see what to enter. This type of selection is made by moving the cursor horizontally from the Drawing Window into the Status Window. When the cursor enters the Status Window, it highlights the option in reverse video. This marker is then moved vertically until the desired option is highlighted. The option is then selected by pressing the mouse button.

The Status Window is a quick and easy way to make changes in your part while you are constructing it. The status can be changed on many of these items while you are in the middle of the function. Many of the options found in the Status Window are also available as Immediate Mode commands.

Status Window Options	
HELP	brings up the help window
VIEW = 1	current view assigned (TOP)
ACTVINST	active instance
CPLANE = 1	current construction plane
PAP LIMITS	paper limits
ALEV = 1	current working level
MLEV = 0	current masking level
COLOR = 1	present color assigned (green)
CONST: 3D	status of the 2D/3D construction switch
COORD: VW (View)	current coordinate system (World or View)
D = 0.000 (Depth)	current working depth
S = 1.000 (Scale)	current working scale
GRID: OFF	current grid status
SNAP: OFF	status of the snapping feature
VERSEL: OFF	verify selection
L-TYPE = 1	current line type (solid)
L-WID = 1	current line width
PEN = 1	current pen number assigned for plotting
DB: FWD	current setting for direction of data base search
NH = 0.300	current note height
DH = 0.300	current dimension height
AS = 0.500	current text aspect ratio
RT = 0.000	current note text angle
NF = 1	current note text font
DF = 1	current dimension text font
WIT: BOTH	status of witness lines currently displayed
ARR: IN	status of arrow direction
LDR: BOTH	status of leader line display
QTRIM: OFF	status of the Quick Trim option
HTRIM: ON	current status of the Hatch trim option
SELLSET: ON	status of the selection set option
HTCH_UP: OFF	status of the Hatch update option
REGEN	will regenerate your entire file
—more—	toggle to return to the first page of the Status Window

HELP

The first option found on the Status Window is the **HELP** option. When you select this option, a Pop-Up Window will appear, providing a written explanation of the command that you are currently using. You can query the window for more information if needed.

VERSEL (Verify Selection)

If verify selection is ON every time the user makes a selection, a **NO/YES** menu will appear. The user must then verify whether or not the highlighted entity is the correct one by responding **NO** or **YES**. It will continue to search for the next closest entity until you respond **YES**.

SYSTEM MODE INDICATOR

The system mode indicator can be found above the Menu Options Window. It indicates what mode CADKEY is currently running. When you are operating CADKEY it will display **CADKEY**. If you are recording a macro it will display **RECORD**. If you are running a CADL program it will display **CADL**. If you are creating a layout drawing it will display **LAYOUT**.

POP-UP WINDOWS/MENUS

Pop-Up Windows appear over the top of the viewports for the user to read or select a displayed option. The **FILES-PART-LIST/LD** option is an example of a Pop-Up Window.

THE CURSOR TRACKING WINDOW

The Cursor Tracking Window is in the lower left corner of your display monitor. Cursor tracking allows you to see the current location of the screen cursor. In the Cursor Tracking Window, "MENU AREA" or "STATUS WINDOW" will be displayed depending on where the cursor is located. When the crosshaired cursor is in a viewport, the X, Y, and Z locations will be displayed when you are using the World Coordinate system. If the View Coordinate system is being used, the XP and YP coordinates are displayed in the Cursor Tracking Window, with the Z found as D (depth) in the Status Window. Each time the cursor is moved the X, Y, and Z coordinates are updated to the new location of the cursor. This status is very helpful when you are creating or positioning entities using the **CURSOR** option from the position menu. The cursor tracking option, like most of the other options in the Status Window, is in the OFF position when you begin creating a part. Cursor tracking can be turned ON two different ways. From the Menu Options Window you would select **DISPLAY-CURSOR** and then **TRACKING**. These steps have been combined into one Immediate Mode command sequence, **CTRL-T**. This command can be invoked by pressing and holding the Ctrl key and pressing the "T" key at the same time. A menu with three options appears. Select **OFF**, **VIEW** (to track using the View Coordinates system) or **WORLD** (to track the cursor using the World Coordinates system).

Once the first position is indicated by the mouse button, the coordinate position is displayed:

WORLD	VIEW
X = 1.000	XP = 1.250
Y = 2.625	YP = 2.500
Z = 0.875	

COORDINATE SYSTEMS

CADKEY offers two coordinate systems, World and View. These two coordinate systems can be used interchangeably within one part. You may start constructing the two-dimensional shape of a part in View Coordinates and then change to World Coordinates when the part is to be transformed into a three-dimensional part.

View Coordinates

The View Coordinate system is relative to the screen with the X, Y, and Z axes positioned as in Figure 3-4. No matter which view you are in, the axes will always be found as listed below:

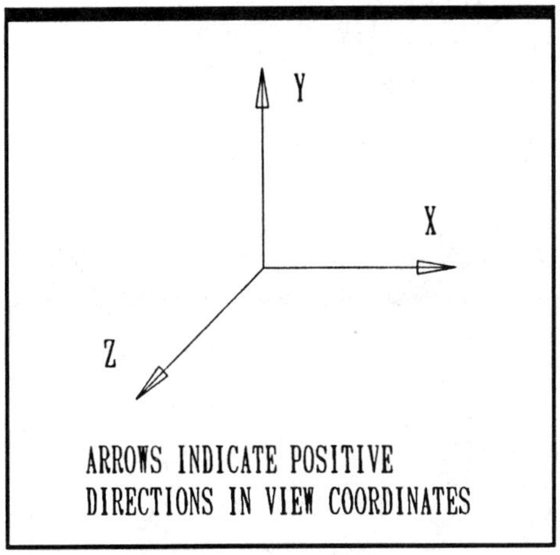

Figure 3-4 CADKEY's view coordinate system

Axis	Position	Positive Direction
X	horizontal	to the right
Y	vertical	toward the top
Z	perpendicular	toward you from the screen

World Coordinates

World Coordinates are relative to the TOP view, view 1. In view 1, the coordinate axes are the same for both World and View. In World Coordinates the axes of the object remain attached to the object so that as you look at different views of the object the direction of the axes is moving relative to the top view of the object. Refer to Figure 3-5 and Figure 3-6 to help visualize the World Coordinate system.

3D VIEWS

CADKEY has eight predefined views. The first six are the regular 2D orthographic views; the next two are isometric and axonometric views. The user-definable view is of great assistance when you are finding the true size and shape of an auxiliary surface that is not parallel to one of the regular orthographic planes of projection. There are an unlimited number of user definable views. When you change views,

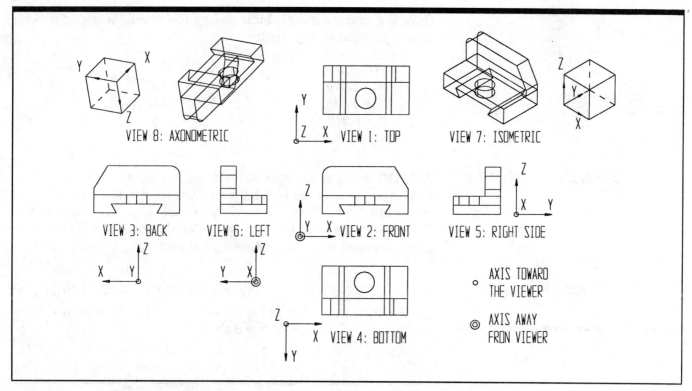

Figure 3-5   CADKEY's predefined views and world coordinate system

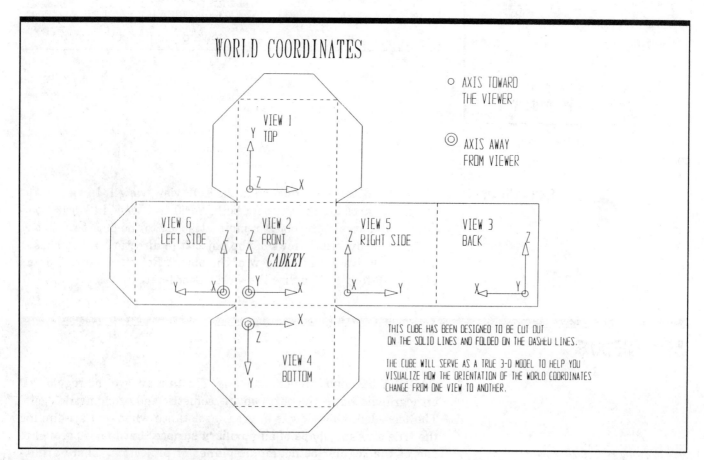

Figure 3-6   CADKEY's world coordinate cube

CADKEY only changes the way you see the object; nothing in the data base is changed.

The nine defined views in CADKEY are:

1. **TOP**
2. **FRONT**
3. **BACK**
4. **BOTTOM**
5. **RIGHT SIDE**
6. **LEFT SIDE**
7. **ISOMETRIC**
8. **AXONOMETRIC**
9. **SYSTEM VIEWS** (unlimited number of user definable views)

VIEWPORTS

In previous versions, when CADKEY was first initialized, the TOP view (view 1) was the view to which the system defaulted. Starting with CADKEY version 3.5, the user now has the ability to configure CADKEY to have multiple views of a part displayed at the same time. The window where a view is displayed is referred to as a VIEWPORT. A viewport has many important features associated with it. It can be active, re-sized, the view changed, or a new view defined. Viewports allow the CAD user to better visualize a part during the 3D modeling process. They give the user multiple points of view at one time. If you are using CADKEY for 2D drafting work, only one viewport need be defined. For more information on viewports see Chapter 14, Display Options.

Since many people are used to creating drawings that have the primary shape description in the front view, it is suggested that you create and save a "blank" file that has the view changed to view 2, the front view. Include in this file any other default changes you wish to make, such as a change in note text height or anything else found in the Status Window. In the future when you initialize the system, retrieve this file and you will be ready to start drawing.

CONSTRUCTION PLANES

The construction plane switch (**CPLANE**) allows the user to define a construction plane or surface which is not normal or parallel to the screen in the construction view. The options for selecting the construction plane include the following:

3 PTS—indicate three points; the first two points determine the X axis and the third point defines the Y axis. DO NOT use the **CURSOR** option when selecting the points. **ENDENT** generally works the best.

PT/LINE—select a point or position in space and two end points of a line to define a new plane. DO NOT use the **CURSOR** option when selecting the end points or position.

2 LINES—select the end points of two lines which bound the desired plane.

VPORT—assigns the plane associated with a specific viewport.

ENTITY—select an existing planar arc, conic, or 2D spline and the new plane is automatically defined in the viewports.

KEYIN—define a plane by selecting a concatenated series of rotations from the rotation menu. Choose **DONE** after rotations have been selected.

CP=DV—automatically sets the construction plane equal to the display view assignments.

SAVE—stores the construction plane defined in the data base. The next available number in the data base is assigned to this new plane.

See Figure 11-1.

2D/3D CONSTRUCTION SWITCH

The 2D/3D construction switch allows the user to work in two different modes. The 2D mode can be used in cases where there is no depth or when there is only the need to work in a defined plane at a given depth. Note the 2D switch uses the depth located in the Status Window. The 3D switch is used in all other 3D construction.

ANGLE CONVENTION

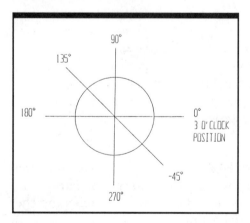

Figure 3-7 CADKEY's standard angle convention

In many options you are requested to enter an angle value. The angle may be specified in two ways: positive or negative numbers of degrees. The zero (0) degree angle is located at 3 o'clock. All (+) positive numbers of degrees will move in a counter-clockwise direction from this origin. All negative (–) numbers of degrees will move in a clockwise direction from this origin. Figure 3-7 illustrates the angle convention used in CADKEY.

For rotation angles, the sign convention (+ or –) obeys the "right-hand" rule:

Positive angles rotate drawings about axes in the direction that your fingers curl when your thumb is pointing along the axis in the positive direction (using your right hand).

TITLE BLOCKS AND BORDERS

CADKEY offers ten different ANSI (American National Standards Institute) standard title blocks in both English and metric modes. These borders are stored as **PATTERN** files and can be installed when CADKEY is installed. These files are not automatically installed in the system; therefore, you must choose to install them during the installation procedure.

Filenames for the title blocks are as follows (all filenames end with the .PTN extension to identify them as pattern files):

English	Metric
BORDA	BORDA0
BORDB	BORDA1
BORDC	BORDA2
BORDD	BORDA3
BORDE	BORDA4

REVIEW QUESTIONS

1. What are the nine areas that the display screen is divided into?
2. When you need help to get through an option, where can you look?
3. What is a default value and how can it be recognized?
4. If you wanted to know what options you had previously selected, where could you find this information?
5. What are the two ways that you can select options from the Menu Options Window?
6. If you wanted to know what view you were in or whether the **SNAP** or **GRID** were on, where could this information be found?
7. What is the function of the Cursor Tracking Window?
8. What is the cursor and what is its function?
9. What is the difference between View and World Coordinates?
10. Draw the positive axes for the View Coordinate system.

CADKEY 2D Drawing and Dimensioning Tutorial

OBJECTIVES

After completing this chapter, the user will be able to:

- Create a title block and border line with text.
- Save a drawing as a part and pattern file.
- Draw lines and rectangles.
- Add text to a drawing.
- Change text height and font.
- Change pens and levels.
- Change the display viewports.
- Change the window.
- Create horizontal, vertical, radial, and diametral dimensions.
- Retrieve a pattern file.

The best way to learn CADKEY is to start using it as soon as possible. This chapter is a series of tutorials that can be followed to learn the basic operations of CADKEY and create simple drawings. Some topics are introduced without much explanation but can be studied further in later chapters. The tutorials in Chapter 4 are an introduction to some of CADKEY's most often used 2D drawing commands. The tutorials are meant to get the user comfortable with CADKEY. The best way to use this chapter is to create the drawings with CADKEY as you read the tutorial.

LOADING CADKEY

Typically CADKEY is loaded by changing to the directory that it is stored in, such as **CD\CADKEY**. The display driver for the display device on your workstation may have to be loaded if it is not done at boot-up from the Autoexec.bat file. Entering **CADKEY** at the prompt will start the executable file that loads the CADKEY software. After the software is loaded a prompt reads: **Enter filename ( )**. To load an existing file enter the name and press **RETURN**. Pressing **ESC** also can be used to load CADKEY without entering a filename. Figure 4-1 shows how the screen display might look after loading CADKEY using **ESC**.

INTERACTION HINTS   Use the following guidelines to help you through this hands-on tutorial.

1. Follow each task step by step in the order that it appears, and refer to the corresponding diagrams on each page.

2. Take the time to read each prompt and review the screen display as you move along. This will help you understand what is going on in the function.

3. When invoking any Immediate Mode commands, remember to hold down the **ALT** or **CTRL** key and simultaneously press the assigned letter.

GETTING STARTED

When you first enter the system, you are prompted for a part filename:

Enter part filename ( ):

Type the filename **blank** and press **RETURN**. For convenience we will store this "empty" part file for use later in the system. Invoke the Immediate Mode sequence **CTRL-F** (hold down the **CTRL** key and simultaneously press the letter **F**) to store the current display. You are prompted:

Enter part filename (blank):

Accept the default filename in parentheses **(blank)** by pressing **RETURN** only. The empty part is automatically stored and we

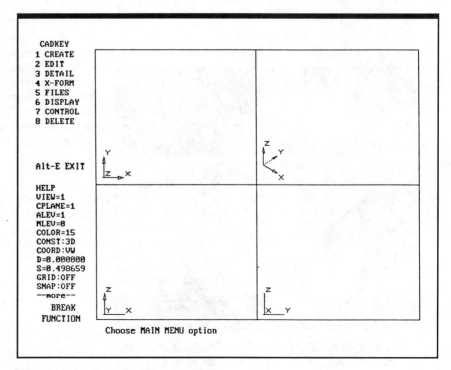

Figure 4-1   Default viewport settings at start-up

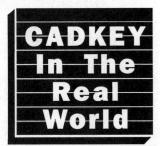

New Product Package Features
3-D Model of Mount Shasta

When a serious bicyclist buys an Avocet Altimeter 50™, he or she purchases a miniaturized computer, designed for bicycle use, that calculates altitude, trip climb and descent, total climb and descent, current speed, average speed, maximum speed, trip distance, and total distance. The Altimeter 50 also serves as a clock and a stopwatch. However, the buyer is buying not only a high-tech product, but also high-tech packaging. When Avocet, Newark, California's well-known manufacturer of equipment for bicycling enthusiasts, planned to introduce its Altimeter 50, the company wanted a unique sales package that would reflect the distinctiveness of the product, as well as provide concerned users with a convenient place to store the manual and the hardware for attaching the Altimeter 50 to a bicycle.

Avocet chose Mount Shasta in northern California as the theme, and Geri Engineering, Inc. of San Carlos, California, as the designer of the Altimeter 50's package. Don Geri, president of Geri Engineering, obtained computerized topographical data of the Mount Shasta area from the United States Coastal and Geodetic Survey (C&GS), and through Geographic Information Systems of Houston, Texas, who converted the C&GS data from their format on magnetic tape into ASCII format on disk.

Replica of Mount Shasta in California makes a dramatic product package for the Avocet Altimeter 50.

A Curious Anomaly

Don discovered an anomaly in the C&GS data: the x and y data were in metric measurements; the z data were in English measurements. He wrote a CADL program to convert the metric measurements into English measurements, and to read the data into CADKEY. The reason for choosing English measurements (rather than metric) for entering the data into CADKEY was that the model of Mount Shasta which Avocet wanted for the Altimeter 50's package was going to be a 3-D topographical model, in which each contour line would represent 100 feet in height.

The C&GS data, now in CADKEY, consisted of a 3-D array of points in space. Don wrote another CADL program to mesh the data points in the array. From this mesh of data points Don created 2-D cubic splines in CADKEY to represent the contours of the topography, using linear interpolation between the data points.

Modeling a Mountain

Don defined the elevation planes in CADKEY as 100-foot-high topographical steps. Using CADKEY's sectioning and level-defining capabilities, he marked all of the points at which the elevation planes intersected the linear meshing. He then sorted the topographical data to levels by elevation step, and joined the data points on each elevation by 2-D cubic splines to form the continuous topography for the model. Mindful that the purpose of this model was to serve as a display package for the Altimeter 50, Don completed the model by constructing four elevated walls supporting the model mountain, and a mounting fixture for the product. Creating this mounting fixture and scaling it to the packaging size required some minor modifications to the sloping contours of the mountain. Designing the corners of the package to snap-fit into a clear plastic calibrated cover also required some other minor modifications to the topography. In most respects, the topography of Mount Shasta in the Avocet model is accurate to scale.

—*Continued, next page*

CADKEY In The Real World (Continued)

Making a Mold

Don's 3-D model of the mountain went to Kattenhorn Tool and Die Company in San Lorenzo, California. Don Kattenhorn used CADKEY and a third-party CAM product to generate the numerical-control tool path to manufacture the mold components directly from the CADKEY part file of the mountain. Don Kattenhorn used drawings only to determine the outline and orientation of the model. In manufacturing the mold, he did not use detailed drawings for any topography.

Don Kattenhorn manufactured five graphite electric-discharge-machine (EDM) electrodes to produce the mold's cavity in steel. An electrode is a positive model made of electrically conductive material, most often graphite, used to create a negative cavity in a mold. The electric arc erodes the steel from which the mold is being made, so that the electrode creates the cavity by quite literally sinking into the steel. Consequently, the electrode is called a *sinker*. Four of the graphite sinkers served for rough-cutting the steel. The fifth electrode provided the final finish on the cavity.

Don Kattenhorn used multiple graphite sinkers to produce the mold's cavity because, as the electrical arcs are formed and broken, the electrode not only erodes the material from which the mold is being made, it also consumes a little of itself. Thus, sharp edges become rounded. The mountain-topography design required some distinctly sharp edges. Using more than one CNC-generated identical graphite electrode was an economical way to produce the mold's cavity with sharp edges.

Henry Plastics of Fremont, California, manufactures the injection-molded plastic packaging for Avocet's Altimeter 50.

Adapted with permission from **3-D World**, *v. 5, no. 4, July./Aug., 1991.*

are ready to begin the new part construction. (If the given filename exists in the part directory, you are prompted to replace the existing file with this new one. Choose **F2 YES** and continue.)

In this first section of the tutorial we will review the various methods of choosing or selecting options from the screen display. You should follow one method throughout the section. You can repeat these same steps using the other method given when completed. For example, if you have a tablet or mouse as your input device, follow the instructions on how to use only these. You will find this an efficient means for understanding the interaction of the system as you progress.

THE WORKPLACE

THE CURSOR   A cross-like symbol called the *cursor* (+) lies within the boundaries of the Drawing Window to make selecting and positioning entities easy and precise. When the keyboard is used as an input device, the cursor is found in the bottom left corner of the screen.

The intersection point of the two lines that make up the crosshair cursor is used as the selection and position area. The closest entity to this area, within the boundaries of the perimeter of the cursor, is the first to be considered for selection. Place the cursor directly upon a desired entity or position for best results.

SELECTING FROM MENUS

The Menu Options Window is located in the upper left corner of the screen. Various menus are displayed here each time a function is activated. Each option is numbered, with the numbers corresponding to the function keys located to the left of the keyboard. The Main Menu includes:

> 1 CREATE
> 2 EDIT
> 3 DETAIL
> 4 X-FORM
> 5 FILES
> 6 DISPLAY
> 7 CONTROL
> 8 DELETE

The **BACKUP** and **ESCAPE** options are always displayed at the same locations in this area. These allow you to cancel the present operation or return to the Main Menu, respectively, at any time. The system's command structure is designed so that you are always asked *what* you want to do, then *how* you want to do it.

Menu options can be selected in one of two ways. The function keys are always available, and the tablet or mouse select area is available whenever you have a tablet or mouse installed. You can switch between them at any time, and in any function.

To begin creating a part, choose the **CREATE** option from the Main Menu using one of two methods.

1. Using the function keys is the quickest access to the system's powerful functions without the need to move the cursor from the work area. Locate the function keys (**F1-F10**) to the left of your keyboard. The **F10** key will always return you to the previous menu (or back one step). Press the **F1** function key to choose the number 1 option, **CREATE**, from the Main Menu displayed.

2. Using a tablet or mouse allows you to select functions using the cursor. Move the cursor control to the upper left side of the work area (tablet/mouse) until two markers appear on each side of a menu option.

Create

Press the first button on the cursor control. The option is selected with the CREATE menu displayed.

Locating Position with the Cursor

Throughout your interaction with the system you are requested to indicate a position or select entities using the cursor. The cursor is easily manipulated using one of two methods.

1. Using a tablet or mouse, move the cursor on the display by sliding the cursor control about the tablet or pad.

2. Using a keyboard, the cursor can only be moved when you are requested to indicate a position. For now, locate the arrow keys to the right of the keyboard. With the **NUM LOCK** key toggled off, the arrows move the cursor vertically (top to bottom) and horizontally (side to side). As we move through the tutorial you will be able to move the cursor about the screen.

The **PgUp** and **PgDn** keys are used to change the increment in which the cursor moves. When the system is initialized the cursor moves at a one-half inch (.5) increment. Each time the **PgUp** key is pressed, this increment doubles. When the **PgDn** key is pressed, the current increment is divided by two. You will find these keys useful when positioning the cursor on desired parts of the drawing later in this tutorial.

INTERACTING WITH THE STATUS WINDOW OPTIONS

The Status Window is found directly below the Menu Options Window. The Status Window offers information about the current part displayed. You will find these options extremely useful for quick changes or alterations needed when constructing your part. As we move forward in the tutorial, a few of these options will be activated using either a keyboard or tablet/mouse using one of the following methods.

1. Using the keyboard–Immediate Mode commands via the keyboard.

Any option found in the Status Window, along with commonly used display and drawing functions, can be accessed using Immediate Mode commands from within any function of the system.

Immediate Mode commands are accessed only through the keyboard, even when a tablet or mouse is attached. Each command is invoked using the **CTRL** or **ALT** key and an assigned letter key. To invoke an Immediate Mode command you must hold down the **ALT** or **CTRL** key while simultaneously pressing the assigned letter.

For example, the key sequence **ALT-V** invokes the **VIEW** function. This automatically prompts for a new view assignment for the part displayed. The Immediate Mode commands available in the system are found in Chapter 9.

If you wish to use a tablet or mouse, consider another convenient method for changing the options in the Status Window area as described next.

2. Using a tablet or mouse allows you to move through a circular stack of windows by choosing the **—more—** option. To choose any of the displayed options, move the cursor into the Status Window area by sliding the cursor control to the left of the work space on the tablet or mouse. Slide the cursor control forward and backward until one marker appears next to a function displayed in this area. For example:

SNAP:OFF

Press the mouse button to activate the function. Each time the mouse button is pressed, the **SNAP** option is activated. Turn **SNAP ON** and **OFF**.

To move to the next window, slide the marker down to the —**more**— option in the same manner:

—more—

and press the space bar or mouse button. A new set of options is displayed. Choose the —**more**— option to review each window until you are returned to the first window in the stack.

HELP
VIEW = 1
CPLANE = 1
ALEV = 1
MLEV = 0
COLOR = 1
CONST: 3D
COORD: VW
D = 0.000
S = 1.000
—more—

INSTRUCTIONS AND INFORMATION

The Prompt Line appears across the bottom of the screen. This provides information and instructions to aid you in choosing menu options, entering data, indicating positions, and selecting entities.

The system automatically assigns defaults to some of its functions. These defaults usually appear within parentheses on the Prompt Line. Press **RETURN** to accept these values.

When a position is requested, the Prompt Line displays two prompts: one for the function and one for the position method chosen. For example,

Indicate start point/cursor-indicate position

requests the start position of an entity using the cursor positioning method.

Once a position is designated, the next prompt or instruction appears.

Keeping Your Place in the Menu Structure

Across the top of the screen the History Line displays the options chosen as you move along in the menu structure. This is a very convenient way to keep your place in the command sequence, especially when you are momentarily interrupted in the course of your work. You will find that each option, as it is selected, will appear on this line.

SPECIAL MENUS

You will see the following menus frequently as you use the system. These menus allow great flexibility in part construction and are reviewed briefly here. As you move through the interactive part of this tutorial, you will be able to choose from each of these menus.

The Selection Menu

The Selection Menu offers different methods of identifying and operating on existing entities.

SINGLE—allows single entities to be specified using the cursor.

CHAIN—allows the user to select a series of connected lines and arcs by defining the start and end entities in the chain.

WINDOW—enables the user to specify entities using a window (rubberbox). The cursor is used to specify two corners of this box. Options include **ALL IN, ALL OUT,** or **PARTIALLY IN** or **OUT.**

POLYGON—allows the user to define a line string or irregular polygon boundary to select entities. Options include **ALL IN, ALL OUT,** or **PARTIALLY IN** or **OUT.**

PLANE—entities are selected by identifying a plane.

GROUP—allows the user to select grouped entities.

ALL DSP—allows the user to select all entities displayed on the screen, only those specified, or excluding all those specified.

A selection symbol appears on all entities selected. The **F10** key may be used to reject those entities selected before **RETURN** is pressed.

The Position Menu

A standard Position Menu allows you to indicate the position (or location) of an entity in 3D space. You will review this menu throughout the tutorial since it is an integral part of the system.

An asterisk (*) always appears next to the active position method (the CURSOR option is the default). It is not necessary to select from this menu more than once within a function unless you wish to change the method of positioning.

Whenever this menu appears, two prompts appear on the Prompt Line, divided by a slash (/). The first prompt requests information on the current function, while the second prompt requests positioning information.

These are the nine position methods available:

1 **CURSOR**
2 **POINT**
3 **ENDENT** (endpoint of entity)
4 **CENTER**
5 **INTRSC** (intersection)
6 **ALONGL** (along a line)
7 **POLAR**
8 **DELTA**
9 **KEYIN**

COORDINATE SYSTEMS

Two types of coordinate systems are available: World and View.

View Coordinates

We will begin the part construction in View Coordinates. The system provides eight defined views (as listed), along with an unlimited number of user-defined views. View Coordinates are relative to the screen; the X axis is horizontal, the Y axis is vertical, and the Z axis points toward you. Each of the eight views is assigned a number for easy access:

1 TOP
2 FRONT
3 BACK
4 BOTTOM
5 RIGHT
6 LEFT
7 ISOMETRIC
8 AXONOMETRIC

The TOP view is the first view assigned when the system is initialized.

World Coordinates

World Coordinates are the units in which geometric entities are stored. Think of them as a stationary coordinate system in which objects are defined when viewed from any position. In this system, World Coordinates are relative to View 1, the top view, sometimes referred to as Model Space.

EXITING THE SYSTEM

NOTE: Do not exit the system if you wish to continue with the tutorial at this time.

To exit from the tutorial and return to DOS at any time, press **ESC** and choose the **EXIT** option from the Break Area or **ALT-E**. If you wish to store the current part displayed, choose **F2 YES** from the exit menu displayed. Enter a part filename (for example, SECTION) and press **RETURN**. If you do not wish to store the current part displayed, choose **NO** to exit the system without storing the part.

CREATING A DEFAULT DRAWING FILE

CADKEY has many default settings. For example, the default setting for the number of viewports is 1, as shown in Figure 4-2. The default text height is .3 and the pen number is 1. Many of the default program settings can be changed so they are set to your needs at startup. For example, you might want most text to be .125 high instead of .3 or the default viewport to be 1 instead of 4. Setting your own defaults can be done after CADKEY is loaded. The following steps are a tutorial in creating a default drawing file.

1. After CADKEY is loaded, change the viewport setting by entering **ESC-F6-F5-F5 (DISPLAY-VWPORTS-AUTOSET)**.

2. The viewport's pop-up menu is displayed. With the cursor, pick the single viewport. The display changes to a single viewport screen (Figure 4-2).

3. The default text height is changed by entering **ESC-F3-F9-F4-F2 (DETAIL-SET-TEXT-NOTE HT)** or selecting **NH** from the Status Window. If **NH** is not displayed in the Status Window, move the cursor to point at the word **—more—** and click to change to another page of the Status Window. In version 6 a dialog box will appear.

4. A prompt reads: **Enter note text height (current=0.3) =**. Enter the new text height of **.125** and press **RETURN**.

5. Change the dimension text height by entering **F1-DIM HT** from the current Dimension Menu or **DH** from the Status Window.

6. A prompt reads: **Enter dimension text height (current=0.3) =**. Enter the new height of **.125** and press **RETURN**.

    Many other detail settings can be changed at this time if desired. For example, the units can be changed from inch to metric. The section view tutorial in Chapter 5 describes how to make a "seed" file, which is similar to creating a default file described here.

7. After the file is created use **CTRL-F** to name and save the file. Enter the filename **BLANK**.

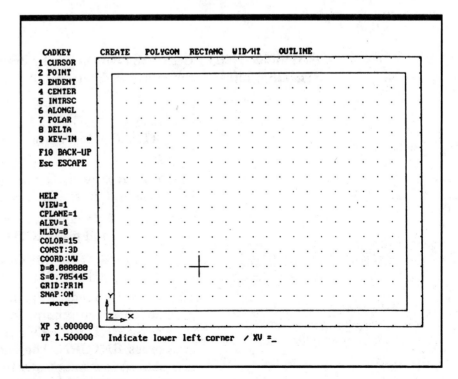

Figure 4-2   The border line

CREATING A TITLE BLOCK AND BORDER LINE

The creation of a title block and border line uses many of CADKEY's important functions, such as drawing lines, rectangles, display options, grid and snap, entering labels, and saving part and pattern files. The following steps are a tutorial that describes how to create a title block and border line. Refer to Figure 4-27 at the end of the chapter when creating the border line and title block. Whenever you see the word "click" in the instructions, use the select button on your mouse or other cursor input device. If you make an error in placing the lines, use the Immediate Mode command **CTRL-Q** to delete the line by moving the cursor over the incorrect line, clicking with the input device, and entering **RETURN**.

SETTING DRAWING PARAMETERS

1. Turn on the grid by pressing **CTRL-G** or selecting **GRID** in the Status Window. A prompt reads: **Choose option for grid display (current=OFF)**. Select **F3-ALL-DSP** to display the grid in the current viewport.

2. Turn the snap on by pressing **CTRL-X** or picking **SNAP** in the Status Window.

3. Turn on the cursor tracking by pressing **CTRL-T**. A prompt reads: **Choose option for tracking coordinate display (current=OFF)**. Select **F2 VIEW** to display View Coordinates. The lower left corner of the display should display x and y coordinates. Moving the cursor will change the coordinate value.

DRAWING THE BORDER LINE

Create the border line by drawing a rectangle. Place the lower left corner at coordinates **0,0,0**.

1. Select **ESC-F1-F1-F7-F2** (**CREATE-LINE-RECTANG-WID/HT**) to create a rectangle of a specified width and height.

2. A prompt reads: **Enter width of rectangle (0.5) =**. Enter **10.5** and press **RETURN**.

3. A second prompt reads: **Enter height of rectangle (0.5) =**. Enter **8** and press **RETURN**.

4. A prompt reads: **Indicate lower left corner**. Select **F9 KEYIN** to locate the lower left corner of the rectangle.

5. A prompt reads: **XV=**. Enter **0** and press **RETURN**. **YV=** is added to the prompt. Enter the Y coordinate position **0** and press **RETURN**. **ZV=** is added to the prompt. Enter **0** and press **RETURN**. The border line is drawn on screen. Enter **ALT-A** (automatic screen scaling) to display all of the rectangle on screen (Figure 4-2).

DRAWING THE TITLE BLOCK

Add the title block to the border line by using the line option with grid and snap. The grid and snap method of geometric construction is a beginner's technique that is not widely used by advanced CADKEY users.

1. Enter **ESC-F1-F1-F1-F1 (CREATE-LINE-ENDPOINTS-CURSOR)**. Draw a line parallel to and ½" from the bottom line of the border line. With grid and snap on, click on coordinate point **0, .5**. Use the cursor tracking read-out located at the lower left corner of the menu area to determine the coordinate position of your cursor.

2. Move the cursor to coordinate point **10.5, .5** and click.

3. Draw the vertical lines in the title block by moving the cursor to **10, .5** and clicking the selection button on the input device.

4. Move the cursor to **10, 0** and pick the point by clicking the mouse or other input device. The first vertical line is added to the drawing.

5. Move the cursor to **7.5, 0** and click.

6. Move the cursor to **7.5, .5** and click. The second vertical line is added to the drawing.

7. Move the cursor to **4, .5** and click.

8. Move the cursor to **4, 0** and click. The third vertical line is added to the drawing (Figure 4-3). This would be a good time to save the drawing in case of error or computer malfunction. A quick way to save a drawing is to use the Immediate Mode command **CTRL-F**. A prompt reads: **Enter filename ( ):**. Enter the name of this drawing, such as **TITLEBLK**, preceded by **A:** to save on the floppy disk, and **RETURN**. Filenames are limited to eight characters. Do not use spaces or periods. Drawings should be saved every ten or fifteen minutes on your data disk and your back-up data disk. Use **CTRL-R (REDRAW)** to remove any markers.

    Add one more horizontal and vertical line to the title block to complete drawing of lines. Before drawing those lines, enlarge the area of the title block on screen by using the Immediate Mode zoom display command.

9. Enter **ALT-W** to create a viewing window. Move the cursor to approximate coordinate position **6, -.5** and click.

10. Stretch the window around the area to enlarge by moving the cursor to approximate coordinate position **11, 1** and click. The area is automatically enlarged on screen, which will make it easier to add the remaining lines.

11. Enter **ESC-F1-F1-F1-F1 (CREATE-LINE-ENDPOINT-CURSOR)**.

12. Select **PgDn** (page down) once on the computer keyboard to halve the snap increment from ½ to ¼. On some keyboards it is necessary to make sure the number lock key is off to activate the **PgDn** key. The **PgDn** and **PgUp** (page up) keys are used to halve (**PgDn**) or double (**PgUp**) the current snap

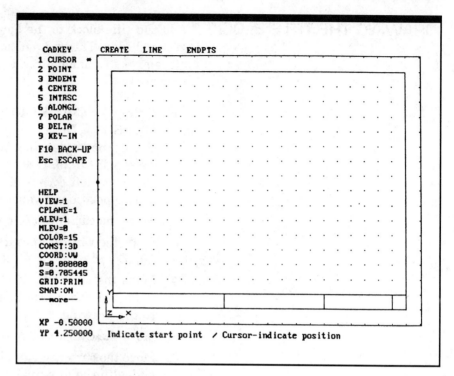

Figure 4-3 Border line and title block

increment. The snap increment can also be changed by using the **DISPLAY** option **SET**.

13. Move the cursor to coordinate position **10, .25** and click.

14. Move the cursor to coordinate position **7.5, .25** and click to draw the horizontal line.

15. Draw the short vertical line by moving the cursor to coordinate position **8.75, 0** and click.

16. Move the cursor to **8.75, .25** and click to draw the last line (Figure 4-4).

ADDING TEXT TO THE DRAWING

Text is added to a drawing by using the **NOTE** option. **NOTE** is a **DETAIL** option that adds text of a defined height, slant, angle, aspect ratio, and filled or unfilled. These **NOTE** settings are called *attributes*. Before adding text to a drawing, set the attributes.

1. Select **NH** from the Status Window. A prompt reads: **Enter note text height (current=0.3)**. Enter a new note text height of **.125** and **RETURN**.

   Before adding the text it is usually good planning to group different types or kinds of entities by level. A *level* is a CAD term used to describe different drawing sheets and can be used to separate entities in a drawing. For the title block, the text is to be placed on a separate layer.

2. Enter **ESC-F6-F6-F1 (DISPLAY-LEVELS-ACTIVE)**. A prompt reads: **Enter new level number from 1-256**

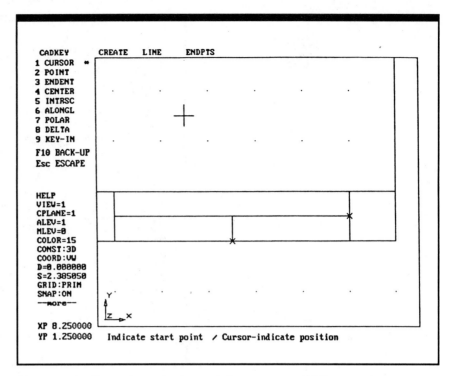

Figure 4-4 Zoom in to the lower left corner to add lines and text.

(**current=1**). Enter the number **2** and **RETURN** to assign the text to be added later to level number 2. CADKEY has 256 levels available to the user, but level 256 is invisible. All entities added to the drawing from this point are assigned to level 2. After entering the level number, a prompt reads: **Enter level descriptor:**. Names can be assigned to every level used in a drawing. Level descriptors are limited to thirty characters. For this example, enter **Title Block Text** as the level descriptor and press **RETURN**. By entering **F7 NAME**, **F4 LIST** a list of levels and their descriptors will be displayed on screen (Figure 4-5).

If a plotter is to be used for hardcopy output, different pens can be assigned to entities. For the title block and border line, use a heavy black line. For the text, use a thin line so it is more readable. Different colors could also be assigned to the pen if you are using a color display device.

3. To change the pen number assigned to the text, use Immediate Mode command **ALT-Z** or select **PEN** from the Status Window. A prompt reads: **Enter new pen number from 1-8 (current=1) =**. Enter the number **2** and **RETURN**. All entities added to the drawing from this point are automatically assigned pen 2. Use **ALT-Z** to change back to pen 1 when needed.

4. If you are using a color display, change the color assigned to text by entering Immediate Mode Command **ALT-X** or selecting **COLOR** from the Status Window. The menu is replaced with a color bar pop-up menu (Figure 4-6) and a prompt reads: **Cursor**

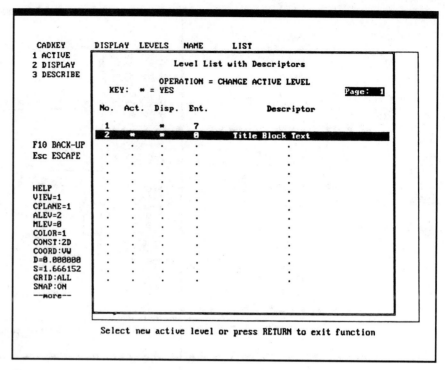

Figure 4-5   Level List with Descriptors pop-up menu

select new color or press RETURN to key-in color number. Pick the color from the color bar with the cursor or enter a number.

5. Enter **PgDn** twice to decrease the snap increment to **.0625**.

6. To enter the text, select **ESC-F3-F2-F1 (DETAIL-NOTE-KEYIN)**. A prompt reads: **Enter text:**. Enter the word **SCALE:** and press **F6 (SAVETX)**.

7. Place the text string by moving the cursor to coordinate position **7.5625, .0625** and click. The text string is drawn on screen (Figure 4-7). Be careful not to click on the input device again because you will add another string of text over the first one without knowing it. This may cause you problems later when deleting or plotting. To erase a line of text from a drawing, select Immediate Mode command **CTRL-Q**, click on the lower left corner of the text string, and press **RETURN**.

8. Select **F10 (BACKUP)** to enter the next line of text. A prompt reads: **Enter text:**. Enter **DATE:** and press **F6 (SAVETX)**.

9. Move the cursor to coordinate position **8.8125, 0.625** and click to place the second line of text (Figure 4-7).

10. Select **F10 (BACKUP)** to enter the next line of text. A prompt reads: **Enter text:**. Enter **YOUR NAME** and press **F6 (SAVETX)**.

11. Move the cursor to coordinate position **7.5625, 0.3125** and click to place the text (Figure 4-7).

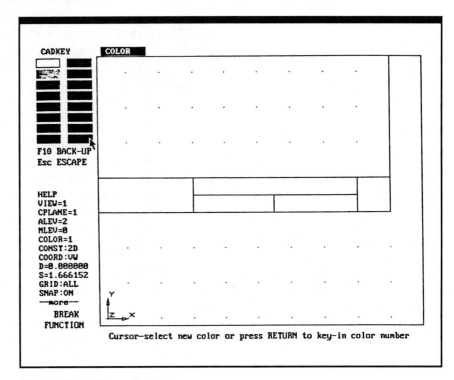

Figure 4-6   The color bar menu

Two more lines of text have to be added to the title block to complete the drawing. Before that is done, the full title block must be displayed and the text attributes changed.

12.   To display all the title block, enter the Immediate Mode command **ALT-A** (automatic scaling) or **ALT-B** (back one).

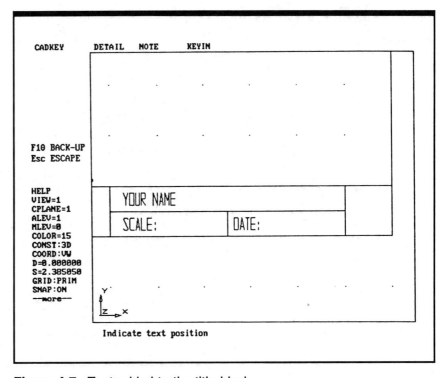

Figure 4-7   Text added to the title block

13. Change the text height by selecting **NH** from the status window. A prompt reads: **Enter note text height (current=.125) =**. Enter **.25** and press **RETURN**.

14. Change the text font by selecting **NF** from the status window, **F2-BOLD**. A prompt reads: **Enter text slant angle =**. Enter **-10** to create a ten degree slant of the text characters. The next prompt reads: **Fill font?** Enter **F2-YES** for filled text.

15. Change pens by entering **ALT-Z**. Enter **3** and **RETURN** to change the pen setting.

16. Enter **ESC-F3-F2-F1 (DETAIL-NOTE-KEYIN)**. A prompt reads: **Enter text:**. Enter the **TITLE** of the drawing and press **F6 (SAVETX)**.

17. Move the cursor to coordinate position **4.1875, 0.125** and click.

18. Enter **F10-BACKUP** once to display the prompt **Enter text:**. Enter the name of **YOUR SCHOOL** or **COMPANY** and press **F6 (SAVETX)**.

19. Move the cursor to coordinate position **.125, .125** and click. The text is added to the drawing (Figure 4-8).

20. Save the drawing as a part file by pressing **ESC** and **CTRL-F**. Enter the same filename used earlier and press **RETURN**. Replace or update the file saved earlier by entering **F2-YES** when prompted if you want to replace the file.

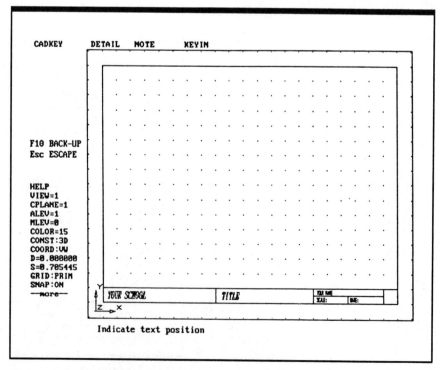

Figure 4-8 Completed title block

CREATING A PATTERN FILE

The border line and title block is completed and saved as a part file. A border line and title block are used many different times. Instead of redrawing them for each new drawing or printing copies to be used in your plotter, make a pattern file. A pattern file creates a template of a drawing that can be merged or added to any other CADKEY drawing. The following steps show how to create a pattern file.

1.   Select **ESC-F5-F2-F1 (FILES-PATTERN-CREATE)**.

2.   Select all the entities by entering **F7-ALL DSP** and **F1-ALL**.

3.   A prompt reads: **Indicate pattern base position**. The point chosen for the base position is very important when making patterns. For example, the base position for an end view of a hex head bolt would be the center of the hex. The bolt could then be accurately placed at the intersection of the center lines of the hole. For this example, move the cursor to the middle of the border lines, approximate coordinate position **5.5** and **4**, then click.

4.   A prompt reads: **Enter pattern filename ( ):**. Enter **TITLE-A** and press **RETURN**. The pattern file is saved on disk. Pattern file names are limited to eight characters with no spaces or periods.

CREATING A 2D DRAWING

This tutorial demonstrates how to create a simple two-dimensional drawing with CADKEY. Lines, arcs, rectangles, circles, and fillets are drawn to create the object. Dimensions are added and the edit function TRIM is demonstrated. The title block pattern is retrieved to complete the drawing. Some of the methods used to create the 2D object may not be the most efficient method of using CADKEY. However, the methods are used to demonstrate important features of CADKEY necessary to create drawings on your own. The drawing to be created is shown in Figure 4-27 (page 73).

PREPARING CADKEY FOR THE DRAWING

Start the drawing by loading the blank file created earlier.

1.   To load a file select **ESC-F5-F1-F2-F1 (FILE-PART-LOAD-NO)**.

2.   A prompt reads: **Enter part filename ( ):**. Enter **BLANK** and press **RETURN**. The default file and its settings are loaded.

3. Before starting the drawing it would be helpful to move the co-ordinate point **0,0,0** closer to the center of the screen instead of the extreme lower left. Use the pan option by entering Immediate Mode command **ALT-P.**

4. A prompt reads: **Cursor indicate new window position.** Moving the cursor moves a rectangle that represents the pan window. With the cursor move the rectangular window to the position shown in Figure 4-9 and click. Press **ESC** to cancel the pan command.

5. Check the new 0,0,0 position by moving the cursor on screen and looking at the cursor display located in the lower left corner of the menu area. If cursor tracking is not on, enter **CTRL-T** and **F2-VIEW** to activate.

6. Make sure that the grid and snap are on. Turn the grid on by picking it in the Status Window or entering **CTRL-G.** Turn the snap on by picking it in the Status Window or entering **CTRL-X.**

DRAWING THE 2D OBJECT

The rectangle command is used to create the 2D object.

1. Select **ESC-F1-F1-F7-F2** (**CREATE-LINE-RECTANG-WID/HT**).

2. A prompt reads: **Enter width of rectangle (0.5) =**. Enter a value of 6 and **RETURN**.

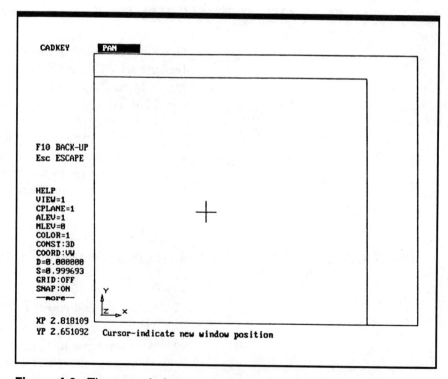

Figure 4-9 The pan window

3. A prompt reads: **Enter height of rectangle (0.5) =**. Enter a value of 4 and **RETURN**.

4. A prompt reads: **Indicate lower-left corner**. Move the cursor to coordinate position **0,0,0** and click. Make sure that the grid and snap are on before placing the rectangle. If you make a mistake, enter **CTRL-Q** and pick each line of the rectangle to erase it.

5. To see the rectangle on screen enter **ALT-A** (automatic scale), then **ALT-H** (automatic half scale; Figure 4-10).

6. Use the line option to create the short lines in the upper left corner of the rectangle. Enter **ESC-F1-F1-F2-F1 (CREATE-LINE-STRING-CURSOR)**.

7. Move the cursor to coordinate **1,4** and click.

8. A rubber-band line stretches from the picked point as the cursor is moved. Move the cursor to coordinate **1,3** and click to draw the first line.

9. Move the cursor to coordinate **0,3** and click to draw the second line (Figure 4-11). The **STRING** option was used to create connected lines, which is faster than using the **ENDPOINT** option, which creates separate lines.

10. The **EDIT** command is used to erase part of the lines from the upper left corner of the rectangle. Enter **ESC-F2-F1-F1 (EDIT-TRM/EXT-FIRST)**.

11. A prompt reads: **Select entity to trim**. Position the cursor on the part of the line that you want to keep. Pick the horizontal line shown in Figure 4-12.

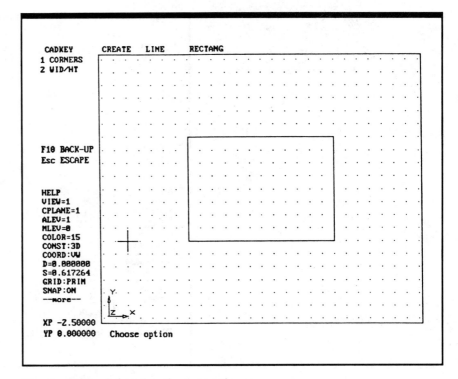

Figure 4-10 A four-by-six rectangle

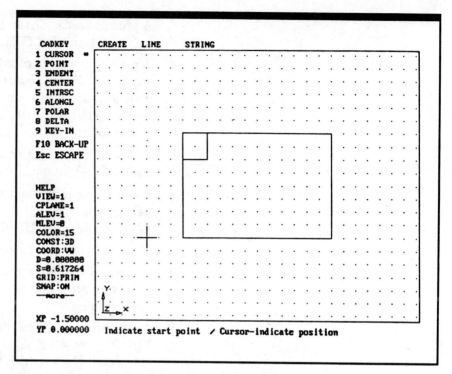

Figure 4-11 Drawing lines with the STRING option

12. A prompt reads: **Select trimming entity (position cursor near intersection)**. Pick the line that will act as the cutting entity. Pick the short vertical line shown in Figure 4-13. The long horizontal line is trimmed.

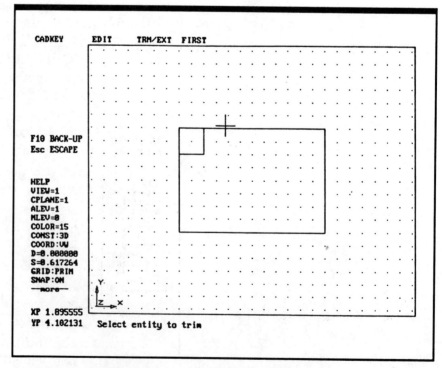

Figure 4-12 Selecting the entity to trim, the "keeper"

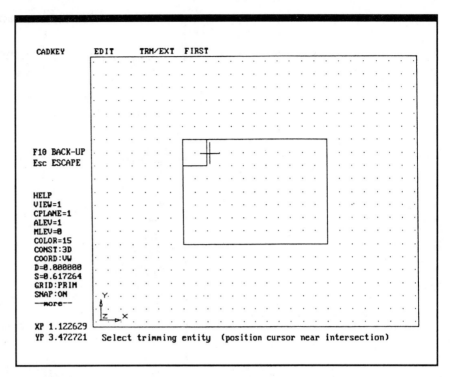

Figure 4-13 Selecting the trimming entity, the "cutter"

13. Trim the vertical line by using the same command. A prompt reads: **Select entity to trim**. Pick the part of the long vertical line that you want to keep.

14. A prompt reads: **Select trimming entity (position cursor near intersection)**. Pick the short horizontal line that will act as the cutting entity. The line is trimmed (Figure 4-14). Use **CTRL-R** to redraw the screen and turn the erased grid points on.

   When using the **EDIT-TRM/EXT-FIRST** option it may be helpful to remember that the part of the first entity to pick is the "keeper" and the second pick is the "cutter."

15. The long vertical line on the right side of the rectangle has to be erased. Press **ESC** and **CTRL-Q** (Single delete) to erase entities by cursor picking.

16. A prompt reads: **Select entity 1 (press RETURN when done)**. Pick the line with the cursor.

17. A prompt reads: **Select entity 2 (press RETURN when done)**. Press **RETURN** to erase the line (Figure 4-15).

18. The arc is added to the drawing by entering **ESC-F1-F2-F6 (CREATE-ARC-BEG+END)**.

19. A prompt reads: **Enter included angle (30) =**. CADKEY is requesting the sweep angle of the arc to be drawn. To draw half a circle enter **180** and press **RETURN**.

20. A prompt reads: **Indicate starting point**. Move the cursor to coordinate position **6,0** or select the end point of the lower line and click.

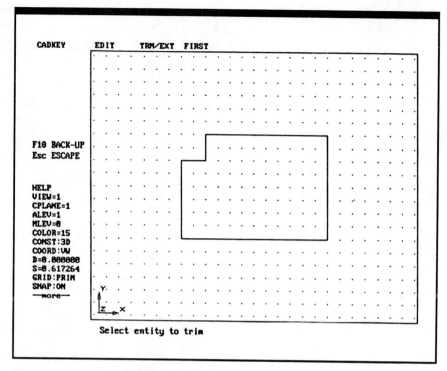

Figure 4-14 Corner after trimming

21. A prompt reads: **Indicate ending point**. Move the cursor to coordinate position **6,4** and click. The arc is added to the drawing (Figure 4-16).

22. The circle is added to the drawing by entering **ESC-F1-F3-F2 (CREATE-CIRCLE-CTR-DIA)**.

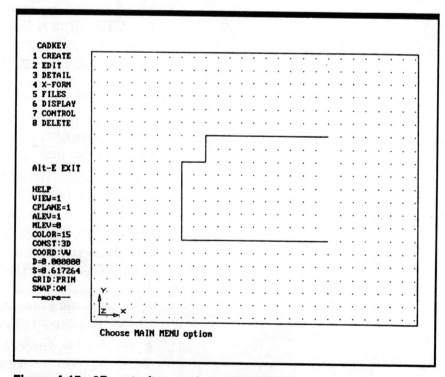

Figure 4-15 2D part after erasing using CTRL-Q

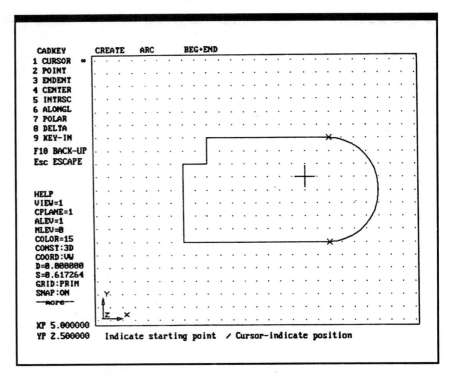

Figure 4-16   Creating an arc using the BEG+END option

23.   A prompt reads: **Enter diameter (1) =**. Enter the diameter of **1.25** and **RETURN**.

24.   A prompt reads: **Indicate center**. Define the center point of the circle by moving the cursor to coordinate position **6,2** and clicking. The circle is added to the drawing (Figure 4-17).

25.   The small round on the lower left corner of the 2D drawing is added by using the **FILLET** option. Enter **ESC-F1-F6-F1-F1 (CREATE-FILLET-ARC-TRIM)**. Make sure **QTRIM** in the status window is **OFF**.

26.   A prompt reads: **Enter radius (0.5) =**. Enter a radius of **.375** and **RETURN**.

27.   A prompt reads: **Select 1st fillet entity**. Pick one of the lines near the corner that is to have the round. Do not put the cursor on the corner.

28.   A prompt reads: **Select 2nd fillet entity**. Pick the other line near the corner with the cursor. The fillet is added to the drawing (Figure 4-18).

29.   Before adding center lines to the drawing, the pen and line type must be changed. Enter **ALT-Z** to change the pen. At the prompt enter the number **2** and **RETURN**. Change the line type by entering **ALT-T**. A prompt reads: **Cursor select new line type or press RETURN to menu-select**. With the cursor pick the third line from the pop-up menu, the center line (Figure 4-19).

30.   Draw the center lines by entering **ESC-F1-F1-F1 (CREATE-LINE-ENDPOINT)**.

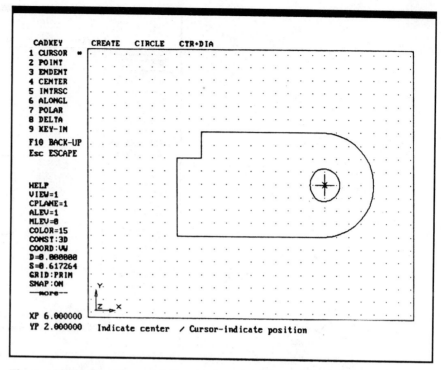

Figure 4-17   Creating a circle using the CTR+DIA option

31.   Select **PgDn** once to decrease the snap increment to **.25**.

32.   Draw the vertical center line by moving the cursor to coordinate position **6, -0.25** and clicking. Move the cursor to **6, 4.25** and click to draw the vertical center line.

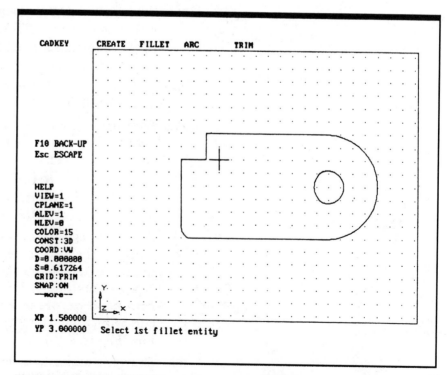

Figure 4-18   The FILLET option used to draw a round corner

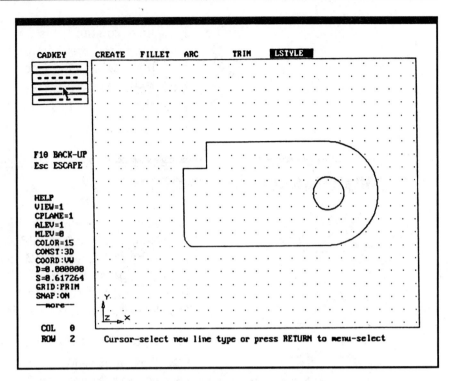

Figure 4-19 Selecting a new line type from the pop-up menu

33. Draw the horizontal center line by moving the cursor to coordinate position **4.5, 2.0** and clicking. Move the cursor to **8.25, 2.0** and click to draw the center line (Figure 4-20). The center lines are not drawn to ANSI standards, which would have the short lines cross at the center of the hole. To create center

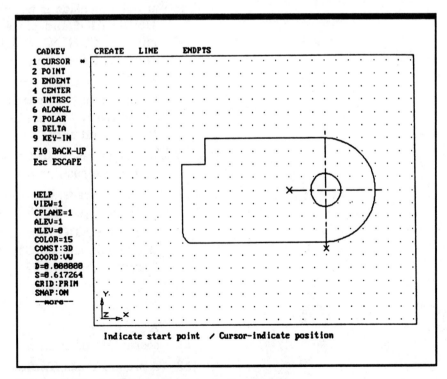

Figure 4-20 Adding center lines to the drawing

lines to standards use **CREATE-LINE-ENDPOINT**, set the snap increment to **.0625** by using PgDn, and draw the center lines using solid lines.

34. Enter **ALT-T** and pick the solid line in the pop-up menu to change back to visible lines.

DIMENSIONING THE DRAWING

Dimensions are added to the drawing using the **DETAIL-DIMEN-SION** options. Before adding dimensions, some parameters must be set by entering **ALT-Z** (pen number). Change the pen number to **3** and press **RETURN**. The dimension text height is changed by selecting **DH** from the Status Window. For the 2D drawing use the default setting of .3 so it is easy to read.

1. Enter **F3-F1-F1 (DETAIL-DIMENSION-HORIZNTL)**.

2. A prompt reads: **Indicate 1st position**. Move the cursor to coordinate position **0, .25** and click.

3. A prompt reads: **Indicate 2nd position**. Move the cursor to coordinate position **6,0** and click.

4. A prompt reads: **Indicate text position**. Move the cursor to coordinate position **3, -1** and click. The dimension is shown on the drawing, but it will not become part of the drawing until the next step is completed.

5. The horizontal dimension is added to the drawing (Figure 4-21). If the dimension is not placed where you want it, simply select **UNDO Alt-I** and place it again. Dimensions are erased by entering **CTRL-Q** (Single delete), picking the dimension text with the cursor, and entering **RETURN**.

6. Another horizontal dimension is added to specify the position of the notch in the upper left corner of the part. A prompt reads: **Indicate 1st position**. Move the cursor to coordinate position **1,4** and click.

7. A prompt reads: **Indicate 2nd position**. Move the cursor to coordinate position **6,4** and click.

8. A prompt reads: **Indicate text position**. Move the cursor to coordinate position **3.5, 5** and click.

9. Click on **UNDO** from the Break Function Window to reposition the dimension. The horizontal dimension is added to the drawing (Figure 4-22).

10. The vertical dimensions are added to the drawing by entering **F10-F2 (BACKUP-VERTICL)**.

11. A prompt reads: **Indicate 1st position**. Move the cursor to coordinate position **.25, 0** and click.

12. A prompt reads: **Indicate 2nd position**. Move the cursor to coordinate position **0, 3** and click.

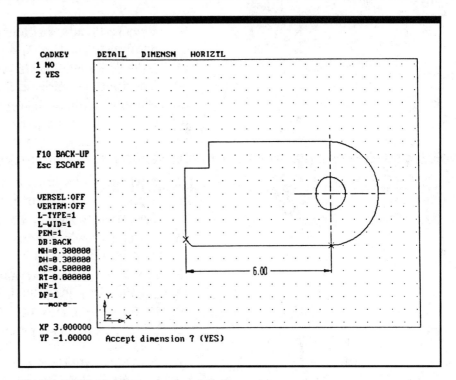

Figure 4-21   Adding a horizontal dimension

13.   A prompt reads: **Indicate text position**. Move the cursor to coordinate position **-1, 1.5** and click.

14.   The vertical dimension is added to the drawing (Figure 4-23).

15.   A prompt reads: **Indicate 1st position**. Move the cursor to coordinate position **-.75, 0** and click.

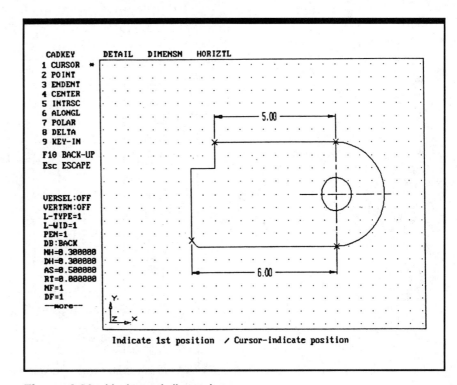

Figure 4-22   Horizontal dimensions

16. A prompt reads: **Indicate 2nd position**. Move the cursor to coordinate position **1, 4** and click.

17. A prompt reads: **Indicate text position**. Move the cursor to coordinate position **-2, 2** and click.

18. The vertical dimension is added to the drawing (Figure 4-23).

19. The radius of the arc is dimensioned by entering **F10-F4-F1 (BACKUP-RADIUS-REGULAR)**.

20. A prompt reads: **Select arc to be dimensioned**. Move the cursor over the arc and click.

21. A prompt reads: **Indicate dimension text position**. Move the cursor to coordinate position **8, 5** and click.

22. A prompt reads: **Choose leader side (current=LEFT)**. Enter **RETURN** or **F1** to select the left option.

23. The dimension will appear (Figure 4-24).

24. The diameter of the circle is dimensioned by selecting **F10-F10-F6-F1 (BACKUP-BACKUP-DIAMETR-REGULAR)**. CADKEY provides two arrow options when placing dimensions: arrows-in or arrows-out. The default setting is arrows-out. For the small diameter in the 2D drawing, the arrows should be set to point in so that the arrow points at the circle from the outside. This is changed by picking **ARR** from the Status Window.

25. A prompt reads: **Select arc to be dimensioned**. Move the cursor over the circle and click on the circle.

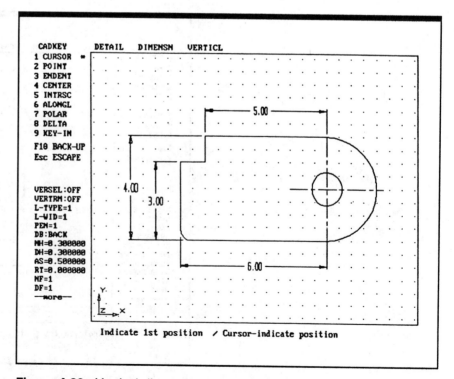

Figure 4-23 Vertical dimensions

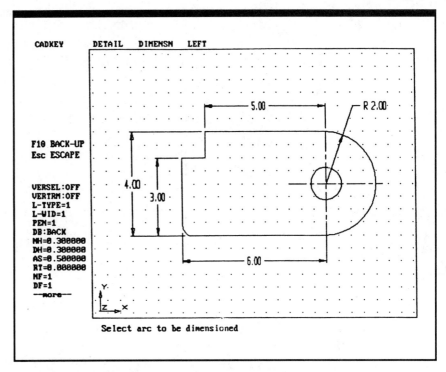

Figure 4-24   Adding a radial dimension

26.   A prompt reads: **Indicate dimension text position**. Move the cursor to coordinate position **8, -1** and click.

27.   A prompt reads: **Choose leader side (current=LEFT)**. Enter **RETURN** or **F1-LEFT** to select the left option.

28.   The dimension will appear (Figure 4-25).

29.   A note has to be added to the drawing to specify the size of the round.   Enter **F10-F10-F2-F1 (BACKUP-BACKUP-NOTE-KEYIN)**.

30.   A prompt reads: **Enter text**. Enter **NOTE: Fillet radius .375**. Press **F6 (SAVETX)** after entering the text.

31.   A prompt reads: **Indicate text position**. Move the cursor to coordinate position **1, -2** and click. The text is added to the drawing (Figure 4-26).

RETRIEVING A PATTERN FILE

To complete the drawing, the border line and title block are added to the drawing. The border line created earlier and saved as a pattern is added to this drawing using the **FILES-PATTERN** option.

1.   Enter **ESC-F5-F2-F2 (FILES-PATTERN-RETRIEV)**.

2.   A prompt reads: **Enter pattern filename ( )**:. Enter **TITLE-A** and **RETURN**.

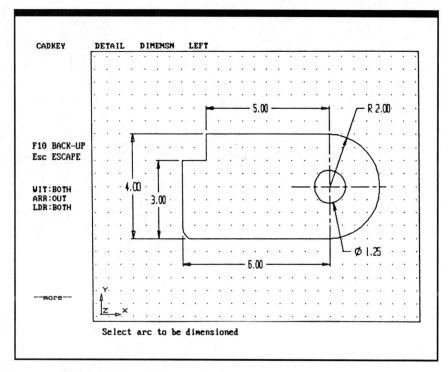

Figure 4-25 Adding a diametral dimension

3. A prompt reads: **Group pattern entities? (YES)**. Grouping entities cause CADKEY to treat the border line and title block as a single entity. By grouping entities it becomes

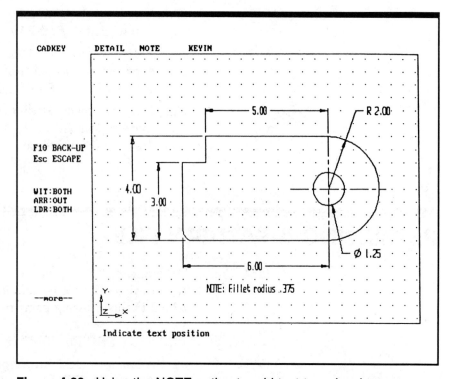

Figure 4-26 Using the NOTE option to add text to a drawing

easier to erase the border and title block if placed incorrectly. Select **F2-YES** to group the entities.

4. A prompt reads: **Enter group name (TITLE-A):**. Use the default name by pressing **RETURN**.

5. A prompt reads: **Choose level option (cur-lev)**. The pattern can be assigned to a separate level if needed. For this example, use the current level by entering **RETURN**.

6. A prompt reads: **Enter scale factor (1) =**. Patterns can be assigned to any scale before placement. For this example, enter a scale of **1.5** to enlarge the pattern.

7. A prompt reads: **Enter rotation angle (0) =**. Press **RETURN** to accept the default setting of zero.

8. A prompt reads: **Indicate pattern base position**. The pattern is placed onto the current drawing by moving the cursor to the center of the drawing (approximately coordinate position **3, 2**) and clicking.

9. Enter **ALT-A** (Automatic scale) to automatically scale the drawing which displays all the entities. If the border line and title block are not positioned accurately, enter **CTRL-Q**, pick any entity in the pattern and press **RETURN** twice to erase. All the entities in the pattern are erased because they were in a group. Move the cursor to a new position and click to place the border and title block again. Figure 4-27 shows the finished 2D drawing with border and title block.

10. To add to or edit the existing text use **ESC F3-F7-F3 (DE-TAIL-CHANGE-EDIT TX)** and select the text to edit.

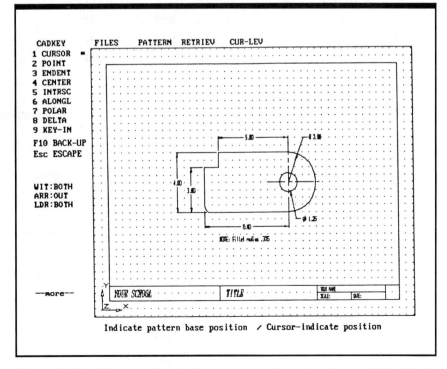

Figure 4-27 The completed 2D drawing with border line and title block

CHAPTER 5

CADKEY 2D Sectional View Tutorial

OBJECTIVES

After completing this chapter, you will be able to:

- Explain the basic operations of the CADKEY software.
- Explain the use of the history line and the prompt line.
- Make option selections from the many CADKEY menus.
- Use and explain the universal Position Menu.
- Use and explain the universal Selection Menu.
- Identify and invoke the Immediate Mode Commands.
- Create lines, arcs, circles, polygons, and fillets.
- Execute the **X-FORM** options **ROTATE**, **COPY**, and **MIRROR**.
- Perform edit options **TRIM/EXTEND**, **FIRST**, and **DIVIDE**.
- Execute the **X-HATCH** option to create section lines.
- Dimension horizontal and vertical linear distances.
- Place the arrows of the dimension lines in and out.
- Dimension circle diameters.
- Key-in and place notes as needed.
- Set detail options for arrowheads, witness lines, and note heights.
- Save and plot the sectional view drawing.

This two-dimensional tutorial has been designed and written to provide the first-time user with step-by-step instruction on how to create a two-dimensional drawing. This tutorial can be used as a training guide for group instruction or as a self-paced guide to be used individually. Of the tasks described, each one is broken down into a step-by-step format. It is to your advantage to review and understand the system's basic operations described on the next few pages so that you may move productively through each operation.

Task 1: Creating a seed file

Command Sequence:

ALT-V (VIEW)
F6 DISPLAY F5 VWPORTS F5 AUTOSET
ESC F5 FILES F2 PATTERN F3 LST/RTV
ESC (ESCAPE)

The first task is the creation of a **SEEDB** file containing a standard size border and the view that you most often begin creating in. In this case we will be creating a two-dimensional drawing, so the view number can be any number between 1 and 6. The front view **(2)** is the suggested view because when creating a three-dimensional model the front view generally contains the best shape description of the object. So, to develop a good habit that will be of use in the future, change to view 2 (front view). Along with your CADKEY software came a library of border files on the Sample Data Disk. For your use, you will use **BORDB.PTN**, which is a "B"-size (11″ × 17″) border. Now, in the future when you begin a new drawing, you can always load part file **SEEDB.PRT** when prompted:

Enter part filename ( ):

When the file comes up, it will have the B-size border displayed and will already be in view 2, the front view.

When you loaded CADKEY, the pattern files for the borders should have been loaded. To begin working in CADKEY, enter SEEDB for the filename when prompted to enter a filename.

Step 1: Press the **ESC** key. Use the Immediate Mode sequence **ALT-V**; or move the cursor into the Status Window area, place the cursor symbol next to the **VIEW:** option, and press the mouse button.

Step 2: You are prompted for a new view number. Enter the value **2** and press **RETURN**. At this point, the status area should read VIEW:2. You have changed your working view from the top to the front.

Step 3: If the border patterns are not loaded on the hard disk drive, press **ALT-H** and go to Step 14.

Step 4: Choose the Main Menu option **FILES (F5)**.

Step 5: Choose to enter the **PATTERN** files (**F2**).

Step 6: Choose the **LST/RTV** option (**F3**).

Step 7: You are prompted to define the path. At this prompt, you must enter the letter **A:** if the border patterns are on floppy diskette, or **C:\CADKEY\PTN** if the patterns are on the hard disk. Press **RETURN**.

Step 8: A list of the pattern files should appear on the screen at this time. Locate **BORDB.PTN**. Move the screen cursor until it highlights **BORDB.PTN** and press the mouse button to select the pattern.

CADKEY Users Design Workstation Furniture and Start New Company

Ronald Ernst, George Moore, and Arlyn Wiesman experienced the restructuring taking place in the American furniture-manufacturing industry first hand. Their initial dismay turned into a resolute decision to put their experience to work in producing high-quality, CAD-workstation furniture that can be adapted easily to meet the real needs of its users. They founded EMW, Inc. in Two Rivers, Wisconsin, in August 1991, during the depths of a recession. EMW uses CADKEY® to design their wood and laminate furniture.

Ease of assembly and disassembly at the customer's site is a major feature of EMW's furniture. Ease of assembly, requiring only a screwdriver (HONESTLY!), is a by-product of tight design and manufacturing control. EMW even includes a flat-head and Phillips-head right-angle screwdriver in the assembly kit for their furniture.

EMW manufactures its furniture from structural panels of wood and high-pressure laminate cut to specific size. The laminate serves particularly well for making resilient top surfaces. A computer-numerical-control manufacturing system does routing, grooving, and point-to-point boring for holes into which metal-to metal joinery is inserted and fastened. "The insertion of the steel-to-steel fasteners into the wood and laminate pieces requires extreme accuracy so that the components fit into a useful and ergonomic piece of furniture," said Arlyn Wiesman, one of the principals of EMW. "This is one place where we find CADKEY especially useful."

"CADKEY allows us to incorporate user-specific needs for work-surface space and, very importantly, for foot space," Arlyn said. "We can include specific needs for the user's line of sight, optimal keyboard height, and angle of viewing the monitor. CADKEY also gives us flexibility in customizing the legs and panels of tables to make them convenient for the customer to use."

"So far," Arlyn added, "the most popular colors for our furniture are natural wood grains, almond, and gray with black trim. However, we can customize our furniture and our colors to match any office decor."

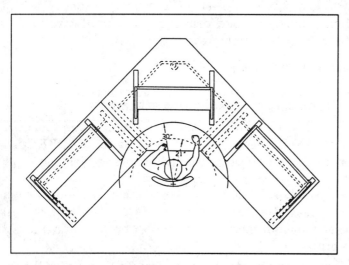

The work area, chair and leg clearance of EMW's CAD workstation furniture. The dashed lines represent the clearances in competing products.

Adapted with permission from **3-D WORLD**, *v. 6 no. 3, 1992.*

Step 9: You are asked if you want to group the pattern entities. Respond **YES (F2)**.

Step 10: When prompted for **Group Name, Level option, Scale factor** and **Angle of Rotation**, press **RETURN** to accept each of the default values from the prompts.

Step 11: You are then prompted to indicate the pattern base position/cursor indicate position. Choose the **KEY-IN (F9)** option for indicating base position for the border pattern.

Step 12:   Enter the value zero (0) three times—once for each of the coordinates—by pressing **RETURN** for each of the prompts.

XV = 0

YV = 0

ZV = 0

This places your base position, which is the lower left corner of the border, at the origin (0,0,0), and all **X** and **Y** points will be relative to this point. At this point, the lower left corner of the border pattern should appear on the screen.

NOTE: Only part of it appears.

Step 13:   Press the Immediate Mode sequence **ALT-A**. This is an autoscaling feature which will scale and display the whole B-size border.

Step 14:   Press the **ESC** key and the Immediate Mode sequence **CTRL-F** to file or save this part file.

Step 15:   When prompted **Enter part filename** ():, enter **A:SEEDB** and press **RETURN** to save it on a floppy disk.

Task 2:   Controlling the display status

Command Sequence:

CTRL-T (CURSOR TRACKING)

CTRL-W (CONSTRUCTION MODE)

CTRL-G (GRID)

CTRL-X (SNAP)

CTRL-L (LEVEL)

Step 1:   **CTRL-T** will turn the cursor tracking display on so that each time the screen cursor moves, the new X and Y coordinates will be updated and displayed. Select **F2 VIEW** from the Menu Window.

Step 2:   **CTRL-W** will toggle the 2D/3D construction switch so that you are working in the 2D mode. In the Status Window **CONST: 2D** should appear.

Step 3:   **CTRL-G** will turn the grid on. The default grid size that will be displayed is .5 on the X and Y axis increments. From the Menu Window select **F2 PRIM DSP** (Primary Display). If you wish to set your grid value to a smaller increment, choose **DISPLAY-GRD/SNAP**. The Status Window should read **GRID:PRIM**.

Step 4:   **CTRL-X** will turn cursor snapping on so that the screen cursor is moving a fixed increment each time it moves, causing a "snapping" effect. The **SNAP** can also be turned on by selecting it from the Status Window.

Step 5: Press the Page Down (**PgDn**) key once to cut the snap increment to .25 or use the **DISPLAY-GRD/SNAP-SNAP INC (F6-F7-F6)** so that you are snapping to the nearest quarter inch. When prompted for the X snap increment enter **.25** and **RETURN**. For the Y increment enter **.25** and **RETURN**.

Step 6: **CTRL-L** will change the current working level to keep the border on level 1 and the drawing on level 2 when prompted for a new level number. Later in this tutorial you may want to turn level 1 off so that it is not displayed. This will increase your redraw time on the screen. When prompted to enter new level number, enter the digit **2** and **RETURN**. When prompted for level descriptor, enter **GEOMETRY**.

Step 7: Press Immediate Mode Command **ALT-X**. Select cyan blue from the color icons in the Menu Window. The Status Window should read **COLOR=5**.

Step 8: Press **CTRL-F** to save. Filename (**A: SEEDB**) will appear. Press **RETURN** to accept the default filename. When prompted, "File already exists, replace it?" respond **F2 YES**.

Task 3:   Creating the basic geometry

Command Sequence:

> **ESC (ESCAPE)**
> **F1 CREATE**
> **F3 CIRCLE**
> **F2 CTR+DIA**
> **F9 KEY-IN**
> **F10 BACKUP**

In this task, you will create five circles and use three different Position Menu options to indicate their center points.

Step 1: Press **ESCAPE (ESC)** to return to the Main Menu.

Step 2: Choose the **CREATE (F1)** option from the Main Menu.

Step 3: Choose to create a **CIRCLE (F3)**.

Step 4: Choose to create the circle by the **CTR+DIA (F2)** method.

Step 5: You are prompted to enter the diameter of the circle. Enter the value **4** and **RETURN** (Figure 5-1, page 79).

Step 6: You are prompted this time to indicate the center. You will notice that the current menu displayed is the Universal Position Menu. Choose the **KEY-IN** option (**F9**). When prompted

> **XV=** enter **5.75** and **RETURN**
> **YV=** enter **5.5** and **RETURN**
> **ZV=** enter **RETURN**

At this point, the first circle should appear.

Step 7: Choose to **BACKUP (F10)**.

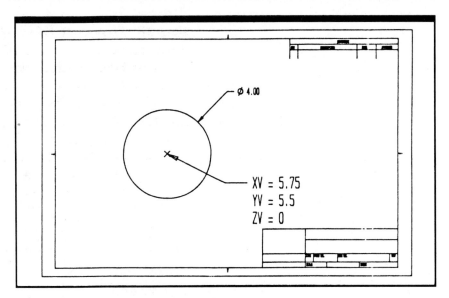

Figure 5-1   Enter the diameter of the circle.

Step 8:   You are prompted to enter a new diameter. Enter the value **3** and **RETURN** (Figure 5-2).

Step 9:   You are prompted to indicate the center. This time use the **CURSOR** option (**F1**) from the Position Menu. You can move the screen cursor to the center mark left from the previous circle and check your cursor tracking status. When the coordinates in the status area read **X = 5.75 Y = 5.5**, press the mouse button.

Step 10:   Choose to **BACKUP (F10)**.

Step 11:   You are prompted to enter a new diameter. Enter the value **2** and **RETURN**.

Step 12:   You are prompted to indicate the center. This time, choose the Universal Position Menu option **CENTER (F4)**, select any point on either of the two previous circles, and press the mouse button. The two-inch diameter circle will appear.

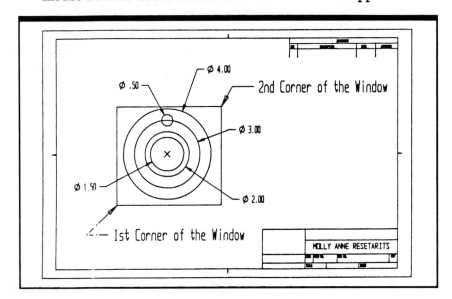

Figure 5-2   Enter a new diameter.

Step 13: Choose to **BACKUP (F10)**.

Step 14: You are prompted to enter a new diameter. Enter the value **1.5** and **RETURN**.

Step 15: You are prompted to indicate the center. Use one of the previously described positioning methods of indicating center (**KEY-IN, CURSOR,** or **CENTER**).

Step 16: Choose to **BACKUP (F10)**.

Step 17: You are prompted to enter a new diameter. Enter the value **.5** and **RETURN**.

Step 18: You are prompted to indicate the center. Choose **KEY-IN (F9)** from the Position Menu. Enter the values
XV = 5.75 and **RETURN**
YV = 7.00 and **RETURN**
ZV= RETURN
The drill hole at 12 o'clock will appear. Stop here.

Press the Immediate Mode command **CTRL-F** to save what you have created to this point. When prompted **Enter part file-name:**, enter the name **SECTION** and press **RETURN**.

Task 4:   A two-dimensional rotation

Command Sequence:

> **ESC (ESCAPE)**
> **ALT-W (WINDOW)**
> **F4 X-FORM**
> **F3 ROTATE**
> **F2 COPY**
> **F1 SINGLE**

Rather than creating and positioning the three remaining circles, the **ROTATE** command will copy and rotate the .5 diameter circle about the center point. The number of copies and the degrees between each rotation will have to be specified.

Step 1: Press **ESCAPE (ESC)** to return to the Main Menu.

Step 2: Press the Immediate Mode command **ALT-W** to create an alternate window. This allows you to zoom in on just the front view of the drawing.

Step 3: You are prompted to pick the first point of the window. Place the screen cursor near the lower left corner of the large circle (Figure 5-2) and press the mouse button.

Step 4: You are prompted to indicate the second point of the window. Move the cursor diagonally toward the upper right until the whole 4.0 diameter circle is within the window, and press the mouse button. The screen will look like Figure 5-3.
NOTE: Make sure that all of the desired entities are within the window.

Step 5: Choose the **X-FORM** option (**F4**).

Step 6: Choose the **ROTATE** option (**F3**).

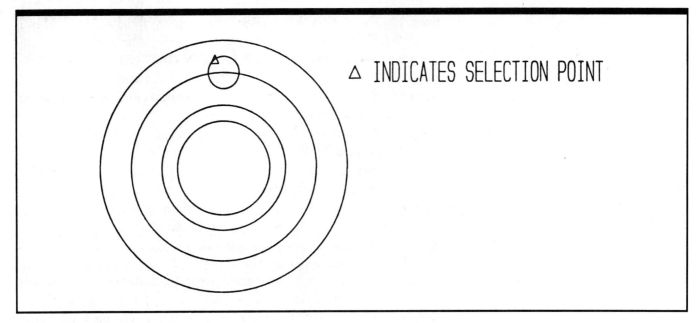

Figure 5-3 Enter the number of copies.

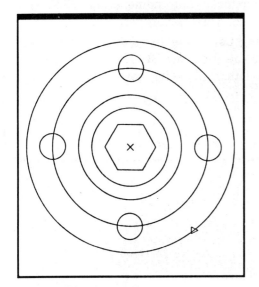

Figure 5-4 Choose the **CENTER** (F4) option.

Step 7: Choose the **COPY** option (**F2**).

Step 8: Choose the **SINGLE** selection method (**F1**) from the Selection Menu.

Step 9: You are prompted to select the first entity. Move the screen cursor until the crosshairs are clearly on the .5 diameter circle, and press the mouse button followed by **RETURN** (Figure 5-3).

Step 10: You are prompted to enter the number of copies. Enter the value **3** and **RETURN**.

Step 11: You are prompted to indicate the first point on axis. Use the position menu choice **CENTER** (**F4**) and select the 4.0 diameter circle as the reference entity (Figure 5-4).

Step 12: You are prompted to indicate second point on axis or press **RETURN**. Press **RETURN**.

Step 13: You are prompted to enter the rotation angle. Enter a value of **90** and **RETURN**.

The remaining three circles should be displayed very quickly.

Step 14: Use **CTRL-F** to save. Use the filename **SECTION** or **A:SECTION** to save to disk.

Task 5: Creating a polygon

Command Sequence:

ESC (ESCAPE)
CTRL-R
F1 CREATE
F1 LINE
F8 N-GON
F1 CORNERS

Step 1: Press **ESCAPE (ESC)** to return to the Main Menu. Press **CTRL-R** for redraw.

Step 2: Choose the **CREATE-LINE-N-GON (F1-F1-F8)** options from the Main Menu.

Step 3: You are prompted to enter the number of sides. Enter the value **6** and **RETURN**.

Step 4: You are prompted to enter the rotation angle. Enter a value of **90** and **RETURN**.

Step 5: You are prompted to enter the radius. The default radius is **.5**, so simply press **RETURN**.

Step 6: You are prompted to choose the option for specifying the radius. Choose distance across **CORNER (F1)**.

Step 7: You are prompted to indicate the center. Choose the **CENTER (F4)** option from the Position Menu. Select the 4.0 diameter circle as the reference entity and press the mouse button (Figure 5-4).

Step 8: You are asked if the entities which make up the polygon should be grouped. Respond **YES (F2)**.

Step 9: You are prompted to enter a group name; enter **HEX**. The polygon will be displayed.

Step 10: **SAVE** your drawing using the Immediate Mode command **CTRL-F** and enter the filename **SECTION**. If you are prompted, "File already exists, replace it?" respond **YES** to update your file.

Task 6: Creating the sectional view

Command Sequence:

> **ESC (ESCAPE)**
> **ALT-A (AUTO SCALE)**
> **F1 CREATE**
> **F1 LINE**
> **F2 STRING**
> **F9 KEY-IN**
> **ALT-W (WINDOW)**

You can see that the sectional view is symmetric, so only the top half of this view needs to be drawn. The other half will be created using a **MIRROR** command.

Step 1: Press **ESCAPE (ESC)** to return to the Main Menu.

Step 2: Press the Immediate Mode sequence **ALT-A**. This is the **AUTO SCALE** option which adjusts the size of your drawing so that it is fully displayed on the screen (Figure 5-5).

Step 3: Choose to **CREATE (F1)** a **LINE (F1)** by the **STRING (F2)** method, which will allow you to draw a continuous series of lines.

Step 4: The Position Menu is displayed and you are prompted to indicate the start point. Choose the **KEY-IN** option **(F9)**.

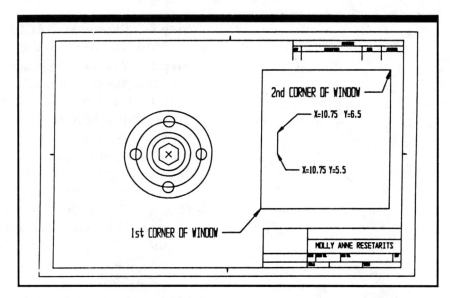

Figure 5-5 The AUTO SCALE option adjusts the size of drawing

When prompted

>**XV=** enter **10.75** and **RETURN**
>**YV=** enter **5.5** and **RETURN**
>**ZV=** enter **0** and **RETURN**

Then enter

>**XV= 10.75** and **RETURN**
>**YV= 6.5** and **RETURN**
>**ZV= RETURN**

A vertical line will be drawn.

Step 5: Press the Immediate Mode command **ALT-W** and zoom a window around the right side of the drawing where the sectional view will be created. You will be prompted to select the first and then the second corners of the window. Note the window size in Figure 5-5. Note also that the string command does not have to be canceled. It is just bypassed temporarily.

Step 6: Change your Position Menu option to **CURSOR (F1)**. Now as you move your mouse/cursor control device you will see a "rubberbanding line." Move the line to the coordinates below and select the points needed to finish the top view (Figure 5-6). Check the Cursor Tracking Window for the X and Y locations. If the X and Y values do not change when you move the cursor, press the Immediate Mode command **CTRL-T**, and select **VIEW (F2)** from the menu.

Step 7: To complete the top half of the sectional view use these X and Y coordinates.

>**X= 12.25 Y= 6.5**
>**X= 12.25 Y= 7.5**
>**X= 13.25 Y= 7.5**
>**X= 13.25 Y= 5.5**

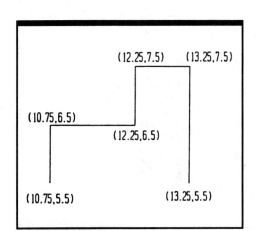

Figure 5-6 Coordinates of top half

Step 8: Choose to **BACKUP (F10)**. This terminates the line string but leaves you in the STRING method.

Step 9: You are prompted to indicate the start point. From the Position Menu, choose **ALONGL (F6)** (along a line). This option allows you to pick a point at a fixed distance from the endpoint of a selected line. The endpoint which the cursor is closest to is selected.

Step 10: You are prompted to select a reference line. Select the lower right portion of the vertical line A (Figure 5-7). When prompted for the distance, enter the value **.75** and **RETURN**.

Step 11: From the Position Menu, choose the **CURSOR (F1)** mode to finish the counterbore line string. Move the cursor to the left horizontally a distance of .5 and press the mouse button. You may have to cut the snap increment in half to move horizontally.

Step 12: Move the cursor vertically down a distance of .75 and press the mouse button.

Step 13: Choose **BACKUP (F10)** twice. You will be prompted to choose line method. This time choose **ENDPTS (F1)**, which will create one line at a time instead of stringing them.

Step 14: Choose the **CURSOR (F1)** option and use **X= 12.25, Y= 7.25** as the start point of the line.

Step 15: You are prompted for the endpoint of the line. Cursor-locate **X= 13.25, Y= 7.25**.

Step 16: You are ready to create another line by endpoints. Cursor-locate **X= 12.25, Y= 6.75** as the start point of the line, or use the grid to locate a line parallel to the previous line at a distance of **.5**.

Step 17: When prompted for the endpoint of the line, cursor-locate **X= 13.25, Y= 6.75**.

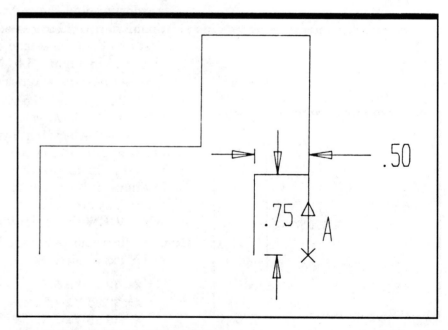

Figure 5-7   Reference line selection

Step 18: The next line by endpoints represents the top of the hexa-gon and will go between the first vertical line and the counter-bore. The start point of the line is located at **X= 10.75, Y= 6**.

Step 19: The endpoint of the line is located at **X= 12.75, Y= 6** (Figure 5-8).

Step 20: Press **CTRL-R** to redraw the screen.

Task 7:   Creating a fillet/round

Command Sequence:

> **ESC (ESCAPE)**
> **F1 CREATE**
> **F6 FILLET**
> **F1 ARC**
> **F1 TRIM**

You will round one corner with a .25 radius using the **FILLET** with **TRIM** option. Make sure **QTRIM** is **OFF** in the Status Window.

Step 1: Press **ESCAPE (ESC)** to return to the Main Menu.

Step 2: Choose to **CREATE (F1)** a **FILLET (F6) ARC (F1)** with **TRIM (F1)**.

Step 3: You are prompted to enter the radius. Enter the value **.25** and **RETURN**.

Step 4: You are then prompted to select the first fillet entity. Se-lect the first vertical line near where the arrow is pointing or the wrong side of the line will be trimmed. Then you are prompted to select the second fillet entity. Select horizontal line A (Figure 5-9).

Step 5: At this point, press the Immediate Mode command **CTRL-F** and save the drawing under the name **SECTION**; if the

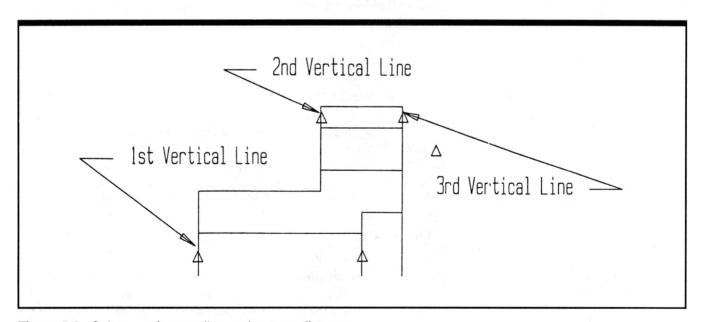

Figure 5-8   Select a reference line and enter a distance.

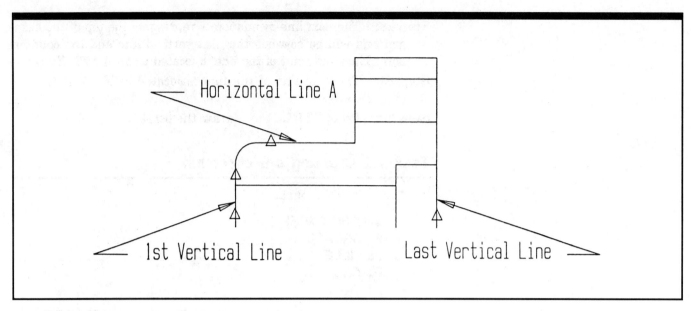

Figure 5-9   Select the first fillet entity.

name **SECTION** appears in parentheses, you can press **RE-TURN** to accept the default name without retyping the name. In response to the prompt **"File already exists replace NO/YES?"** choose **YES (F2)** because you want to update the file, replacing the old file with the new additions. This step should be repeated after every major change that you make to a drawing or, in the case of this tutorial, at the end of every task.

Task 8:   Mirroring about an axis

Command Sequence:

   ESC (ESCAPE)
   F4 X-FORM
   F5 MIRROR
   F2 COPY
   F3 WINDOW

When an object is mirrored, a mirror or reverse image copy of the object is created. In the cases of drawing symmetric objects, only one half of the object needs to be created.

Step 1:   Press **ESCAPE (ESC)** to return to the Main Menu.

Step 2:   Choose the **X-FORM** option (**F4**).

Step 3:   Choose the **MIRROR** option (**F5**).

Step 4:   Choose the **COPY** option (**F2**).

Step 5:   From the Selection Menu, choose the **WINDOW (F3) ALL IN (F1)** option. You are prompted for the first point of the window. Be sure to select a point that will include the whole top half of the sectional view. You are then prompted for the second point of the window. Move diagonally until all of the entities in the top half are in the window. You must press the mouse button for the first point and again for the second point.

Step 6:   You are prompted to choose the method for defining the mirror plane. Select **2 PTS (F3)**. From the Position Menu, choose **ENDENT (F3)**. Then select the bottom portion of the first vertical line, followed by selecting the bottom portion of the last vertical line (Figure 5-9).

Step 7:   **CTRL-F (SAVE)**.

Task 9:   Editing the lines

Command Sequence:

> **ESC (ESCAPE)**
> **F2 EDIT**
> **F1 TRM/EXT (TRIM or EXTEND)**
> **F4 DIVIDE**

To fill the areas with the sectional lines, you need to create an island to be filled with the **X-HATCH** pattern. If there are lines running between the areas to be X-HATCHed, the pattern will fill more than the desired areas. To delete the lines which connect the island, you will edit using the **FIRST** and **DIVIDE** methods. If a line does not trim properly, select **UNDO** from the Break Function window.

Step 1:   Press **ESCAPE (ESC)** to return to the Main Menu. If your side view is off the screen, press **ALT-A** and then zoom a new window with **ALT-W**.

Step 2:   Choose the **EDIT** option (F2).

Step 3:   Choose the **TRM/EXT** option (F1).

Step 4:   Choose the **DIVIDE** option (F4). When using the **DIVIDE** method, you select the line you want to divide by touching the portion you want to remove and then selecting the first and second entities that you are trimming to. Refer to the example below. You are prompted to:

> **Select entity to trim (A)**
> **Select first trimming entity (B)**
> **Select second trimming entity (C)**

Repeat this procedure as illustrated in Figure 5-10.

Step 5:   Choose to **BACKUP (F10)**.

Step 6:   Choose the **FIRST (F1)** method.

Because the sectional view was mirrored, the middle lines are two separate lines and the **DIVIDE** command will not work, so the **FIRST** method must be used.

Continue to trim the lines as needed. When trimming using the **FIRST** method, select the entity you want to trim by picking the portion of the entity (F) you want to remain. The second

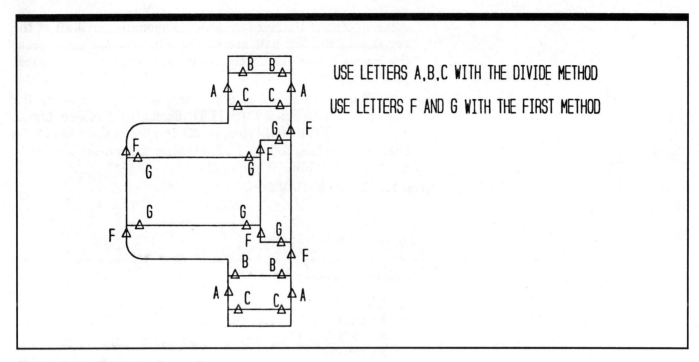

USE LETTERS A,B,C WITH THE DIVIDE METHOD

USE LETTERS F AND G WITH THE FIRST METHOD

Figure 5-10 Trimming the entity

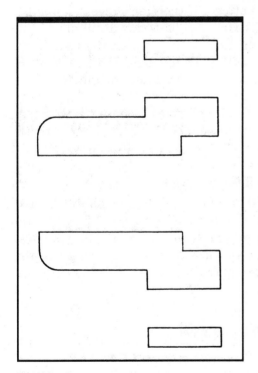

Figure 5-11 Trimmed entity ready to be filled

selection is the entity that you are trimming to (G). When you finish trimming you should have four islands to fill (Figure 5-11).

Task 10: Creating the cross-hatch

Command Sequence:

> **ESC (ESCAPE)**
> **CTRL-L (LEVEL)**
> **F3 DETAIL**
> **F5 X-HATCH**
> **F3 WINDOW**

Step 1: Press **ESCAPE (ESC)** to return to the Main Menu.

Step 2: Press the Immediate Mode sequence **CTRL-L** to change the current level to level 3. When prompted to enter the new level, enter the value **3** and **RETURN**. For the level descriptor, enter **X-HATCH**.

Step 3: Press the Immediate Mode sequence **ALT-X (COLOR)** or select **COLOR** from the Status Window. The color menu is displayed. Select the color purple by placing the cursor on top of the color and pressing the mouse button or by entering the value **6** and **RETURN**.

Step 4: Choose the **DETAIL** option **(F3)** from the Main Menu.

Step 5: Choose the **X-HATCH** option **(F5)** (version 6 **CREATE**).

Step 6: Choose **STEEL** as the pattern option from the pop-up window (Figure 5-12).

Step 7: You are prompted for the angle of the hatch lines. Accept the default of **45** by pressing **RETURN**. You are prompted to

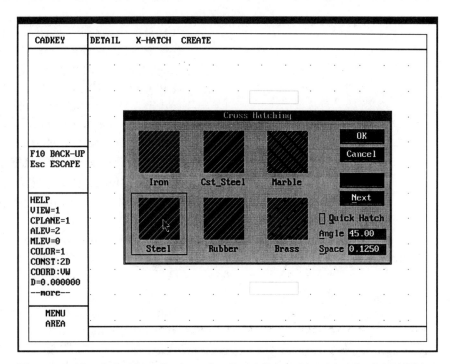

Figure 5-12   Selecting the icon for Steel Cross Hatching in CADKEY 6

enter the distance between hatch lines. The value **.125** is the default; press **RETURN** to accept it.

Step 8:   From the Selection Menu, choose **WINDOW (F3) ALL IN (F1)**.

Step 9:   Create a window around the whole sectional view (Figure 5-11). Pick the lower left corner of the window by pressing the mouse button, then move the cursor diagonally until the entire sectional view is in the window and press the mouse button or space bar.

Step 10:   You are asked if the selection is complete. Respond **YES (F2)** and **RETURN**. The section lines will be placed.

If the prompt **INVALID DATA** appears, repeat the above steps but select the islands one at a time to find out which one has the invalid data (usually a double line). If you use **SINGLE** selection on the island with invalid data, it may work. Also try the **TRM/EXT-BOTH** option to assure a continuous connection in the corners.

Task 11:   Extend the lines that were previously trimmed

Command Sequence:

> **ESC (ESCAPE)**
> **CTRL-L (LEVEL)**
> **ALT-X (COLOR)**
> **F2 EDIT**
> **F1 TRM/EXT**
> **F1 FIRST**

Step 1: Press **ESCAPE (ESC)** to return to the Main Menu.

Step 2: Press the Immediate Mode command sequence **CTRL-L** to change back to level 2. When prompted for a new level number, enter the value **2** and **RETURN** twice.

Step 3: Press the Immediate Mode command sequence **ALT-X** to change the color back to cyan blue (color number 5).

Step 4: Choose the **EDIT** option **(F2)** from the Main Menu.

Step 5: Choose the **TRM/EXT** option **(F1)**.

Step 6: Choose the **FIRST** method **(F1)**.

Step 7: Select the line that you wish to extend. In Figure 5-13, this is line A.

Step 8: You are prompted to select the reference entity to which line A will be extended. In Figure 5-13, this is line B.

Task 12: Creating the lines for the hexagon and placing the center lines

Command Sequence:

    ESC (ESCAPE)
    F1 CREATE
    F1 LINE
    F1 ENDPT

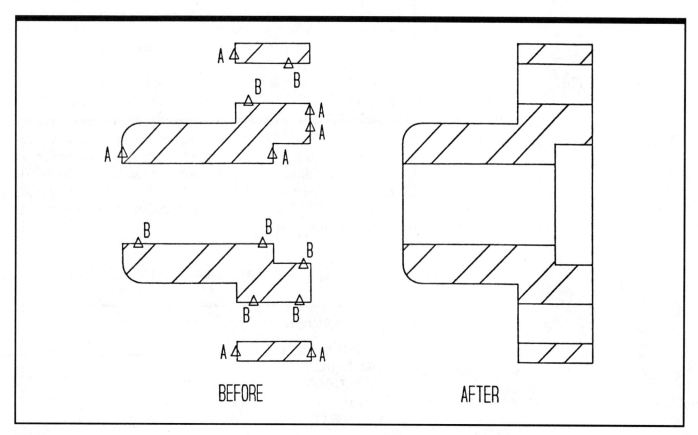

BEFORE   AFTER

Figure 5-13 Reference entity

Step 1:   **ESCAPE (ESC)** to return to the Main Menu. Make sure the snap is **ON**.

Step 2:   **CREATE (F1)** a **LINE (F1)** by the **ENDPT** method **(F1) KEY-IN (F9)**.

Step 3:   The two lines being created represent the hexagon in the sectional view. The coordinates for the first line are:

> **XV= 10.75 YV= 5.75**
> **XV= 12.75 YV= 5.75**

Step 4:   The coordinates for the second line are:

> **XV= 10.75 YV= 5.25**
> **XV= 12.75 YV= 5.25**

Step 5:   Press the Immediate Mode commands **ALT-A** to autoscale, and **ALT-X** to change the color to yellow number 4. Use **ALT-T** to change the line type to **CENTER**. Select the **CENTER** line when displayed on the screen.

Step 6:   Place a horizontal center line through the center of the front view (Figure 5-14).

Step 7:   Place a center line through the two drill holes.

Step 8:   Place a center line through the counterbore.

Step 9:   **CTRL-F (SAVE)**. Replace? **YES (F2)**.

Task 13: Placing the cutting plane line and attaching arrowheads

Command Sequence:

> **ESC (ESCAPE)**
> **ALT-T (LINE TYPE)**
> **ALT-Y (LINE WIDTH)**

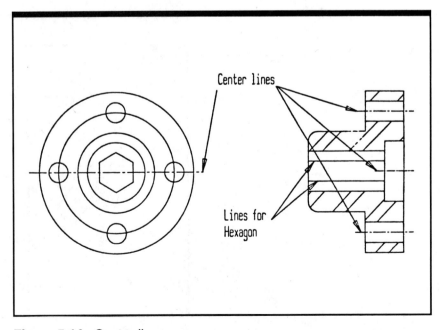

Figure 5-14   Center lines

CTRL-X (SNAP)
F1 CREATE
F1 LINE
F1 ENDPTS
ESC (ESCAPE)
ALT-T (LINE TYPE)
ALT-Y (LINE WIDTH)
F3 DETAIL
F4 ARROW/WIT
F1 ARROWS
F1 ENDPTS
F3 ENDENT
F1 CURSOR

The cutting plane line will be drawn vertically through the front view with a width value of 3. After the cutting plane line is drawn, arrows will be attached at a right angle to the cutting plane line (Figure 5-15).

Step 1: Press **ESCAPE (ESC)** to return to the Main Menu.

Step 2: Press the Immediate Mode command sequence **ALT-T** or select **L-TYPE** from the Status Window. Change the line type to **PHANTOM** by moving the cursor on top of the fourth line displayed on the line menu and pressing the mouse button.

Step 3: Press the Immediate Mode command sequence **ALT-Y** or select **L-WIDTH** from the Status Window. Change the line width to a value of **3**, the second icon on the Line Menu.

Step 4: Check your status area to see if the snap is on. If not, the Immediate Mode command **CTRL-X** will turn it on.

Step 5: **CREATE (F1)** a **LINE (F1)** by the **ENDPTS** method **(F1)**.

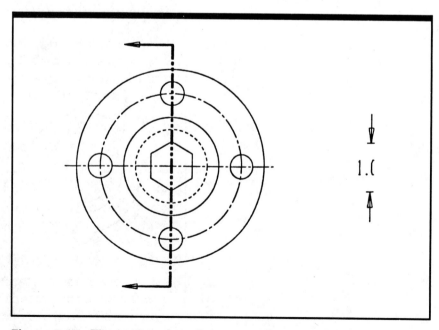

Figure 5-15 The cutting plane line

Step 6: Start the line **.5** above the 4.0 circle and end it **.5** below the 4.0 circle.

Step 7: Press **ESCAPE (ESC)** to return to the Main Menu.

Step 8: Press the Immediate Mode command **ALT-T**. Change the line type to **SOLID** by selecting the first line in the line menu with the cursor.

Step 9: Press the Immediate Mode command **ALT-Y** and enter the value **1** to change the line width back to **1**.

Step 10: Choose **DETAIL (F3)**.

Step 11: Choose **ARROW/WIT (F4)**.

Step 12: Choose **ARROWS (F1)**.

Step 13: Choose **ENDPTS (F1)**.

Step 14: You are prompted to indicate the first point of the arrow. From the Position Menu, choose **ENDENT (F3)** and attach the arrow to the end of the cutting plane line.

Step 15: Change the positional method to **CURSOR (F1)**, move horizontally to the left about **.5**, and press the mouse button. The arrow will appear.

Step 16: Repeat Steps 14 and 15 to attach an arrow to the other end.

Task 14: Controlling (changing) attributes

Command Sequence:

    ESC (ESCAPE)
    F7 CONTROL
    F2 ATTRIBUTES
    F2 L-TYPE
    F2 CHANGE
    F3 CENTER
    F1 SINGLE

In this task, we will change the line type of the 3.0 diameter circle from solid to center and the 1.5 diameter circle from solid to dashed. The current line type is changed from solid to center and the needed center lines are created.

Step 1: Press **ESCAPE (ESC)** to return to the Main Menu.

Step 2: Choose **CONTROL (F7)** from the Main Menu.

Step 3: Choose to control **ATTRIBUTES (F2)**.

Step 4: Choose **L-TYPE (F2)** and then **CHANGE (F2)**.

Step 5: From the **L-TYPE** menu select the **CENTER** line by moving the cursor on top of the third line and pressing the mouse button.

Step 6: From the Selection Menu choose **SINGLE (F1)**.

Step 7: Select the 3.0 diameter circle by placing the cursor clearly on the circle and pressing the mouse button or space bar, followed by a **RETURN**. This circle is the center line for the four .5 diameter holes.

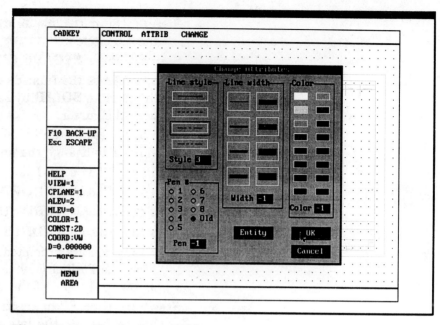

Figure 5-16   The Attributes dialog box

Step 8:   You are prompted to choose a line type. Select the second line on the line menu, the **DASHED** line.

Step 9:   You are prompted to choose a selection method. Choose **SINGLE (F1)**.

Step 10:   Select the 1.50-diameter circle by placing the screen cursor anywhere on the circle and pressing the mouse button or space bar, followed by a **RETURN**. The hidden circle represents the counterbored hole from the back side.

Step 11:   Choose to **BACKUP (F10)**.

NOTE: If you have a color monitor, you may want to change the color attributes of the different line types. This would be done with **CONTROL-ATTRIBUTES-COLOR-CHANGE**. Use the following standards:

CADKEY Line Color Standards

Line Type	CADKEY Color/Number	Plotter Pen #/Color	Width
Border Lines	Green #1	Pen #1....Black	P.7
Object Lines	Blue #5	Pen #1....Black	P.7
Text	White #15	Pen #2....Black	P.3
Dimensions	Magenta #13	Pen #2....Black	P.3
Hidden Lines	Light Red #12	Pen #3....Red	P.3
Center Lines	Yellow #4	Pen #4....Blue	P.3
Section Lines	Purple #6	Pen #5....Green	P.3
Unassigned		Pen #6....?	P.3

Task 15:   Horizontal dimensioning

Command Sequence:

> **ESC (ESCAPE)**
> **CTRL-L (LEVEL)**
> **F3 DETAIL**
> **F1 DIMENSION**
> **F1 HORIZONTAL**
> **F3 ENDENT**
> **ALT-W (WINDOW)**

Step 1:   Press **ESCAPE (ESC)** to return to the Main Menu.

Step 2:   Press the Immediate Mode command **CTRL-L** to change levels. You are prompted to enter a level number. Enter the value **4** and **RETURN**. For the level description enter **DETAILS**.

Step 3:   Press the Immediate Mode sequence **CTRL-F** to save your drawing at this time.

Step 4:   From the Main Menu, choose **DETAIL (F3)-DIMENSION (F1)-HORIZONTAL (F1)**.

Step 5:   From the Position Menu choose **ENDENT (F3)**.

Step 6:   Use **ALT-W** to zoom window in on the sectional view to begin to dimension.

Step 7:   Press **CTRL-X** to turn off the snap.

Step 8:   You are prompted to indicate position for the start of the first witness line. Select the top side of the first vertical line, being careful not to select the fillet (Figure 5-17).

Step 9:   You are prompted to indicate position for the start of the second witness line. Choose the top side of the last vertical line to get an overall dimension.

Step 10:   You are prompted to indicate dimension text position. Move the screen cursor about 3/4 inch away from the object and press the mouse button.

Step 11:   If you don't like the position, select **UNDO** from the Break Function window. Move the screen cursor to a new position and press the mouse button.

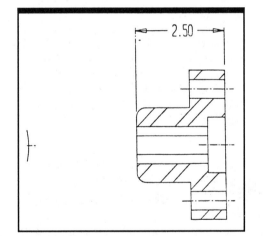

Figure 5-17   Horizontal dimensioning

Task 16:   Horizontal dimensioning with the arrows outside

Command Sequence:

> **ESC (ESCAPE)**
> **CTRL-A (ARROWS IN/OUT)**
> **CTRL-B (WITNESS LINES)**
> **F3 DETAIL**
> **F1 DIMENSION**
> **F1 HORIZONTAL**
> **F3 ENDENT**

Step 1:   Press **ESCAPE (ESC)** to return to the Main Menu.

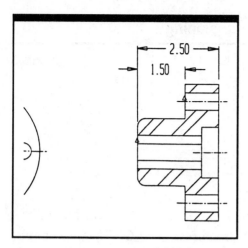

Figure 5-18 Horizontal dimensioning with arrows outside

Step 2: Press the Immediate Mode sequence **CTRL-A** to change the direction of the arrowheads relative to the witness lines.

Step 3: The Status Window should read **ARR: OUT**.

Step 4: Press the Immediate Mode sequence **CTRL-B** to change the displayed witness lines to display only the second witness line. When prompted, choose the **SECOND** option (**F3**).

Step 5: Press the Immediate Mode command **CTRL-R** to redraw the screen and remove the temporary markers.

Step 6: From the Main Menu, choose **DETAIL (F3)-DIMENSION (F1)-HORIZONTAL (F1)**.

Step 7: From the Position Menu, choose **ENDENT (F3)**.

Step 8: You are prompted to indicate the position for the start of the first witness line. Select the top side of the first vertical line, being careful not to select the fillet (Figure 5-18).

Step 9: When prompted to indicate the position for the second witness line, select the top side of the second vertical line.

Step 10: You are prompted to indicate the dimension text position. Move the screen cursor to a distance about .375 above the sectional view, and place the dimension text by pressing the mouse button.

Task 17:   Vertical dimensioning

Command Sequence:

> **ESC (ESCAPE)**
> **F3 DETAIL**
> **F1 DIMENSION**
> **F2 VERTICAL**
> **F3 ENDENT**
> **CTRL-B (WITNESS LINES)**
> **F1 BOTH**

In this task, the dimension across the corners of the hexagon will be placed using a vertical dimension.

Step 1: Press **ESCAPE (ESC)** to return to the Main Menu.

Step 2: From the Main Menu choose **DETAIL (F3)-DIMENSION (F1)-VERTICAL (F2)**.

Step 3: From the Position Menu choose **ENDENT (F3)**.

Step 4: Press the Immediate Mode command **CTRL-B** for witness lines and choose **BOTH (F1)**.

Step 5: You are prompted to indicate the position of the first witness line. Select horizontal line A, as shown in Figure 5-19, being careful not to choose the fillet or the X-hatch. If a wrong selection is made, **BACKUP (F10)** will cancel it.

Step 6: You are prompted to indicate the position of the second witness line. Move the cursor in to select horizontal line B, again being careful not to select the fillet or the X-hatch.

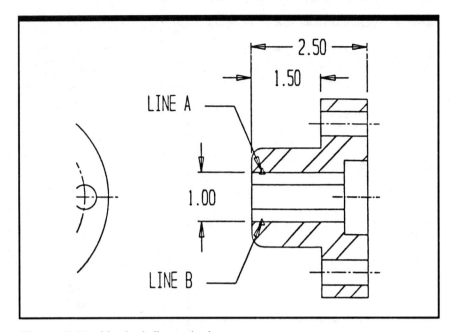

Figure 5-19 Vertical dimensioning

Step 7: You are prompted to indicate the position for the dimension text. Move the cursor far enough away so that the dimension text does not conflict with the center line.

Task 18:   Dimensioning the diameters

Command Sequence:

> **F10 BACKUP**
> **F6 DIAMETER**
> **ALT-A (AUTO SCALE)**
> **ALT-W (WINDOW)**

Step 1: Choose **BACKUP (F10)**, which brings you to the **DIMENSION** menu.

Step 2: From the **DIMENSION** menu, choose the **DIAMETER** option **(F6)**.

Step 3: Use the Immediate Mode commands **ALT-A** to auto scale and **ALT-W** to zoom a window around the front view.

Step 4: You are prompted to select the arc to be dimensioned. Move the screen cursor to the edge of the 4.00 diameter circle, where the leader line will begin, and press the mouse button (Figure 5-20).

Step 5: You are prompted to indicate the text position next. Select an area where the leader line and dimension text will not interfere with other entities.

Step 6: You are prompted to choose which side of the text to place the shoulder of the leader line. Choose **RIGHT (F2)**.

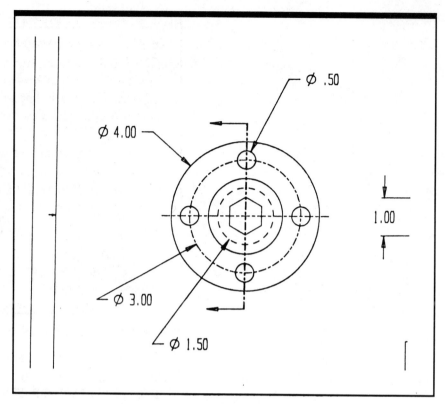

Figure 5-20 Dimensioning the diameter

Step 7:   You are returned to the prompt found in Step 3. Just keep following steps 3 through 6 until the diameters for all circles have been dimensioned.

Task 19:   Placing notes

Command Sequence:

      ESC (ESCAPE)
      F3 DETAIL
      F2 NOTE
      F1 KEYIN
      F6 SAVE TX
      F10 BACKUP

Step 1:   Press **ESCAPE (ESC)** to return to the Main Menu.

Step 2:   You will find it necessary to zoom in for correct placement of the notes. Just use the Immediate Mode commands **ALT-W (WINDOW)**, **ALT-B (BACKUP 1)** and **ALT-A (AUTO SCALE)** as needed to get around the drawing.

Step 3:   Check to make sure that your note height **(NH)** is equal to your dimension height **(DH)** by selecting —**more**— twice from the Status Window. They should both be set to **.300**.

Step 4:   From the Main Menu, choose **DETAIL (F3)**.

Step 5:   From the **DETAIL** Menu, choose **NOTE (F2)**.

Step 6:   When prompted for the method, choose **KEYIN (F1)**.

Step 7:   You are prompted to enter text. Key in all the information for the first line of text and **RETURN** to go on to the next line of text. If there is not another line below the first line of text, press **SAVE TX (F6)**. In this case, key in **SECTION A-A** and press **RETURN** and **SAVE TX**.

Step 8:   You are prompted to indicate text position. Place the screen cursor under the section view and press the mouse button.

Step 9:   Press **F10** to **BACKUP**. You are prompted to enter the text for your next note.

Notes to be placed include:

A.   4 HOLES
B.   CBORE
C.   .50 DEEP
D.   NOTE: FILLETS AND ROUNDS R .25
E.   Enter your name in the title block.

The completed drawing is shown in Figure 5-21.

Step 10:   If you wish to move text after it has been placed, use **DETAIL-CHANGE-TXT POS (F3-F7-F2)**.

Task 20:   Saving the drawing and exiting the system

Command Sequence:

ESC (ESCAPE)
ALT-E (EXIT)
F2 YES

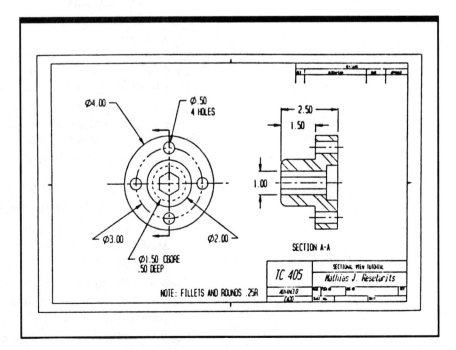

Figure 5-21   The finished drawing

If a plotter is attached to your workstation, then you will most likely want to reverse Tasks 20 and 21.

Step 1:   Press **ESCAPE (ESC)** to return to the Main Menu.

Step 2:   From the Break Area, select **EXIT** or **ALT-E**.

Step 3:   You are asked if you wish to save the file before exiting; respond **YES (F2)**.

Step 4:   You are prompted to enter the filename under which the drawing is to be saved. The filename **SECTION** should be the default name in parentheses. Press **RETURN** to accept the default or type in the filename.

Step 5:   If you have saved this file before (hopefully, you have been using the Immediate Mode command **CTRL-F** to save the file as you have been creating it), you will get the prompt **File already exists do you want to replace it?**; respond **YES (F2)**. The file will be saved and you will be exited from CADKEY.

Task 21:   On-line plotting

Command Sequence:

> **ESC (ESCAPE)**
> **F7 CONTROL**
> **F7 PLOT**
> **F1 PLOT**
> **F1 AUTO**
> **F1 NO**
> **F1 PEN#**
> **F2 COLOR**
> **F6 ALL DSP**
> **F1 ALL**

If a plotter is attached to your workstation, then you will most likely want to reverse Tasks 20 and 21. The following method of plotting is an online method which requires you to have your drawing displayed on the screen while you are plotting. If your drawing is not presently displayed, you can load it using the Escape Code **ESC-F5-F1-F2-F2 FILES-PART-LOAD-YES**. You are then prompted to enter the filename (**SECTION**) and press **RETURN**.

Step 1:   Press **ESCAPE (ESC)** to return to the Main Menu.

Step 2:   From the Main Menu, select **CONTROL (F7)**.

Step 3:   From the menu of **CONTROL** options, select **PLOT (F7) PLOT (F1)**; if you are using a printer instead of a plotter, select **PRINT (F6)**.

Step 4:   When prompted for method, select **AUTO (F1)**, which will automatically scale and center your drawing to fit on the paper.

Step 5:   When you are asked whether or not to display the scale and offsets, respond **NO (F1)**.

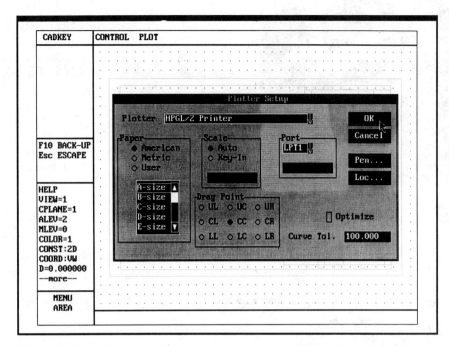

Figure 5-22   Making the proper settings in the Plotter Setup dialog box.

Step 6:   For the last prompt, you will be asked whether to plot by using pen numbers that you assigned to entities when they were created or the colors that are on the screen (if you have a color monitor). The choice is yours! Try **COLOR (F2)**.

Step 7:   The Universal Selection Menu will be displayed. Select **ALL DSP (F6) ALL (F1)**.

Step 8:   Finally, you are all set and ready to begin plotting. Make sure that the paper is in the plotter and that the plotter is turned on. Press **RETURN** and your drawing will begin plotting.

3D Dovetail Tutorial

OBJECTIVES

After completing this chapter, you will be able to:

- Create a three-dimensional model of a part.
- Create a three-view orthographic drawing from the model.
- Create lines, rectangles, and circles.
- Edit lines and circles.
- Group entities.
- Make display changes.
- Use the **X-FORM** commands to create a 3D model.
- Add 3D fillets.
- Define a new view.
- Change levels and turn them on or off.
- Define a new construction plane.
- Dimension an orthographic drawing.
- Add notes to a drawing.
- Change dimensions or text.
- Retrieve a title block pattern.
- Retrieve orthographic view patterns.

This tutorial has been designed to familiarize you with the world of three-dimensional CAD. You will be introduced to the basic building blocks of this three-dimensional design system. Understanding these building blocks will give you the ability to construct three-dimensional parts of your own. This hands-on session will take you through the construction of a three-dimensional part, detailing procedures, and the creation of a layout drawing by placing multiple views of the three-dimensional part in one drawing. The part you are about to create is referred to as the dovetail part, shown in Figure 6-1.

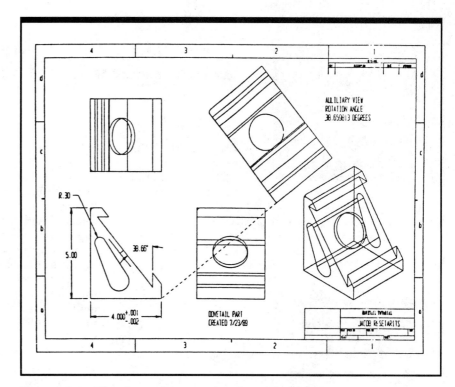

Figure 6-1   The dovetail layout drawing

Task 1:   Setting up the display

Command Sequence:

F6 DISPLAY
F5 VWPORTS
F5 AUTOSET
PLANE (Construction Plane)

When beginning a new part, it is advantageous to set up the screen display with the parameters and attributes needed for creation. The options displayed in the Status Window allow you to view the current status of some of these parameters and attributes. Changes may be made to any of those options listed using the cursor control of any tablet or mouse device attached, or using the assigned Immediate Mode sequence. First, make sure your display has four Viewports.

Step 1:   Select **DISPLAY-VWPORTS-AUTOSET** and select the 4 Viewport icon.

Step 2:   Select the **CPLANE** option. Move the cursor into the Status Window area, place the cursor symbol next to the **CPLANE:** option, and press the mouse button.

Step 3:   You are prompted for a new view assignment. Enter the value **2** and press **RETURN**.

Step 4:   The construction plane is changed to the Front Viewport (**2**). The Status Window displays the new view assignment: **CPLANE: 2**.

Designed Physical Fitness Equipment With CADKEY 1!

Steve Tritter's first exposure to CAD-KEY® true, 3-D mechanical-design software occurred in 1986-1987, while he was serving as head of the Department of Electrical and Mechanical Technology at New Hampshire Vocational/Technical College, Nashua, New Hampshire. Steve learned to use CADKEY 1 (the educational version) on an 80286

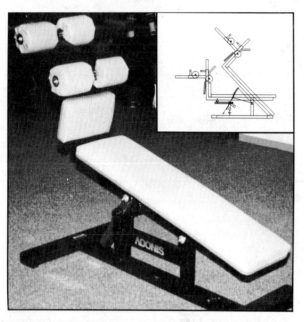

Bent-leg, adjustable, decline/flat bench designed by Steve Tritter, President of Adonis Fitness Equipment, using CADKEY 1

PC-AT compatible without a math co-processor or a mouse. In 1988, Tritter purchased Adonis® Fitness Equipment, Inc. in Hudson, New Hampshire, which he heads as President. He has used CADKEY in the design of several of his company's products. He recently designed and added to the Adonis product line a bent-leg, adjustable decline/flat bench.

The Adonis bench is, so far, unique in the world of fitness equipment because it can be raised from flat to a true 45-degree angle while still maintaining a strong, wobble-free integrity. This feature comes from a unique, telescopic, swivel-joint which doubles as a horizontal rest. "Without true 3-D," Steve said, "it would not have been possible to design this joint with its close tolerances and complex motions." The bench is built of 11-gauge steel tubing (⅛-inch wall thickness), ¾-inch plywood, and upholstered with the highest quality Naugahyde.

Sophistication of design and ruggedness of construction have made Adonis' benches and fitness equipment, designed or redesigned on CADKEY, into prominent items in corporate fitness centers across the country. U.S. government agencies have been purchasing Adonis fitness equipment on the GSA schedule for three years. Tritter directly credits the better design and stronger products that CADKEY helped him to achieve for his increased sales.

On September 6, 1989, Steve Tritter donated his personal CADKEY-designed Adonis bench to Cadkey, Inc. to express how delighted he is with Cadkey's software.

Adapted with permission from **3-D World,** *v. 4, no. 2, Mar./Apr., 1990.*

Task 2:   Create a rectangle (4 × 5 inches)

Command Sequence:

F1 CREATE
F1 LINE
F5 RECTANG
F2 WID/HT
F9 KEY-IN

Step 1:   Choose to create a rectangle, **CREATE-LINE-RECTANG (F1-F1-F7).**

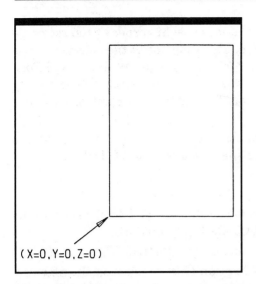

(X=0,Y=0,Z=0)

Figure 6-2 Creating a rectangle with the base position 0,0,0

Step 2: Choose the **WID/HT** (width/height) method (**F2**) to enter in values for the rectangle size.

Step 3: You are prompted for the width value of the new rectangle. Enter in the number **4** and press **RETURN**.

Step 4: You are prompted for the height value of the new rectangle. Enter in the number **5** and press **RETURN**.

Step 5: Choose the **KEY-IN** method (**F9**) from the Position Menu displayed. This method allows you to determine a position for the new rectangle by keying in the coordinate values of its lower left corner.

Step 6: Enter a zero (**0**) value for each coordinate by pressing **RETURN** at each prompt (the zero value is the default assigned by the system). This positions the bottom left corner of the rectangle at the "home" position.

> **XV = 0**
> **YV = 0**
> **ZV = 0**

Note that this positioning method causes only part of the rectangle to appear on the screen. Once we display the complete part within the drawing window (Figure 6-2), the 0,0,0 position will be at the bottom left corner of the rectangle.

Step 7: To display the complete part within the Viewports, invoke an Auto-Scale using the Immediate Mode command **ALT-A**. Press **RETURN**, then **ESCAPE**.

Task 3:   Creating line entities

Command Sequence:

F1 CREATE
F1 LINE
F1 ENDPTS
F3 PARALEL
F3 ENDENT

Lines with Endpoints

Step 1: Press **ESCAPE** to return to the Main Menu.

Step 2: Choose to **CREATE (F1)** a **LINE (F1)** by specifying **ENDPTS (end points, F3)**.

Step 3: The Position Menu is displayed offering nine methods for indicating the start and end points of the new line. Choose the **ENDENT** method (**F3**) to locate the endpoints of existing lines. An asterisk (*) appears next to the selected menu option.

Step 4: Move the cursor to the upper left corner of the rectangle and press the mouse button. The endpoint of the selected line is designated as the start point of the new line. Note that when the cursor is moved, a "rubberband" line is attached until an endpoint is designated.

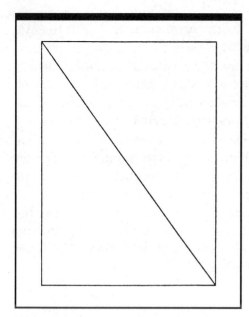

Figure 6-3 A diagonal line created using the endpoints of an entity

Step 5: Move the cursor to the lower right corner of the rectangle and press the mouse button. The endpoint of the selected line is designated as the end point of the new line. With a modal Position Menu, one positioning method can be selected for a variety of tasks; in this case, it was finding the endpoint of an entity (Figure 6-3).

NOTE: If an unwanted entity is created by mistake, use the Immediate Mode command single delete option **CTRL-Q**.

Parallel Lines

Step 1: To create a line parallel to the diagonal line, return to the **LINE** Menu using the **BACKUP** option (**F10**).

Step 2: At the **LINE** Menu, choose the **PARALEL** option (**F3**).

Step 3: Choose the **AT DIST** option (**F2**) to position the new line a specified distance from the reference line selected.

Step 4: Enter in a distance value of **.5** and press **RETURN**. When the parallel line is created it is placed one-half inch (**.5**) from the line referenced.

Step 5: Move the cursor to line 1 and press the mouse button (Figure 6-4).

Step 6: Indicate a position for the new parallel line by moving the cursor to the right of the line referenced and pressing the mouse button. The new line is created .5 inches from the reference line on the side selected.

Note that it is the same length as the line referenced.

Step 7: Create another parallel line by selecting line 2 as the reference line, and positioning the new line to its left.

Step 8: Select line 3 as the third reference line and position the new line below it. A total of three new lines is created, parallel to lines 1, 2, and 3.

Step 9: Invoke the redraw command. Hold down the **CTRL** key, press the letter **R** and press **RETURN**.

Step 10: Press **CTRL-F** to save. Use the filename **DOVE**.

Task 4:   Trimming lines

Command Sequence:

F2 EDIT
F1 TRM/EXT
F2 BOTH
F4 DIVIDE
F5 MODAL

Trimming Both Lines

Step 1: Return to the Main Menu using the **ESC** key.

Step 2: Choose to **EDIT** (**F2**), and **TRM/EXT** (trim/extend, **F1**) designated lines. The lines have been numbered for your reference.

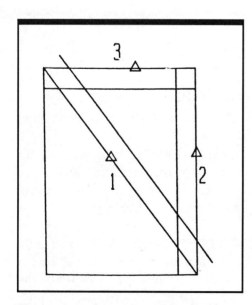

Figure 6-4 Creating parallel lines

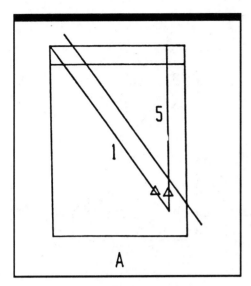

Figure 6-5 Both lines 1 and 5 trimmed

Step 3: Choose to trim **BOTH (F2)** lines selected.

Step 4: When selecting lines for trimming, it is important to position the cursor near to the desired trimming intersection. In the diagrams that follow, select each line at the selection symbol.

Move the cursor to line 1, indicating the first line selection. Press the mouse button. Select line 5 in the same manner. (See diagram A, Figure 6-5.)

Both lines are trimmed at their intersection point.

Step 5: Trim lines 2 and 4 in Figure 6-6 (diagram B), 3 and 4 (diagram C), and 1 and 6 (diagram D) in the same manner.

Step 6: To clear the screen of temporary markers, invoke the Immediate Mode command **CTRL-R** (hold down the **CTRL** key and simultaneously press the letter **R**) and press **RETURN**.

Modal Trimming

Step 1: Return to the trim menu using the **BACKUP (F10)** option.

Step 2: Choose the **MODAL** option **(F5)**. This allows you to continually trim selected entities to a reference entity.

Step 3: Move the cursor to line 1, as shown in Figure 6-7, page 108. Press the mouse button. This is the reference entity to which lines will be trimmed.

Step 4: Select the two lines which intersect this line (lines 2 and 3 of Figure 6-7) on the portion of the line to keep. Remember to press the mouse button to indicate each line.

Step 5: Lines 2 and 3 are trimmed to their intersection points with line 1.

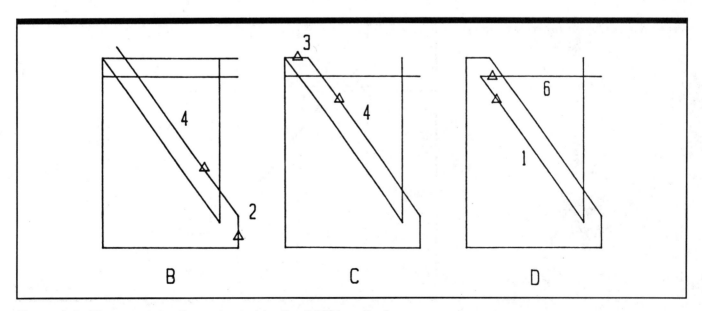

Figure 6-6 The remaining lines trimmed by the BOTH method

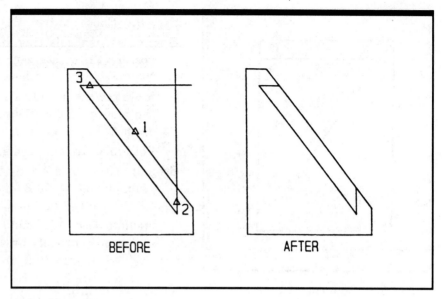

Figure 6-7 Modal trimming

Divide Trimming

Step 1: Once again, return to the trim menu (press **F10** twice).

Step 2: Choose the **DIVIDE** option **(F4)**. This function divides a selected line at its intersection points and removes the middle entity.

Step 3: Move the cursor to the center of line 1, as shown in Figure 6-8, and press the mouse button.

Step 4: Next, select the lines (2 and 3 of Figure 6-8) which intersect this diaagonal line by moving the cursor to each line and pressing the mouse button. Trimming is automatically completed.

Step 5: Redraw the screen using **CTRL-R** and **RETURN**.

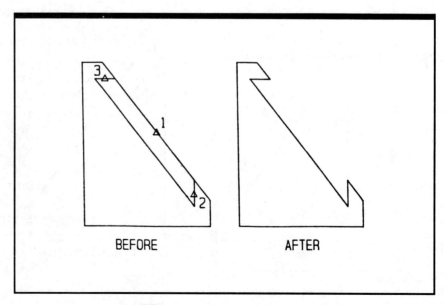

Figure 6-8 Trimming with the DIVIDE option

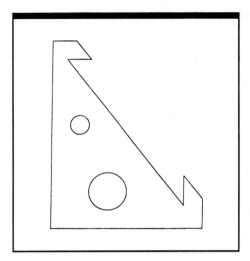

Figure 6-9   Creating a circle with the CTR + DIA option

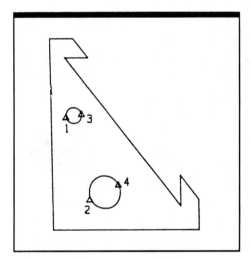

Figure 6-10   Creating tangent lines

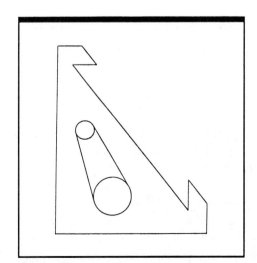

Figure 6-11   Lines tangent to circles

Task 5:   Creating circles

Command Sequence:

ALT-X (Color)
F1 CREATE
F3 CIRCLE
F2 CTR + DIA
F9 KEY-IN

Step 1:   Return to the Main Menu **(ESCAPE)** and change the color to blue by selecting **ALT-X**.

Step 2:   Choose to **CREATE (F1)** a **CIRCLE (F3)** using the **CTR + DIA** method **(F2)** and **KEY-IN**.

Step 3:   When prompted for the Diameter enter the value of .5 and press **RETURN**, for the center location KEY-IN X = .75, Y = 2.75 and Z = 0. Select **BACKUP (F10)**. When prompted for the Diameter of the second circle enter the value of 1 and press **RETURN**, for the center location KEY-IN X = 1.5, Y = 1 and Z = 0.

Task 6:   Creating tangent lines

Command Sequence:

F1 CREATE
F1 LINE
F4 TAN/PRP
F3 TAN TAN
F2 2 ARCS

Step 1:   Return to the Main Menu (press **ESCAPE**).

Step 2:   Choose to **CREATE (F1)** a **LINE (F1) TAN/PRP (F4) TAN TAN (F3)**.

Step 3:   The two circles on the part will be selected as the two arcs for the tangent lines. Move the cursor to the left side of the first circle (point 1 in Figure 6-10) and press the mouse button.

Step 4:   Move the cursor to the same side of the second circle (point 2) and again press the mouse button. A tangent line is created on the left side of the circles.

Step 5:   Select **BACKUP (F10)** and repeat steps 3 and 4; designate the right side of each circle for the tangent line (Figure 6-11).

Press **CTRL-R** and **RETURN** to redraw the screen.

Task 7:   Trimming circles

Command Sequence:

F2 EDIT
F1 TRM/EXT
F3 DOUBLE

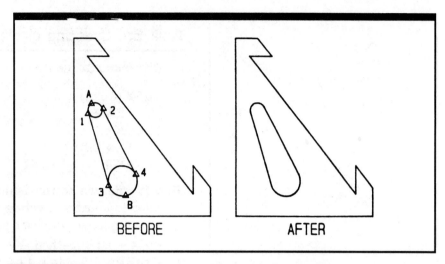

Figure 6-12   Trimming circles with the DOUBLE option

Step 1:   Return to the Main Menu **(ESCAPE)**.

Step 2:   Choose to **EDIT (F2)** the circles using the **TRM/EXT-DOUBLE (F1-F3)** option to trim the circles.

Step 3:   Zoom a Window **ALT-W** around both circles by pressing the mouse button to select opposite corners of the window.

Step 4:   Move the cursor to the outer section of the top circle (point A) and press the mouse button (Figure 6-12).

Step 5:   Next, move the cursor to the point on the circle where a tangent line touches (point 1) and press the mouse button. Repeat this for the opposite side of the circle (where the other tangent line touches, point 2). The circle is trimmed to these points.

Step 6:   Repeat steps 4 and 5 for the bottom circle using point B as the reference position. Both circles are trimmed.

Clean up the screen using the Immediate Mode command **ALT-A** and **RETURN**.

Task 8:   Grouping entities

Command Sequence:

F2 EDIT
F5 GROUP
F1 MAKE

Grouping allows you to link together entities so that they may be selected as a complete unit. For this part, let's group the inner portion of the dovetail for easier manipulation.

Step 1:   Return to the **EDIT** menu (press **F10** twice).

Step 2:   Choose the **GROUP** option, **F5**.

Step 3:   Choose the **MAKE** option, **F1**.

Step 4:   Enter the name **SLOT** for the entities to be grouped and press **RETURN**.

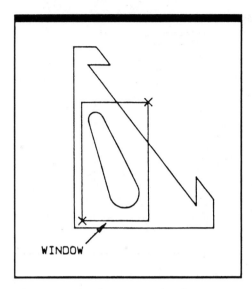

Figure 6-13 Group entities in the window

A subgroup is created for the selected entities and is assigned the number 1. Up to 256 subgroups may be created for any one group.

Step 5: Select the entities to group using the **WINDOW (F3) ALL IN (F1)** method from the Selection Menu displayed. This allows you to select only those entities surrounded by the rubberbox window.

Step 6: Move the cursor to the lower left corner, as shown in Figure 6-13, by the **X**. Press the mouse button. Next, move the cursor to the upper right corner of the desired entities. As you move the cursor, the rubberbox moves along with it. Only those entities which lie completely within this window are selected for grouping; therefore, the inner portion of the part must be completely surrounded. Press the mouse button.

If the selection window overlaps any portion of these desired entities, press **F10** and designate each corner of the window once again. From this point on, all entities chosen from this group are considered as a whole unit and may be selected or altered as such.

Task 9:   Making changes to the display

Command Sequence:

DISPLAY
AXES
DISP VW
OFF

Throughout the creation of a part, you may find it convenient to re-organize the screen display by clearing out temporary symbols, altering the part size, changing the view, reassigning a new coordinate system, adding levels, etc. Most of these changes may be made without leaving your place in the menu structure via Immediate Mode commands or the Status Window. Let's make some changes to the current part before continuing.

Step 1: The coordinate axes, which appear in the lower left corner of all of the viewports, are not needed. Turn them off using **DISPLAY-AXES-DISP VW-OFF**. Now turn on the Construction View axes which coincide with the current Construction Plane by selecting **CONST VW-ALL DSP**. The construction axes should appear in the upper right corner of all of the viewports.

Step 2: Invoke an **AUTO SCALE** to fit the completed part on the screen (**ALT-A**) or **DISPLAY-ZOOM-AUTO** and press **RETURN**.

Step 3: Reduce the drawing to half its size using the Immediate Mode command **ALT-H** or **DISPLAY-ZOOM-HALF**. Then select viewport 7 to reduce the scale (Figure 6-14).

Step 4: **SAVE** your part file by pressing **CTRL-F**, entering the name **DOVE**, and pressing **RETURN**.

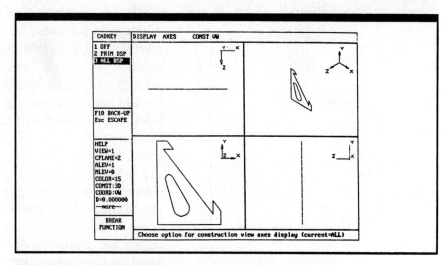

Figure 6-14 Making changes to the Display Viewports.

With any of these new assignments you are always returned to your place in the current menu structure. Let's continue on.

Task 10: Creating the third dimension

Command Sequence:

F4 X-FORM
F1 DELTA
F3 JOIN
F2 CHAIN
F4 GROUP

Step 1: Press **ESCAPE** to return to the Main Menu.

Step 2: Transform the part using the **X-FORM** function (**F4**).

Step 3: Choose to transform the drawing relative to its current position, **DELTA (F1)**, and **JOIN (F3)** the endpoints.

Step 4: Choose the **CHAIN** method, **F2**, from the Selection Menu displayed to select a series of connected lines.

Move the cursor to the left-most vertical line of the part and press the mouse button. This line is the starting point for the selection chain and is marked with a marker (Figure 6-15).

Next, define the direction of the chain by moving the cursor upwards, and press the mouse button. (No marker will appear at this position.)

Finally, to define an endpoint for the chain, select all connected lines up to the starting point of the chain by pressing **RETURN** only. All connected lines up to the starting point are selected and marked with selection markers. Hence, a selection chain is created or turns white and your part looks something like Figure 6-15.

You are then prompted:

Selection o.k. ?

Choose **YES (F2)** to continue.

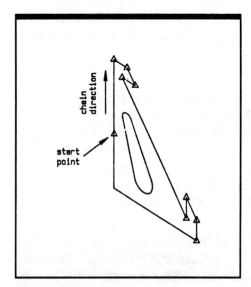

Figure 6-15 Chain selecting entities to be transformed

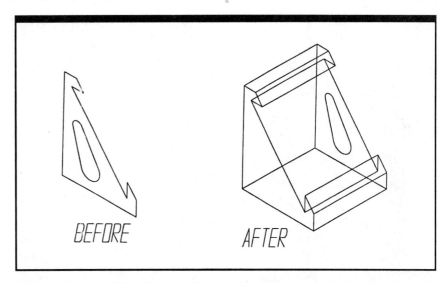

Figure 6-16   Adding the 3rd dimension, depth

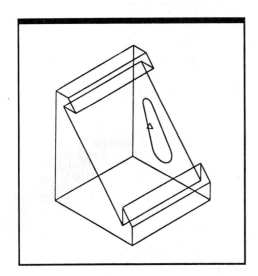

Figure 6-17   Transforming the grouped entities

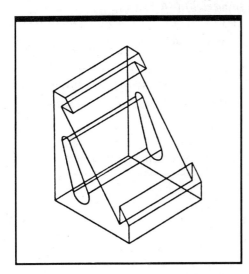

Figure 6-18   The round end slot transformed

Step 5:   When prompted for the number of copies, enter the number **1** and press **RETURN**.

Step 6:   Position a copy of the part at the following location (remember we are in World Coordinates). Press **RETURN** after each number entered:

> **dxv = 0   dyv = 0   dzv = 4**

Use **F10** to back up if necessary. The outer portion of the part is transformed and joined right before your eyes (Figure 6-16)! Select **UNDO** if it is not correct.

Step 7:   Return to the Selection Menu using the **F10** key twice. Press **ALT-A** and **RETURN**.

Step 8:   Choose the **GROUP** method, **F4**. This allows you to select entities that are grouped, such as the inner portion of the part.

Step 9:   Choose **F1**, **SELECT**, from the menu displayed.

Step 10:   Position the cursor on the grouped portion of the part, as in Figure 6-17 (previously named **GROUP1**), and press the mouse button.

Having grouped these entities previously, this portion of the part is treated as a complete unit no matter where it is selected.

Step 11:   When prompted for the number of copies, enter the number **1** and press **RETURN**.

Step 12:   Position a copy of the part at the following location (remember we are in VIEW Coordinates):

> **dxv = 0   dyv = 0   dzv = 4**

Press **RETURN** after each number entered. Use **F10** to back up if necessary.

The inner portion of the part is now transformed and joined right before your eyes (Figure 6-18)! Select **UNDO** if it is not correct.

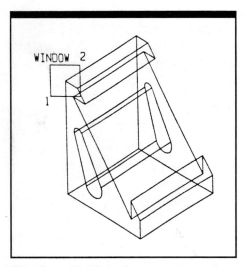

Figure 6-19 Zooming an alternate window

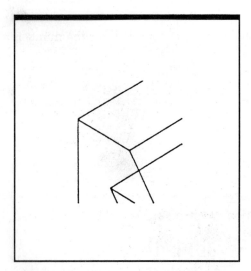

Figure 6-20 The part after zooming

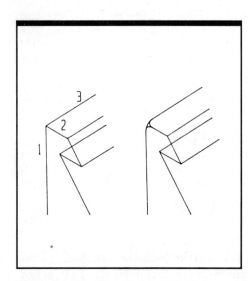

Figure 6-21 Creating a 3D fillet

Task 11: Adding 3D fillets

Command Sequence:

F1 CREATE
F6 FILLET
F1 ARC
F1 TRIM

A fillet allows you to connect potentially intersecting line and arc entities that lie in the same or a parallel plane at the same depth. The fillet takes on the appearance of an arc.

Step 1: Return to the Main Menu **(ESC)**.

Step 2: Choose to **CREATE (F1)** a **FILLET (F6) ARC (F1)**.

Step 3: Choose to **TRIM (F1)** the selected intersection of the new fillet.

Step 4: Enter a radius value of **.2** for the arc fillet and press **RETURN**.

Step 5: To enlarge a corner of the part for easier manipulation, use the zoom-window option, Immediate Mode command **ALT-W**. This allows you to create a "rubberbox" window to surround and enlarge a desired part of the entity. You are prompted:

Indicate position for 1st window corner

Step 6: Move the cursor near to the upper left corner of the part (point 1 in Figure 6-19) and press the mouse button. The **F10** key allows you to back up and re-select a new position if necessary.

Step 7: With the box positioned, move the cursor upwards and to the right of the part's left corner until the "rubberbox" completely surrounds the desired corner (point 2). Press the mouse button.

The upper left corner of the part is enlarged to the boundaries of the Drawing Window (Figure 6-20).

Step 8: Select the designated lines for filleting. Move the cursor to line 1 (Figure 6-21) and press the mouse button, then to line 2 and press the mouse button once again. A fillet is created for one corner of the box.

Step 9: Select lines 2 and 3, and then 3 and 1, using the same procedures described in **Step 8**.

Step 10: Return the part to its previous size using the Immediate Mode command **ALT-B (BACK-1)**.

Task 12: Defining a new construction plane

Command Sequence:

COORD: (View Coordinates)
CPLANE:
F2 2 LINES

Step 1: Return to the Main Menu **(ESC)** and invoke an Auto Scale on the part **(ALT-A)** and **RETURN**.

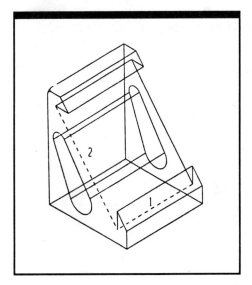

Figure 6-22 Identifying a construction plane by two lines

For color monitors only: To keep track of the **CIRCLE** entity we are about to create, let's change its **COLOR**. As you already know, there are a few ways to access options in the system, so choose one of these methods for assigning a new color.

1. Use the Immediate Mode command **ALT-X**.
2. Invoke the **COLOR** option from the Status Window area.
3. Use the **ESCAPE** codes **ESC-F7-F2-F1-F1**.

For any of these options, cursor-select the color red from the color table and press the mouse button.

Step 2: From the Status Window, check **COORD:** to make sure the View Coordinate system (**VW**) is active.

Step 3: Select **CPLANE:** from the Status Window. Use the **2 LINE** option to identify the construction plane.

Step 4: When selecting the two lines for a construction plane, select the end of the line opposite to the positive axis direction, where the vertex of the axes will be found. After defining the construction plane choose **DISPLAY-AXES-CONST VW-ALL DSP** to display the construction plane axes in all of the viewports. If the axes arrowheads point in the proper positive direction, select **CPLANE** from the Status Window, and **SAVE** from the Menu Window.

Task 13:   A new circle

Command Sequence:

F1 CREATE
F3 CIRCLE
F2 CTR+DIA
CTRL-X (Snap)

Step 1: Return to the Main Menu by pressing **ESC**.

Step 2: From the Main Menu, choose to **CREATE (F1)** a **CIRCLE (F3)** by designating its center and diameter (**CTR+ DIA, F2**).

Step 3: Enter a value of **2** for the circle's diameter and press **RETURN**.

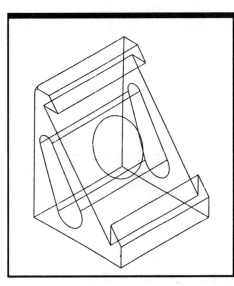

Figure 6-23 Creating a circle on the inclined surface

Step 4: Choose **DELTA-ENDENT** and select the lower end of line 2 in Figure 6-22 as your delta reference position. This will allow you to accurately locate the center of the circle relative to the corner of the inclined surface. When prompted for **dXV** enter **2**, for **dYV** enter **2.25**, and for **dZV** press **RETURN**. The circle is automatically created on the face of the part at the proper depth (Figure 6-23).

Step 5: To give the hole a .25 depth use **X-FORM-DELTA-JOIN-SINGLE (F4-F1-F3-F1)**. Select the circle and press **RETURN**.

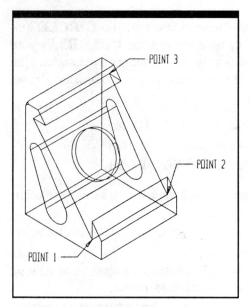

Figure 6-24 Defining a new view

Step 6: Press **RETURN** for each of the following to make (1) copy:

dXV=0
dXY=0

Step 7: When prompted for **dZV=**, enter the value **–.25**. Note the depth in the various Viewports.

Step 8: Use **CTRL-F** to save.

Task 14: Defining a new view

Command Sequence:

F6 DISPLAY
F4 VIEW
F3 NEW
F2 3-PTS

Step 1: Return to the Main Menu by pressing **ESC**.

Step 2: Choose **DISPLAY (F6)** a **VIEW (F4)**. You are prompted to select a Viewport. Select the view **7** Viewport. Then choose **NEW (F3)** by the **3-PTS (F2)** method. Use **ENDENT (F3)** from the Position Menu.

Step 3: A new view is defined by indicating a series of three points. Choose the **ENDENT** method **(F3)** from the Position Menu to locate these three points.

Step 4: Use the cursor to select the following endpoints (1,2,3, Figure 6-24) for the new view.

Step 5: When all three points are selected, as in Figure 6-24, a new view is displayed in the Viewport. This is the auxiliary view showing the inclined surface and circle in true size and shape (Figure 6-25).

Step 6: If you wish to save this view for future reference select the Viewport and **SAVE**. The view will be assigned the next available view number, in this case view **9**. Version 6 allows a name to be entered.

Step 7: To change the Viewport back to view **7** select **VIEW** from the Status Window, then select the Viewport and enter the number **7**.

Task 15: Changing levels

Command Sequence:

F6 DISPLAY
F6 LEVELS
F1 ACTIVE

Before dimensioning the existing part, let's change the current active level. This will allow us to create dimensions on a different layer than the part construction itself. This offers us the flexibility to view the part, with or without dimensioning, when desired.

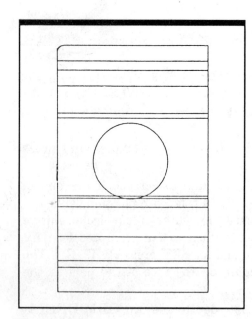

Figure 6-25 The new auxiliary view—number 9

The following steps describe how to change the current active level of 1 to 2. The active level is displayed in the Status Window as the **ALEV** option. Note that the system allows you to build up to 256 different levels!

Step 1: Return to the Main Menu by pressing **ESC**. Change the current color assignment to yellow (value number 4) using **ALT-X**. All text created from this point on will be yellow. **CTRL-F** to save your part file. Enter the name **DOVE** and **RETURN**.

Step 2: Choose the **DISPLAY** option **(F6)**.

Step 3: Choose the **LEVELS** option **(F6)** and a menu is displayed offering four methods of manipulating levels.

Step 4: Choose the **ACTIVE** option **(F1)**. This option allows you to add a new level to the part. From this point on, all dimensioning will be stored on this new level.

Step 5: Enter in the level number **2** and press **RETURN**. For the level description enter **DETAILS**. This level is displayed next to the **ALEV:** prompt in the Status Window.

Step 6: Reduce the part to one-half its size with **ALT-H**. Select the Front Viewport.

Task 16:   Dimensions

Command Sequence:

F3 DETAIL
F1 DIMENSN
F2 VERTICL
F3 ENDENT
F1 HORIZTL
F3 ENDENT

Step 1: Return to the Main Menu **(ESC)** and choose the **DETAIL** option **(F3)**.

Step 2: Choose to dimension **(DIMENSN, F1)** selected points of the part.

Step 3: Choose to select **VERTICL (F2)** points as your first dimension.

Step 4: Locate dimension points by choosing the **ENDENT** method **(F3)** from the Position Menu displayed.

Step 5: For this next step, use the "zoom" feature to enlarge the upper left corner of the part for easier selection. This feature is invoked using the Immediate Mode command **ALT-W**. Indicate each corner of the window to surround the upper left corner of the part. To select the first point of the dimension, move the cursor near an endpoint of line 1 (Figure 6-26) and press the mouse button once again. The endpoint of the line is located and marked with an **X** symbol. If the fillet's endpoint is selected by mistake, press **F10 (BACKUP)** and reselect the line entity.

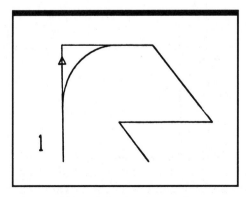

Figure 6-26 Locating witness lines

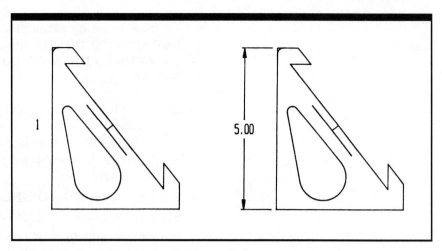

Figure 6-27 Positioning the dimension text

Step 6: Use the Immediate Mode command **ALT-B (BACK-1)** to return the part to its previous size.

Step 7: Select the opposite endpoint of the same line (at the lower left corner of the part) by moving the cursor near the endpoint and press the mouse button.

(**NOTE:** The **ENDENT** method chosen is still in effect.)

Step 8: When both points are located you are prompted for a dimension text position. Move the cursor to the left of the part, between the two marked points, and press the mouse button.

Step 9: The dimension text appears with arrows, witness lines and text (Figure 6-27). The **UNDO** option allows you to redo the dimension.

Step 10: Press **F10** to return to the **DIMENSN** Menu and choose **HORIZTL** dimensioning (**F1**). Choose the same positioning method, **ENDENT (F3)**. Select the endpoints of line 2 (Figure 6-28) for horizontal dimensioning. Position the dimension text underneath this line.

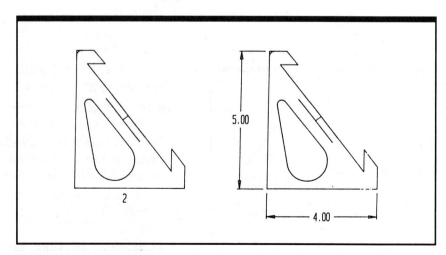

Figure 6-28 Positioning a horizontal dimension

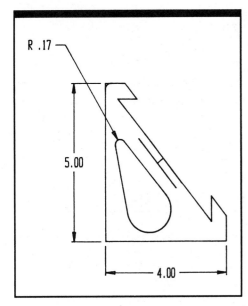

Figure 6-29   Dimensioning a radius

Task 17:   Dimensioning a radius

Command Sequence:

F3 DETAIL
F1 DIMENSN
F4 RADIUS
F2 RIGHT

Step 1:   Press **ESC** and select **DETAIL-DIMENSN (F3-F1)**.

Step 2:   Choose the **RADIUS** option **(F4)**. This automatically dimensions the radius of a selected arc or circle.

Step 3:   Move the cursor to an arc (one of the trimmed circles) and press the mouse button.

Step 4:   The system automatically sets the arrows of a dimension to appear inside the witness lines displayed or within an arc or circle. Change this default assignment by invoking the Immediate Mode command **CTRL-A**. The last page of the Status Window should read **ARR: OUT**. You are returned to your place in the menu structure.

Step 5:   Position the dimension text for the radius by moving the cursor to the outer left portion of the part (Figure 6-29). Press the mouse button.

Step 6:   Choose the **RIGHT** option **(F2)** from the menu displayed to indicate on which side of the dimension text the leader line will appear.

Step 7:   The radius is automatically calculated and an **R** appears to indicate this as a radius dimension. (**NOTE:** Radius values will not be consistent with those given here.) Use the **UNDO** option if you do not like the placement of the dimension.

Task 18:   Adding notes

Command Sequence:

F3 DETAIL
F2 NOTE
F1 KEYIN

Step 1:   Return to the **DETAIL** menu by selecting **F10** twice.

Step 2:   Choose to create a **NOTE (F2)**.

Step 3:   There are two methods of creating notes: reading data in from disk, or typing a note using the keyboard. Choose the **KEYIN** option **(F1)** to type a note.

Step 4:   You are prompted to enter the text. Type the following and press **RETURN** after the first line of text.

Dovetail Part
Created: 5/15/86

To correct mistakes, use the left arrow key or backspace key before pressing **RETURN**. You can also press the **F10** key and

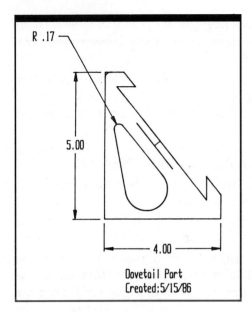

Figure 6-30 Placing notes

Figure 6-31 Selection point for a note

retype the text. If text has already been placed, use **DETAIL-CHANGE-EDIT TX**.

Step 5: When the note is completed, select **SAVE TX (F6)**.

Step 6: Use the cursor to position the note text underneath the completed part (Figure 6-30). Text is displayed to the right of the cursor's center point. Press the mouse button to indicate this position.

Step 7: Save **(CTRL-F)**.

Task 19:   Changing dimensions or text

Command Sequence:

F3 DETAIL
F7 CHANGE
F1 TXT ATT
F4 ASPECT
F7 DECIMAL
F8 TOLER

Various parts of a dimension or note (e.g., attributes, position, decimal precision, etc.) may be altered or changed using the CHANGE function. Only a few of these options are described here. Feel free to try others on your own.

Changing Attributes

Step 1: Use the Immediate Mode command **ALT-A** and **RETURN** to AUTO SCALE all of the Viewports. Return to the **DETAIL** Menu using **F10** and choose to **CHANGE (F7)** text attributes **(TXT ATT=F1)**.

Step 2: Four attributes are offered for change. To change the character aspect ratio of the note, choose the **ASPECT** option **(F4)**.

Step 3: Enter a character aspect ratio of **1** (one inch) and press **RETURN**. Characters are assigned a one-to-one ratio of width to height (in this case, 1 by 1).

Step 4: No changes are made to text until the **TXT ATT** menu is exited and entities are selected. Choose the **DONE** option **(F8)** from this menu.

Step 5: The Selection Menu appears. Since only a single note has changed, choose the **SINGLE** option **(F1)**.

Step 6: Move the cursor near the lower left-hand corner of the upper leftmost character of the text (in this case, the D in Dovetail in Figure 6-31) and press the mouse button. A selection symbol appears at this location.

Step 7: Press **RETURN**. The text is redrawn and takes on the new character aspect ratio assigned.

Changing Decimal Precision

Step 1: Return to the **CHANGE** menu (press **F10** twice) and choose the **DIM REP-DECIMAL** option.

Step 2: A menu appears offering different decimal precisions (where **n** represents any value). Change the decimal precision to three places, **n.nnn (F4)**.

Step 3: Choose the **SINGLE** option **(F1)** from the Selection Menu displayed. Move the cursor to the lower left corner of the horizontal dimension text (4.00) and press the mouse button.

Step 4: Press **RETURN**. The dimension is redisplayed with the new decimal assignment of three places.

Changing Tolerances

Step 1: Return to the **CHANGE** menu by selecting **F10** twice. Change tolerance **(TOLER, F8)** values.

Step 2: Alter the **TYPE (F2)** of tolerance used to the **+/- (F2)** option. This displays both the negative and positive tolerance values with a dimension.

Step 3: Select the same horizontal dimension **(4.000)** using the **SINGLE** option **(F1)** from the Selection Menu. Move the cursor to this dimension and press the mouse button.

Step 4: Press **RETURN** and the default tolerance value of **.010** is displayed, positive over negative, with the dimensioned part.

Step 5: To change this value, return once again to the **TOLER** menu by selecting **F10** twice. This time select the **VALUES** option **(F1)**.

Step 6: Change **BOTH (F3)** tolerances. Enter a positive value of **.001** and press **RETURN**. Then enter a negative value of **-.002** (negative .002) and press **RETURN**.

Step 7: Once again select the same horizontal value **(4.000)** by choosing the **SINGLE (F1)** method from the Selection Menu using the cursor.

Step 8: The dimension is redisplayed with the new tolerance values.

Task 20:   Repositioning a dimension

Command Sequence:

F3 DETAIL
F7 CHANGE
F1 TXT POS

Step 1: Return to the **CHANGE** menu. Alter the text position **(TXT POS, F2)** of a selected note or dimension.

Step 2: Move the cursor to the lower left corner of the radius **(R)** dimension text and press the mouse button.

Step 3: A text box appears around the dimension text. Use this as your cursor for new text placement. Move the text box to the right of the part.

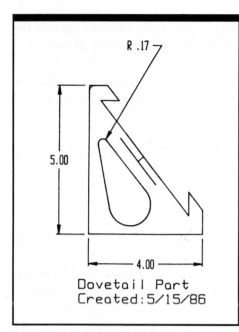

R .17

5.00

4.00

Dovetail Part
Created:5/15/86

Figure 6-32 Changing the
dimension tolerance

Step 4: When the text box is positioned (Figure 6-32), press the mouse button. The dimension text appears at this new location. (The text may be repositioned until the **F10** option is chosen to back out of the function.)

Press **F10** to return to the **CHANGE** menu.

Step 5: Press **CTRL-F** to save.

Task 21: Turning levels on and off

Command Sequence:

F6 DISPLAY
F3 LEVELS
F4 LIST

Up to now we have created entities on two different levels. The main body of the part is on level 1, while all dimensioning is on level 2. To view the part once again without the dimensioning, level 2 can be turned off.

Step 1: Return to the Main Menu **(ESC)**.

Step 2: Choose to **DISPLAY (F6)** the **LEVELS (F6)** options.

Step 3: Choose the **LIST** option **(F5)** to list and turn desired levels on or off.

Step 4: The first page of levels lists numbers 1-128 with a reverse video box surrounding levels 1 and 2. These boxes represent the levels which are currently turned on.

Step 5: Turn off level 2 by selecting **TURN OFF (F2)**, moving the pointer arrow to the number **2**, and pressing the mouse button.

Step 6: Level 2 is no longer highlighted in reverse video and is turned off.

Step 7: Press **RETURN** to return to your part. The dimensions are no longer displayed.

Turning Levels On

To turn levels on select **DISPLAY-LEVELS-LIST-TURN ON** and point to the levels you wish to turn on, in this case, level 2.

Task 22: Creating a layout drawing from the 3D model

Command Sequence:

F3 DETAIL
F6 LAYOUT
F5 INSTANCE
F1 CREATE
F3 ALIGN
F1 AUTO

This method of creating a layout drawing no longer requires the user to create patterns for each of the views. Instead, a layout mode is linked directly to the part file. When changes are made to the part file, the layout is automatically updated. If at any time you wish to get out of the layout mode, use **DETAIL-LAYOUT-EXIT (F3-F6-F8)**.

Step 1: Select **DETAIL-LAYOUT (F3-F6)**.

Step 2: A pop-up window will appear in the middle of the drawing window. First you must enter a name, **Dovetail**, for the layout type. Select **PAPER SIZE** from the window. Change the paper size from A to C by selecting the letter **C** from the list in the window. Check the drawing scale to make sure it is **1:1**. Then select **OK** from the window. The System mode indicator will now read **Layout**.

Step 3: From the Menu Options Window select **INSTANCE-CREATE (F5-F1)**. An instance is a view of your 3D model.

Step 4: A list of views of your model which have been defined will appear. Select the view which you would like to CREATE or add to your layout. Start by selecting the Front view from the pop-up menu.

Step 5: You are prompted **Enter drawing instance rotation angle.** For all of the views except the auxiliary view number 9 the rotation angle will be **0**. The rotation angle for the auxiliary view is **38.659813**. Press **RETURN** after you have entered the angle.

Step 6: Use the mouse to position the views as they appear in Figure 6-33.

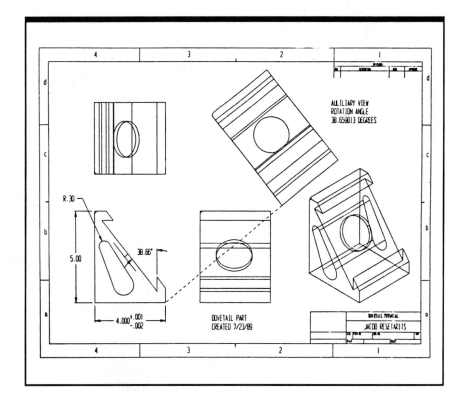

Figure 6-33 Completed part assembly with border.

Step 7: Repeat Steps 3 through 6 for each of the views in the layout drawing. The views in the drawing includes the **Front Top, Right Side, Isometric,** and **System View (Auxiliary 9).**

Step 8: Once the views are placed in the paper boundary you can align the Top, Right Side, and Auxiliary views with the front view using the options **ALIGN-AUTO (F3-F1).**

Step 9: You are prompted **Select the fixed drawing instance.** This is the base view to which you align another view. In this case select the front view.

Step 10: You are then prompted **Select drawing instance to align.** Select the view which you would like to align to the front view. Move the view into the desired position (notice how you can only move it one direction because it is aligned with the fixed instance) and press the mouse button to align the view.

Step 11: Repeat Steps 8 through 10 until the views are aligned.

Step 12: Use the Immediate Mode command **CTRL-F** to save your part file. Use the same filename, **DOVE,** and replace the existing file. The changes you now make on your original part file will automatically appear on the layout, because the part file and the layout are the same file, displayed two different ways. If you should want to get back to the 3D model (we do not), use **DETAIL-LAYOUT-EXIT.**

At this point you can add a border pattern and more detail features to your layout. They will appear only in the layout mode and not on the 3D model unless you choose to modelize the features. See Chapter 16, Modelizing, for more information.

Task 23: Retrieving the title block

Command Sequence:

F5 FILES
F2 PATTERN
F2 RETRIEV
F9 KEY-IN

A title block may be used when creating a drawing layout (displaying a part in different views) and may be thought of as a "frame" for your part.

The Student Work Disk holds two title blocks stored as pattern files for your access. These pattern files are not copied onto your hard disk during installation procedures so you must copy them into your pattern directory yourself or retrieve them directly from the diskette as needed. The title blocks supplied follow ANSI standards and are presented in English and Metric modes. Filenames are as follows:

English
BORDB
BORDC

These names represent the name of the pattern file (BORD) along with the title block size (A,B,C, etc.).

Press **ESC** to return to the Main Menu.

Step 1:   Call up the **PATTERN:RETRIEV** function using the **ESCAPE** code: **ESC-F5-F2-F2**. At the prompt:

Enter pattern filename ( ):

type the following:

BORDC

or **A:BORDC** (if the Student Work Disk is in drive a:) and press **RETURN**.

Press **RETURN**. Remember that you must retrieve the border pattern files from the diskette or copy them into the pattern file directory in order to access them through this option.

Step 2:   Group the entities in the pattern being retrieved. Choose the **YES** option, **F2**.

Step 3:   You are prompted to enter the group name, level assignment, and the scale factor of the retrieved pattern. Accept the default values by pressing the **RETURN** key only.

Step 4:   At the next prompt:

Enter rotation angle (0) =

accept the current rotation angle of **0** by pressing **RETURN** only.

Step 5:   Indicate a base position for the border by choosing the **KEY-IN** option, **F9**, from the Position Menu displayed. This option allows Cartesian coordinates to be entered from the keyboard to indicate a desired position.

When this pattern was stored, the bottom left corner was chosen as the base position. When the pattern is retrieved, this same corner is used as the reference point of the new base position. Assigning zero (0) coordinate values for the title block location will position this reference point in the lower left corner of the screen (the cursor's home position).

At the first **KEY-IN** option, **XV**, press **RETURN** to accept the View Coordinate default value of **0**. Repeat this procedure for each coordinate:

XV = 0   YV = 0   ZV = 0

Step 6:   Use **CTRL-F** to save as **DOVE**.

Task 24:   Deleting unwanted entities

This function is useful only if you have created unwanted entities on the screen.

If you wish to edit the individual patterns it is best to **DEGROUP** the patterns first. They were only grouped to allow for repositioning and deleting. To degroup the patterns select **EDIT-GROUP-DEGROUP-ALL DSP-ALL (F2-F5-F3-F7-F1).**

Single Entities:

Step 1: Invoke the Immediate Mode command **CTRL-Q**.

Step 2: Move the cursor to each entity you wish to erase and press the space bar or mouse button. If you make a wrong selection, press **F10, BACKUP**; this causes a # marker to appear on the canceled selection.

Step 3: Press **RETURN** to activate the deletion process of these entities.

Deleting Entity Groups

To delete more than one entity:

Step 1: Return to the Main Menu (by pressing **ESC**).

Step 2: Choose the **DELETE** option **(F8)**.

Step 3: Choose to **SELECT** entities **(F1)**.

Step 4: Choose an option from the new menu displayed:

1 SINGLE—allows you to indicate each entity for deletion (same as **CTRL-Q**).

Press **RETURN** when the selection process is completed.

2 CHAIN—selects a "chain" of entities.

3 WINDOW—erases entities that are surrounded by a "rubberbox." Use the space bar or mouse button to indicate each corner of the window. The entities desired for deletion must be completely surrounded by this "rubberbox."

4 POLYGON—allows the user to define a line string or irregular polygon boundary to select entities. Options include ALL IN, ALL OUT, or PARTIALLY IN or OUT.

5 GROUP—selects grouped entities.

6 PLANE—entities are selected by identifying a plane.

7 ALL DSP—deletes everything on the display screen as indicated.

Step 5: When the deletion process is complete, activate a screen **REDRAW (CTRL-R)** to clean up the screen. The screen is refreshed with all temporary markers erased.

Press **ESC** to exit this function. If necessary, reconstruct desired entities and continue on.

Deleting Dimensions and Text

Select the dimension or text for deletion by placing the cursor at the bottom left corner of the furthest left line of text or dimension,

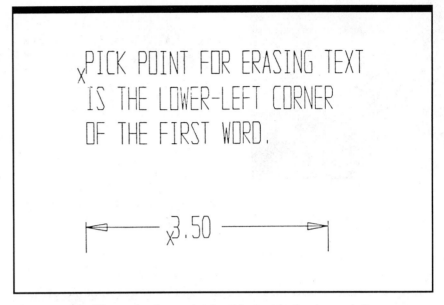

Figure 6-34   The selection point for this text is the lower left corner represented by the X.

as shown in Figure 6-34. If you do not select a point near the lower left corner, CADKEY may not find the text and the Prompt Line will display an error message.

CHAPTER 7

Deleting Entities

OBJECTIVES

After completing this chapter, you will be able to:

- **DELETE** entities using the Immediate Mode command.
- **DELETE** entities by level.
- **DELETE** entities by chaining.
- Use **WINDOW** to delete entities.
- Use **GROUP** to delete entities.
- Use **PLANE** to delete entities.
- Describe how truncated entities are handled by the **WINDOW** delete option.

The ability to delete entities quickly and in a variety of ways is a very important feature of a CAD system. CADKEY has a powerful delete function that allows the user to delete entities one at a time, by group, level, window, all entities displayed, and chain, and by other methods to be described in this chapter. An Immediate Mode delete allows the user to selectively erase entities at any time in the program. CADKEY's delete function can save you much time when erasing unwanted entities or modifying an existing drawing. The **RECALL** function located under the Main Menu item **EDIT** can be used to retrieve previously deleted entities, so if you made a mistake and deleted the wrong entity, the **RECALL** function can be used. The deleted entities can also be recalled using the Immediate Mode command **CTRL-U**. **DELETE** has two options: **SELECT** and **LEVEL** (Figure 7-1).

IMMEDIATE MODE DELETE

DELETE is Main Menu option 8. Single entity delete can also be accessed through the Immediate Mode command **CTRL-Q**. The Immediate Mode delete can only be used to selectively delete entities by digitizing each entity that you want to erase, and it cannot be used to erase entities by window, level, or some other methods which are available under **F8 DELETE** in the Main Menu. If the

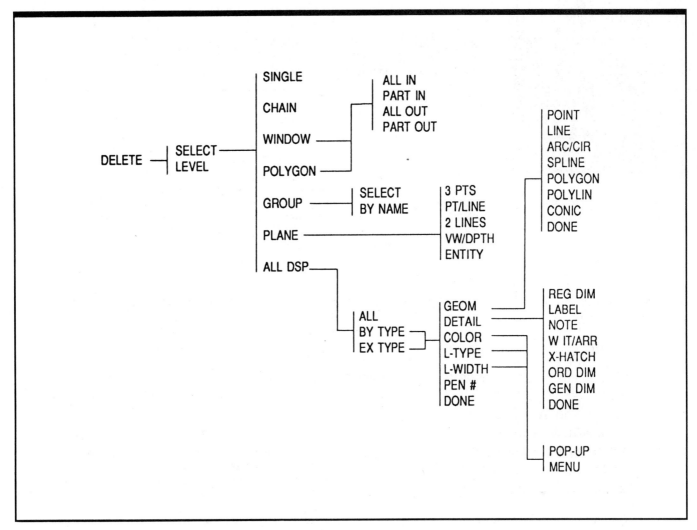

Figure 7-1   The **DELETE** menu

entity selected is part of a group, you are prompted to select from the Group Menu display. However, the main advantage of the Immediate Mode delete is that it allows you to quickly delete a few entities without interrupting your current function.

After the Immediate Mode delete is accessed, a prompt reads: **Select entity 1 (press RETURN when done).** Move the cursor over the entity to be deleted and press the space bar or the mouse button on the input device. For text, pick the placement or pick point. A small triangle will be placed over the entity that the system will delete, or the entity will be highlighted, depending on the configuration choices. The prompt now reads: **Select entity 2 (press RETURN when done).** If for some reason the system places the triangle over the wrong entity, press **F10 BACKUP**. This will cause the system to place a small "#" sign over the triangle, indicating that that item will not be deleted. The prompt will now read: **Select entity 1 (press RETURN when done).** You can now attempt to pick the correct entity.

Amdahl Corporation Uses CADKEY® to Design Mainframe Computers!

For many years, personal computers and their compatible software were designed and manufactured with the help of mainframe computers in conjunction with minicomputers. As technology allows more computing power to be packed into smaller machines, PCs have now grown into the eventual replacement for the older, and traditionally more powerful, minicomputers.

Almost four years ago, however, a new scenario began to unfold around Amdahl Corporation. The Sunnyvale, California-based manufacturer of mainframe computers had realized that the minicomputer-based design and manufacturing system, which it had purchased four years earlier, and the computer-aided design (CAD) software running on it were limiting their design efforts. In addition to not being user-friendly, the system failed to offer the same high level of user-productivity and performance when compared with a personal-computer-based system. Furthermore, the cost-per-user for the minicomputer-based system was five times greater than the cost-per-user for the PC-based system, and the cost to upgrade the minicomputer system was estimated then at approximately $400,000.

The minicomputer-based system presented the company with other problems, as well. "If the system went down, every terminal linked to it went down, too," explained Tim O'Rourke, Amdahl's senior mechanical engineer and CAD systems support manager. "Using personal computers as stand-alone workstations eliminated the problem because each workstation was independent of the host system."

Besides the hardware issue, Amdahl needed to identify a PC-based, three-dimensional, CAD software package that would offer the same electronic-design advantages that their engineers were getting from the mini-based package then in use. Amdahl had already weighed the differences between drafting and design systems. The company had determined that continuing with a three-dimensional system was their only real option. Amdahl required CAD software sophisticated enough to be used both by O'Rourke's mechanical-design team, and by the company's manufacturing team that designs special fixtures, tools, and equipment used to manufacture and assemble Amdahl's mainframe computers. Amdahl called on Mike Poelman of Poelman Design in Campbell, California, among several other local value-added resellers (VAR) of CAD software, to suggest the best hardware/software configuration to do the job.

Poelman recommended CADKEY®. At the same time, Poelman encouraged Amdahl to conduct an extensive evaluation of several products, including CADKEY, before making a final decision to standardize.

Amdahl's test results showed that "CADKEY is fully three-dimensional, very user-friendly, and it can support mice and/or tablets, high-resolution monitors, and a variety of printers and plotters," O'Rourke added.

Amdahl based its choice on CADKEY (Version 2.11), first introduced five years ago. Since then, CADKEY's upgrades include such enhancements as the user's access to multiple viewports simultaneously, and the ability to maneuver rapidly from wire-frame modeling to solid modeling, as well as advanced types of curves, including cubic parametric splines, ellipses, parabolas, and hyperbolas.

Admahl has already begun incorporating these enhanced versions into their current installations to increase productivity beyond their original expectations. A key feature of prime interest to Amdahl's engineering team is CADKEY's ability to let users test the mechanical parts and assemblies in real-life situations, even before building a prototype. This lowers the overall cost of development and virtually eliminates the unwieldy process of going through several design and prototype stages.

*Adapted with permission from **3-D WORLD**, v. 5, no. 2, Mar./Apr., 1991.*

 POSSIBLE USE

The Immediate Mode delete is an option that you will find very useful and one that is used quite often, though one hopes not too often! One possible use is to correct a mistake immediately without interrupting your current place in the menu structure. Another possible use would be to immediately erase entities used for construction purposes. Suppose you had to create an auxiliary view of a part to show the true size of a surface. If the part was not too complicated, you could use the Immediate Mode delete to erase the construction lines used to create the view.

HINTS FOR IMPROVING THE ENTITY SELECTION PROCESS

If you are having difficulty deleting entities, try these suggestions. If many items are on the screen, window-in to the entity to be deleted so the system does not have to search so long to find the desired entity. Picking a unique point on the entity that is as far as possible from other entities will improve chances of the system selecting the correct entity. Do not pick the point where two entities intersect. This is not a unique point and the system slows down when searching for the correct entity to delete. Also, try to plan ahead and place groups of entities on different levels. For example, place all construction lines on a separate level so they can be deleted or masked by level. A mask can be set with any selection method simply by pressing **ALT-M** and choosing the entities to mask.

A common problem for many beginning CAD operators is that deleted entities keep coming back after using **REDRAW**. This occurs because two entities have been drawn directly over one another. When one of the entities was deleted it temporarily blanked out the other one. Selecting **REDRAW** will refresh the screen and the blanked entity returns. Although it appears that the entity was not erased, one of the entities was erased but the other one was not. This can also happen when trimming entities. If a trimmed entity returns after selecting **REDRAW**, two or more entities may be drawn on top of each other.

LEVEL DELETE

Choosing Main Menu item **F8 DELETE** will give the user two choices: **SELECT** and **LEVEL**. The level function deletes all entities on each level specified. By planning ahead and placing groups of entities on different levels, it is possible to delete large numbers of entities very quickly using the **DELETE** by **LEVEL** function. For example, all construction lines placed on a drawing could be drawn on a level separate from other entities. All entities on the level containing the construction lines could be deleted using the **LEVEL** option.

Figure 7-2, page 132, shows a simple part with dimensions placed on level 2. Level 1 contains the lines and circle. All the dimensions and the cross-hatching could be quickly deleted using the **LEVEL** option. This is demonstrated in the following steps.

1. Press **ESC, F8 DELETE, F2 LEVEL**.
2. A prompt reads:

   Enter level to delete entities from =.

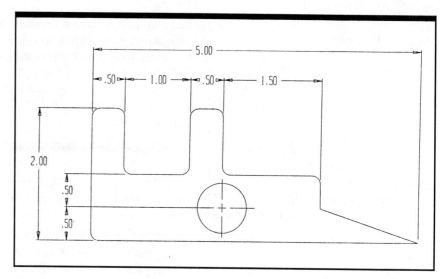

Figure 7-2 Part with dimensions on level 2

CADKEY uses 256 levels; choose the level(s) which contains the entity(ies) to delete. Input the level to be deleted, which in this example is 2. Only one level can be deleted at a time. Press **BACKSPACE** to remove an incorrect level number.

3. Press **RETURN** if the desired entity level is displayed in the prompt. In this example, the number 2 should be displayed in the prompt. After **RETURN** is pressed, all entities on that level will immediately be erased (Figure 7-3).

4. Press **ESC** to return to the Main Menu.

SINGLE DELETE

When the **SELECT** delete option is picked, the Selection Menu appears on screen. The first option listed is **SINGLE**. The **SINGLE** delete option is used to cursor-select entities to be erased by entering **F8 DELETE, F1 SELECT, F1 SINGLE**. With this option you move the cursor over the entity to be deleted and press the space

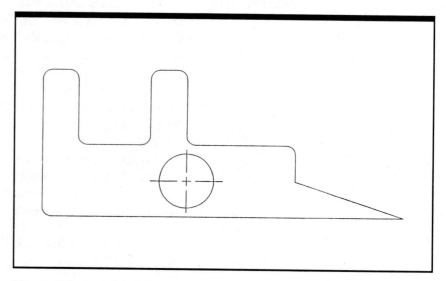

Figure 7-3 Level 2 deleted

bar or the mouse button on the cursor control device. A small triangle is placed on the entity that the system has selected to delete. If that selection is incorrect, simply press **F10 BACKUP**. A small "**#**" sign is placed over the triangle drawn on the entity and you can try to select the correct entity. For drawings with many entities displayed, it is better to use the **DISPLAY** function to window-in on the entity to be deleted so that the system can more readily find the desired entity. This option is also available with the Immediate Mode command **CTRL-Q**.

When using the single delete option you are prompted to cursor-select the entities. After the entities have been selected, press **RETURN** and the entities are erased. The small markers or triangles are left on screen after the delete, but are easily removed using the **REDRAW** option under the **DISPLAY** function or using the Immediate Mode command **CTRL-R**.

Of course, many times you will be using the single delete option. Figure 7-4 shows an object with two width dimensions (4.50). The one shown at the bottom of the part will be deleted using the single delete option using the following steps.

1. Press **ESC, F8 DELETE, F1 SELECT, F1 SINGLE**.

2. A prompt appears on screen:

   Select entity 1 (press RETURN when done).

3. Cursor-select the dimension to be deleted. A small triangle appears near the dimension figure.

4. A prompt reads:

   Select entity 2 (press RETURN when done).

5. Only one entity is to be deleted, so press **RETURN**. The entity is immediately deleted from the screen.

6. Press **CTRL-R** to refresh the screen and remove any markers (Figure 7-5 on page 134).

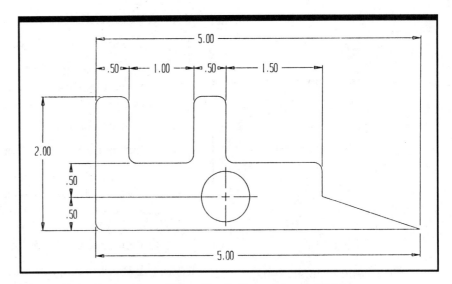

Figure 7-4 Part with a dimension that must be deleted

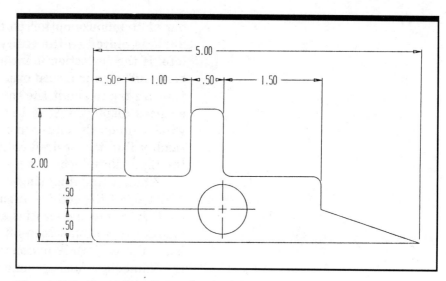

Figure 7-5 Part after dimension is deleted

CHAIN DELETE

The **CHAIN** delete option allows the deletion of connected entities, such as lines and arcs. With this option, some or all of the entities in the chain are deleted. Only connected lines, arcs and/or splines are available with this function. The direction from the start of the chain that the delete is to travel is selected with the cursor. The last entity in the chain that is to be deleted is then picked. If all entities in the chain are to be deleted, press **RETURN** and the system automatically identifies the entities included in the chain.

Figure 7-6 shows a part with an irregular shape punched in one end. This irregular shape is made from a series of lines and arcs. The entities could be deleted one at a time using the **SINGLE** option, but they could more effectively be deleted using the **CHAIN** option. The following steps demonstrate how to use the **CHAIN** delete option.

1. Press **ESC, F8 DELETE, F1 SELECT, F2 CHAIN**.

2. A prompt reads:

    Select start entity in chain.

    Cursor-select one of the entities in the chain. A small triangular marker identifies the entity selected. If this is not the one you want, press **F10 BACKUP** and try again.

3. A prompt reads:

    Indicate chaining direction.

    Pick a point in the general direction that the chain delete is to travel. If all the entities in a closed chain are to be deleted, either direction from the first entity picked will work. However, this step becomes more important if only a few of the entities in the chain are to be deleted.

4. The prompt reads:

    Select termination entity in chain or press RETURN.

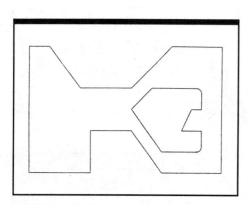

Figure 7-6 Part with an irregular shape

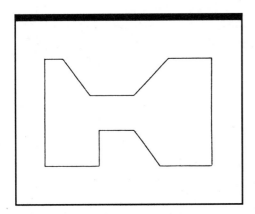

Figure 7-7 **CHAIN** selected lines erased

For this example, **RETURN** could be used to automatically have the entities in the chain selected. However, if only part of the chain is to be deleted, then the last entity in the chain has to be picked. In either case, the system marks those entities to be deleted with a small triangle after this step is executed.

5. A prompt reads:

   Selection OK?

   Carefully check that only those entities that you want to be deleted are marked with a small triangle. A **NO-YES** menu will appear on screen. If the chained entities are incorrect answer **F1 NO** and the system returns to the prompt shown in step 2. If you answer **F2 YES**, all entities making up the defined chain are deleted.

6. Press **CTRL-R** to refresh the screen and remove the triangular markers (Figure 7-7).

WINDOW DELETE

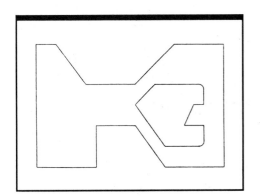

Figure 7-8 The Window **DELETE** option

Window **DELETE** defines a window that includes the entities to be deleted in much the same way a window is defined using the Window Display function. Window **DELETE** is a quick method of defining a group of objects to be deleted. There are four Window **DELETE** options: **ALL IN**, **PART IN**, **ALL OUT**, and **PART OUT**. The **ALL IN** option will delete all the entities that are totally within the defined window. Truncated entities (those not entirely enclosed by the window) are not deleted. The **PART IN** option will delete every entity that is totally or partially within the window. Truncated entities will be deleted. The **ALL OUT** option will delete all entities that are entirely outside of the defined window. Truncated entities are not deleted. The **PART OUT** option will delete all entities outside the defined window, including those partially within. Truncated entities will be deleted.

Figure 7-8 is the same object used in the previous example for chain delete. The irregular punched shape in the part will be deleted using the Window **DELETE** function.

1. Press **ESC, F8 DELETE, F1 SELECT, F3 WINDOW F1 ALL IN**.

2. A prompt reads:

   Indicate position for 1st selection window corner.

   Cursor-select a corner for the window that completely encloses the entities to be deleted. If that point proves to be unsatisfactory, simply press **F10 BACKUP** and select again.

3. A prompt reads:

   Indicate position for 2nd selection window corner.

   Move the cursor to the corner of the window located diagonally from the first corner. Notice that a rubberband window is formed as the cursor is moved to assist you in properly locating the window (Figure 7-9). Carefully position the window

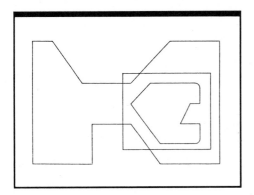

Figure 7-9 Rubberband box surrounding entities

completely around those entities to be deleted, then press the mouse button on the input device. The entities completely enclosed within the window are deleted (Figure 7-10). Truncated entities are not deleted.

GROUP DELETE

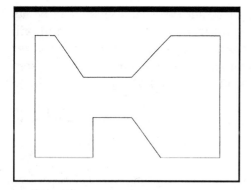

Figure 7-10 Lines erased

Any entity that can be placed into a defined group can be deleted as a group. This option can save considerable time, especially when cleaning up a drawing that has just been completed. For example, suppose that a mechanical drawing was created using a large number of construction lines. If you wanted to delete these construction lines quickly, they could be placed in a group and deleted using the Group **DELETE** option. The Group **DELETE** option is used by identifying the group name or picking any one of the entities making up the group. When you select the Group **DELETE** option, two menu choices will appear, **SELECT** and **BY NAME**. Each method will be explained.

Figure 7-11 shows an object with a number of dimensions that have been placed in a group. These dimensions are deleted using the following steps.

1. Press **ESC, F8 DELETE, F1 SELECT, F5 GROUP, F1 SELECT**.

2. A prompt appears:

   Cursor-select group.

   Pick any one of the entities making up the group. In this example picking any one of the dimension notes causes all

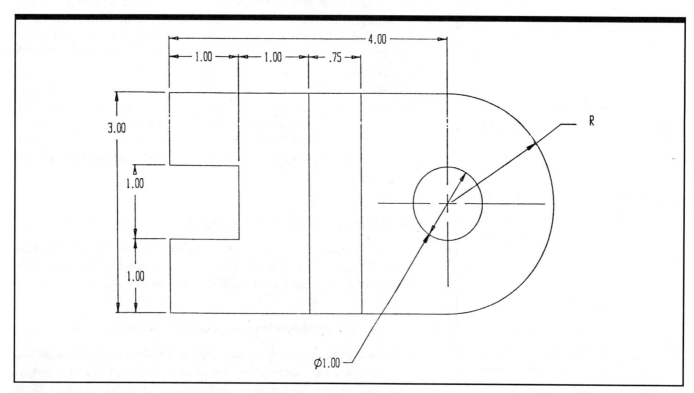

Figure 7-11 Part with dimensions to be deleted

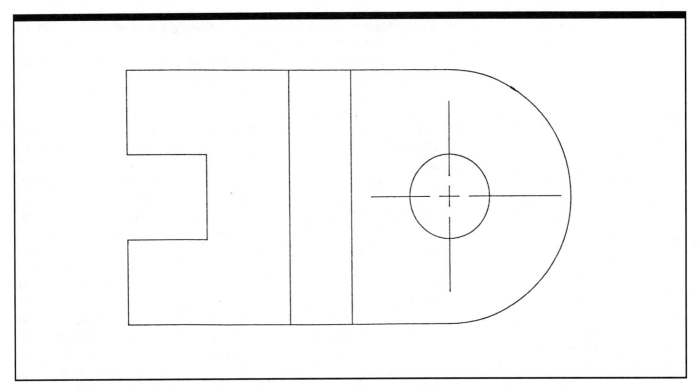

Figure 7-12   Part after dimensions deleted

the dimensions to immediately be deleted from the screen (Figure 7-12).

The dimensions can be deleted by typing in the name given to the group. The following steps describe how this is accomplished using the same Figures 7-11 and 7-12.

1.   Press **ESC, F8 DELETE, F1 SELECT, F5 GROUP, F2 BY NAME**.

2.   A prompt reads:

   Enter group name ( ):

   Type in the name given to the group to be deleted and press the **RETURN** key. All entities in the group are deleted from screen.

PLANE DELETE   The **PLANE** option allows entities to be deleted by defining a plane. Five options are given: **3 PTS, PT/LINE, 2 LINES, VW/DPTH,** and **ENTITY**. The selection process is the same as is described in Chapter 11. After the plane is selected, all entities in the plane are deleted.

ALL DISPLAYED DELETE   The **DELETE** All Displayed option deletes everything that is displayed on the screen or in the current window. It also has a masking delete option that works with the **ALL DSP** option. This allows the deletion of all the entities displayed **BY TYPE** or **EX**

TYPE (except type). These two options allow masking on any entity or combination of entities.

When using the **ALL DSP** and **ALL** function together, every entity displayed is deleted. But what if you are in a zoom window at the time you select these options? What happens to those entities lying outside of the currently displayed window? Only those entities displayed are deleted, including truncated or partially displayed entities. Be careful that only those entities to be deleted are displayed when this option is selected.

The **ALL DSP** option is chosen by pressing **F8 DELETE, F1 SELECT, F7 ALL DSP**. A menu displays the different options available with this function: **ALL, BY TYPE,** and **EX TYPE**. Choosing **F1 ALL** deletes all the displayed entities, including those that are truncated. Choosing **F2 BY TYPE** allows the setting of a mask. A prompt appears asking: **Choose option for masking**. The Masking Menu is displayed and those entities to be deleted are selected from the menu. The menu number is displayed in the prompt. For example, if dimensions and cross-hatching are to be deleted from Figure 7-13, select **F2 DETAIL** and select the numbers 1 and 5. These numbers will appear after **(current=)** in the prompt like this: **(current1,5)** because **REG DIM** is menu item number 1 and **X-HATCH** is number 5. To remove an entity from the list, simply select that entity from the menu again and it is removed from the prompt. After the entities to be deleted are selected press **F8 DONE** and all the entities selected are immediately deleted (Figure 7-14). In CADKEY 6, the hierarchial style of masking menu has been replaced by a much more understandable and easier to use Set Mask dialog box (Figure 7-15). In this dialog box the user simply selects all of the desired attributes and entity types on which to mask. If the user is

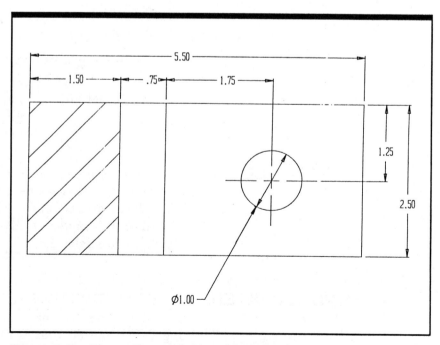

Figure 7-13 Dimensions and cross-hatching to be deleted

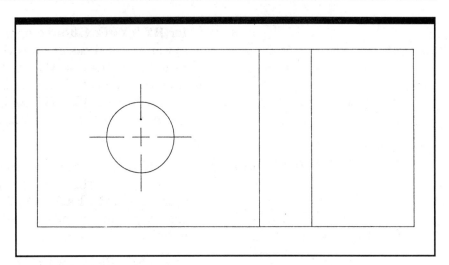

Figure 7-14   Dimensions and cross-hatching erased

masking **BY TYPE,** all of the selected entities will be affected by
the command. If the user is masking **EX TYPE,** all of the selected
entities will not be affected by the command.

If more than one viewport is displayed when **ALL DSP** is selected a prompt reads: **Indicate viewports to select all entities
in.** Move the cursor into the viewport that displays all the entities
to be deleted and press the mouse button. All the entities displayed in the selected viewport will be deleted.

Another method of using the **ALL DSP** option is to mask those
entities that are not to be deleted. This is done by choosing **F3 EX
TYPE.** Selecting this option displays the same prompt described

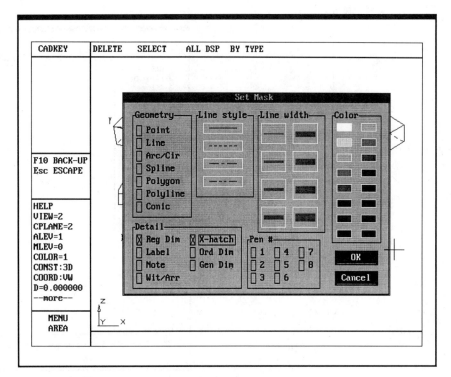

Figure 7-15   The Set Mask Dialog Box for Version 6

for **BY TYPE: Choose option for masking:**. For the **EX TYPE** option select those entities that are not to be deleted from the current display. Choose the entities from the selection masking menu. Remove an entity from the mask by selecting that entity from the menu. Press **F8 DONE** to delete those entities displayed, except those that were masked using the **EX TYPE** option.

POLYGON DELETE

The **POLYGON** option is used to define a polygon shape to delete entities. With this option you can create a multisided polygon to delete **ALL IN**, **PART IN**, **ALL OUT**, or **PART OUT**. The following steps and Figures 7-16 and 7-17 describe the **POLYGON** option.

1. Select **ESC, F8 DELETE, F1 SELECT, F4 POLYGON, F2 PART IN**.
2. A prompt reads: **Indicate position for polygon vertex (or RETURN to finish)**. Use the cursor to draw a polygon which will define the area from which entities are deleted.
3. Press **RETURN** to close the polygon (Figure 7-16).
4. After **RETURN** is pressed, the entities totally or partially within the defined polygon are deleted (Figure 7-17).

SUMMARY

You can take advantage of CADKEY's convenient deletion options if you plan carefully before beginning your drawing. And if for some

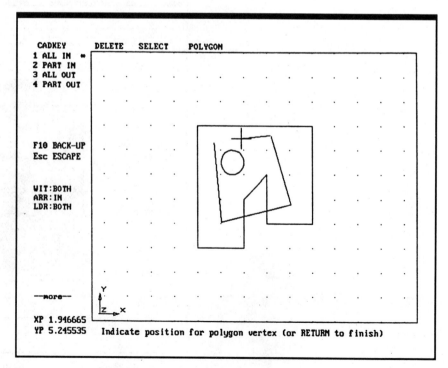

Figure 7-16 Defining the polygon area

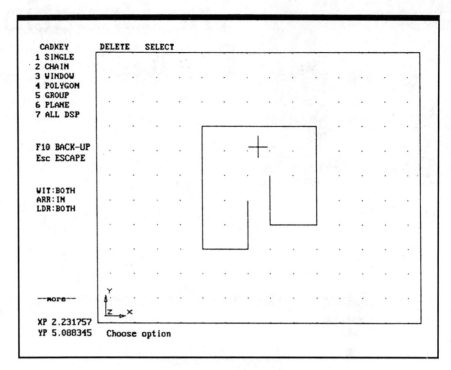

Figure 7-17   Entities deleted using **POLYGON**

reason you should make a mistake (hard to believe but possible), use the **RECALL** option under the **EDIT** function to recover entities that you deleted. This option is explained further in Chapter 15.

REVIEW QUESTIONS

1. What Immediate Mode command is used for **SINGLE DELETE?**
2. List the entities that are deleted using the **CHAIN** option.
3. Will truncated entities be deleted when using the Window **DELETE ALL-IN** option?
4. What is the name of the option that is **used** to retrieve previously deleted entities?
5. List some of the hints that will improve the selection process.

CHAPTER 8

The Creation of Entities

OBJECTIVES

After completing this chapter, you will be able to:

- Create lines using endpoints, string, parallel, perpendicular, tangent, horizontal, vertical, rectangle, polygon, and mesh.
- Create arcs using one of the seven CADKEY options.
- Create a circle using one of the six CADKEY options.
- Create a polyline using one of the five CADKEY options.
- Create fillets and rounds between lines, arcs, and circles.
- Create a chamfer on a corner.
- Create a conic section.
- Create regular polygons.
- Create spline curves.
- Create a helix.

Entities are the basic building blocks used to create a geometric model. Every geometric figure created using CADKEY can be made of a single entity or hundreds of entities; the number will vary depending on the complexity of the part or drawing. Lines, arcs, circles, splines, conics, polygons, and points are examples of entities (Figure 8-1). There are many different methods to create these and other entities when using CADKEY. This chapter will discuss those methods that are found under the **CREATE** function on the Main Menu. Figure 8-2 illustrates the **CREATE** menu.

CREATE OPTIONS

When you create entities in CADKEY you produce more than a picture; you are developing a true 3D data base. This data base can

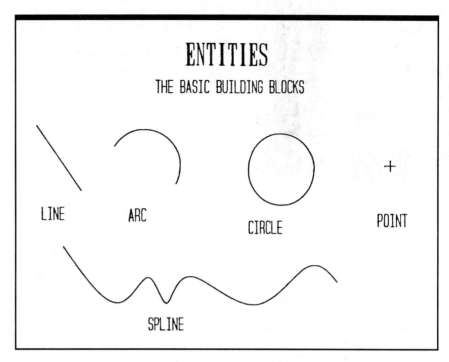

Figure 8-1 Entities are the basic building blocks.

then be used for other applications in designing and manufacturing a part.

When you are using the options found under **CREATE** several parameters (**2D/3D** construction switch, indicating position, depth, and escaping a function) are shared by most of the functions.

Whenever you are prompted to enter data (a filename or increment value), the data in parentheses on the prompt line is the default value. If the default value is the same as the desired value you need not enter the value again; you can simply accept the default by pressing **RETURN**.

When you are requested to indicate the position of a point (start point, endpoint, center, etc.) you will always have the opportunity to use the universal Position Menu (see Chapter 10) to aid you in accurately positioning the point. It is strongly suggested that you become skilled in the use of the Position Menu to create the most accurate parts with the greatest ease.

Whenever you are prompted to select an entity you will always have the benefit of the universal Selection Menu (see Chapter 11) to aid you in selecting the desired entity. Again, because this is a universal menu that is used in many options, it is to your benefit to become familiar with all of the options.

CADKEY is a true 3D system, so that whenever you indicate the position of an entity, you have to consider what the current depth is and whether or not it is the desired depth. Most entities, when they are created, are positioned at the current depth. The current depth can be found in the Status Window next to the letter **D**. The depth can be changed by using the Immediate Mode command **CTRL-D**, which will prompt you to enter a new depth value. The **DEPTH-POSITN** command will allow you to change the current depth to

Buell Motorcycles Make It in the '90s with CADKEY!

Erik Buell started his motorcycle manufacturing company in Mukwonago, Wisconsin, in the Horatio Alger tradition of American entrepreneurs. Find yourself a niche in the market that satisfies professional and personal goals, and utilize your skills to start a company, preferably in the family garage. In the spring of 1987, with experience at America's premiere motorcycle manufacturer, Harley-Davidson, Erik found his niche market in high-performance motorcycles. Not the big and beautiful Harleys made famous in Hollywood, but sleek sporty machines that combined aerodynamics with ergonomics and a sense of the rider. Fast and functional, powerful and precise.

Erik was convinced that there was a void in the market between the economical Japanese models and the luxury Harleys, so he founded the Buell Motor Company Inc. and began to design the dream machine.

Creating a New Motorcycle

Creating the kind of motorcycle that he wanted required three things: light weight, chassis rigidity, and high-quality components. Erik developed a unique drivetrain mounting system that allows the engine to be a stressed member of the chassis without transmitting any of the vibration to the rider.

Buell Motorcycle, Model RS 1200, designed with CADKEY, displayed at AUTOFAC '90.

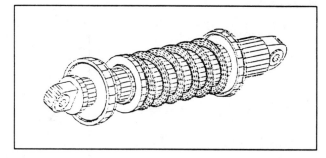

Shock-absorber assembly of the Buell Motorcycle, Model RS1200.

The Uniplaner™ system results in a motorcycle that offers chassis rigidity that may be the best in the world while weighing only 450 pounds. This is accomplished despite an engine/transmission unit that weighs 205 pounds. The high-quality components are produced specifically for Buell by the best names in the business: White Power, Performance Machine, Works Performance, etc. Casting, machining, and fabricating are done by small, precision firms in the Milwaukee area. "I produce the engineering drawings that all these firms need using CADKEY," says Erik.

"I had worked with other CAD programs, but I needed to work in 3D," Erik recalls. "I was more interested in conceptual design." Erik, like many designers, finds three dimensions easier because people conceptualize as they see, in 3D, so that the translation between idea and design is less convoluted. He also wanted something that afforded him design flexibility but wouldn't inflate his expenses.

Evolutionary, Modular Development of Motorcycles

Buell Motor Company's motorcycles have already evolved from the initial R1000 series, through the RR1200, into the current RS 1200 model. Erik uses part files from older motorcycles to evolve new bikes. He reworks them to current specifications, and imports them into contemporary designs. Erik designs Buell motorcycles in modular fashion. Each component occupies a single level. Erik begins with an x,y plane, on Level 1, representing the

—Continued, next page

CADKEY In The Real World—Continued

ground on which the motorcycle stands. The front wheel occupies Level 2. The rear wheel occupies Level 3. And so on ... As he completes the geometry of each component, on its own level, in the master part file, Erik also saves that component as a CADL file.

Erik prefers to save components as CADL files rather than as pattern files because the geometry in a CADL file is not related to a specific view. He sets the origin of each component's CADL file at x=0,y=0,z=0. That makes it easier for him to integrate the CADL files of the geometry of individual components into the master part file containing the

subassemblies and assemblies of the entire motorcycle. Erik also details the CADL file of each component for use by machine shops that manufacture parts for Buell Motor Company. If the master part file becomes unwieldy, he can delete the geometry on particular levels, and continue working, secure in the knowledge that reassembling the assembly of components is a matter of calling up CADL files.

Adapted with permission from 3-D WORLD, v. 4, no. 6, Nov./Dec., 1990.

---

the depth of a position that you select. If you had an entity or surface at an unknown depth, you could select a point on that surface and the depth value of that point would be assigned to the current depth.

If the **2D** construction switch is on, any entities created are only able to be positioned at the current working depth in a plane parallel to the display plane. If the construction switch is set to **3D**, the points can be positioned at varying depths and the entity created does not have to lie in a plane parallel to the display plane. In cases where it is needed, a new view or construction plane can be defined for entities that are created in new planes.

When you have completed a function, you will always be prompted with the first prompt of the function. This allows you to repeat the function as many times as needed without having to re-invoke the function each time. When you are done with a function or if you only wish to use the function once, press **ESCAPE** to exit the function and return to the Main Menu. If you wish to **BACKUP** one step in a function to change a value or option, press **F10**.

The **CREATE** function has nine options for creating different types of entities:

LINE
ARC
CIRCLE
POINT
POLYLIN (Polyline)
FILLET
CONIC
POLYGON
SPLINE

All of the options on this menu are discussed in depth in this chapter.

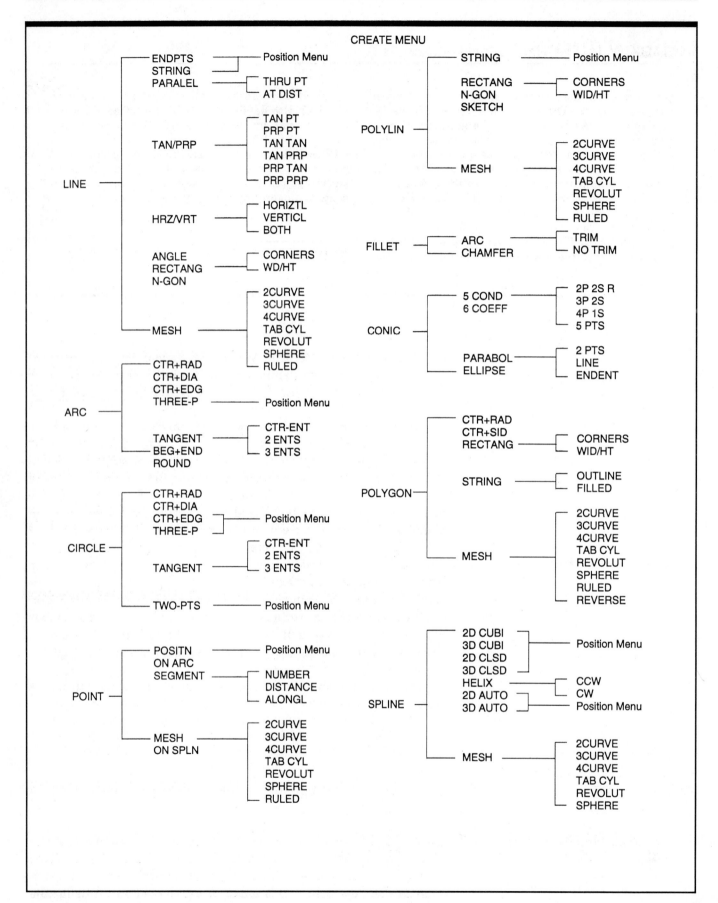

Figure 8-2 The CREATE Menu

LINE OPTIONS

There are nine different methods of creating a line in CADKEY. Each method has its own use for which it was developed. The nine methods are:

> **ENDPTS (Endpoints)**
> **STRING**
> **PARALEL (Parallel)**
> **TAN/PRP (Tangent/Perpendicular)**
> **HRZ/VRT (Horizontal/Vertical)**
> **ANGLE**
> **RECTANG (Rectangle)**
> **N-GON**
> **MESH**

LINE OPTION: ENDPTS (Endpoints)

The **ENDPTS** option allows you to create an individual line segment by selecting the two endpoints of the line. The Position Menu can be used to assist you in selecting the endpoints. To access this command quickly, you can use the Escape Code **F1-F1-F1 (CREATE-LINE-ENDPTS)**

You are prompted to indicate a start point for the segment. Once the start point has been selected, you are then prompted to indicate the endpoint. As you move the cursor away from the start point you will see the line from the start point to the current cursor position; this is called *rubberbanding*. Rubberbanding allows the user to check the fit and visualize the appearance of a line before it is positioned.

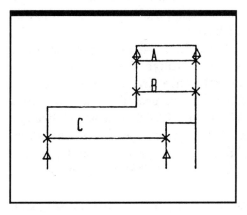

Figure 8-3   Lines created using the ENDPOINT option

 POSSIBLE USE

In many cases, you will need to draw a series of line segments. The ENDPTS option allows you to draw one line at a time. This is a great advantage over a line command that will only draw continuous line strings that require you to terminate and re-invoke the command to draw another line segment. In Figure 8-3, from the 2D sectional view tutorial, the horizontal lines A, B, and C for the drill holes were drawn using the **ENDPTS** option.

LINE OPTION: STRING

The line **STRING** option allows you to draw a continuous chain or "string" of line segments by indicating successive endpoints. Each time a new point is selected, a line is drawn from the previous endpoint to the new point. This function can be activated using the Escape Code **F1-F1-F2 (CREATE-LINE-STRING)**.

You are prompted to indicate the start point of the first line and then to indicate the endpoint for all of the lines that follow. The rubberbanding effect occurs when the cursor is moved to select an endpoint. To terminate the line string, press **F10 BACKUP** or **ESCAPE**.

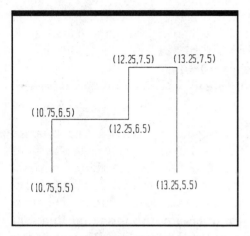

Figure 8-4 Profile using a STRING

POSSIBLE USE

The line **STRING** option is very useful where defining the perimeter of an object. Once the start point has been identified, the subsequent points along the perimeter are then selected to form a line string of multiple line entities that represent the shape of the object. In Figure 8-4, the top half of the sectional view was created using a line string. Starting from point **10.75, 5.5**, each point was indicated in order. When you reach the final point of the line string you can press **F10 BACKUP** if you wish to start another line string or **ESCAPE** to exit this function.

**LINE OPTION:
PARALEL (Parallel)**

The **PARALEL** option copies or creates a line entity parallel to an existing line. The two methods that can be used when creating a parallel line are **THRU PT** (Through a Point) and **AT DIST** (At a Distance). The Escape Code for using the parallel line option is **F1-F1-F3 (CREATE-LINE-PARALEL)**.

The **THRU PT** option creates a line parallel to an existing line with the parallel line passing through an indicated point. You are prompted to select the reference line to which the new line will be drawn parallel. You are then prompted to indicate a reference point (with the aid of the Position Menu) through which the parallel line is drawn. The new line will be the same length as the reference line. You can create additional parallel lines by selecting additional points. Press **ESCAPE** to exit the function.

The **AT DIST** option creates a line parallel to an existing line at a specified distance. You are prompted to enter the distance away from the reference line opposite which you wish to have the parallel line created. You can accept the default value in parentheses by pressing **RETURN** or you can enter a new distance value. Next you are prompted to select the reference line to which the new line is to be parallel. Finally you are prompted to indicate the side of the reference line where the parallel line is to be created. Once this is done a line is created parallel to the reference line on the side indicated and at the distance specified.

POSSIBLE USE

Once the perimeter of a part has been defined, frequently the interior details require lines that are parallel to one of the edge lines. This can be done quickly using either of the two methods of creating parallel lines. In Figure 8-5 the line for the top of the dovetail was created using the **CREATE-LINE-PARALEL-AT DIST** function. The line for the edge above it was created using the **CREATE-LINE-PARALEL-THRU PT** function. A point a half-inch away was indicated.

LINE OPTION: TAN/PRP (Tangent/Perpendicular)

The **TAN/PRP** line option allows you to create a line tangent and/or perpendicular to two entities. Lines can be created tangent to arcs, conics, splines, and points. They can be created perpendicular to lines, arcs, conics, splines, and points. If you are constructing in 3D and the entities are not co-planar, an error message will be displayed. When the **L-LIMIT** switch is set to **FUNCTN**, the line is created using the points of tangency and/or perpendicularity as the endpoints of the line. The Escape Code to access this option quickly is **F1-F1-F4 (CREATE-LINE-TAN/PRP)**. The six different options available for creating tangent and perpendicular lines are:

> **TAN PT**
> **PRP PT**
> **TAN TAN**
> **TAN PRP**
> **PRP TAN**
> **PRP PRP**

TAN PT (Tangent Point)

The **TAN PT** option creates a line tangent to a selected entity and through a reference point. Select the desired entity near the

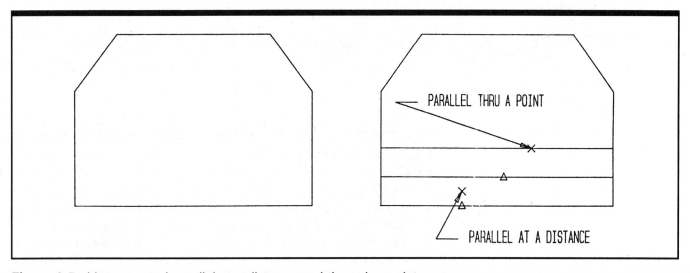

Figure 8-5   Lines created parallel at a distance and through a point

point of tangency. Then indicate the reference point that the tangent line will pass through, using any option from the Position Menu.

PRP PT (Perpendicular Point)

The **PRP PT** option will create a line which is perpendicular to the reference line and passes through a specified point. You are prompted to select a reference line to which the new line is to be created perpendicular (intersect at a 90 degree angle).

Next you are prompted to indicate the point from which the line will originate. Once this point is indicated a line is drawn from the point to the reference line. If the reference line does not lie below the point, the perpendicular line will be created as if the reference line were extended to meet it.

TAN TAN (Tangent Tangent)

The **TAN TAN** option creates a line tangent to two entities. You are prompted to select an existing entity to which the line will be drawn tangent. You are then prompted to select a second entity to which the line will be drawn tangent. The line is drawn between the tangent points.

TAN PRP (Tangent Perpendicular)

The **TAN PRP** option will create a line that is tangent to the first entity selected and perpendicular to the second entity selected. One entity can be a point, but both entities cannot be points. Select the first entity near the point of tangency and the second entity near the point of perpendicularity.

PRP TAN (Perpendicular Tangent)

The **PRP TAN** option creates a line perpendicular to the first selected entity and tangent to the second selected entity. Simply select the two entities in the correct order.

PRP PRP (Perpendicular Perpendicular)

The **PRP PRP** option creates a line perpendicular to two selected entities. Simply select each entity near the point of perpendicularity.

 ## POSSIBLE USE

When you need to create a line perpendicular to another line (reference line) from a specific point of origin, the **PRP PRP** option will allow you to create the line quickly and easily. The lines perpendicular to the base of Figure 8-6 were created using this option.

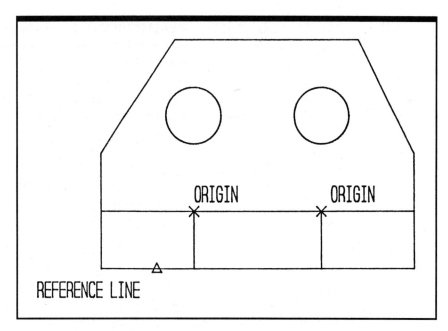

Figure 8-6   Lines created using PRP PRP perpendicular to two entities.

POSSIBLE USE

When a line is required to be drawn tangent to an arc/circle, the **TAN TAN** option allows you to accurately find the point of tangency rather than eyeballing it or leaving it to dumb luck. In Figure 8-7 a line was created tangent to two arcs by selecting the two arcs. Two other tangent lines were created tangent to an arc and a point by using **TAN PT** and selecting the arc and indicating the point.

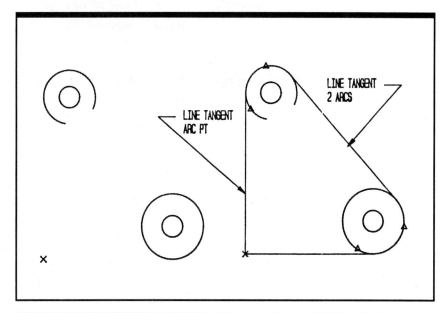

Figure 8-7   Lines created TAN TAN to 2 arcs and a TAN PT

LINE OPTION: HRZ/VRT (Horizontal/Vertical)

The **HRZ/VRT** option creates a horizontal line, vertical line, or both a horizontal and a vertical line entity through a specified reference point. The Escape Code for this function is **F1-F1-F5** (**CRE-ATE-LINE-HRZ/VRT**).

When the **HRZ/VRT** option is selected, you are given three options.

> **HORIZTL (Horizontal)**—creates a horizontal line through a specified point.
>
> **VERTICL (Vertical)**—creates a vertical line through a specified point.
>
> **BOTH**—creates both a horizontal and a vertical line through a specified point.

Once you have selected the type of line you wish to create, you are prompted to indicate the point where the line is to be drawn. The line(s) is (are) drawn to the length of the screen through the point that was specified.

 POSSIBLE USE

The **HRZ/VRT-BOTH** option is a great help in placing center lines in circles. Zoom a window **(ALT-W)** equal to the width and height of the center lines around the circle for which you wish to create the center lines, as in Figure 8-8. Remember that the lines will be created equal to the width and height of the screen. This is why you need to zoom in. Change the line type to center lines and then, using this option and the **CEN-TER** option from the Position Menu, select the circle for which you wish to create the center lines. When this is done, the center point of the circle is located and both a horizontal and a vertical center line are drawn through the circle's center.

LINE OPTION: ANGLE

The **ANGLE** option creates a line at a specified angle from another line. To access this function quickly, use the Escape Code **F1-F1-F6** (**CREATE-LINE-ANGLE**).

You are prompted to enter the angle at which you wish the new line to be drawn with reference to the line you will select. If you enter a positive angle, the new line will be rotated about the endpoint of the reference line in a counterclockwise direction. If a negative angle is entered, the new line is rotated about the endpoint of the reference line in a clockwise direction. The same angle convention that applies to other options applies here.

You are prompted to select the reference line. Next, you are prompted to indicate the new line's endpoint with the aid of the Position Menu. This is the point from which you wish to have the angled line start. A new line is created which is equal in length to the reference line at the specified angle. If the **LINE LIMITS** switch is

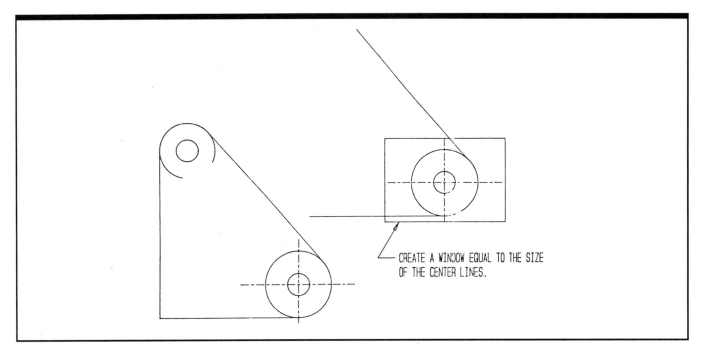

CREATE A WINDOW EQUAL TO THE SIZE OF THE CENTER LINES.

Figure 8-8   Horizontal and vertical lines created using the HRZ/VRT-BOTH option

set to **SCREEN**, the new line will be created at the full length of the screen.

 POSSIBLE USE

In any case where a line needs to be created at an angle, the **LINE-ANGLE** option can be used. The angle specified is a rotation angle. The reference line is being copied and created from a specified point such as an endpoint, as illustrated in Figure 8-9. If the rotation is in a clockwise direction, a negative angle should be specified. The length of the angle line is equal to the reference line. In many cases, the line which is created may be too long or too short. By using the **TRIM/ EXTEND** function you can either trim the excess line or extend the angled line to the desired length.

LINE OPTION: RECTANG (Rectangle)

The **RECTANG** option gives you two methods of creating a rectangle: **CORNERS** and **WID/HT** (Width/Height).

CORNERS

The **CORNERS** option allows you to create a rectangle by indicating two opposite corners. The rectangle is comprised of four separate line entities rather than being a single entity. The Escape Code for this option is **F1-F1-F7-F1** (**CREATE-LINE-RECTANG-CORNERS**).

You are prompted to indicate the position of the first corner, which is the lower left corner. Then you are prompted to indicate the

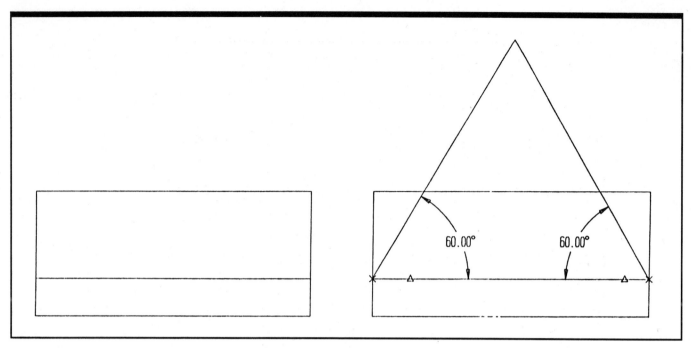

Figure 8-9 Lines created using the LINE-ANGLE option

position of the second corner, or the upper right corner. As you move the cursor away from the first corner a rubberbox will appear. The rubberbox allows you to see the size of the rectangle before the second point is selected.

 POSSIBLE USE

This option can be used to fit a rectangle in a specified area (Figure 8-10). By selecting corners that do not interfere with any other surfaces the rectangle can be accurately placed. If you want to quickly define the approximate height and width of a view you can arbitrarily select two diagonal corners of a rectangle and one will be created. If SNAP and Cursor Tracking are on you can accurately indicate the corner positions of the rectangle. The lines that are created are not grouped so that they can be trimmed or deleted as needed.

WID/HT (Width/Height)   The **WID/HT** option allows you to specify the width and the height of the rectangle and then to indicate the position of the lower left corner. The rectangle is comprised of four separate line entities rather than being one complete entity. The Escape Code for this option is **F1-F1-F7-F2 (CREATE-LINE-RECTANG-WID/HT)**.

You are prompted to enter the width value for the rectangle and press **RETURN**. The default value can be accepted by pressing **RETURN** only. Next, you are prompted to enter the height value for the rectangle and press **RETURN**. Last, you are

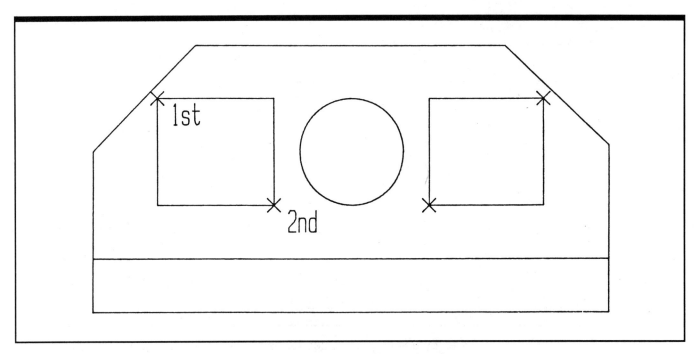

Figure 8-10   A LINE-RECTANG created by indicating two diagonal corners

requested to indicate the position of the lower left corner of the rectangle. The rectangle is now created.

 POSSIBLE USE

This function is very useful when you are creating a view and you know the width and height ahead of time. You can simply key in the value for the width and then the value for the height. Once those values are entered indicate the base position and the rectangle representative of the overall dimensions (Figure 8-11). This rectangle can be thought of as a sculptor's block which is used to create the view by cutting away, editing and creating entities within and on the rectangle.

LINE OPTION: N-GON

The **N-GON** option allows you to create a regular polygon, which is made up of individual line entities that can be deleted or edited independently. *The lines in the polygons created using the CREATE-POLYGON function cannot be edited; only whole polygons can be deleted.* The Escape Code for this option is **F1-F1-F8** (**CREATE-LINE-N-GON**).

You are prompted to enter the number of sides (3 to 100) that you wish to have in the polygon and press **RETURN**. Next you are requested to enter the rotation angle and press **RETURN** if you wish to have the polygon rotated to a special angle.

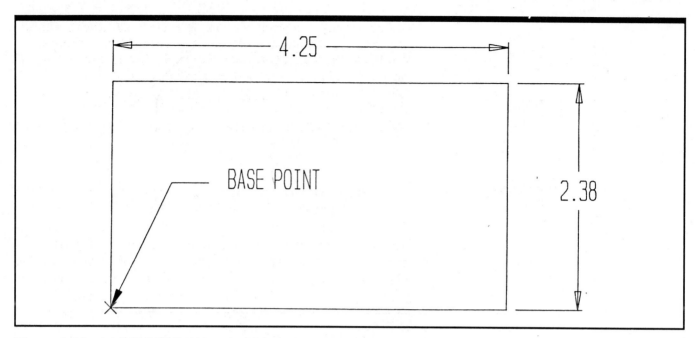

Figure 8-11   A LINE-RECTANG created by entering the values for height and width, and then indicating a
base point

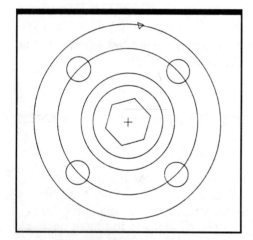

Figure 8-12   Creating a POLYGON
distance across the corners

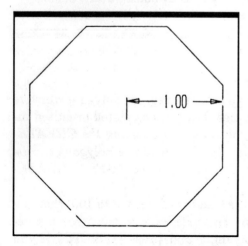

Figure 8-13   A line POLYGON flat

For the next prompt you are asked to choose a method to specify the polygon radius. The two options that are displayed on a menu are:

F1 CORNER (Distance across corners, inscribed)
F2 FLAT (Distance across the flats, circumscribed)

CORNER measures the radius from the center of the polygon to the intersection of two sides of the polygon (Figure 8-12).

FLAT measures the radius of the polygon from the center point to a point on the side which forms a perpendicular line from the side to the center (Figure 8-13).

Once you have selected the radius method you are then prompted to enter the radius value and press **RETURN**. Next you must indicate the position of the center point.

After the center point is positioned you are asked whether or not to group the entities. The **NO/YES** menu is displayed. If **F1 NO** is selected each side of the polygon will be a separate **LINE** entity. If **F2 YES** is selected all of the **LINE** entities which make up the polygon will be grouped (you are prompted to enter a group name) and treated as a single entity.

At this point, the polygon is created with the current attributes.

POSSIBLE USE

The **N-GON** option is a favorite function for most CADKEY users. When polygons were created by manual drafting techniques they required many laborious steps and the steps were different from one polygon to the next. Now, with CADKEY, you can create a polygon with as many sides as needed with only a few quick steps which are applicable to all polygons. The hexagon in the 2D tutorial (Figure 8-12) was created using the **CREATE-LINE-N-GON-CORNER** function. The radius from the center to the corner is .5, which gives you a distance across corners of 1.00. This option is a real time-saver.

LINE OPTION: MESH

NOTE: These **MESH** options also apply to the **POINT, POLYLIN, POLYGON**, and **SPLINE** options.

The **MESH** option allows you to generate a mesh that approximates a two-curve, three-curve, or four-curve swept surface, a tabulated cylinder, a surface of revolution, a sphere or a ruled surface. Meshes can be generated of line, point, spline, polyline, or polygon entities. These meshes can then later be used in a finite element analysis package to analyze structural properties of a part design.

The **MESH** option allows the user to define a surface and generate a mesh that approximates the surface to a user-specified level of accuracy. When creating a mesh, the **DIRECTOR** curve is the curve along which the generator curve is swept. The **GENERATOR** curve is the curve that is swept by the director curve. The methods of generating a mesh include the following:

> **2CURVE**
> **3CURVE**
> **4CURVE**
> **TAB CYL (Tabulated Cylinder)**
> **REVOLUT (Surface of Revolution)**
> **SPHERE**
> **RULED**

The **2CURVE** method does not require the endpoints of the curves to intersect. However, the **3CURVE** and the **4CURVE** methods require that the endpoints of the curves intersect to form a closed boundary.

MESH Option: 2CURVE

This option generates a swept surface, an ordered surface of entities between selected director and generator entities. This surface represents the sweep of the profile defined by the generator entity. The profile moves along the trajectory (path through space) defined by the director entity. The two methods of defining the sweep of the profile are:

> **TRANSL (Translate)**
> **ROTATE**

TRANSL—maintains the generator curve's orientation as it moves along the trajectory of the director curve, relative to World Coordinates (Figure 8-14).

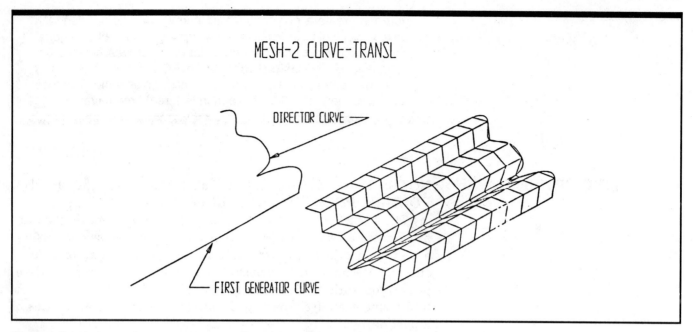

MESH-2 CURVE-TRANSL

DIRECTOR CURVE

FIRST GENERATOR CURVE

Figure 8-14 A mesh created using 2CURVE and the Translate option

ROTATE—maintains the generator curve's orientation to the director curve as it moves along the trajectory of the director curve.

Select the line, arc, conic, or spline to be used as the generator curve that forms the edges of the surface mesh. When prompted, enter the number of segments desired for the generator curve. The greater the number, the better the surface will look and the slower to generate and draw. Then select the director entity near the end where you wish the sweep of the generator curve to start. Then enter the number of segments desired for the director entity to be broken into and press **RETURN**.

A temporary mesh will be displayed in the primary viewport along with the **NO/YES** menu. You are prompted: **Are the curves connected correctly?** Choose the appropriate response. When the line mesh has been accepted, the mesh will be displayed in the other Viewports. You are given the opportunity to group the line mesh entities. Choose from the **NO/YES** menu.

MESH Option: 3CURVE

The **3CURVE** option creates a mesh defined by the combination of either two generator curves and a director curve or by a generator curve and two director curves. The resulting mesh approximates the blending of the two-curve surfaces. The two methods of defining the sweep of the profile are:

TRANSL—maintains the generator curve's orientation as it moves along the trajectory of the director curve, relative to World Coordinates (Figure 8-15).

ROTATE—maintains the generator curve's orientation to the director curve as it moves along the trajectory of the director curve (Figure 8-16).

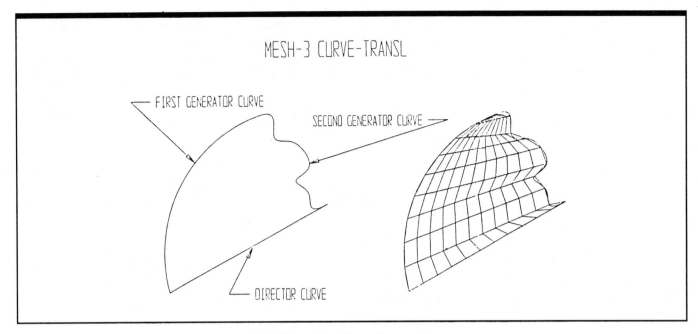

Figure 8-15   A mesh created using 3CURVE and the Translate option

Select the line, arc, conic, or spline to be used as the first and second generator curves which are the edges of the surface mesh. If you do not wish to use two generator curves, press **RETURN**. When prompted, enter the number of segments that the generator curve(s) are to be broken into. The greater the number, the better the surface will look. If you have selected two generator curves, select one director curve. If you have selected one generator curve then you must select two director curves. Select the director entity near the end where you wish the sweep of the generator curve to start. Then enter the number of segments desired for the

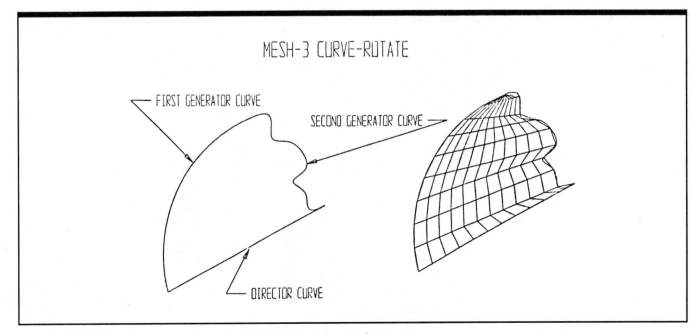

Figure 8-16   A mesh created using 3CURVE and the Rotate option

director entity to be broken into and press **RETURN**, or press **RETURN** to accept the default value.

A temporary mesh will be displayed in the Primary Viewport along with the **NO/YES** menu. You are prompted: **Are the curves connected correctly?** Choose the appropriate response. When the line mesh has been accepted, the mesh will be displayed in the other Viewports. You are given the opportunity to group the line mesh entities. Choose from the **NO/YES** menu.

MESH Option: 4CURVE

The **4CURVE** option creates a mesh defined by the combination of two generator curves and two director curves. The resulting mesh approximates the blending of the two-curve surfaces. The two methods of defining the sweep of the profile are:

> **TRANSL**—maintains the generator curve's orientation as it moves along the trajectory of the director curve, relative to World Coordinates (Figure 8-17).
>
> **ROTATE**—maintains the generator curve's orientation to the director curve as it moves along the trajectory of the director curve (Figure 8-18).

Select the line, arc, conic or spline to be used as the first and second generator curves which are the edges of the surface mesh. When prompted, enter the number of segments that the generator curves are to be broken into. The greater the number, the better the surface will look. Select two director curves or select one director curve and press **RETURN** to use the same entity as the second director curve. Select the director entity near the end where you wish the sweep of the generator curve to start. Then enter the number of segments desired for the director entity to be broken into and press **RETURN**, or press **RETURN** to accept the default value.

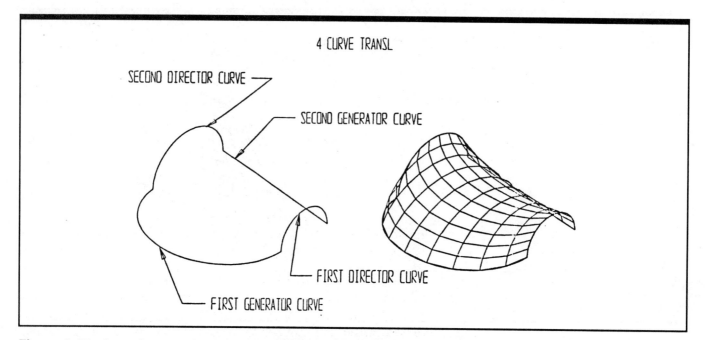

Figure 8-17   A mesh created using 4CURVE and the Translate option

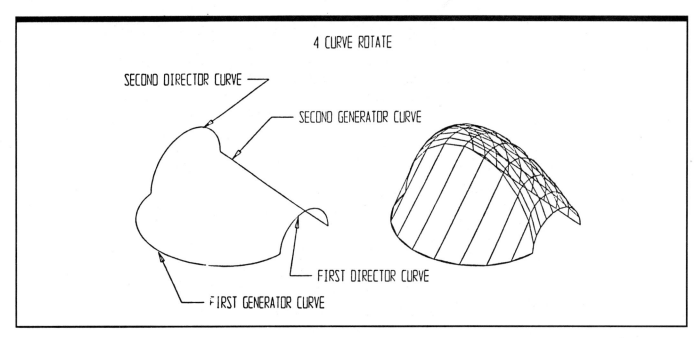

4 CURVE ROTATE

SECOND DIRECTOR CURVE

SECOND GENERATOR CURVE

FIRST DIRECTOR CURVE

FIRST GENERATOR CURVE

Figure 8-18   A mesh created using 4CURVE and the Rotate option

A temporary mesh will be displayed in the primary viewport along with the **NO/YES** menu. You are prompted: **Are the curves connected correctly?** Choose the appropriate response. When the line mesh has been accepted, the mesh will be displayed in the other Viewports. You are given the opportunity to group the line mesh entities. Choose from the **NO/YES** menu.

MESH Option: TAB CYL (Tabulated Cylinder)

The **TAB CYL** option creates a mesh that approximates the surface that is defined by sweeping a generator curve along an axis vector (the director curve) of a specified length (Figure 8-19).

Begin by selecting the line, arc, conic, or spline to be used as the generator curve. Enter the desired number of segments into which to break the generator curve and press **RETURN**.

To identify the cylinder axis vector, choose from the following menu choices:

2 PTS—allows you to define the cylinder axis vector by selecting a start point and an endpoint.

LINE—select a line entity to define the cylinder axis vector. The axis vector is directed from the end of the line closest to the selection point to the opposite end of the line.

END ENT—select a line, arc, conic, polyline, or spline to define the cylinder axis vector.

KEYIN—the cylinder axis is defined by entering the x,y,z vector components, in either World or Construction Coordinates.

Enter the length of the cylinder and press **RETURN**. If a negative length value is entered, the direction of the axis vector will be reversed. Enter the number of segments that are to be generated in the axis direction and press **RETURN**.

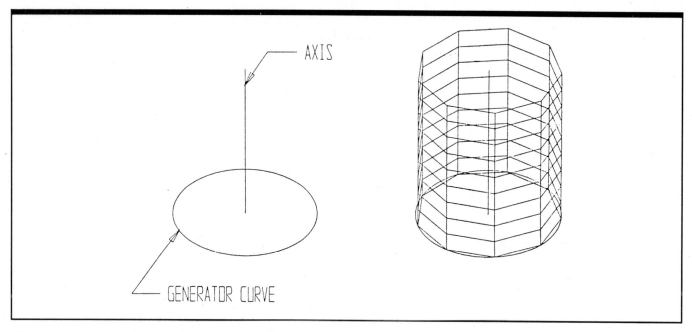

Figure 8-19 A tabulated cylinder mesh using a line as the cylinder axis

A temporary mesh will be displayed in the Primary Viewport along with the **NO/YES** menu. You are prompted: **Are the curves connected correctly?** Choose the appropriate response. When the line mesh has been accepted the mesh will be displayed in the other Viewports. You are given the opportunity to group the line mesh entities. Choose from the **NO/YES** menu.

MESH Option: REVOLUT (Revolution)

The **REVOLUT** option creates a mesh approximating the surface formed by rotating a generator curve about an axis vector (the director curve) of a specified angular distance (Figures 8-20 and 8-21).

Begin by selecting the line, arc, conic, or spline to be used as the generator curve. Enter the desired number of segments into which to break the generator curve and press **RETURN**.

To identify the rotation axis vector for the mesh of revolution, choose from the following menu choices:

2 PTS—define the rotation axis vector by selecting a start point and an endpoint.

LINE—select a line entity to define the rotation axis vector. The axis vector is directed from the end of the line closest to the selection point to the opposite end of the line.

END ENT—select a line, arc, conic, polyline, or spline to define the rotation axis vector.

KEYIN—enter the x,y,z vector components, in either World or Construction Coordinates to define the rotation axis.

Enter the total angular distance in degrees that the generator curve is to be rotated about the axis and press **RETURN**. Enter the number of segments that are to be generated in the rotation

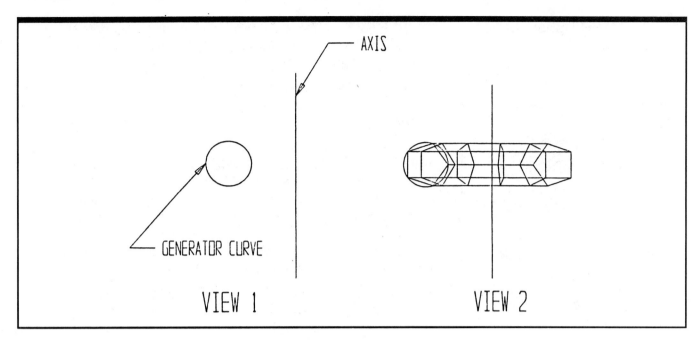

Figure 8-20   A mesh revolution as seen in view 2

direction and press **RETURN**. A temporary mesh will be displayed in the primary viewport along with the **NO/YES** menu. You are prompted: **Are the curves connected correctly?** Choose the appropriate response. When the line mesh has been accepted, the mesh will be displayed in the other Viewports. You are given the opportunity to group the line mesh entities. Choose from the **NO/YES** menu.

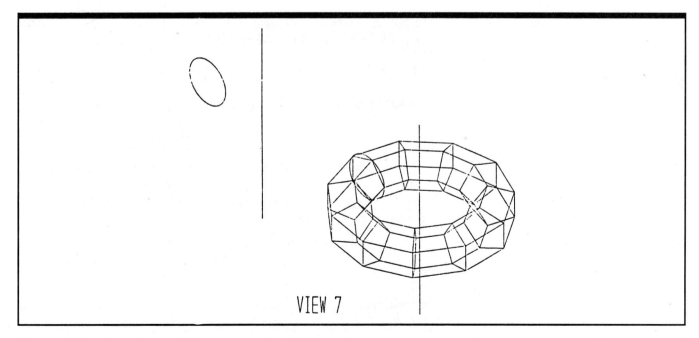

Figure 8-21   A mesh revolution as seen in view 7

MESH Option: SPHERE

The **SPHERE** option creates a line mesh that approximates a sphere. The two methods of creating a sphere are: **CTR+RAD** and **CTR+DIA** (Figure 8-22).

> **CTR+RAD**—creates a mesh by specifying the center and the radius of the sphere, and the number of latitude and longitude lines in the sphere.

Locate the center of the sphere using one of the Position Menu options. Enter a sphere radius value of greater than 0.0 and press **RETURN**. Enter the number of longitude and latitude lines to be generated in the sphere and press **RETURN** after each entry.

You are prompted as to whether or not to group the mesh entities. Choose **NO** to display the sphere mesh without grouping the entities or **YES** to display the sphere mesh as a group. You will be prompted to enter a group name if **YES** is selected.

> **CTR+DIA**—creates a mesh by specifying the center and the diameter of the sphere, and the number of latitude and longitude lines in the sphere.

Locate the center of the sphere using one of the Position Menu options. Enter a sphere diameter value of greater than 0.0 and press **RETURN**. Enter the number of longitude and latitude lines to be generated in the sphere and press **RETURN** after each entry.

You are prompted as to whether or not to group the mesh entities. Choose **NO** to display the sphere mesh without grouping the entities or **YES** to display the sphere mesh as a group. You will be prompted to enter a group name if **YES** is selected.

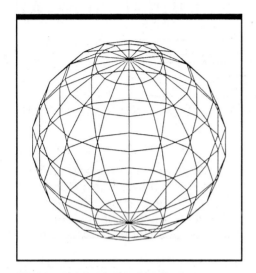

Figure 8-22 A mesh sphere

MESH Option: RULED

This option creates a line mesh between entities by specifying **NUMSEGS (Number of Segments)** or **ARCLNTH (Arc Length)**.

> **NUMSEGS**—divides an edge entity into a specific number of equal segments.
> **ARCLNTH**—divides an edge curve into segments of a specified length. The length of the last segment will be equal to or less than the specified arc length (Figure 8-23).

You are first prompted to enter a value for the number of segments or length of the arcs into which you wish to divide the line or arc, conic or spline. You are then prompted to select the first director curve near the endpoint for the mesh to begin, and the second director curve near the endpoint where you wish to begin creating the mesh lines. Be sure to select both entities near the same end or else the mesh lines will be twisted.

The lines will be generated quickly and you will be prompted: **Curves connected correctly?** Respond **YES** to accept the line mesh or **NO** to select the first and second endpoints again and regenerate the line mesh.

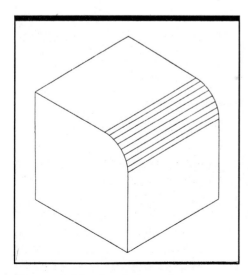

Figure 8-23 A ruled mesh

When the line mesh has been accepted you are prompted as to whether you wish to generate the line mesh in the cross direction. Choose from the **NO/YES** menu. If **YES** is selected you are then prompted with the **NUMSEGS/ARCLNTH** menu. To generate a mesh in the cross direction use the same procedure as you did to create the previous mesh.

Finally, you are given the option to group the line mesh entities. Choose from the **NO/YES** menu.

POSSIBLE USE

The primary use of the line **MESH** option is to generate a meshed surface containing a series of nodes or knot points that are used by a third party FEA (Finite Element Analysis) program such as ALGOR's SUPERSAP. Another possible use is to generate the appearance of a surface model by using the line mesh option.

ARC OPTIONS

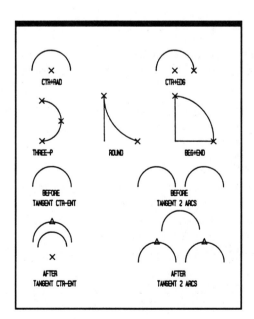

Figure 8-24 The ARC creation option

There are seven different options for creating an **ARC** (Figure 8-24). Each of these seven options allows you to create an arc by specifying different points that are included in the arc entity. The seven **ARC** options are:

> **CTR+RAD (Center and Radius)**
> **CTR+DIA (Center and Diameter)**
> **CTR+EDGE (Center and Edge)**
> **THREE-P (Through 3 Points)**
> **TANGENT**
> **BEG+END (Beginning and Endpoints)**
> **ROUND**

When specifying the start and end angles requested by several of the options, the normal angle conventions are in effect. Zero is the origin at 3 o'clock; positive angles are measured in the counterclockwise direction and negative angles in the clockwise direction.

To get some hands-on experience with the arc options, set up the axes in Figure 8-25. To create the center lines use the **CREATE-LINE-ENDPTS** function. For positioning the endpoint of the angled line use the **POLAR** option or **CREATE-LINE-ANGLE**.

CREATE-ARC OPTION: CTR+RAD (Center and Radius)

The **CTR+RAD** option allows the user to create an arc by entering the radius, the start angle and the end angle and indicating the center point. The Escape Code is **F1-F2-F1 (CREATE-ARC-CTR+RAD)**.

You are prompted to enter the radius of the arc you wish to create and to press **RETURN**. The radius must be a value between .00005 and 10,000.

Next you must enter the start and end angles for the beginning and the end points of the arc. When specifying the angles the normal angle conventions are in effect. Zero is the origin at 3 o'clock; positive angles are measured in the counterclockwise direction and negative angles in the clockwise direction.

Last, you are prompted to indicate the center point of the arc. Once the center point is indicated, the arc will be drawn starting at the angle specified and ending at the end angle. The radius of the arc will equal that of the radius specified earlier.

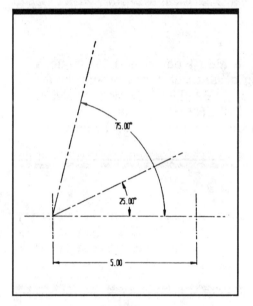

Figure 8-25 This axis will be used to create arcs by several different methods.

POSSIBLE USE

When the radius and the start and end angles of an arc are known, this method of creating an arc is very accurate. In Figure 8-26, the arcs were drawn using this method. The start angle is 75 degrees and the end angle is 255 degrees for the R.625 and the R.250 arcs. The angle number of 255 was determined by adding the included angle of 180 degrees to the start angle of 75 degrees. For the other R.250 arc, the start angle is 205 degrees and the end angle is 25 degrees because arcs are drawn in a counterclockwise direction.

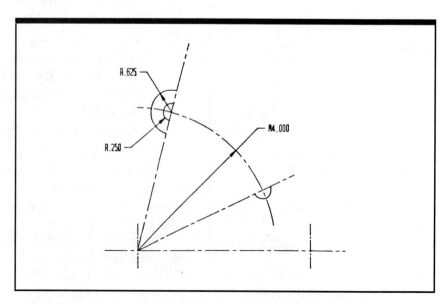

Figure 8-26 Arcs created using CTR+RAD option

CREATE-ARC OPTION: CTR+DIA (Center and Diameter)

The **CTR+DIA** option allows you to create an arc by specifying the diameter, the start and end angles, and the center point. The Escape Code for this option is **F1-F2-F2 (CREATE-ARC-CTR+DIA)**.

This option is the same as the **CTR+RAD** option except, in this case, you are entering the diameter of the arc instead of the radius. You are prompted to enter the diameter of the arc that you wish to create.

Once again you are prompted to enter the start angle and press **RETURN**. Then enter the end angle and press **RETURN**. Finally, you are prompted to indicate the position of the center point. The Position Menu is displayed to aid you in indicating the center point.

When the center point is indicated, the arc will be drawn. If the arc is not visible press the Immediate Mode command **ALT-A** to autoscale your part so that the arc is displayed on the screen.

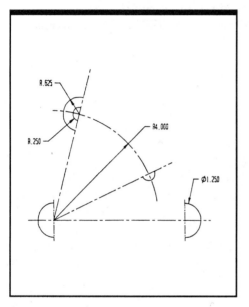

Figure 8-27 Arcs created using the CTR+DIA option

 POSSIBLE USE

The possible use of the command is somewhat limited because, in most cases, arcs are dimensioned by their radius rather than their diameter, which results in the previous option being used more often. The two commands are interchangeable in use. In Figure 8-27, the two arcs drawn on the base were created using the **ARC-CTR+DIA** option. For the arc on the left, the diameter is 1.25, the start angle is 90 degrees, and the end angle is 270 degrees. For the arc on the right, the start angle is 270 degrees and the end angle is 90 degrees.

CREATE-ARC OPTION: CTR+EDG (Center and Edge)

The **CTR+EDG** option allows the user to create an arc by specifying the start and end angles and a point at the center and the edge for the radius. The Escape Code for this option is **F1-F2-F3 (CREATE-ARC-CTR+EDG)**.

To start, you are prompted to enter the start angle and press **RETURN**, then to enter the end angle and press **RETURN**. Next, the Position Menu is displayed and you are prompted to indicate the center point and then the edge point which the arc will pass through.

POSSIBLE USE

When a part is being designed, in some cases the radius may not be known yet: the center point and an edge are known. By

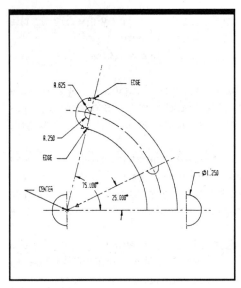

Figure 8-28 Arcs created using the CTR+EDG option

using this option an arc can be drawn to fit the given situation. The two large arcs in Figure 8-28 are created using CADKEY's **CTR+EDG** option. The start angle for these arcs is zero (0) degrees and the end angle is 75 degrees. When you are prompted to select the center point for the arc, the Position Menu is displayed. Select the **ENDENT** option. Then, to indicate the center point, select the 25 degree center line near the end closest to the center point. An X marks the center. Next, indicate the edge point again using the Position Menu option **ENDENT**. Select the R.625 arc near the endpoint where the **CTR+EDG** arc will meet it. Don't select the intersection because the center line may be mistakenly selected. To be sure you are getting the arc you may have to zoom a window using **ALT-W**. Repeat the same steps for the other **CTR+EDG** arc.

CREATE-ARC OPTION: THREE-P (Through Three Points)

The **THREE-P** option allows the user to create an arc by indicating three points. CADKEY will calculate the radius and center point needed to create an arc that passes through the three points. The Escape Code for this option is **F1-F2-F4** (**CREATE-ARC-THREE-P**).

When using this option you are prompted to indicate three points. The points can be positioned with the aid of the Position Menu. The first two points specify the direction in which the arc will be drawn. Once all of the points are indicated, the arc is immediately drawn through the points.

When working in the 3D mode, the points can be placed at different depths. This will create an arc that appears elliptical in shape. If viewed in the new reference view that is created it will appear as a regular arc.

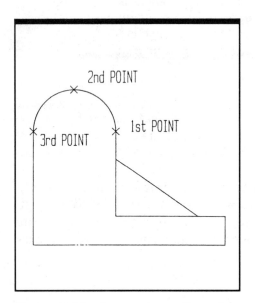

Figure 8-29 An arc drawn through three points

 POSSIBLE USE

This possible use is not included in the part being drawn by the other arc options. As a part is being designed an arc often needs to be created that passes through three known points, as in Figure 8-29. Previously in manual drafting techniques this was done by bisecting the distance between the first and second and the second and third points to find a common center point. Now, by just indicating the three points, the arc is automatically drawn.

CREATE-ARC OPTION: TANGENT

The **TANGENT** option allows you to create an arc that is tangent by three different methods: **CTR-ENT, 2 ENTS** and **3 ENTS**.

CTR-ENT (Center-Entity)

The **CTR-ENT** option allows the user to create a tangent arc by indicating the arc's center and the entity to which it is to be tangent. The Escape Code for this option is **F1-F2-F5-F1** (**CREATE-ARC-TANGENT-CTR-ENT**).

You are prompted to enter the start angle and press **RETURN**, and then to enter the end angle and press **RETURN**. You are then prompted to select the entity to which the new arc will be tangent. Finally, you are requested to indicate the center point of the arc. At this time the arc is displayed.

2 ENTS (Two Entities)

The **2 ENTS** option is used to create an arc tangent to two existing entities. The Escape Code for this option is **F1-F2-F5-F2** (**CRE-ATE-ARC-TANGENT-2 ENTS**).

You are prompted to enter the start angle and press **RETURN**, then to enter the end angle and press **RETURN**. Next you are requested to enter the radius of the arc and press **RETURN**. Last you are prompted to select the two entities that you wish to create the arc tangent to (Figure 8-30).

3 ENTS (Three Entities)

The **3 ENTS** option is used to create an arc tangent to three existing entities on a drawing. The Escape Code for this option is **F1-F2-F5-F3** (**CREATE-ARC-TANGENT-3 ENTS**).

You are prompted to enter the start angle and press **RETURN**, then to enter the end angle and press **RETURN**. Next, you are requested to enter the radius of the arc and press **RETURN**. Last, you are prompted to select the three entities that you wish to create the arc tangent to (Figure 8-31).

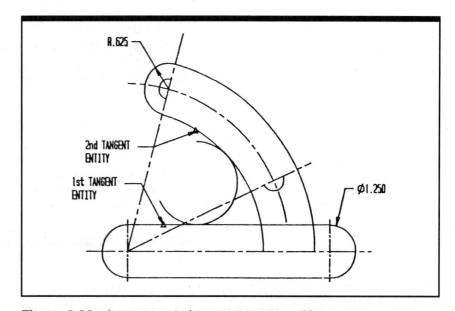

Figure 8-30 An arc created tangent to two entities

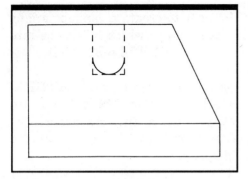

Figure 8-31 An arc created tangent to three entities

 POSSIBLE USE

This option can be used when you need to create an arc tangent to **2 ENTS**—in Figure 8-30, the arc and the line. To place the two lines use the function **CREATE-LINE-ENDPTS** and the position menu option **ENDENT**. Then select the 1.25 diameter arcs to create the line between them.

The start angle for the tangent arc is **205** degrees and the end angle is **90** degrees. The radius is **1.00**. After entering this data, select the arc and line to which you wish to create the new arc tangent. The arc which is drawn can be trimmed to its points of tangency using the function **EDIT-TRM/EXT-FIRST**.

CREATE-ARC OPTION: BEG+END (Beginning and Endpoints)

The **BEG+END** option allows you to specify the start and end points and the included angle for the arc to be created. The Escape Code for this option is **F1-F2-F6** (**CREATE-ARC-BEG+END**).

With the **BEG+END** option you are prompted to enter the included angle which will determine the length of the arc. Enter the included angle in degrees and press **RETURN**.

Next, you are prompted to indicate (with the aid of the Position Menu) the beginning and the endpoints of the arc. Once the two points have been indicated, the arc is drawn. The center point or radius is determined by the size of the included angle and the distance between the two points selected. The direction in which the arc is drawn is determined by which point is selected first.

 POSSIBLE USE

The **BEG+END** option can be used to create an arc where the included angle and the beginning and end points are known. The arcs for the center slot in Figure 8-32 are a good example of this. The included angle is 50 degrees (75 − 25 = 50). When you are prompted to enter the begin point, the Position Menu is displayed. Choose the **ENDENT** option. Remembering that arcs are drawn in a counterclockwise direction, select the lower R.250 arc near the begin point. The **ENDENT** option will position the begin point at the end of the R.250 arc. Then indicate the endpoint by selecting the other R.250 arc near the endpoint, being careful not to select the wrong entity. The arc will be created with the correct radius. Repeat these steps for the other arc on the center slot.

CREATE-ARC OPTION: ROUND

The **ROUND** option allows you to create an arc tangent to the endpoint of an existing line or arc entity and ending at a

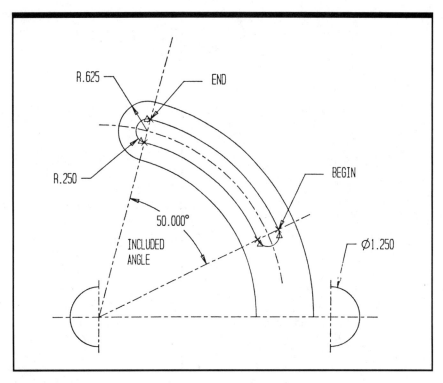

Figure 8-32 Arcs created using the BEG+END option

specified point. The Escape Code for this option is **F1-F2-F7** (**CREATE-ARC-ROUND**).

You are prompted first to select the entity that you wish to have the arc tangent to. The selection should be made nearest to the endpoint from which you wish the arc to begin. You are then prompted to indicate the termination or endpoint for the arc.

When both of the points have been indicated, a new menu is displayed which prompts you to indicate the direction in which the arc is to be created. The menu choices are:

F1 CLKWSE (Clockwise)
F2 CCLKWSE (Counterclockwise)

When this direction has been selected, the arc is created and displayed.

 POSSIBLE USE

For this option, first delete the two 1.25 diameter arcs on the ends of the base in Figure 8-33. Use the Immediate Mode command **CTRL-Q (DELETE)**, select both arcs and press **RE-TURN**. You are left with two lines. On the left side you will create an **ARC-ROUND-COUNTERCLOCKWISE** by selecting the top line as the tangent entity, using the **ENDENT** option from the Position Menu and selecting the bottom line to indicate the termination point. For the right end, the arc must be created in the clockwise direction if the top line is again used as the

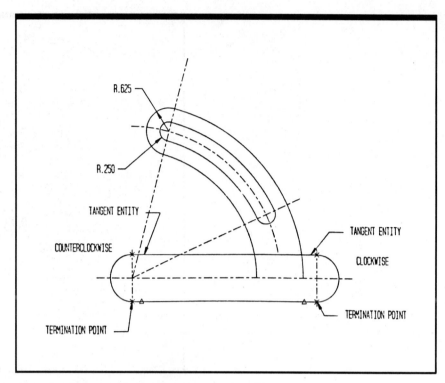

Figure 8-33 Arcs created using the ROUND option

tangent entity and the **ENDENT** of the bottom line is used to indicate the termination point. If the bottom line were used as the tangent entity and the top line for the termination point, the arc would have to be created in the counterclockwise direction.

CIRCLE OPTIONS

There are six methods of creating a circle (Figure 8-34). The six methods are:

> **CTR+RAD (Center and Radius)**
> **CTR+DIA (Center and Diameter)**
> **CTR+EDG (Center and Edge)**
> **THREE-P (Three Points)**
> **TANGENT**
> **TWO-PTS (Two Points)**

If a circle is created but it is not visible on the screen it may have been created outside of the Drawing Window. To place in the Drawing Window everything that has been created use the Immediate Mode command **ALT-A**, Autoscale.

In most cases, when a circle is created in the 2D construction mode, the circle is created at the current depth. When the circle is created in the 3D construction mode it can be placed at different

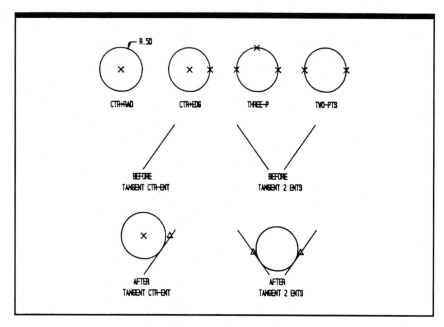

Figure 8-34 The CIRCLE creation options

depths, depending on the option. In most cases it is created at the depth position of the center point.

CREATE-CIRCLE OPTION: CTR+RAD (Center and Radius)

The **CTR+RAD** option allows you to create a circle by specifying a radius value and indicating the center position. To invoke the **CTR+RAD** option quickly, use the Escape Code **F1-F3-F1** (**CREATE-CIRCLE-CTR+RAD**).

You are prompted to enter the value for the radius and press **RE-TURN**. The next prompt requests that you indicate the center position. The Position Menu is displayed to aid in positioning the center point. If the circle does not appear on the screen once the center is indicated, press the Immediate Mode command **ALT-A**, Autoscale, to place the circle within the Drawing Window.

 POSSIBLE USE

The **CTR+RAD** option can be used to create a circle entity with a radius which ranges in size from .00005 to 10,000 units. The circles in Figure 8-35 were created by specifying a radius of .50. The center point was indicated using the cursor option from the Position Menu.

CREATE-CIRCLE OPTION: CTR+DIA (Center and Diameter)

The **CTR+DIA** option allows the user to create a circle by specifying the diameter and indicating the center point. To access this option using the function keys, use the Escape Code **F1-F3-F2** (**CREATE-CIRCLE-CTR+DIA**).

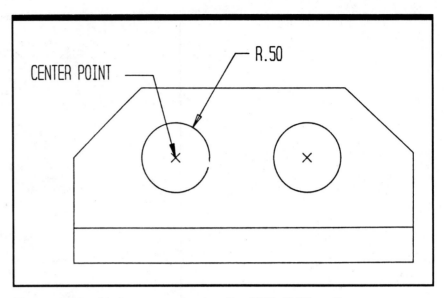

Figure 8-35 Circles created using the CTR+RAD option

You are prompted to enter the diameter value and press **RE-TURN**. To accept the default value in the parentheses press **RE-TURN**. You are then prompted to indicate the center point. Once the center point has been positioned, the circle is drawn.

 POSSIBLE USE

The **CTR+DIA** option is the most popular option for creating circle entities because most circles are dimensioned by their diameter rather than their radius. In Figure 8-36, the circles were created using this option. The diameter is entered and then the center point must be indicated for the first circle. The **KEY-IN** option from the Position Menu allows you to precisely position the center point. The center point for the remaining concentric circles can be located by using the **CENTER** option from the Position Menu and selecting the first circle drawn. This will search for the center of the first circle and use it for the center of the concentric circles.

CREATE-CIRCLE OPTION:
CTR+EDG (Center and Edge)

The **CTR+EDG** option allows you to create a circle by indicating the center point and one point on the edge of the circle. The Escape Code for this option is **F1-F3-F3 (CREATE-CIRCLE-CTR+EDG)**.

The first prompt requests that you indicate the center point. Next, you are prompted to indicate the edge point. The Position Menu is available to aid you in positioning both points. Once both points have been indicated, the circle is displayed.

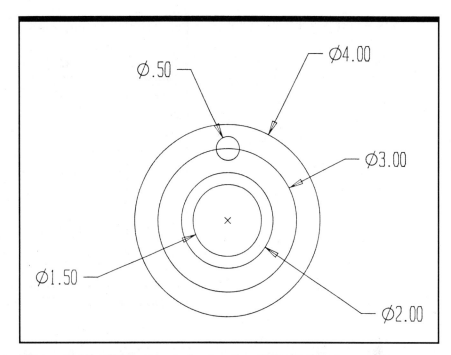

Figure 8-36   Circles created using the CTR+DIA option

 POSSIBLE USE

When designing a part you have to fit a drill hole (circle) in a specific space. So that the circle does not exceed the area that you have to work with, the center can be specified and then an edge that will fit in the space provided. In Figure 8-37 the drill hole had to be fit between two other holes.

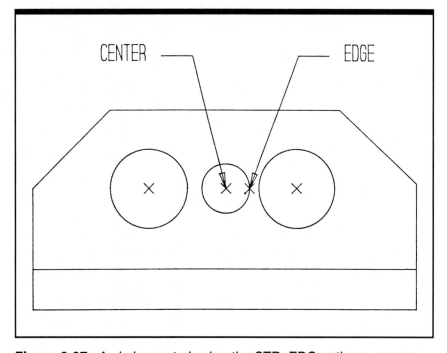

Figure 8-37   A circle created using the CTR+EDG option

CREATE-CIRCLE OPTION:
THREE-P (Three Points)

The **THREE-P** option allows the user to specify three points that the circle will pass through. When the circle is created, the points will all lie on the circumference of the circle. The Escape Code for this option is **F1-F3-F4 (CREATE-CIRCLE-THREE-P)**.

You are prompted for the three points with three separate prompts. For each point that you are requested to indicate, the Position Menu is displayed to aid you in positioning the points. Once the three points have been indicated, the circle is drawn.

 POSSIBLE USE

This option gives the designer the opportunity to fit a circle between three points. The three points or positions may already exist, as in Figure 8-38. The points can also be indicated with the use of the Position Menu.

CREATE-CIRCLE OPTION:
TANGENT

The **TANGENT** option allows you to create a tangent circle by three different methods: **CTR-ENT, 2 ENTS and 3 ENTS**.

CTR-ENT (Center-Entity)

This method allows you to create a circle by indicating the position of the center point and selecting an entity to which the circle will be tangent. The Escape Code for this option is **F1-F3-F5-F1 (CREATE-CIRCLE-TANGENT-CTR-ENT)**.

You are prompted to select the entity to which the circle will be tangent. Then you are requested to indicate the center point. The circle is then displayed.

2 ENTS (Two Entities)

This method allows you to create a circle tangent to two entities. The Escape Code for this option is **F1-F3-F5-F2 (CREATE-CIRCLE-TANGENT-2 ENTS)**.

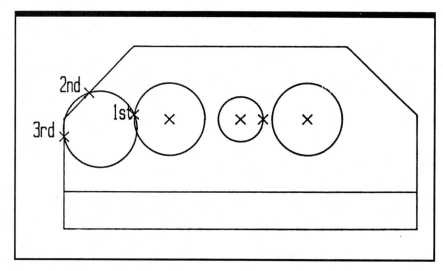

Figure 8-38 A circle created by indicating three points

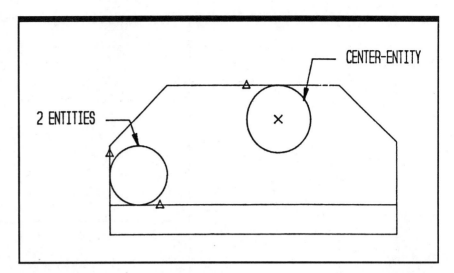

Figure 8-39 Circles created tangent to two entities and one entity after indicating a center point

You are prompted to enter the radius value for the tangent circle and press **RETURN**. You are then prompted to select the two entities to which the circle is to be created tangent. The circle will be drawn (Figure 8-39).

3 ENTS (Three Entities)

This method allows you to create a circle tangent to two entities. The Escape Code for this option is **F1-F3-F5-F3** (**CREATE-CIRCLE-TANGENT-3 ENTS**).

You are prompted to enter the radius value for the tangent circle and press **RETURN**. You are then prompted to select the three entities to which the circle is to be created tangent. The circle will be drawn.

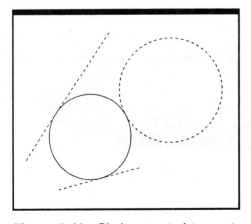

Figure 8-40 Circles created tangent to three entities

💾 POSSIBLE USE

When the center point is a known location, the center point can be indicated, the tangent entity selected, and a circle with the correct diameter created (Figure 8-40). If a circle must be created tangent to two entities, use the **2 ENTS** option and select the two entities. Then a circle of the proper diameter will be created.

CREATE-CIRCLE OPTION: TWO-PTS (Two Points)

The **TWO-PTS** option allows you to create a circle by specifying two points. The distance between the two points is used as the diameter of the circle. The Escape Code for this option is **F1-F3-F6** (**CREATE-CIRCLE-TWO-PTS**).

You are prompted to indicate the position for the first point and then the second point. The circle is drawn with the two points on the circumference.

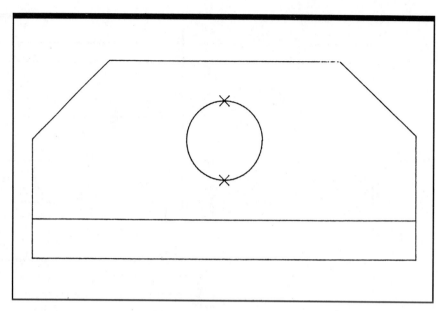

Figure 8-41 A circle created by indicating 2 points

POSSIBLE USE

The circle in Figure 8-41 was created using the **TWO-PTS** option. Like the **3-PTS** option, this option can be used to fit a drill hole between two objects. The two points for the diameter can be indicated with the aid of the Position Menu.

POINT OPTIONS

You can create point entities using any one of five options. The options are:

> **POSITN**
> **ON ARC**
> **SEGMENT**
> **MESH**
> **ON SPLN**

CREATE-POINT OPTION:
POSITN

The **POSITN** option allows you to create point entities on the screen at an indicated position. The Escape Code for this option is **F1-F4-F1** (**CREATE-POINT-POSITN**).

The Position Menu is displayed to aid you in indicating the position for the point entities. Each time a position is indicated, a point entity is created and a plus sign (+) appears on the screen to mark the point.

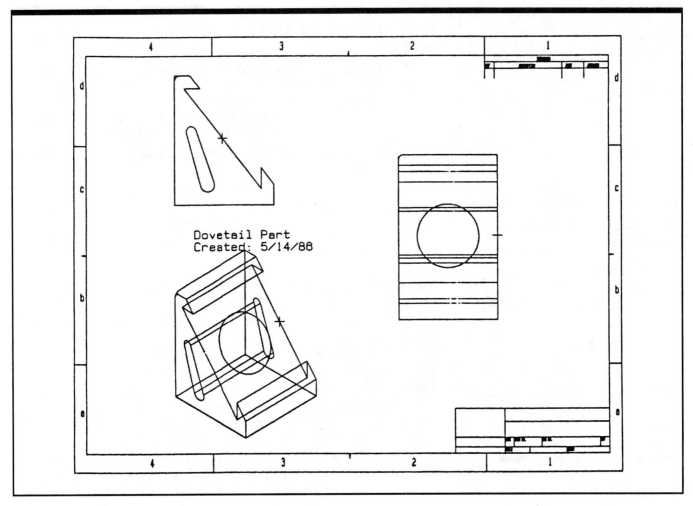

Figure 8-42   A point placed at a strategic position

POSSIBLE USE

Many times a point entity is positioned in a strategic location to aid in positioning other entities. In Figure 8-42 a point is positioned on the center of a line on the inclined surface as a reference entity. When the active depth is changed, the point can be used as a reference for the new active depth.

CREATE-POINT OPTION: ON ARC

The **ON ARC** option allows you to select an arc or circle entity and to create a point entity at a specific location relative to or on the arc. The Escape Code for this option is **F1-F4-F2** (**CREATE-POINT-ON ARC**).

You are prompted to select the desired arc or circle. Then you must enter an angle which will place the point relative to the arc or circle's 3 o'clock position (zero degrees). The angle specified does not have to be between the start and endpoints of the arc or circle. The point is placed at the depth of the arc or circle.

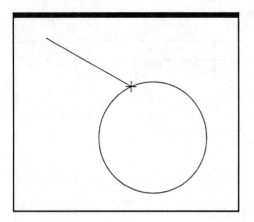

Figure 8-43 A point placed on a circle at 115 degrees

 POSSIBLE USE

If you need a reference entity at a specific location on an arc you can use this option. In Figure 8-43 a point was placed at the 115 degree point on the circle. This point was then referenced as an end point for the line that was drawn next to the circle using the **POINT** option from the Position Menu.

CREATE-POINT OPTION: SEGMENT

The **SEGMENT** option allows you to create point entities on a line or arc at specific intervals by segmenting the line or arc. The Escape Code for the option is **F1-F4-F3** (**CREATE-POINT-SEGMENT**).

You are prompted to enter the number of segments into which you wish to divide the line or arc and press **RETURN**. The value that you enter (n) equals the number of segments that are created. The number of points that are created is equal to (n + 1). If a line were divided into two segments, three points would be created: one at each endpoint and one in the middle.

You are asked whether to group the entities. The **NO/YES** menu is displayed. If you select **YES** you are requested to enter a group name and press **RETURN**. If **NO** is selected the next prompt is displayed. You are requested to select the line, arc/circle or spline which results in the points being displayed at the segment points.

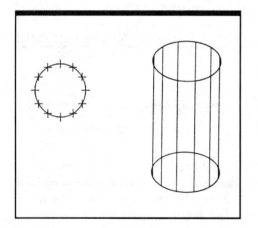

Figure 8-44 Points placed at each segment point on a circle for visual reference

 POSSIBLE USE

When the **AUTOSEG** option is used you cannot visually see the segment points. By using the **POINT-SEGMENT** option you can place a point entity along the segmented entity as a visual reference to where the segments begin and end (Figure 8-44).

CREATE-POINT OPTION: MESH

The **MESH** option allows you to create a mesh of points between two line or arc entities, similar to the mesh option for **LINE**. The Escape Code for this option is **F1-F4-F4** (**CREATE-POINT-MESH**).

The options for creating a **POINT MESH** are the same as a **LINE MESH**. The options include:

> **2CURVE**
> **3CURVE**

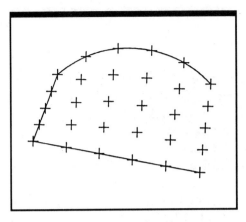

Figure 8-45   A POINT-MESH generated between an arc and a line

**4CURVE
TAB CYL
REVOLUT
SPHERE
RULED**

Refer to the **LINE MESH** option in this chapter for the operation of the **MESH** option.

POSSIBLE USE

The **POINT-MESH** can be utilized as node points for FEA (Finite Element Analysis). In Figure 8-45, a point mesh has been generated between the arc and the base line.

**CREATE-POINT OPTION:
ON SPLN**

The **ON SPLN** option allows you to recreate the knot points on the spline in the form of point entities. The Escape Code for this option is **F1-F4-F5 (CREATE-POINT-ON SPLN)**.

You are prompted to select an existing spline entity to create the knot points on. The Selection Menu is displayed. When the spline is selected, the points are created where the original knot points were indicated. The points will take on the current color and level.

POSSIBLE USE

Once a spline has been created, the knot points used to calculate the curve fit are no longer displayed. If you wish to keep the knot points as a visual or positional reference, they can be regenerated and point entities placed at each knot point, as in Figure 8-46.

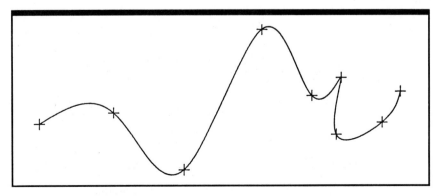

Figure 8-46   Points created at the knot points on a spline

POLYLINE OPTIONS

A polyline differs from a regular line entity in that a polyline is one continuous series of grouped line segments. A polyline cannot be broken, edited or have a single segment deleted from it. For most normal construction where editing is involved you should use the **LINE** function, not the **POLYLIN** function.

The **POLYLIN** function should be used when creating lines with widths greater than one because the corners or intersection points of the segments will be filled and blend nicely together. The great advantage of a polyline is that it has the ability to be filled with a solid color/pattern using **CONTROL-ATTRIB-OUT/FIL**. The requirements are that the polyline must be closed and have fewer than 51 sides. The polyline gives you the ability to fill areas that are regular (equal length sides) polygons. The options found under **POLYLIN** are similar to **LINE**. They are:

> **STRING**
> **RECTANG**
> **N-GON**
> **SKETCH**
> **MESH**

POLYLINE OPTION: STRING

The line **STRING** option allows you to draw a continuous chain or "string" of line segments by indicating successive endpoints. Each time a new point is selected, a line is drawn from the previous endpoint to the new point (Figure 8-47). This function can

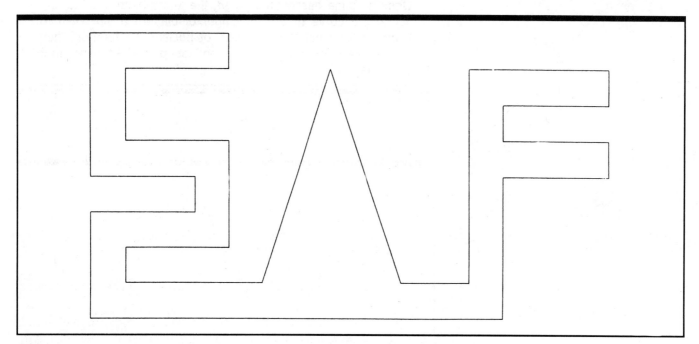

Figure 8-47 A polyline string

be activated using the Escape Code **F1-F5-F1** (**CREATE-POLYLIN-STRING**).

You are prompted to indicate the start point of the first line and then to indicate the endpoint for all of the lines that follow. The rubberbanding effect occurs when the cursor is moved to select an endpoint. To terminate the line string, press **F10 BACKUP** or **ESCAPE**.

POLYLINE OPTION: RECTANG (Rectangle)

The **RECTANG** option gives you two methods of creating a rectangle: **CORNERS** and **WID/HT**. A polyline rectangle is a grouped polygon that cannot be de-grouped and is therefore unable to be edited.

CORNERS

The **CORNERS** option allows you to create a rectangle by indicating two opposite corners. The rectangle is comprised of four grouped line entities and is considered a single entity. The Escape Code for this option is **F1-F5-F2-F1** (**CREATE-POLYLIN-RECTANG-CORNERS**).

You are prompted to indicate the position of the first corner, which is the lower left corner. Then you are prompted to indicate the position of the second corner or the upper right corner. As you move the cursor away from the first corner, a "rubberbox" will appear. The rubberbox allows you to see the size of the rectangle before the second point is selected.

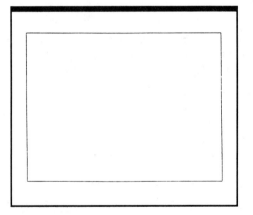

Figure 8-48   A polyline rectangle using the CORNERS option

 POSSIBLE USE

This option can be used to fit a rectangle in a specified area (Figure 8-48). By selecting corners that do not interfere with any other surfaces the rectangle can be accurately placed. If you want to quickly define the approximate height and width of a view you can arbitrarily select two diagonal corners of a rectangle and one will be created. If **SNAP** and Cursor Tracking are on you can accurately indicate the corner positions of the rectangle. The lines that are created are grouped so that they cannot be trimmed or deleted.

WID/HT (Width/Height)

The **WID/HT** option allows you to specify the width and the height of the polyline rectangle and then to indicate the position of the lower left corner. The rectangle is comprised of four grouped line entities and is therefore one complete entity. The Escape Code for this option is **F1-F5-F2-F2** (**CREATE-POLYLIN-RECTANG-WID/HT**).

You are prompted to enter the width value for the rectangle and press **RETURN**. The default value can be accepted by pressing **RETURN** only. Next, you are prompted to enter the height value

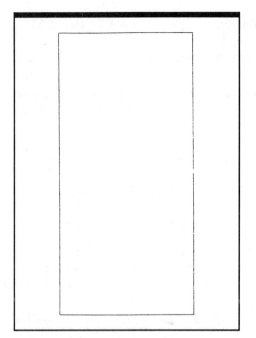

Figure 8-49 A polyline rectangle using the WIDTH and HEIGHT options

POLYLINE OPTION: N-GON

for the rectangle and press **RETURN**. Last, you are requested to indicate the position of the lower left corner of the rectangle. The rectangle is created.

 POSSIBLE USE

This function is very useful when you are creating a view or layout size and you know the width and height of the area you wish to work within. You can simply key in the values for the width and the height. Once those values are entered indicate the base position and the rectangle representative of the overall dimensions (Figure 8-49).

The **N-GON** option allows you to create a regular polygon which is made up of individual line entities that can be deleted or edited independently. The lines in the polygons created using the **CRE-ATE-POLYGON** function cannot be edited; only whole polygons can be deleted. The Escape Code for this option is **F1-F5-F8** (**CRE-ATE-POLYLIN-N-GON**).

You are prompted to enter the number of sides that you wish to have in the polygon and press **RETURN**. Next you are requested to enter the rotation angle and press **RETURN** if you wish to have the polygon rotated to a special angle.

For the next prompt you are asked to choose a method to specify the polygon radius. The two options that are displayed on a menu are:

F1 CORNER (Distance across corners, inscribed)
F2 FLAT (Distance across the flats, circumscribed)

CORNER—measures the radius from the center of the polygon to the intersection of two sides of the polygon.

FLAT—measures the radius of the polygon from the center point to a point on the side which forms a perpendicular line from the side to the center.

Once you have selected the radius method you are then prompted to enter the radius value and press **RETURN**. Next, you must indicate the position of the center point. After the center point is positioned you are asked whether or not to group the entities. The **NO/YES** menu is displayed. If **NO** is selected each side of the polygon will be a separate **LINE** entity. If **YES** is selected all of the **LINE** entities which make up the polygon will be grouped (you are prompted to enter a group name) and it is treated as a

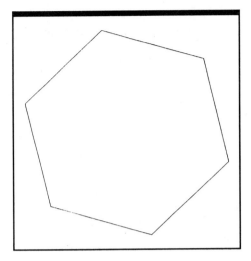

Figure 8-50   A polyline N-GON

single entity. At this point the polygon is created with the current attributes.

POSSIBLE USE

The **N-GON** option is a favorite function for most CADKEY users. When polygons were created by manual drafting techniques they required many laborious steps and the steps were different from one polygon to the next. Now, with CADKEY, you can create a polygon with as many sides as needed using only a few quick steps which are applicable to all polygons. The hexagon in Figure 8-50 was created using the **CREATE-POLYGON-CORNER** function. The radius from the center to the corner is .5, which gives you a distance across corners of 1.00. This option is a real time-saver.

POLYLINE OPTION: SKETCH

The **SKETCH** option allows you to draw a freehand line by holding down the mouse control button and moving the mouse/puck. The result is a series of connected line entities. The Escape Code for the **SKETCH** option is **F1-F5-F4** (**CREATE-POLYLIN-SKETCH**).

When the **SKETCH** option is selected you are prompted with the option to group the entities. If you choose **YES**, all of the line entities that make up the sketched line are grouped and take on the characteristics of a single entity. If **NO** is chosen, each line entity within the sketched line is treated as a single entity.

At this point you can start sketching by pressing down on the mouse control button to indicate the start point and moving the mouse as if it were a pencil. The screen cursor will disappear while you are sketching and will reappear when you release the mouse control button.

To begin sketching in a new position press **F10 BACKUP** and indicate a new start point.

NOTE: The configuration option "minimum increment for sketching" is located under option 1 (Select Graphics Device) in the CONFIG program. This dramatically affects the resolution and appearance of sketched polylines. The default value is .5 inches (very coarse); use .01.

POSSIBLE USE

The **SKETCH** option can be used to draw freehand break lines in a broken-out sectional view (Figure 8-51). Any other

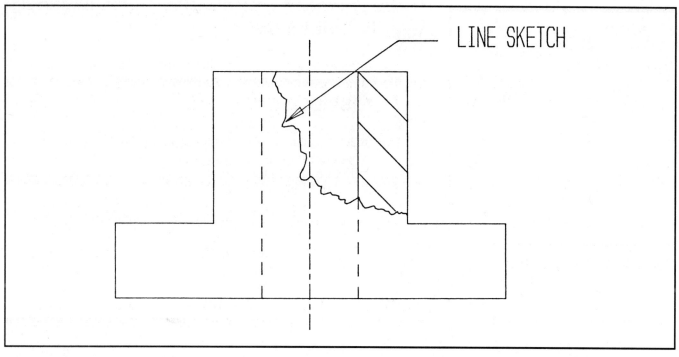

Figure 8-51 A freehand break line drawn using the POLYLIN-SKETCH option

freehand features that you wish to create so that they do not appear as though they came off a plotter can be done with the **SKETCH** option.

POLYLINE OPTION: MESH The **MESH** option allows you to generate a wire mesh on a surface between two, three, or four entities. This can then later be used in a finite element analysis package to analyze structural properties of a part design. The Escape Code for this option is **F1-F5-F5** (**CREATE-POLYLIN-MESH**).

The options for creating a polyline mesh are the same as a line mesh. The options include:

> **2CURVE**
> **3CURVE**
> **4CURVE**
> **TAB CYL**
> **REVOLUT**
> **SPHERE**
> **RULED**

Refer back to the **LINE MESH** option for the operation of the **MESH** option.

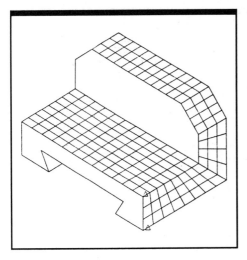

Figure 8-52 A polyline mesh generated on surfaces of a part

POSSIBLE USE

The primary use of the line **MESH** option is to generate a meshed surface that contains a series of nodes or knot points that are used by a third party FEA (Finite Element Analysis) program, such as ALGOR's SUPERSAP. In Figure 8-52 a **LINE-MESH** has been created on several surfaces.

Another possible use is to generate the appearance of a surface model by using the line mesh option.

The two options for generating a mesh are **RULED** and **GENERAL**.

FILLET OPTIONS

The **FILLET** option allows you to create an arc of a specified radius tangent to two entities, or a chamfer at specified distances between two lines.

FILLET OPTION: ARC

The entity combinations can be:

1. Two lines
2. A line and an arc/circle
3. An arc/circle and arc/circle

The entity combinations above can be used if they meet the following requirements:

1. The selected entities must lie in the same or parallel planes at the same depth.
2. The selected lines must intersect or have the potential to intersect if they were extended.
3. When one or both of the selected entities are arc/circle entities, the minimum distance between the entities must be less than or equal to the diameter of the fillet.

The direction of the fillet is determined by the order in which the entities are selected. When a fillet is created, it will take on the current construction attributes (color, line type, pen number, line width, etc.; Figure 8-53).

CREATE-FILLET-ARC OPTION: TRIM

The **FILLET-TRIM** option allows you to create an arc tangent to two entities and have the excess portion of the entities trimmed to the tangent/intersection point where the fillet connects the

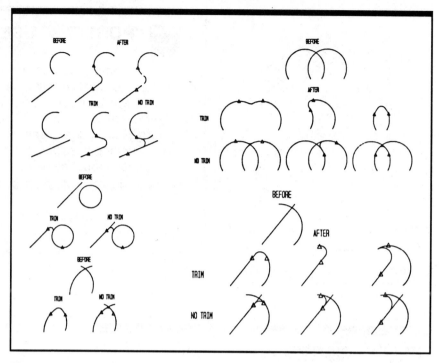

Figure 8-53 The FILLET options

entities. The Escape Code for this option is **F1-F6-F1-F1** (**CREATE-FILLET-ARC-TRIM**).

You are prompted to enter a radius value for the fillet and press **RETURN**. To accept the default value press **RETURN**. Next, you are prompted to select the first fillet entity and then the second fillet entity.

 ## POSSIBLE USE

This is a very efficient method of creating a fillet, but be careful when placing fillets between arcs. The order in which the arcs are selected will make a difference in the arc that is trimmed and the direction of the fillet. The result is not always what you wanted. To protect yourself against having the wrong portion of the arc entity trimmed away, it is recommended that you save your part first using the Immediate Mode command **CTRL-F SAVE**. If the wrong arc is trimmed you can then just go to **FILES** and **LOAD** the part in the state that you saved it and try again. In Figure 8-54 the horizontal line was divided using **TRM/EXT-DIVIDE** so that a fillet could be placed on both sides of the line. A one-inch fillet was placed on the left side selecting the line first and then the arc. On the right side a .438 radius fillet was created selecting the arc and the line. In both cases the arc and the line were trimmed to the point of tangency with the fillet. In many cases the fillet will be created between two lines, as in the 2D sectional view tutorial.

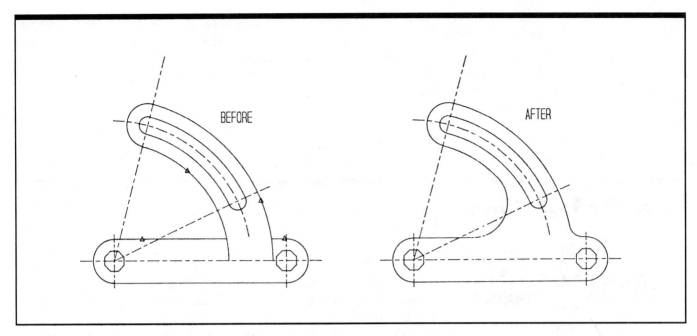

Figure 8-54   A fillet created with the TRIM option

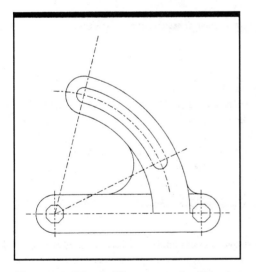

Figure 8-55   A fillet created without the TRIM option

POSSIBLE USE

In some of the cases where you may have trouble with the **TRIM** option you may be better off with the **NO TRIM** option. If you use this option you can always go to the **EDIT-TRIM** option to trim the portion of the desired arc or line quickly and accurately. Using the same radii for the fillets from the example above, the fillets were created without trimming their tangent entities in Figure 8-55. The arc and line could then be trimmed back to the point of tangency using the **EDIT-TRM/EXT-FIRST** function.

If the radius selected is more than half of the furthest distance between the two selected entities, the error message **No solution...press RETURN to continue** is displayed.

When this happens press **RETURN**, enter a new radius, press **RETURN**, and select the two entities again.

Once both entities are selected, the fillet is drawn and the entities are trimmed to the tangent points unless the entities are circles. Circles cannot be trimmed.

CREATE-FILLET-ARC OPTION: NO TRIM

The **NO TRIM** option allows you to create an arc tangent to two entities without trimming the selected entities. This option can be accessed using the Escape Code **F1-F6-F1-F2** (**CREATE-FILLET-ARC-NO TRIM**).

You are prompted to enter the radius value for the fillet and press **RETURN**. You are then prompted to select the first fillet entity and then to select the second fillet entity. If the radius entered fits between the two entities selected, the fillet is drawn without trimming the entities. If there is no solution, an error message appears and you are prompted through the option again.

CHAMFER OPTIONS

CREATE-FILLET OPTION: CHAMFER

The **CHAMFER** option allows you to create an angled line that intersects two other lines. The intersection point on each line is identified and the remaining portion of the line is trimmed back to the chamfer line. The Escape Code to invoke the **CHAMFER** option is **F1-F6-F2 (CREATE-FILLET-CHAMFER)**.

 POSSIBLE USE

The **CHAMFER** option allows you to create inclined lines/surfaces by specifying the distance along the first line where the chamfer should start. Then the distance along the second line is specified. In Figure 8-56, the distance along the first line is .50 and the distance along the second line is .75. If the distances were equal the inclined line would be at a 45-degree angle.

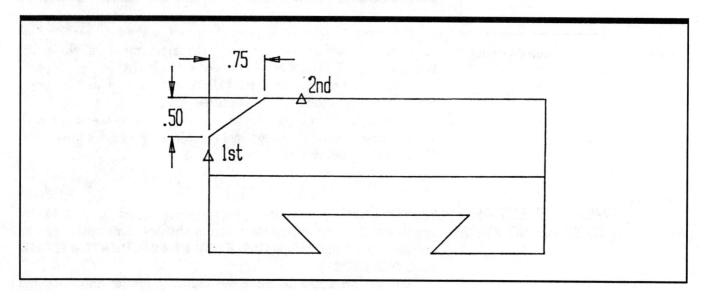

Figure 8-56   Creating a chamfer

You are prompted to enter the distance along the first line that you wish the chamfer to intersect and press **RETURN**. The distance will be measured from the endpoint of the line closest to the cursor when the line is selected. The distance value must be between .00005 and 10,000. You are then prompted to enter the distance along the second line and press **RETURN**. The final two prompts request that you select the first and second lines closest to the end where you wish the chamfer to be created. The **CHAM-FER** line is created as specified and the selected lines are trimmed or extended to meet the new chamfer line with the current attributes (color, line type, etc.).

CONIC OPTIONS

A conic section is defined by Webster's Ninth New Collegiate Dictionary (1983) as "a plane curve, line, or point that is the intersection of or bounds the intersection of a plane and a cone with two nappes." The conic option allows you to create an ellipse, parabola or hyperbola, all curves generated by the intersection of a plane and a cone. The Escape Code for this option is **F1-F7** (**CREATE-CONIC**). The options found under **CONIC** include:

> **5 COND (Five Condition)**
> **6 COEFF (Six Coefficients)**
> **PARABOL (Parabola)**
> **ELLIPSE**

5 COND (Five Condition)

The **5 COND** option gives you the ability to generate a conic section (ellipse, parabola, or hyperbola) by identifying five conditions (points, slopes, or rho). This can be done by indicating two, three, four, or five points the conic entity will pass through. The Escape Code for this option is **F1-F7-F1** (**CREATE-CONIC-5 COND**).

For all of the options except **5 POINT**, your definition of the desired conic also includes the slope of the conic at one or two of the indicated points. Four options define slope: **2 PTS, ENDENT, KEYIN,** and **ANGLE**.

> **2 PTS**—defines the slope of the conic at the selected point with the tangent vector that passes through two indicated positions.
>
> **ENDENT**—defines the slope of the conic at the selected point with the tangent vector at the end point of a selected entity.
>
> **KEYIN**—defines the slope of the conic at the selected point with keyed-in delta X, Y, and Z values.

ANGLE (2D construction mode only)—defines the slope of the conic at the selected point with a keyed-in value for the angle of the tangent vector. The value for the angle (in degrees) moves in a counterclockwise direction from the construction view X axis. The options for creating a conic are:

2P 2S R (2 Points, 2 Slopes, Rho)
3P 2S (3 Points 2 Slopes)
4P 1S (4 Points 1 Slope)
5 PTS (5 Points)

2P 2S R (2 Points, 2 Slopes, Rho)

The **2P 2S R** option allows you to generate a conic section by specifying the following conditions (Figure 8-57).

1. Identify the start and end points with the aid of the Position Menu.

2. Enter a rho value (0.0 rho 1.0). The rho value determines the conic type.

rho = 0.0 to 0.5 Ellipse
rho = 0.5 Parabola
rho = 0.5 to 1.0 Hyperbola

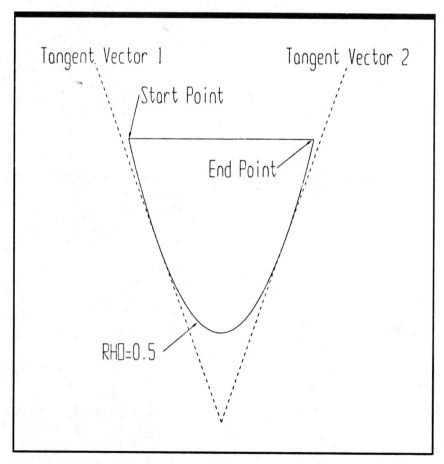

Figure 8-57 A five condition conic created using two points, two slopes, and rho, 2P 2S R

3. Define the slope of the conic at the start and end points.

3P 2S (3 Points, 2 Slopes)

In the **3P 2S** option you must identify three points (start, second, end) with the aid of the Position Menu. Then you must assign the slope to two of the three points.

4P 1S (4 Points, 1 Slope)

The **4P 1S** option generates a conic by indicating the start, second, third, and end points using the Position Menu and assigning the slope to one of the four points.

5 PTS (5 Points)

The **5 PTS** option allows you to generate a conic by identifying the start, second, third, fourth, and end points using any option from the Position Menu.

6 COEFF (Six Coefficients)

The **6 COEFF** option allows you to create a conic section by entering values for the six coefficients of the following equation, which is the general quadratic representation of a conic:

$$A*x + B*x*y + C*y + D*x + E* + F = 0$$

Your definition of the desired conic also must include the start point and the end point of the entity. The Escape Code for this option is **F1-F7-F2 (CREATE-CONIC-6 COEFF)**.

When prompted, enter an appropriate value for each of the six coefficients. Then with the aid of the Position Menu, indicate the start and end points of the conic section. An error message will be displayed if the conic cannot be created.

PARABOL (Parabola)

The **PARABOL** option is used to create a parabola by defining the directrix, the focus, and the start and end points for the entity. A parabola is a set of points in a plane, with the condition that the distance from any point in the set of a given line (directrix) is equal to the distance from that point to a given point (focus). The Escape Code for this option is **F1-F7-F3 (CREATE-CONIC-PARABOL)**.

First you are given three options for defining the directrix.

2 PTS (Two Points)—using the Position Menu options, indicate two positions as the endpoints for the directrix.

LINE—cursor-select an existing line as the directrix (Figure 8-58).

ENDENT—define the directrix as the line that is tangent to a selected curve entity at the end of the curve that is closest to the selection point.

Once the directrix has been defined, the focus and the start and end points of the parabola must be defined using an option from the

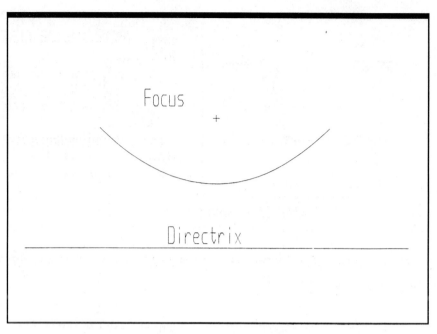

Figure 8-58 A conic parabola using a line as the directrix

Position Menu. If the parabola cannot be generated, an error message will be displayed.

ELLIPSE   The **ELLIPSE** option creates an elliptical arc by defining the length and orientation of the major and minor axes, and the start, end and center points of the entity. This option works similar to the creation of an approximate ellipse in traditional drafting. The Escape Code for this option is **F1-F7-F4** (**CREATE-CONIC-ELLIPSE**).

To begin you are prompted to enter the length of the semi-major axis. This value is equal to half of the major axis or the distance from the center to one edge along the major axis. The major axis is sometimes referred to as the major diameter. Enter a value and press **RETURN**.

Next, enter a value for the semi-minor axis and press **RETURN**. This value should be equal to half of the minor axis (minor diameter). Then enter a value for the angle between the major axis and the construction view X axis and press **RETURN**.

You are then prompted for the start and end angles for the ellipse. Normal angle conventions apply with 0 at the 3 o'clock position. For a full ellipse the start and the end angle must be the same. Last, you are prompted to indicate the center point with the aid of the Position Menu. The ellipse will be created in the current construction view at the current depth, with the center projected to the current construction plane.

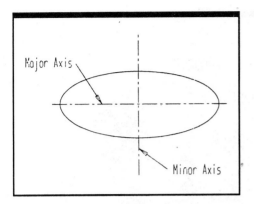

Figure 8-59   A conic ellipse

 POSSIBLE USE

The **ELLIPSE** option can be used to create an ellipse that is the result of the intersection of a plane and a cone or cylinder (Figure 8-59), or to create an ellipse found in other parts, such as a gasket.

POLYGON OPTIONS

The **POLYGON** option allows you to create a regular polygon with equal angles and equal-length sides. These sides cannot be trimmed or deleted; if you want a polygon that can be edited use the **CREATE-LINE-N-GON** option. The number of sides can range from three to eight. The polygons can be either filled or non-filled. The Escape Code for this option is **F1-F8 (CREATE-POLYGON)**.

The options found under **POLYGON** are:

CTR+RAD (Center and Radius)
CTR+SID (Center and Side)
RECTANG (Rectangle)
STRING
MESH

POLYGON OPTION:
CTR+RAD

You are prompted to enter the number of sides that you wish to have in the polygon and press **RETURN**. Next, you are requested to enter the rotation angle and press **RETURN**, if you wish to have the polygon rotated to a special angle. Enter a value for the polygon's radius and press **RETURN**.

For the next prompt you are asked to choose a method to specify the polygon radius. The two options that are displayed on a menu are:

F1 CORNER (Distance across corners, inscribed)
F2 FLAT (Distance across the flats, circumscribed)

CORNER—measures the radius from the center of the polygon to the intersection of two sides of the polygon.

FLAT—measures the radius of the polygon from the center point to a point on the side which forms a perpendicular line from the side to the center.

Once you have selected the radius method you are then prompted to choose whether to have an **OUTLINE**(d) or **FILLED** polygon

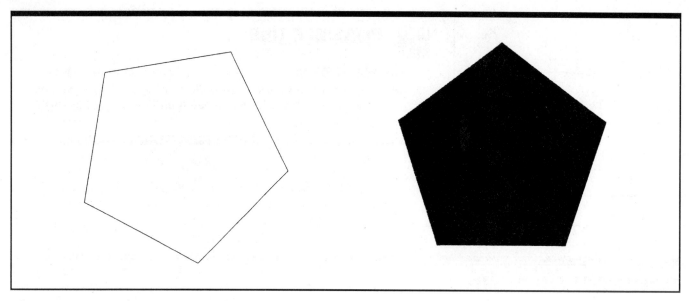

Figure 8-60 A polygon outline and a filled polygon

(Figure 8-60). Next you must indicate the position of the center point. The polygon is now created with the current attributes.

POLYGON OPTION: CTR+SID

The **CTR+SID** option allows you to create a polygon entity by specifying the number of sides, rotation angle, side length, fill type, and center position. You are prompted to enter the desired number of sides up to eight, and the rotation angle in degrees. Positive angles rotate the polygon's base in a counterclockwise direction. Then enter the length of the side and press **RETURN**. Choose whether the polygon should be an **OUTLINE** or **FILLED**. Last, indicate the center of the polygon.

POLYGON OPTION: RECTANG (Rectangle)

The **RECTANG** option gives you two methods of creating a rectangle: **CORNERS** and **WID/HT**.

CORNERS

The **CORNERS** option allows you to create a rectangle by indicating two opposite corners. The rectangle is a single entity. The Escape Code for this option is **F1-F8-F3-F1** (**CREATE-POLYGON-RECTANG-CORNERS**).

You are prompted to indicate the position of the first corner, which is the lower left corner. Then you are prompted to indicate the position of the second corner or the upper right corner. As you move the cursor away from the first corner, a rubberbox will appear. The rubberbox allows you to see the size of the rectangle before the second point is selected.

POSSIBLE USE

The **CORNERS** option can be used to fit a rectangle in a specified area. By selecting corners that do not interfere with any other surfaces, the rectangle can be accurately placed. If you want to quickly define the approximate height and width of a view, you can arbitrarily select two diagonal corners of a rectangle and one will be created. If **SNAP** and Cursor Tracking are on, you can accurately indicate the corner positions of the rectangle. The lines that are created are grouped so that they cannot be trimmed, deleted, or degrouped as needed.

WID/HT (Width/Height)

The **WID/HT** option allows you to specify the width and the height of the rectangle and then to indicate the position of the lower left corner. The rectangle is one complete entity. The Escape Code for this option is **F1-F8-F7-F2 (CREATE-POLYGON-RECTANG-WID/HT)**.

You are prompted to enter the width value for the rectangle and press **RETURN**. The default value can be accepted by pressing **RETURN**. Next, you are prompted to enter the height value for the rectangle and press **RETURN**. Last, you are requested to indicate the position of the lower left corner of the rectangle. The rectangle is created.

 POSSIBLE USE

This function is very useful when you are creating a view and you know the width and height ahead of time. You can simply key in the value for the width and then the value for the height. Once those values are entered, indicate the base position and the rectangle representative of the overall dimensions. This rectangle can be thought of as a sculptor's block, which is used to create the view by cutting away, editing and creating entities within and on the rectangle.

POLYGON OPTION: STRING

The line **STRING** option allows you to draw a continuous chain or "string" of line segments by indicating successive endpoints. Each time a new point is selected, a line is drawn from the previous endpoint to the new point. All of the line segments are grouped as a polygon and cannot be edited. This function can be activated using the Escape Code **F1-F8-F4 (CREATE-POLYGON-STRING)**.

You are prompted to indicate the start point of the first line and then to indicate the endpoint for all of the lines that follow. The rubberbanding effect occurs when the cursor is moved to select an endpoint. To terminate the line string, press **RETURN,** which will automatically close the polygon. Up to eight sides (lines) can be indicated.

POLYGON OPTION: MESH

The **MESH** option allows you to generate a wire mesh on a surface between two, three, or four entities. This can then later be used in a finite element analysis package to analyze structural properties of a part design. The Escape Code for this option is **F1-F8-F9** (**CREATE-POLYGON-MESH**).

The options for creating a **POLYGON MESH** are the same as a **LINE MESH**. The options include:

> **2CURVE**
> **3CURVE**
> **4CURVE**
> **TAB CYL**
> **REVOLUT**
> **SPHERE**
> **RULED**
> **REVERSE**

Refer to the **LINE MESH** option in this chapter for the operation of the **MESH** option.

 POSSIBLE USE

The primary use of the line **MESH** option is to generate a meshed surface that contains a series of nodes or knot points that are used by a third party FEA (Finite Element Analysis) program such as ALGOR's SUPERSAP. Another possible use is to generate the appearance of a surface model by using the line mesh option.

REVERSE

This process reverses the normals of grouped polygons in the geometric data base. This process is helpful in situations where shading is used.

SPLINE OPTIONS

A **SPLINE** is an irregular curve that is comprised of successive segments of geometric curves. A spline is a curve fit through several points known as knot points. In manual drafting, an irregular curve was created by tracing a french curve (a plastic template containing many irregular curves).

Splines are used widely in design fields dealing with aerodynamics where a surface with a smooth curve passing through a defined set of knot points is created, providing a path of least resistance on the surface. The spline has continuity of slope and curvature at all points, which results in a curve of minimum strain energy. Splines are used in the design of automobiles, aircraft, and boats. All of these vehicles move through air or water. When designing their outside shape a smooth curve or spline gives the vehicles less resistance and allows them to move faster with less energy.

There are several different classifications of splines. The splines created in CADKEY are cubic parametric splines. In a cubic parametric spline, the spline passes directly through the knot points, as opposed to a B-spline, which passes through the midpoint between the knot points. The cubic parametric spline gives you the greatest control of the irregular curve. CADKEY generates splines exactly to IGES standards.

There are eight types of splines that can be created using CADKEY. They are:

2D CUBI (2D Cubic)
3D CUBI (3D Cubic)
2D CLSD (2D Closed)
3D CLSD (3D Closed)
HELIX
2D AUTO
3D AUTO
MESH

**SPLINE OPTION:
2D CUBI (2D Cubic)**

The **2D CUBI** option allows you to create a 2D cubic parametric spline through a series of knot points. This type of spline lies within only one plane. The Escape Code for this option is **F1-F9-F1 (CREATE-SPLINE-2D CUBI)**.

You are prompted to indicate the position for the knot points through which the spline will be created. The Position Menu can be used and points can be attached to existing entities.

When you are in the 2D construction mode, the knot points indicated are always at the current working depth and the spline is created at that depth. When you are in the 3D construction mode, the knot points may not lie in the current plane. If this is the case, you are given the opportunity to change to the splines definition plane. The **NO/YES** menu is displayed for you to choose from.

Next, a menu of Start and End Condition options is displayed. Use this menu to choose the Start and End Conditions for the spline. The Start/End Condition options are:

NATURAL
2 PTS (2 Points)
ENDENT (End of Entity)
TANGENT
ANGLE

NATURAL | This option automatically assigns a curvature of zero to the start or end of the spline. This is equivalent to the spline being attached to the endpoint without any external forces being applied.

2 PTS (2 Points) | This option allows you to specify a tangent vector at a spline endpoint by indicating two points equal to the vector length or magnitude. The position of the vector is the same as the position of the two points. You are prompted to indicate the position of the first point and then the position of the second point. The next prompt requests that you enter the desired magnitude and press **RETURN** or accept the default magnitude by pressing **RETURN**. When a negative magnitude value is entered, the direction of the vector will be reversed.

ENDENT (End of Entity) | This option allows you to specify a tangent vector by using the endpoint of the spline and the endpoint of an existing entity. You must first select the line or arc near the desired endpoint. This point is marked with a temporary (X). Enter the desired vector magnitude and press **RETURN**. The positive vector direction is pointing from the end of the selected entity closest to the cursor away from the entity.

TANGENT | This option allows you to enter the X and Y vector components (used to determine the slope X/Y) and the magnitude of the desired tangent vector. The default values can be accepted by pressing **RETURN**. You are prompted to enter the X component value and press **RETURN**, then to enter the Y component value and press **RETURN**.

ANGLE | This option allows you to determine the direction of a tangent vector by setting the angle between the tangent vector and the positive X axis. Enter the angular value in degrees and press **RETURN**. The angle is measured counterclockwise from the positive X axis of the spline's definition view. Enter the tangent vector magnitude and press **RETURN**. If a negative value is entered, the direction of the vector will be reversed.

 POSSIBLE USE

A **2D CUBIC SPLINE** can be used for two-dimensional drafting work where the spline will connect with another entity, such as the pattern development for a truncated cylinder (Figure 8-61).

SPLINE OPTION: 3D CUBI | The **3D CUBI** option allows you to create a three-dimensional cubic parametric spline with knots that lie in more than one plane. The result is a true 3D curve. The Escape Code for this option is **F1-F9-F2 (CREATE-SPLINE-3D CUBI)**.

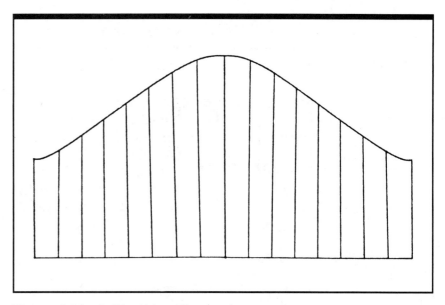

Figure 8-61   A 2D cubic spline on the top edge

You are prompted to indicate the position for the knot points through which the spline will be generated. To erase the last knot point selected press **F10 BACKUP** and indicate a new position. Press **RETURN** when you have finished indicating knot points. The start point is temporarily marked with an asterisk (*).

A new menu is displayed containing different methods for selecting the start and end conditions for the spline. The options are the same as the **2D CUBIC** spline except there is no **ANGLE** option. The options are:

> **NATURAL**
> **2 PTS (2 Points)**
> **ENDENT (End of Entity)**
> **TANGENT**

Refer to the **2D CUBI** option for explanations of the Start and End Conditions.

You are prompted to choose the the Start Condition and then to choose the End Condition.

 POSSIBLE USE

A **3D CUBIC SPLINE** is the equivalent of a rope hanging in 3D space. The rope can twist and bend in any direction, always forming a smooth curve. A tool like this is extremely important in the design of automobiles and aircraft. The edges on the car in Figure 8-62 are an example of a 3D cubic parametric spline.

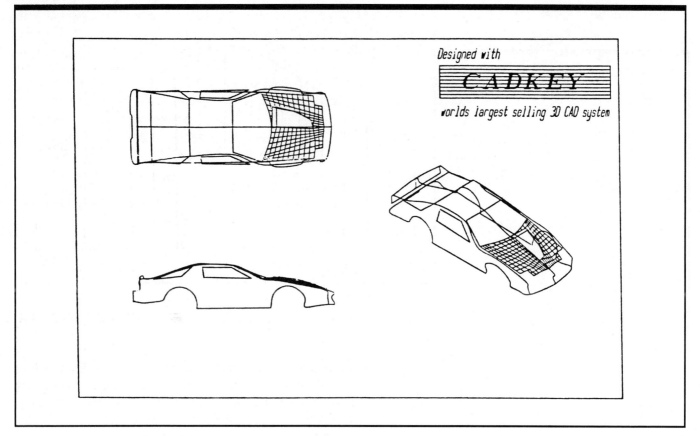

Figure 8-62   A 3D cubic spline (Courtesy of Cadkey, Inc.)

SPLINE OPTION: 2D CLSD

The **2D CLSD** option allows you to create a 2D cubic parametric spline that starts and ends at the first point selected. The spline will lie in a single plane and appear as an unbroken or closed loop. The slope and curvature are continuous at all points on the spline. The Escape Code for this option is **F1-F9-F3** (**CREATE-SPLINE-2D CLSD**).

You are prompted to indicate the knot points which will be used to define the path of the spline. The Position Menu is displayed to aid in indicating the position of the knots. Once you have indicated all of the desired knots press **RETURN** and the spline is generated through the knots.

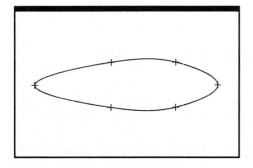

 POSSIBLE USE

The **2D CLOSED SPLINE** can be used for cross-sectional work with irregular objects. This option can also be used for any drafting techniques that previously required a french curve. A 2D closed spine is pictured in Figure 8-63.

Figure 8-63   A 2D closed spline

SPLINE OPTION: 3D CLSD

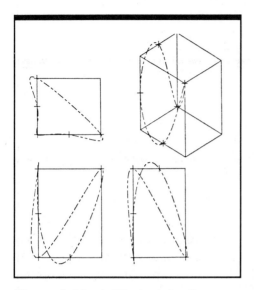

Figure 8-64   A 3D closed spline

The **3D CLSD** option allows you to create a cubic parametric spline that is three-dimensional. The closed spline will begin and end at the first point indicated. The points can be indicated at various depths and a true curve fit will be generated. The Escape Code for this option is **F1-F9-F4** (**CREATE-SPLINE-3D CLSD**).

You are prompted to indicate the knot points for the 3D spline. With the aid of the Position Menu, knot points can be attached to existing geometry. Once all of the desired knot points are indicated press **RETURN** and the 3D spline is displayed.

 POSSIBLE USE

The **3D CLOSED SPLINE** can be used in the creation of such objects as an aircraft wing or any other closed irregular surface (Figure 8-64).

SPLINE OPTION: HELIX

The **HELIX** option allows you to create a 3D spline or splines, which represent a helix curve of arbitrary orientation. The Escape Code for this option is **F1-F9-F5** (**CREATE-SPLINE-HELIX**).

You are prompted to select the direction for the helix to turn. Select from the two menu options displayed:

F1 CCW (counterclockwise)
F2 CW (clockwise)

Next, you are prompted to enter the helix's pitch value. This value is equal to the number of turns per unit length.

You must now define the extent or length of the helix curve. A menu with three options for defining the length of the curve is displayed. The options are:

ROTS—allows you to enter the number of rotations for the
(Rotations) helix. Fractional values may be used.
ARCLEN—allows you to define the arc length of the helix
(Arc Length) curve.
AXIS—allows you to define the size of the helix along the major axis.

Enter the appropriate value for whichever option you select. You will then be prompted to indicate two points along the major axis which will serve as the center/axis of the helix.

You must now indicate the start point. This point determines the radius, the direction of the X axis, and the base point on the major axis from which the helix is created.

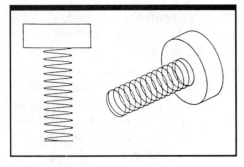

Figure 8-65 A helix

The helix is created using the information you have entered. It will begin at the start point and move along the major axis in specified direction with the appropriate screw.

POSSIBLE USE

The **HELIX** option can be used to create screw threads and springs or any other coiled object (Figure 8-65).

SPLINE OPTION: 2D AUTO

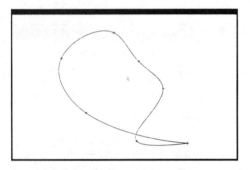

Figure 8-66 A spline created using the 2D AUTO option

The **2D AUTO** option automatically creates a 2D spline through a path of minimum distance point entities that have been projected to the current construction plane. The Escape Code for this option is **F1-F9-F6** (**CREATE-SPLINE-2D AUTO**).

The first position/point, second position/point, and optional end-point position determine the start, end, and direction of the automatic spline. After all of the positions for the spline have been defined, press **RETURN** and the spline will be created passing through a maximum of up to 76 entities. If desired, the start and end positions can be identified with the use of the Start/End Condition Menu (Figure 8-66).

SPLINE OPTION: 3D AUTO

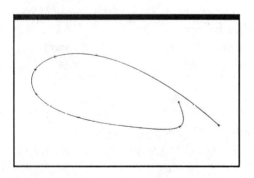

Figure 8-67 A spline created using the 3D AUTO option

The **3D AUTO** option automatically creates a 3D spline through a path of minimum distance point entities in 3D space. The Escape Code for this option is **F1-F9-F7** (**CREATE-SPLINE-3D AUTO**).

The first position/point, second position/point, and optional end-point position determine the start, end, and direction of the automatic spline. After all of the positions for the spline have been defined, press **RETURN** and the spline will be created passing through a maximum of 76 entities. The start and end positions can be identified with the Start/End Condition Menu (Figure 8-67).

SPLINE OPTION: MESH

The **MESH** option allows you to generate a wire mesh on a surface between two, three, or four entities. This can then later be used in a finite element analysis package to analyze structural properties of a part design. The Escape Code for this option is **F1-F9-F8** (**CREATE-SPLINE-MESH**).

The options for creating a **SPLINE MESH** are the same as a **LINE MESH**. The options include:

2CURVE
3CURVE
4CURVE
TAB CYL
REVOLUT
SPHERE
RULED
REVERSE

Refer to the **LINE MESH** option in this chapter for the operation of the **MESH** option. (Refer to the **POLYGON MESH** option for **REVERSE**.)

REVIEW QUESTIONS

1. What is an entity?
2. If you needed to create a single line segment, which **LINE** option would you use?
3. How can center lines for a circle be created quickly and placed accurately at the center of a circle?
4. How can a freehand line be drawn?
5. When specifying start and end angles for arcs, at what position is zero degrees?
6. Which option for creating circles is most commonly used?
7. What are point entities used for?
8. What are the two methods of creating a rectangle?
9. What is a **FILLET**?
10. What values must be entered to create a **CHAMFER**?
11. What is the difference between a polygon created using the distance across corners and a polygon with the distance across the flats?
12. What is a spline?
13. What are splines used for?
14. What is a helix used for?

The Immediate Mode

OBJECTIVES

After completing this chapter, you will be able to:

- Access all of CADKEY's Immediate Mode commands.
- Escape CADKEY's Immediate Mode commands.
- Use the on-line calculator for solving simple mathematical problems.

Nearly all CAD systems have a menu structure that can be compared to the structure of a tree. That is, to access an option within the software it is necessary to locate the option by following the branches that make up the program. For example, to draw a line it may be necessary to choose the correct option from the Main Menu, such as **CREATE**. Invoking this command will reveal another list of options to create entities. On this menu list there will probably be an option called **LINE**. **LINE** is selected, which will probably reveal another menu providing the user with various methods of placing the line on the drawing. A typical option from this branch of the menu might be "end of an entity." From this example, you get the idea of how the tree structure is used to create a drawing. From the Main Menu, various options are selected by following the branches of the software.

The problem with this type of menu structure is that it is difficult and time consuming to go from one unrelated menu option to another. For example, if you were in the middle of drawing lines and decided to erase one of the lines, you would be forced to leave the line drawing option and find the erase option. If the erase option was not under the **CREATE** menu branch it would be necessary to return out of the **LINE** function until you reached the Main Menu and choose the commands necessary to reach the **ERASE** option. Few CAD systems have a method of reaching frequently used commands immediately without having to page through menus and interrupt your current operation. Fortunately for you, CADKEY has one of the most powerful Immediate Mode commands available.

THE IMMEDIATE MODE COMMANDS

CADKEY's Immediate Mode allows you to easily access frequently used commands regardless of your position in the menu. Although all of the Immediate Mode commands can be located through the normal menu system, it is much quicker and more convenient to use the Immediate Mode. You must remember that a CAD system is only as efficient as the operator using the software. To become an effective user of CADKEY you should learn the Immediate Mode commands as soon as possible.

CADKEY has three basic methods of changing the current status or changing the active menu item. The first method is to choose the menu items displayed on the left side of the screen, using the function keys. The number to the left of the displayed command corresponds to the function key located on the left side of the IBM PC or compatible keyboard. This method involves many keystrokes and it takes some time to become proficient. The second method of choosing menu items is to move the cursor over the desired menu item and pick it using a cursor control device such as a mouse. This method takes time because the cursor must be moved about the screen and you must wait for each new menu to be displayed before continuing. The third method is to use the Immediate Mode commands. This is by far the most effective and efficient method of choosing some commands with CADKEY.

Using the Immediate Mode does not involve moving the cursor or invoking menu options using the function keys. The Immediate Mode is invoked by holding down the **CTRL** or **ALT** key and simultaneously pressing an assigned letter key. Prompts are then used for any additional information to complete the command. After the command is completed, you are automatically returned to your place in the menu before the Immediate Mode command was selected. Immediate Modes can be interrupted by using the **ESC** key, which will automatically return you to your place in the menu. Thus, you can easily move to frequently used commands without interrupting your current position in the menu or the position of your cursor on screen. Items listed in the Status Window can also be activated by using the Immediate Mode. For example, the color assignment for entities can be changed by moving the cursor into the Status Window and over the command **COLOR**. The Immediate Mode command used to change the color is **ALT-X**.

The Immediate Mode command list follows. These commands are listed on a plastic card provided by CADKEY and shown in the front of this text.

Function	Immediate Mode Command
ARROWS IN/OUT	**CTRL-A**
AUTO SCALE	**ALT-A**
BACK-1 window	**ALT-B**
CALCULATOR	**CTRL-I**
change DEPTH	**CTRL-D**
change VIEW	**ALT-V**
COLOR bar	**ALT-X**
cursor SNAPPING on/off	**CTRL-X**
CURSOR TRACKING display	**CTRL-T**
DATA BASE SEARCH (forward/backward)	**ALT-R**

Dunlop Motorsport Wins with CADKEY

CADKEY In The Real World

Dunlop Motorsport helped make another piece of racing history in June 1991 when the company joined with Mazda to produce the first-ever Japanese victory in the world's most famous race: the Le Mans 24 Hours. The win creates a record which cannot be equalled by any other tire manufacturer well into the 21st century. It was Dunlop Motorsport's second 24-hour race in a week, following its Group N victory in partnership with Nissan at Nurburgring, Germany. The winning Mazda 787B used Dunlop radial tires in its sensational defeat of the Mercedes, Jaguar, and Peugeot factory teams, as well as 19 private Porsches.

SP Tyres UK Limited of Fort Dunlop, Birmingham, England, designs, develops and builds Dunlop competition tires. SP Tyres' Motorsport Development Department relocated and combined its design operations with SP Tyres' UK Motorsport Division in January 1990. The move prompted a need for a self-contained design tool to be located in the center of teh development office, which is some distance away from the company's existing large computer facility.

Ease of Use — Key Feature

SP Tyres chose CADKEY for their design and drafting requirements, and the system now sees regularly increasing use by more and more members of the design staff. Matthew Simpson, a Senior Tire Engineer at Fort Dunlop, explained how CADKEY helped Dunlop to help Mazda make it in the '90's. "Most people were self-trained on the system, acquiring skills through work-related exercises," Matthew Simpson said. "There was never the need to provide a formal training program as the package is so user friendly." "CADKEY enables me to alter things on the run, sit and sketch concepts on mold shapes, tire constructions, tread patterns, and all the associated aspects of the design process for Group C racing tires, before committing myself to final drawings," Matthew added.

Tread Patterns — Critical for Racing

Late in 1990, the size parameters and wheel widths for the tires to be used on the Mazda racing car were defined. Initial work involved the design of the mold profile for the front tires. Working within the dimensional constraints of the vehicle, and using knowledge acquired from previous years of developing radial tires with Jaguar and Nissan sports prototypes, the design group finalized the profile in January 1991. "During our development period, work progressed on new tread patterns using CADKEY, both for light and for heavy rain conditions. Tread-pattern development is an amalgamation of style, and performance," Matthew said. "We produce a lot of preliminary sketches with the CAD system to assess how the various elements of the pattern relate to one another, and to achieve the requisite relationship of *land* to *sea* (rubber-to-groove ratios). It is possible to assess new concepts of pattern theory and style by producing templates directly from CADKEY at full size, cutting the grooves out with a scalpel, and spraying paint through onto a plain-treaded tire. The pattern is then *hand-cut* to provide a model for visual appraisal. We produce full sets of test tires in the same way. This method is fairly common within the tire industry. Ultimately, however, we will look toward a solid-modeling system,

Mazda racer with Dunlop tires designed with CADKEY for Le Mans 1991.

—Continued, next page

CADKEY In The Real World—Continued

like CADKEY® SOLIDS, to save time and effort in producing these visual appraisals. CADKEY Version 4 includes such a package, and an upgrade to the new system seems viable in the near future."

SP Tyres provides tires for each race in a variety of tread-compound types to suit prevailing conditions of temperature, track surface, vehicle-handling characteristics, driver preference, etc. Typically, each car will have as many as 20 sets of dry *slick* tires available per race. In sprint races, cars will use only three sets under normal circumstances. At Le Mans 1991, the winning Mazda car used 18 sets of tires during the 24 hours of the race.

Adapted with permission from **3-D WORLD**, *v. 6, no. 1, 1992.*

DELETE (single)	**CTRL-Q**
DIGITIZE	**CTRL-Y**
EXECUTE MACRO	**CTRL-E**
FILE PART (save)	**CTRL-F**
GRID ON/OFF	**CTRL-G**
LAYOUT	**ALT-G**
LEVEL masking	**ALT-N**
LINE LIMITS switch	**ALT-L**
LINE TYPE menu	**ALT-T**
LINE WIDTH menu	**ALT-Y**
PAN	**ALT-P**
PAUSE on/off macro	**CTRL-K**
PEN # assignment	**ALT-Z**
RECALL last (undelete up to 100 entities)	**CTRL-U**
RECORD macro	**CTRL-J**
REDRAW	**CTRL-R**
SCALE with center	**ALT-S**
set working LEVEL	**CTRL-L**
SLIDE file save	**ALT-F**
2D/3D switch	**CTRL-W**
type MASKING	**ALT-M**
UNDO	**ALT-I**
USER prompt macro	**CTRL-O**
VIEW/WORLD coordinates	**CTRL-V**
WINDOW	**ALT-W**
WITNESS LINE	**CTRL-B**
ZOOM-DOUBLE	**ALT-D**
ZOOM-HALF	**ALT-H**

Some of the Immediate Mode commands will be used more than others, even though they all should be mastered as soon as possible. Some of the more important ones are **DELETE, WINDOW, REDRAW, AUTO SCALE, FILE PART, BACK-1 WINDOW, UNDO,** and **RECALL LAST.**

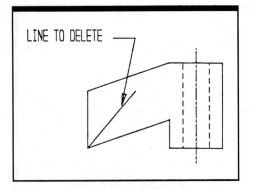

Figure 9-1 Deleting a line using
CTRL-Q

POSSIBLE USE

Figure 9-1 shows a simple drawing that has an incorrect line that must be deleted before the designer can continue drawing. To **DELETE** this line using function keys and the menu would involve the following keystrokes: **ESC, F8, F1, F1** (**DELETE, SELECT, SINGLE**) to delete the line and **ESC, F1, F1, F1, F3** to return to the previous menu position and start drawing lines again. The Immediate Mode can be used to **DELETE** the line by pressing **CTRL-Q**. After the line is deleted, you will automatically be returned to your previous position in the menu. This example would save you six keystrokes and much time.

Notice that many of the Immediate Mode commands use the first letter from the option that it will activate, for example, **CTRL-A** for **ARROW IN/OUT**, **ALT-A** for **AUTO SCALE**, **ALT-W** for **WINDOW**, and **ALT-P** for **PAN**. Most of the Immediate Modes are from Main Menu item **DISPLAY** and from the Status Window. Exceptions are **DELETE-SINGLE, FILES-PART, LEVEL-MASKING, LINE-LIMITS, ARROWS IN/OUT, MASKING** by type, **RECALL,** and **WITNESS LINE**. Shown below are those options from the **DISPLAY** Menu that are Immediate Mode commands.

Display
ZOOM-Double
Half
Window
Auto
Back-1
PAN
LEVEL
GRID
SNAP
REDRAW
VIEW

POSSIBLE USE

The Immediate Mode Command **UNDO Alt-I** can undo actions implemented using the **X-FORM, DIMENSION, TRIM/EXTEND,** and **BREAK** commands. Figure 9-2 shows a series of circles drawn using the **X-FORM** option **C-ARRAY**. Immediately after the circular array is drawn, the **UNDO** command is

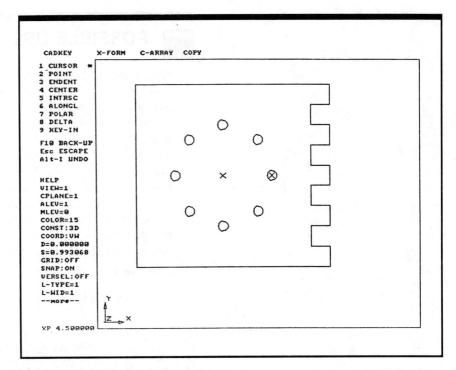

Figure 9-2 Circular array option is one command where UNDO will work

displayed in the Break Area of the screen below **ESC ESCAPE**. Selecting **Alt-I UNDO** from the screen or by entering **Alt-I** from the keyboard will undo the circular command (Figure 9-3).

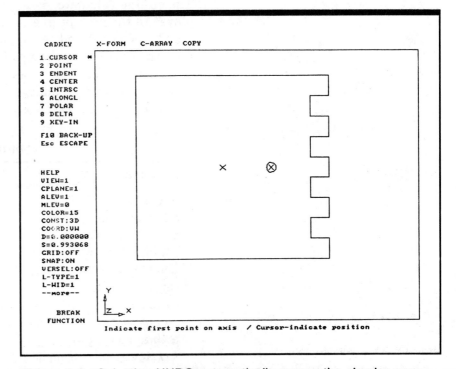

Figure 9-3 Selecting UNDO automatically erases the circular array.

POSSIBLE USE

One possible use for the Immediate Mode on a Status Window command is with the **PEN** command, which controls the assigned pen number given to an entity. It is usually changed before the entity is drawn. You could easily be in the middle of drawing lines when you find it necessary to change the pen number. This would make it very inconvenient to leave your present menu position. To change the pen number using function keys would mean following this sequence: **ESC, F7, F2, F4, F2**. Of course, you could also move the cursor into the Status Window and choose —**more**— and **PEN:** to change the pen number. However, the quickest method would be to use Immediate Mode command **ALT-Z**.

In addition to those commands listed above, other commands listed in the Status Window are activated using the Immediate Mode command. Status Window commands that are activated using the Immediate Mode are **VIEW, ALEV, COLOR, SNAP, TRACKING, CONST, L-TYPE, GRID, L-WID, PEN**, and **DATA BASE SEARCH**.

CALCULATOR

All of the CADKEY commands accessed using the Immediate Mode are explained in other parts of this text, except for **CALCULATOR**. The on-line calculator is used by entering **CTRL-I**. **CALC** is displayed in the prompt line ready for user input of calculations after **RETURN** is entered. After the expression is entered, the results are displayed in the Prompt Line. The precedent of mathematical operations is (1) math functions, (2) exponents, (3) multiplication and division, and (4) addition and subtraction. The math and logical operators that CADKEY recognizes are listed in the "Operators" table below.

OPERATORS

Math Operators	Logical Operators
a+b addition	**a==b** equals comparison
a-b subtraction	**a!=b** not equals comparison
a*b multiplication	**ab** greater than comparison
a/b division	**a** less than comparison
a^n nth power of a	**a=b** greater than or equal
(a) precedence	**a** less than or equal
-a negative a	**a&&b** logical AND operation
+a positive a	**allb** logical OR operation
a%b a mod b	**(a)?x:y** conditional operation
xxx=b variable assignment	

To add two numbers enter **CTRL-I** or press the **TAB** key. The prompt line displays **CALC>**. Enter your problem using the operators shown in the "Operators" table above. For example, to add two numbers, enter **476+924**. Spaces are ignored when entering operations. Press **RETURN** to display the answer, which is displayed in the prompt line as: **CALC. 1400 (Press any key. . .).** Pressing a key will put you back into the calculation mode. Press **ESC** to leave the calculation mode.

Mathematical equations are written similar to the C programming language. The "Calculator Functions" table below shows the mathematical functions that can be performed by the calculator. For example, to find the cosine of an angle, enter **CTRL-I** to start the calculator. Enter **cos(65)** and **RETURN** to determine the cosine of a 65 degree angle. The prompt line displays the answer as: **0:422618 (Press any key. . .).**

CALCULATOR FUNCTIONS

Value	Function
abs(x)	Absolute value of a number
acos(x)	Arc cosine
asin(x)	Arc sine
atan(x)	Arc tangent
atan2(y,x)	Arc tangent y/x
ceil(x)	Integer ceiling
cos(x)	Cosine
cosh(x)	Hyperbolic cosine
exp(x)	Natural exponential function
floor(x)	Integer floor
hypot(x,y)	Length of hypotenuse, sqrt (x*x+y*y)
log(x)	Natural logarithm
log10(x)	Base 10 logarithm
sin(x)	Sine
sinh(x)	Hyperbolic sine
sqrt(x)	Square root
tan(x)	Tangent
tanh(x)	Hyperbolic tangent

SUMMARY

The Immediate Mode is an extremely important method of invoking frequently used options for CADKEY. If you want to become a very efficient operator of CADKEY or any other CAD system, it is important that you learn the most effective method to use the system. Not only will you be an efficient operator, but you will find that you will become a more effective and less frustrated designer/drafter. Master Immediate Mode and you may never go back to using traditional drafting tools.

REVIEW QUESTIONS

1. Describe the meaning of a "tree" menu structure.
2. What problem is associated with the tree menu structure?
3. List the three methods of changing the current status or changing the active menu item on CADKEY.
4. How can an Immediate Mode command be interrupted?
5. For which commands will the **UNDO** option work?

The Position Menu

OBJECTIVES

After completing this chapter, you will be able to:

- Create or position entities using the cursor option.
- Create or position an entity at a point.
- Snap to the end of an existing entity.
- Snap to the center of a circle or arc.
- Snap to the intersection of lines, arcs, and circles.
- Define an entity's position relative to a distance along a line.
- Define a position using polar coordinates.
- Define a position using delta coordinates.
- Define a position using X, Y, and Z coordinate values.

The universal Position Menu gives you the accuracy needed for fine detail work by allowing you to locate entities at precise positions. For example, if a window were zoomed to magnify the intersection point of two entities, no break would be found between them, but the Position Menu options can assist you in positioning entities in exact locations. The Position Menu contains nine different methods or options of indicating the position of a point in space. The Position Menu can be found with many functions and will be displayed whenever you are requested to indicate a position for an entity. It should be noted that it is only a single position point that is being indicated. Generally, the creation of an entity requires several position points to be indicated. To create a line, two positions must be indicated: a beginning and an end point. With the use of the Position Menu these endpoints can be selected with a high degree of accuracy. The positions can be indicated by using existing entities and attaching to the center or the end of the entity or one of the other options. In the 2D sectional tutorial, a review of the different positional methods that can be used for creating a line can be found.

Once a method is selected from the Position Menu, that option will stay active or modal for every point selected in the function. The active method will have an asterisk (*) after it on the Screen Menu. A new method can be selected at any time when the Position Menu is displayed. The nine methods of selecting position are:

1. **CURSOR**
2. **POINT**
3. **ENDENT (End of Entity)**

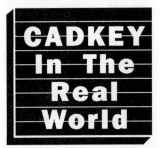

Ergonomics and Speed Skiing!

Ergonomics and CAD

Ueli Thomi and Martin Wyler used ANYBODY™ (Version 3.5) Ergonomic Stencil software to create, in CADKEY 3™ (Version 3.5), an animated, three-dimensional, CAD model of a speed skier wearing the Speed Skiing Performance Engineering Program (SS PEP's) newly designed, aerodynamic, speed-skiing helmet. What's more, they did it *live*, during the NAVY MICRO '90 trade show in San Diego, July 10-13, 1990.

Ueli Thomi and Martin Wyler of Industrial Technical Software, AG (INTESO) of Flamatt, Switzerland are international distributors for ANYBODY software developed by Industrial-Design Somatographic CAD Training (I.S.T.) of Gernsheim-am-Rhein, Germany. Ueli and Martin collaborated with Steve McKinney and Mellissa Dimino of the United States Speed Skiing Team and with Braxton Carter and Stephen Gubelmann of SS PEP to create the animation of the model with helmet, skis, and boots.

I.S.T. developed ANYBODY to work exclusively with CADKEY 3, as a seamless integration written in CADL™ (CADKEY Advanced Design Language).

ANYBODY features CAD models of men and women whose body types measure at the 50th percentile of normal physical types. It also includes models of body types in the 5th percentile and in the 95th percentile of normal physical types.

All of this human-related data incorporated into ANYBODY has produced human-model software modules that clearly tell a mechanical or architectural engineer whether or not the person for whom a product is being designed will be able to use it comfortably. If the man, woman, or child cannot move in a way that a designer would like, ANYBODY only allows the maximum movement that the person can perform in real life. The software also tells the designer what the limitations are for this particular movement by this type of person.

ANYBODY-CADKEY 3: Seamless Integration

After installation, ANYBODY becomes an extension of CADKEY 3's menus and cursor-selectable operations. The human models are CADKEY part files and pattern files. CADKEY macros connect ANYBODY's different modules. ANYBODY works with CADKEY's AutoSwap utility for processing large models.

A user creates an animation with ANYBODY by creating and saving a series of part files or pattern files with variations in body position. The ideal that 3-D CAD models of human beings created on personal computers should move in the same way as humans in real time is not yet possible given the present state of PC CAD technology. However, sequences of variations in the body position of the human model can create animations that very closely approximate actual human motion.

With ANYBODY (Version 3.5), the user can move joints in the human model by making a selection with the cursor in the body-icon menu that appears in the upper left of the screen. The human body's elements reside on different levels in the human model's part file, and parts of the body that are related in real life (e.g., right arm, left leg) are grouped together in the part file. The user can modify, scale, rotate, move, proportion body parts and joints. By using the cursor to select a joint on the body-icon, one can specify the movement not only of the individual joint, but also of the other parts and joints related to it.

Exertion Analysis

ANYBODY also has an exertion-analysis module to verify the level of a person's tolerable strain in the lifting and carrying of loads. The module indicates whether the task is within the capabilities of the human model whose body type and characteristics have been defined by the user.

3-D, wireframe CADKEY/ANYBODY model of speed skier wearing SS PEP's aerodynamic helmet, designed at Navy Micro '90.

Adapted with permission from **3-D WORLD**, *v. 4, n. 5, Sept./Oct. 1990.*

4. **CENTER**
5. **INTRSC (Intersection)**
6. **ALONGL (Along a Line)**
7. **POLAR**
8. **DELTA**
9. **KEYIN**

POSITION MENU OPTION:
CURSOR (F1)

The **CURSOR** option allows you to place an X,Y point that uses the current working depth. The point is placed by moving the screen cursor on the X and Y axes to a desired location and selecting that point. The point is selected by pressing the mouse button if you're using a mouse or digitizer, or by pressing the space bar if you're using the keyboard for input. The **CURSOR** option is the default option if no other Position Menu option is selected.

To know the exact X and Y coordinates for the position of the cursor, invoke the Immediate Mode command sequence **CTRL-T** (cursor tracking). The X and Y values for the position of the cursor will be continually displayed and updated in the lower left corner of the display monitor. The Z value can be found in the Status Window next to the letter "D" for the current working depth.

It is recommended that if you are planning to use the cursor option to indicate position, you should use the **SNAP** and **GRID** options. With the **SNAP** and **GRID** options on you are moving in fixed increments that you can specify, such as .125, rather than moving freely in space.

 POSSIBLE USE

The **CURSOR** option is helpful when creating basic geometry. The **CURSOR** option can be used effectively and accurately to define the outline of an object. By using the **LINE-STRING** option you can cursor-select the endpoints for the lines in the perimeter of the object in Figure 10-1. You can also use the **CURSOR** option to select the center point of a circle or a polygon or the beginning and endpoint for an arc.

POSITION MENU OPTION:
POINT (F2)

The **POINT** option searches out and locates a **POINT** entity that has been previously created. The **POINT** option uses the same X,Y,Z coordinate values as the **POINT** entity to locate the start, end or center point for the entity being created.

To position to a **POINT** move the screen cursor to the **POINT** entity you would like to locate and select it by pressing the cursor button or the space bar.

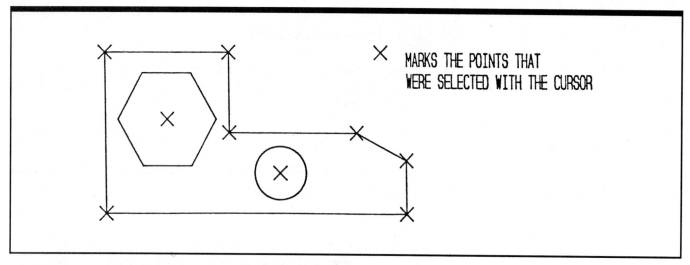

Figure 10-1   Positioning points using the CURSOR option

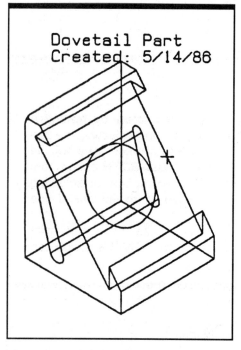

Figure 10-2   Changing the depth to the position of the point

If the construction switch is set to 3D, the X,Y,Z values for the position are returned. When the construction switch is set to 2D, only the X and Y values are returned. The Z value is the same as the current working depth.

▢ POSSIBLE USE

In the 3D dovetail tutorial, the **POINT** option is used to change the current working depth. A point is placed on an inclined surface at an unknown depth in Figure 10-2. By using the coordinate values of that point, the depth is changed to equal the depth of the inclined surface. This is done by using the Immediate Mode command sequence **CTRL-D, POSITN-POINT (F1-F2)**. This can be found in Task 14 in the 3D dovetail tutorial.

POSITION MENU OPTION: ENDENT (End of Entity) (F3)

This option locates the exact X,Y,Z values for the end of an entity and uses them for position.

This is particularly helpful when a line is to be drawn from one line to another or for replacing a deleted line.

When using the **ENDENT** option, select the entity desired near the endpoint that is desired. Once the entity is selected, an X will mark the endpoint chosen. If the wrong end is selected press **F10 BACKUP** to re-select the entity.

If a circle is selected, the endpoint is the same as the beginning point at 0 degrees or 3 o'clock (Figure 10-3).

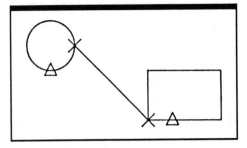

Figure 10-3   A line between the end
of a circle and the end of a line entity

POSSIBLE USE

Many times, you will find the need to draw a line/arc which begins at the endpoint of another entity. By using the **ENDENT** option, you will be sure that the line/arc begins exactly at the end of the entity (Figure 10-4). By simply selecting the entity, the correct position is found, rather than just eyeballing the endpoint and hoping they meet.

**POSITION MENU OPTION:
CENTER (F4)**

The **CENTER** option quickly locates the center point or midpoint of an entity. By selecting the entity desired, the center point is calculated and marked with an X.

When an arc/circle is selected, the center point is marked with an X. When a line is selected, its midpoint is marked with an X.

POSSIBLE USE

A line can be bisected quickly and accurately by placing a **POINT** entity using the **CENTER** option and selecting the line desired to be bisected (Figure 10-5B). Center lines will be placed quickly by using the **LINE-HRZ/VRT-BOTH** option from the Create Menu and using the **CENTER** option for selecting the center of the circle for the position of the horizontal and vertical lines (Figure 10-5A)

When drawing concentric circles, once the center point has been identified for the first circle the **CENTER** option can

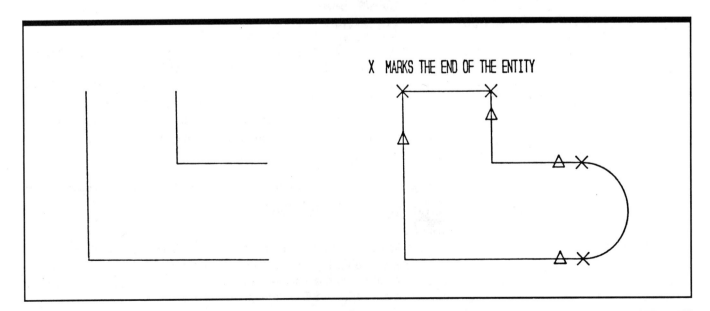

X MARKS THE END OF THE ENTITY

Figure 10-4   Positioning a point using the ENDENT option

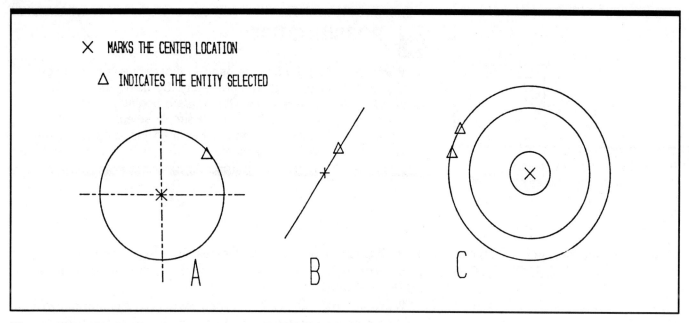

X MARKS THE CENTER LOCATION

△ INDICATES THE ENTITY SELECTED

A   B   C

Figure 10-5   Positioning points using the CENTER option

be chosen and the previous circle selected (Figure 10-5C) for indicating the center of the subsequent circles.

POSITION MENU OPTION: INTRSC (Intersection) (F5)

The **INTRSC** option is used to accurately position a beginning, center, or endpoint at the intersection of two entities. Some of the possible entity intersection combinations are:

> **Arc (Circle)/Arc (Circle)**
> **Arc (Circle)/Line**
> **Arc (Circle)/Spline**
> **Line/Line**
> **Line/Spline**
> **Point/Line**

The intersection of any of the above combinations can be used for position. If the entities do not actually intersect, the intersection point will be projected as if the entities actually did. If the construction switch is set to 2D, the entities selected do not need to lie in the same plane. If the construction switch is set to 3D, the entities do not have to lie in the same plane at the same depth.

 POSSIBLE USE

Positioning a polygon or circle at an intersection point, drawing a diagonal line across the corners of a rectangle, or placing a circle at the intersection of two entities are all possible uses of the **INTRSC** option (Figure 10-6).

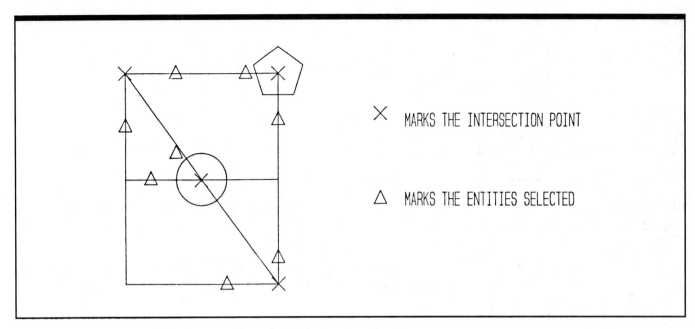

Figure 10-6 Positioning points using the INTRSC option

If there is no true intersection, and if the entities' projections onto the construction plane intersect, choose one of the following menu options to indicate the projections' intersection at a user-defined depth (with respect to the construction view):

> **FIRST**
> **SECOND**
> **CONST**
> **DEPTH**

FIRST—returns a position on the first entity or its logical extension.

SECOND—returns a position on the second entity or its logical extension.

CONST—returns a position on the construction plane (i.e., the same position that is returned when you set the construction switch to 2D).

DEPTH—returns a position at a user-defined depth.

POSITION MENU OPTION: ALONGL (Along a Line) (F6)

The **ALONGL** option is very helpful when drawing a line from a point located at a distance along a selected line entity. To do this, you must select the line nearest to the endpoint from which you wish to reference the distance. An X will mark the endpoint selected. If this is not the desired endpoint, press **F10 BACKUP**, and select again.

The line length will be displayed and you will be prompted to: **Enter distance (n.nn)=**. The value found in parentheses is the actual length of the line you have selected. Respond to this prompt with the value of the distance along the line you wish to locate the point. If a negative value is used, the point will be

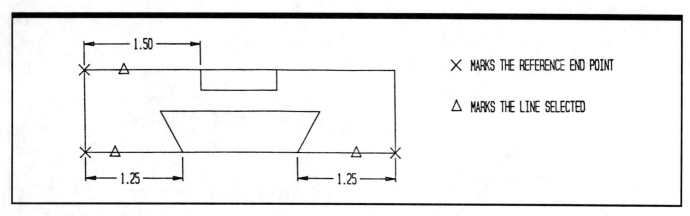

Figure 10-7   Positioning points along a line with the ALONGL option

positioned away from the line at that distance. If the length is longer, the position will be placed at a projected intersection.

POSSIBLE USE

In a 2D drafting mode, a line for the outside edge of a keyway is located at a distance in from an edge. To accurately draw this, the position for the first point on the keyway could be located using the **ALONGL** option. The edge points for the dovetail could also be located in this way (Figure 10-7). After the lines are drawn, the function **EDIT-TRM/EXT-DIVIDE** can be used to trim the middle portion of the line.

POSITION MENU OPTION: POLAR (F7)

The **POLAR** option locates a position at a specified angle and distance from an indicated reference position.

When you are first prompted to indicate the **POLAR** origin, the position menu will be displayed without the **POLAR** or **DELTA** options. You may use any of these options for indicating the **POLAR** origin or reference position.

Next, you will be prompted to enter the angle (n.nn) and to press **RETURN**. When entering the angle, use Figure 10-8 to help specify the angle.

Finally you must enter the distance away from the **POLAR** origin. If the **POLAR** option is being used to draw a line, the distance need not be exact. With the use of the **TRIM/EXTEND** option, the line can be trimmed or extended to provide a clean, accurate intersection.

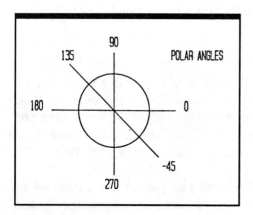

Figure 10-8   The polar angles are the same as the normal angle conventions.

POSSIBLE USE

The **POLAR** option can be used for creating a line at an angle. The dovetail in Figure 10-9 used the **POLAR** option to

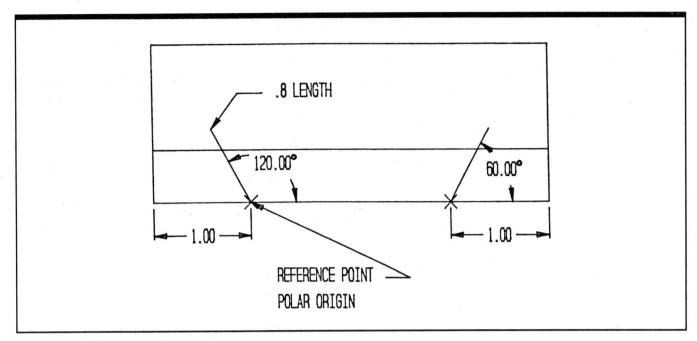

Figure 10-9   The endpoints of the dovetail lines are positioned using the POLAR option.

create the endpoints on the angled lines. The steps for creating a polar line are as follows:

1. Choose **CREATE** a **LINE** by **ENDPOINTS** (**F1-F1-F1**).
2. To indicate the first endpoint, use the **ALONGL** option (**F6**) to place the point one inch from the end.
3. Change the **POSITION** method for the second point to **POLAR** (**F7**).
4. When you are prompted for **POLAR** origin, again select the beginning point one inch from the end. You can use the **ALONGL** option to select this point again. You are selecting the same point that you selected in Step 2. This time, however, you are using this point as the **POLAR** origin to place the endpoint of the line.
5. You are then prompted to enter the angle of the position of the endpoint with reference to the origin. In this case, enter **120** degrees.
6. Next you will be prompted to enter the distance. In this case, enter **.8**. This distance is longer than needed, but you can trim the line later with the **TRIM** option.

POSITION MENU OPTION: DELTA (F8)

The **DELTA** option positions points relative to another reference position. The distance is entered in X,Y, and Z values and can be used in either the World or View Coordinate systems. You will be prompted to indicate a reference position which is similar to the **POLAR** option. The Position Menu will be displayed without the **POLAR** or **DELTA** option to assist you in selecting the reference position. After the reference position has been selected, you are prompted according to your coordinate mode. The prompts will appear as below:

dXV=   dYV=   dZV= for the View Coordinate system

dX=   dY=   dZ=   for the World Coordinate system

Each value is entered in inches or millimeters depending on which units you are using. Each value entered should be followed by a **RETURN**. The values entered should equal the distance away from the reference position that the new point will be on each of the axes.

An X will mark the point that has been located. If the position is undesirable, press the **F10** key to **BACKUP** and enter new values.

POSSIBLE USE

The **DELTA** option could be used for placing the center position of a circle at a desired location from a reference position on a corner (Figure 10-10).

POSITION MENU OPTION: KEY-IN (F9)

The **KEY-IN** option indicates a position relative to the origin, which has an X,Y,Z value of 0,0,0. Using a Cartesian coordinate system, the X,Y,Z position is entered from the keyboard. The X,Y,Z values can be entered using either the World or View Coordinates.

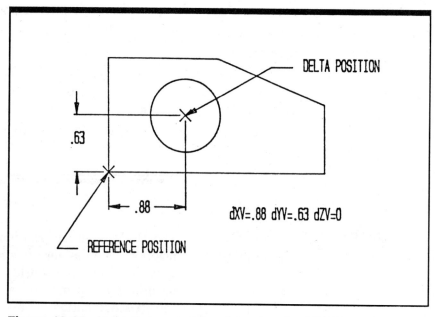

Figure 10-10 Indicating a center point using the DELTA option

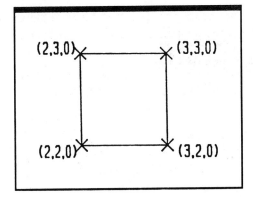

Figure 10-11 Indicating points with the KEY-IN option

POSSIBLE USE

This option can be used for entering and plotting points. To create a one-inch square, as in Figure 10-11, start with the beginning point 2,2,0 in View Coordinates. Key in the following points:

XV=2	XY=2	XZ=0
XV=2	XY=3	XZ=0
XV=3	XY=3	XZ=0
XV=3	XY=2	XZ=0
XV=2	XY=2	XZ=0

To enter the coordinates, either set of numeric keys on the keyboard can be used. You will be prompted to enter the X,Y,Z values for each point. The prompts appear as follows:

XV=	YV=	ZV=	for View Coordinates
X=	Y=	Z=	for World Coordinates

After each value entered, press **RETURN**. If the value is zero you can simply press **RETURN** because zero is the default value.

If a mistake is made when entering numbers, the backspace key may be used to back up one digit, or the **F10 BACKUP** key can be used to back up one coordinate value.

REVIEW QUESTIONS

1. When will the Position Menu be displayed?
2. What is the Position Menu used for?
3. Compare the **DELTA** option with the **KEY-IN** option.
4. Which option would be used if you wanted to indicate the start point of an arc at the endpoint of a line?
5. When would the **CENTER** option be used?
6. Which of the Position Menu options is the default option when no other option is selected?
7. When is the **ALONGL** option used?

CHAPTER 11

Selecting Entities

OBJECTIVES

After completing this chapter, you will be able to:

- Define the term *mask*.
- Define the function of the Selection Menu.
- Use the **UPDATE** function to change attributes.
- Use the Selection Menu to choose entities on a drawing.

For many CAD functions it is helpful to identify existing entities, such as the endpoint of a line. CADKEY gives you seven methods of identifying entities: **SINGLE, CHAIN, WINDOW, POLYGON, GROUP, PLANE**, and **ALL DSP** (all displayed). CADKEY refers to this as the Selection Menu. The Selection Menu (Figure 11-1) is a universal menu that is always available when you must select an entity. For example, when using the **DELETE** command from the menu, it is necessary to identify the entity or entities to be deleted by using the Selection Menu.

THE SELECTION MENU

The Selection Menu offers seven different methods of identifying and operating on existing entities:

SINGLE—allows single entities to be specified using the cursor.

CHAIN—allows the user to select a series of connected lines and arcs by defining the start and end entities in the chain.

WINDOW—allows the user to specify entities using a window (rubberbox).

POLYGON—allows the user to specify entities using a defined polygon up to twenty sides.

GROUP—allows the user to select those entities that have been grouped.

PLANE—allows the user to select entities by defining a plane using orientation and depth.

ALL DSP—allows the user to select all entities displayed on the screen, only those specified, or excluding all those specified.

The **F10** key can be used to reject those entities selected before **RETURN** is pressed.

The Selection Menu allows you to mask or specify precisely those entities to be deleted or to group those entities to be deleted using the **GROUP** or **ALL DSP** options. Some of many functions that use the Selection Menu are **UPDATE, X-FORM, DELETE-SELECT, CONTROL, EDIT-GROUP, FILES-PLOT, LEVELS-MOVE, FILES-PATTERN, SECTION, X-HATCH,** and **CADL-OUTPUT-SELECT.**

SELECTION SET LIST

A new option found on the Status Window of CADKEY version 6 is **SELLSET:ON/OFF**. If this option is turned off selection of entities within CADKEY will be performed as always. If the option is on you will now have the ability to make multiple selections using any of the seven selection methods. You may select with a **WINDOW** and find that not all of the entities you wanted were selected. You could then go to the **SINGLE** method and select the remaining entities. After you have made a selection the following menu will appear:

ADD

allows the user to make additional selections using any of the selection methods

SUBTRACT

allows the user to remove any of the unwanted entities in the current selection set

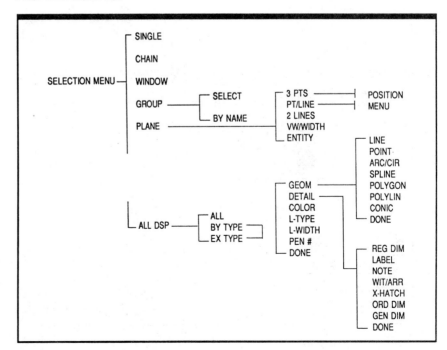

Figure 11-1 The Selection Menu

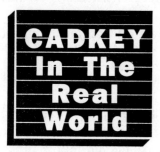

SWECOMEX Uses CADKEY 3™ To Design Heat Exchangers Used Around The World!

SWECOMEX, S.A., of Guadalajara, Jalisco, Mexico, has been designing and manufacturing heat-transfer process equipment for more than thirty years. For more than three years, SWECOMEX's engineers have been using CADKEY® to design their heat exchangers.

A heat exchanger is any device, such as a radiator, for transferring heat energy from a warmer medium to a cooler medium. SWECOMEX custom designs many different types of heat exchangers. They can range from one tube inside another with a small exchange surface to complex condensers for the discharge of steam from turbines in which there can be many square miles of exchange surface. Between these two extremes lies a very wide gamut of shell-and-tube heat exchangers, each identified by distinctive design characteristics, for example, U-shaped tube bundles, finned tubes, fixed tube sheets, floating head, etc.

The design of heat exchangers involves two essential elements: (1) thermodynamic and fluid-processing requirements, and (2) mechanical requirements. The process requirements determine the mechanical requirements. SWECOMEX's Engineering Department is responsible for the mechanical design of all the equipment that the company manufactures.

CADKEY in Heat-Exchanger Design

SWECOMEX typically uses CADKEY in designing its heat exchangers to verify that the number of tubes required by the thermal design will physically fit inside the internal diameter of the heat exchanger's shell. A designer creates a circle to represent the limit of the tubes that the shell can contain and identifies it as the Limit Tube Circle (LTC). He or she displays a grid with one of four standard, circular tube patterns inside the LTC: (1) a triangle with 30-degree angles, (2) a triangle with 60-degree angles, (3) a square with four 90-degree angles, or (4) a diamond (a square rotated 45 degrees). The designer aligns the snap to the grid and begins to add circles representing peripheral tubes into the LTC using the grid. Then he or she verifies that all of the tubes are contained inside the Limit Tube Circle (shell), and that the number of tubes corresponds to the quantity required by the thermal design. This design becomes a template.

After the template is dimensioned, the next step is to design the tube bundles inside the heat exchanger, and to generate, through a computer program, the calculations that yield the thickness and dimensions of all of the heat exchanger's components: shell, channel side, flanges, tube sheets, covers, heads, nozzle, reinforcement, supports, etc. The designer dimensions the equipment using the results of these calculations, and also taking into consideration the standard dimensions imposed by the plant's manufacturing facilities.

Now a draftsman prepares a preliminary bill of materials and uses CADKEY to create all the part files that will become the plant's manufacturing drawings. The draftsman enters all the customer's data, final dimensions and material specifications into CADKEY through a computer

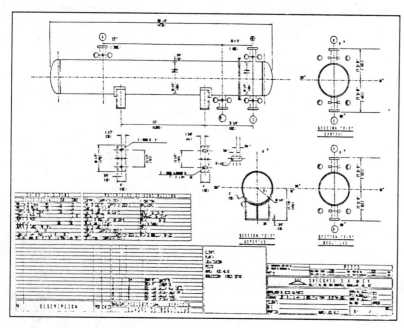

A refrigeration unit to cool sulfuric acid

—Continued, next page

CADKEY In The Real World—Continued

program that creates a CADL file containing all the text, circles, arcs, lines, points, polylines and other primitive entities to be included in the design. He or she then calls up a part file that is a multi-level *master file* for the particular type of heat exchanger being

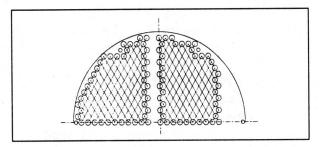

A heat-exchanger tube template

designed, and executes the CADL file containing the specific data for this design. All of the data entities appear in their proper locations in the master file. Using different combinations of levels in a master file, along with the program that generates the CADL file containing all the coordinate data, allows considerable flexibility in the design process.

After the master file has been reviewed and any necessary corrections made, the design is plotted in A or B size on a Houston Instruments DMP-40 plotter. If the draftsman needs to plot all of the master files at one time, he or she saves each one as a pattern file and they are all plotted later, in a batch mode, on a Zericon 4800 A-E plotter that handles paper up to size E.

Adapted with permission from 3-D World, v. 4, no. 4, July/Aug., 1990.

SHOW

will highlight all of the entities in the current selection set

DONE

must be picked when you are finished selecting entities

THE SINGLE OPTION

SINGLE allows the user to select individual entities with the cursor. Each entity selected is marked with a triangle or oval, or it may simply flash, depending on the assigned marker in the **CONFIG** program. The prompt line keeps track of the number of selected entities and will warn you if the entity selected is part of a group. If a group is chosen, you can select all entities in the group, select all entities in a subgroup, or select an individual entity in the group.

When the selection is complete, press **RETURN**. If an entity is chosen by mistake, the **F10** key is pressed to reject the selection. A pound sign (#) marks the canceled selection. If **F10** is selected twice in a row, the **SINGLE** option is canceled. A common use of the **SINGLE** option is with the Immediate Mode command **CTRL-Q**, used to delete entities.

USING THE CURSOR TO SELECT ENTITIES

The screen cursor is used to pick or select entities with CADKEY. The proximity of the entity to the cursor determines whether it will be found and how long it takes. A circular active area surrounds the cursor (Figure 11-2). To increase selection speed, the center of the cursor should be located within .1 inch of the entity.

THE CHAIN OPTION

CHAIN is used to select a series of connected entities, such as lines, splines, or arcs, by defining the start and end entities. The first step in the **CHAIN** option is to select the start entity. The direction of the chain is then selected with the cursor. The end of the chain is then selected by cursor-selecting the end entity and pressing **RETURN** to select all the entities in the chain. The last entity in the chain displayed on screen is automatically selected as the end entity. A **YES/NO** menu is displayed to accept or reject the defined chain.

MASKING

Once the Selection Menu is displayed, a selection mask is automatically set. A *mask* is a method of screening or designating certain entity types when making a selection. For example, if you had a very complicated part and wanted to delete the dimensions from the drawing, this would be a very difficult and time-consuming task if you were to use the **DELETE-SINGLE** function. You would have to selectively pick each dimension to delete it. If you masked the dimensions, you could delete all of them at once. It is important to note that the masking stays in effect only for the function selected and it returns to normal after the masking selection is completed. The mask option is explained in detail in Chapter 12.

**THE UPDATE
SELECTION MENU**

One of the many different functions that use the Selection Menu is the **UPDATE** option found under the **DETAIL** function. **UPDATE** changes certain attributes to the current modal value. For example, if you had a drawing with two or three lines of notes and you wanted to change the height and font of one of the lines, it could easily be accomplished using the **UPDATE** option and the Selection Menu. The alternative would be to erase the note, change the modal values of notes, and then use the **DETAIL-NOTE** function to replace the note on the drawing.

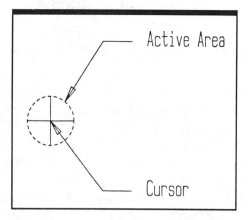

Figure 11-2   The active cursor area

 POSSIBLE USE

Figure 11-3 shows a drawing with two lines of notes. One line must be changed from .30 height to .20 height and the font changed from Box to Slanted Bold.

1. After the note parameters are changed using the **SET** option under the **DETAIL** function, press **ESC**.
2. Select **F3 DETAIL**.
3. Select **F8 UPDATE**.
4. The Selection Menu is now displayed. Select **F1 SINGLE**.
5. Pick the note to be changed, which, in this example, is **1/4" CORK GASKET**.

The note is immediately updated to the current modal values that were set to .20 height and Slanted Bold (Figure 11-4).

**THE X-FORM
SELECTION MENU**

The **X-FORM** function also uses the Selection Menu. A possible use of the **X-FORM** Selection Menu is illustrated in Figures 11-5 and 11-6. The steps to follow in this example are shown below. In this example, the **MIRROR** option is used to demonstrate the Selection Menu.

1. Select **F4 X-FORM**.
2. Select **F5 MIRROR**.
3. Select **F2 COPY**.
4. The Selection Menu is now displayed. Select **F7 ALL DSP (All Displayed)**.
5. Select **F1 ALL**. Two points on the object are picked (Figure 11-5) and the object is mirrored (Figure 11-6).

THE DELETE-SELECT SELECTION MENU

The Selection Menu used with the **DELETE** function gives the user many different options when entities must be erased on complicated parts and drawings. The last thing you want as a CAD operator is to draw entities and then have to unnecessarily delete them because the CAD software is inadequate for some deleting options. **DELETE** is illustrated in the following steps and in Figures 11-6 and 11-7, which demonstrate the **CHAIN** option of the Selection Menu. This option selectively deletes five connected lines from the top of the drawing.

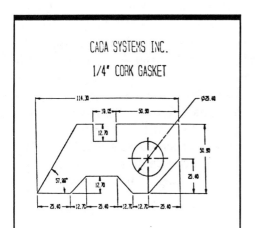

Figure 11-3   Drawing showing two lines of text

1. Select **F8 DELETE**.
2. Select **F1 SELECT**.
3. The Selection Menu is displayed. Select **F2 CHAIN**.
4. When this option is chosen, you are prompted:

   Select start entity in chain.

   Move the cursor to the first entity in the string to delete and press the mouse button. A small triangle will be displayed on the line to verify your choice. If this is not the line you wanted press **F10 BACKUP**, and a small "**x**" will appear over the triangle indicating that the choice has been canceled.

5. After the entity is picked, the system prompts:

   Cursor-indicate chaining direction

   Move the cursor in the direction you want the chaining process to move. In this example we want the chaining process to move to the right of the picked line. Again, to cancel the direction chosen, simply select menu item **F10 BACKUP**.

6. After the direction is specified you are prompted:

   Select termination entity in chain or press RETURN

   Move the cursor to the last entity to be deleted in the chain and press the mouse button. Pressing **RETURN** before selecting the termination entity causes the system to delete all the lines and arcs making up the chain. If an entity is selected for the end of the chain, then the entities making up the defined chain will be deleted. Entities from the start and endpoints of the chain are marked with a small triangle.

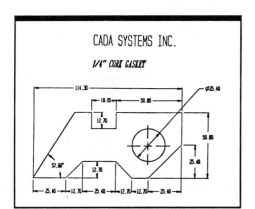

Figure 11-4   Drawing after the notes have been changed

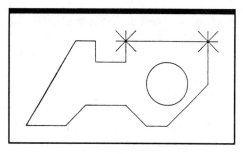

Figure 11-5 Drawing to be mirrored

THE DEGROUP AND PATTERN-CREATE SELECTION MENUS

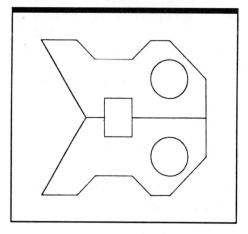

Figure 11-6 Drawing after MIRROR-ALL is used

THE WINDOW OPTION

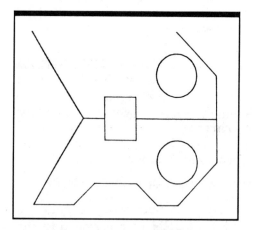

Figure 11-7 Use the DIVIDE mode to enhance the dimensions and text.

7. You are then prompted:

Selection OK?

A **NO/YES** menu appears. Select **YES** to delete those entities marked with the small triangle. A **NO** selection cancels the selected entities and allows you to re-define the chained entities. Figure 11-7 shows the result of the **CHAIN DELETE** function.

8. Press **ESC** to exit this function.

Two other menu options use the Selection Menu: **DEGROUP** and **PATTERN-CREATE**. The Selection Menu used with the **DEGROUP** option allows the user to specify those entities in a group to be removed from the group without removing the group and subgroup references. For example, suppose that a group is created that contained the entities lines, arcs, dimensions, and notes. You then decide that lines and arcs should be removed from the group for a selective delete. The Selection Menu under the **EDIT GROUP DEGROUP** option would allow you to remove the lines and arcs from the group.

In the same way, the Selection Menu can be used to create a pattern file. A part may be created that can be used as a pattern, but it may contain entities that you do not want as part of the pattern file. For example, Figure 11-3, the Cork Gasket, had a 1-inch diameter punched hole. You may decide that this gasket can be used as a pattern file except the hole size may vary for different applications. A pattern file could be created without the hole and dimensions by using the **FILES-PATTERN-CREATE** selective menu function. The Selection Menu function could be used to identify only those entities that should be part of the pattern file.

Up to this point, we have concentrated on those menu options that use the Selection Menu. In the examples given, Selection Menu options **ALL, SINGLE, CHAIN,** and **MASKING** have been illustrated. The **WINDOW** option of the Selection Menu is used in a way similar to the **WINDOW** option under **DISPLAY**. Entities can be selected by defining diagonal corners of a box. All entities falling completely within the defined window will be identified if **ALL IN** is selected.

 POSSIBLE USE

In Figure 11-8 you may decide that the isometric view of the part is unnecessary for the detail drawing. By using the **WINDOW** option with the **DELETE** function you could define a window which includes the isometric view of the part and delete all entities in the window. These steps are illustrated below.

1. Select menu item **F8 DELETE**.
2. Select menu item **F1 SELECT**.

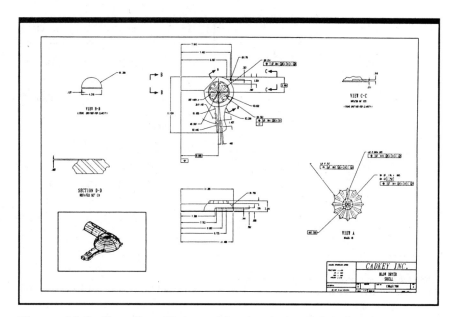

Figure 11-8   Drawing with isometric view to be deleted

3. The Selection Menu will now be displayed. Select menu item **F3 WINDOW, F1 ALL IN**.
4. You are prompted:

Indicate position for 1st selection window corner

Move the cursor to the lower left corner of the window needed to completely surround the isometric view of the part in Figure 11-8 and press the mouse button.

5. You are now prompted:

Indicate position for 2nd selection window corner.

Move the cursor to the upper right corner of the window. A rubberbox will identify the size of the window as you move the cursor to the upper right corner of the window. Press the space bar or the mouse button when the isometric view is within the window. The isometric view is deleted after the second point is picked.

THE GROUP OPTION

The **GROUP** option of the Selection Menu allows the user to select grouped entities. When this option is used, a menu is displayed with two choices: **SELECT** and **BY NAME**. **SELECT** allows for the selection of a subgroup of entities by cursor-indicating any entity belonging to the subgroup. For example, if a subgroup were made of dimensions, to select this group pick any dimension making up the subgroup. The **GROUP** option can be used with any of the functions that use the Selection Menu, such as **DELETE** or **EDIT**.

THE PLANE OPTION

The **PLANE** option defines entities within a plane using orientation and depth for selection. This option is especially useful when selecting entities from a 3D wireframe model. Only those entities

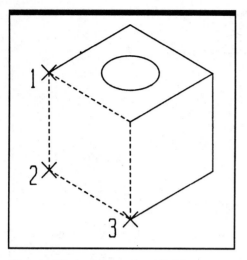

Figure 11-9 Using the 3PTS method to define a plane

that lie completely within the defined plane are selected. CAD-KEY provides five different methods of using the **PLANE** option.

The **3 PTS** method defines a plane using three positions in space (Figure 11-9).

The **PT/LINE** method defines a plane by picking a position and a selected line entity (Figure 11-10).

The **2 LINES** method defines a plane by picking the end-points of intersecting lines (Figure 11-11).

The **VW/DPTH** method defines a plane by selecting a specified view and the depth. The depth is defined by entering a value or picking a position (Figure 11-12).

The **ENTITY** method defines a plane by picking a line, arc, or 2D spline. If a line is picked, it will define the plane as being normal to the screen and at the same inclination as the line projected onto the screen. Picking an arc or spline defines the plane that contains the entity at the depth of the entity (Figure 11-13).

THE ALL DSP (ALL DISPLAYED) OPTION

ALL DSP allows for all the displayed entities to be picked or only those displayed entities you choose. **ALL DSP** offers three options: **ALL**, **BY TYPE**, and **EX TYPE**.

ALL selects all the entities displayed, including those that are partially displayed.

BY TYPE selects only those entities that you select from a menu of seven items.

EX TYPE selects all entities except those that you select from a menu of seven items. The selection type menu used with **BY TYPE** and **EX TYPE** is:

> **GEOM**
> **DETAIL**
> **COLOR**
> **L-TYPE**
> **L-WIDTH**
> **PEN #**
> **DONE**

Each of these items has a submenu (see Figure 11-1) to further specify entities.

In CADKEY 6 the hierarchial style of masking menu has been replaced by a much more understandable and easier to use Set Mask dialog box (Figure 11-14). In this dialog box the user simply selects all of the desired attributes and entity types on which to mask. If the user is masking **BY TYPE** all of the selected entities will be affected by the command. If the user is masking **EX TYPE** all of the selected entities will not be affected by the command.

THE POLYGON OPTION

The **POLYGON** option is used to define a polygon shape to select entities. With this option you can create a multisided polygon to select **ALL IN**, **PART IN**, **ALL OUT**, or **PART OUT**. Figures 7-15 and 7-16 describe how to use the **POLYGON** option.

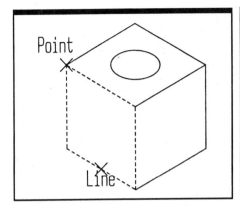

Figure 11-10   Using the PT/LINE method to define a plane

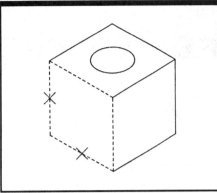

Figure 11-11   Using the 2 LINES method to define a plane

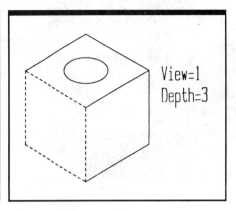

Figure 11-12   Using the VW/ DEPTH method to define a plane

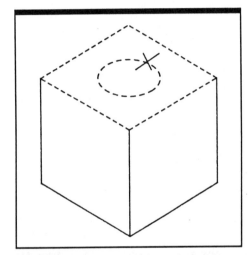

Figure 11-13   Using the ENTITY method to define a plane

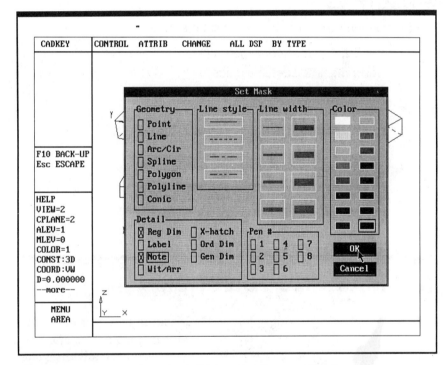

Figure 11-14   The Set Mask Dialog Box for version 6

REVIEW QUESTIONS

1. Define the term "mask" as used by CADKEY.
2. List the three options under **ALL DSP**.
3. List the two options used with **GROUP** to select entities.
4. What option can be used to change the height and font of a line of text without erasing the text?
5. What is the size of the active area around a cursor?
6. List the four options for **WINDOW**.

CHAPTER 12

The Masking Menu

OBJECTIVES

After completing this chapter, you will be able to:

- Set a mask for geometric entities.
- Set a mask for detail entities.
- Set a mask for more than one type of entity.
- Set a mask by color.
- Set a mask by line type.
- Set a mask by line width.
- Set a mask by pen number.

THE MASKING FUNCTION

Masking is a part of CADKEY that some people never take the time to learn. Do not be one of those people! If you don't use masking you are not taking advantage of one of the greatest and most helpful features of a CAD system. Masking is a method of selecting specific entities within a part that will only permit the entities for which the mask is set to be selected. The mask acts as a filter that will sort out and select only the type of entities for which the mask has been set. When entities are being selected by type, this is referred to as *type masking*. Masking should be used in conjunction with the Selection Menu discussed in Chapter 11.

When choosing the **ALL DSP** (all displayed) option from the Selection Menu, you then have the opportunity to mask **BY TYPE** or **EX TYPE** (except type). The **BY TYPE** option will only allow you to select those entities, details, or entities with the attributes that you have selected from the Masking Menu. The **EX TYPE** option will allow you to select any entities *except* those that have been selected from the masking menu. You can mask with any of the Selection Menu options whenever you have to select an entity. A commonly used combination is **CTRL-Q** (Single Delete) and **ALT-M** (masking type), which allows you to select specific entity types individually. You have the option of masking by entity or attribute or a combination of both. To set a type mask you can press the Immediate Mode sequence (**ALT-M**) which will display the Main Masking Menu (Figure 12-2) or Masking dialog box (Figure 12-1).

In CADKEY 6, the hierarchial style of masking menu has been replaced by a much more understandable and easier to use Set Mask dialog box (Figure 12-1). In this dialog box the user simply selects all of the desired attributes and entity types on which to mask. If the user is masking **BY TYPE**, all of the selected entities will be affected by the command. If the user is masking **EX TYPE**, all of the selected entities will not be affected by the command.

ENTITY MASKING

The first two options on the Main Masking Menu, **GEOM** and **DETAIL**, are considered Entity Masks because they mask geometric and detail entities which are generated using the **CREATE** and **DETAIL** options on CADKEY's Main Menu.

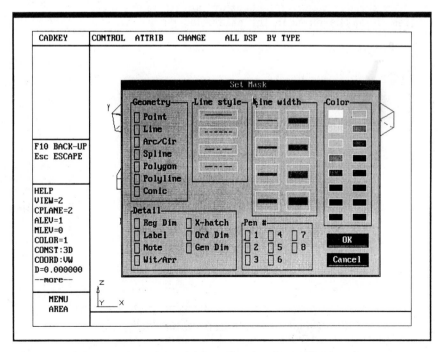

Figure 12-1   The Set Mask Dialog Box for CADKEY Version 6

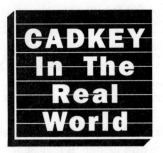

Sound Makes Martin Guitars "the Stradivarius of Guitars"

C.F. Martin Company of Nazareth, Pennsylvania, travels in some very elite circles: historical, musical, and technological. C.F. Martin is one of a handful of companies in the United States that still operates under the active management of its original founding family, after more than 150 years. Its products, acoustic guitars, are renowned around the world for the quality of their sound and craftsmanship. Their sound has given Martin guitars the nickname the "the Stradivarius of guitars." The guitar makers of C.F. Martin Company combine their age-old tradition of high-quality hand craftsmanship with CADKEY to produce even better guitars.

Hand Craftsmanship, CAD and CAM?

"We still make all of our guitars by hand," says Bob Headman, mechanical engineer at C.F. Martin. "However, we use CADKEY to design the tooling to make the parts that go into a guitar." To design the tooling to make the parts requires designing the guitar first. Some ninety parts, not counting ornamentation, go into the making of a guitar. Nearly all of these parts are made of wood. "Our major use for CADKEY is to quantify what has historically been done through hand craftsmanship, and to improve the fit of the parts in the assembled guitar. We use CADKEY's 2-D and 3-D splines a lot. We use CADKEY to define the surface area that we need to cut. Then we use CADKEY to design the tooling, the fixtures, and the cutters that we need to do the cutting." C.F. Martin has its own computer-numerical-control machine shop to make the tooling and fixtures. Their CADKEY files serve as input to the CNC system.

Curves and More Curves

"There are relatively few straight lines on a guitar," Bob added. "Everywhere a guitar has curves which often have very subtle radii, so subtle that the human eye frequently does not perceive them. But, these radii affect the aesthetic appearance of the guitar, its tonality through the stiffness of the wood, and the way it plays in the hands of the musician. Many companies make guitars. The major difference in our guitars is the sound. We sell sound. Our guitars have become the standard definition of the acoustic guitar."

Bob gave examples of the subtle radii which affect the way a guitar sounds. "For example, a flat-topped guitar is not actually flat-topped," he said. "There is an arc to the top whose radius is, in reality, 42 feet long. You need to use a spline to design that. Another example is the back of our Dreadnought guitar. The perimeter of the body on the Dreadnought's back surface is a 62-inch 3-D spline. A third example is the guitar's bridge. The bridge is a matrix combination of 3-D splines both in the X and Y planes."

On an acoustic guitar, the interplay of the vibration of each string when plucked by the musician, the transmission of this vibration through the bridge into the wood of the guitar's top, the vibration of the top of the guitar's body, the reverberation of this vibration inside the guitar's body, the tonal qualities of the wood used in the guitar's body, the shape and depth of the body, the size and location of the bridge and sound hole, even the varnish, all affect the clarity, range, resonance, and volume of the guitar's tone. A deep body tends to emphasize the bass sounds and to increase volume. A shallow body tends to emphasize the treble sounds and to diminish volume. Producing the correct tonal balance in a guitar's performance depends upon the type of music to be played, as well as on the

Martin Dreadnought guitar, crafted by hand, with tooling designed using CADKEY.

—Continued, next page

CADKEY In The Real World—Continued

tastes of the individual musician. This interplay has led C.F. Martin to produce twenty-five different models of guitars made from varying combinations of ten different structurally sound acoustic woods. "We are

working at the extreme high end of woodworking craftsmanship," Bob said.

"We work our wood within plus or minus five one-thousandths of an inch," Bob continued. "We hold the holes and slots to even tighter tolerances. For example, the slots where the frets are inserted into the fingerboard are held to one one-thousandth of an inch. We have special saws to cut these slots, and we check the slots with feeler gauges."

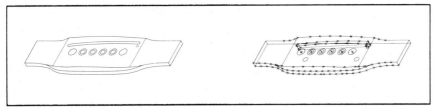

A six-string guitar bridge. A simple part of lines and arcs? No, a complex part of 3-D splines.

Adapted with permission from **3-D WORLD**, *v. 6, no. 2, 1992.*

MASKING OPTION: GEOM (Geometric Entities)

This option will allow you to select one or more geometric entity types at a time for masking. Simply select the entity type(s) for masking from the menu. Press **DONE** when the mask is complete or **RETURN** to display the Main Masking Menu to add other options to your mask. The entity options found under **GEOM** are:

> **POINT**
> **LINE**
> **ARC/CIR (Circle)**
> **SPLINE**
> **POLYGON**
> **POLYLIN**
> **CONIC**
> **DONE**

MASKING OPTION: DETAIL

This option will allow you to select one or more **DETAIL** entity types at a time for masking. Simply select the entity type(s) for masking from the menu. Press **DONE** when the mask is complete or **RETURN** to display the Main Masking Menu to add other options to your mask. The entity options found under **DETAIL** are:

> **REG DIM (Linear, Angular, Radial, and Diameter Dimensions)**
> **LABEL**
> **NOTE**
> **WIT/ARR (Witness lines, Arrows)**
> **X-HATCH**
> **ORD DIM (Ordinate Dimensions)**
> **GEN DIM (Generic Dimensions)**
> **DONE**

ATTRIBUTE MASKING

Options 3 through 6 on the Main Masking Menu are considered Attribute Masks because they represent the four attributes that all entities have, **COLOR**, **L-TYPE**, **L-WIDTH**, and **PEN #**. Attribute masking is a new feature that has added a great amount of power and flexibility to masking.

MASKING OPTION: COLOR This option will allow you to mask entities by color. When selected, the color bar is displayed and you select the color(s) that you wish to mask. Markers will appear next to the colors which are active in the mask. If the color is selected a second time, the marker will no longer be displayed and the color will no longer be active in the mask.

MASKING OPTION: This option will allow you to mask entities by line type. When se-
L-TYPE (Line Type) lected, the line type menu is displayed and you select the line type(s) that you wish to mask. Markers will appear next to the line type(s) which are active in the mask. If the line type is selected a second time, the marker will no longer be displayed and the line type will no longer be active in the mask.

MASKING OPTION: L-WIDTH This option will allow you to mask entities by line width. When you
(Line Width) select it, you must enter the odd number of the line width(s) that

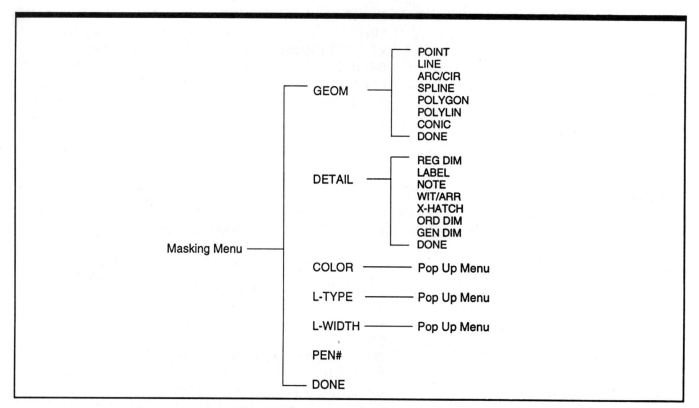

Figure 12-2 The Masking Menu

you wish to mask. A hyphen can be used between numbers to indicate a range of widths, such as 3-9. If the line width is entered a second time, then it will no longer be displayed and the line width will no longer be active in the mask.

MASKING OPTION:
PEN # (Pen Number)
This option will allow you to mask entities by pen number. When you select it, you must enter the pen number(s) that you wish to mask. A hyphen can be used between pen numbers to indicate a range of pens, such as 1-5. If the line width is entered a second time, then it will no longer be displayed and the pen number will no longer be active in the mask.

A single mask can be set to select only one entity type, or a multiple mask can be set that includes a combination of any of the entity type or attribute choices from the Main Masking Menu. To set a mask, simply choose the options from the Masking Menu that you plan to select from your part/drawing. All of the active options selected in the masking sub-menu will be listed on the prompt line. If an option is chosen that is already displayed on the prompt line, that option will become inactive. Once you have completed your masking choices on one of the sub-menus, press **RETURN**. Do not press **DONE** on any of the menus until you wish to execute the current function. If you are using multiple masking options, press **RETURN** at the end of each sub-menu until you have set the desired mask. By pressing **RETURN**, your active masking options are temporarily stored and you are returned to the Main Masking Menu to select another masking option. Once the desired mask is set and you are returned to the Main Masking Menu, then press **DONE** to execute the current function.

Even though all options always appear on the Masking Menu, not all options are available for every function. The mask that you have set will only remain active for the function you are currently using. Once you leave that function the mask is discontinued so that all items again become selectable. In many cases, the Masking Menu can be found within another function; an example is the **DELETE** function.

 POSSIBLE USE

Many times in a 3D part, there are multiple entities very close to or overlapping one another. When you want to select one of these entities it can be difficult without the aid of masking. If there were several lines running through a circle and you wanted to select that circle but not the lines, a mask would be set on the circle and only the circle would be selected. For large parts or parts with many closely placed entities, the speed and reliability of selection are greatly increased by using a mask.

During the construction of a part POINT entities are often used to locate strategic placements for other entities. After the entities have been placed the points must be deleted. To delete the points without deleting the entities on top of them can be time consuming. With the aid of a mask on

POINT entities you can **DELETE**, select **ALL DSP, BY TYPE, GEOM POINT, DONE**.

The masking menu is also very helpful when selecting entities to change their attributes. For instance, you may want to change the color of all dimensions. By invoking the function sequence **CONTROL, ATTRIB, COLOR, CHANGE** and selecting **ALL DSP, BY TYPE, DETAIL** with a mask on **REG DIM** arcs and circles, the color of all of the dimensions will be changed to the color selected.

Another time when type masking is very useful is in the deletion of **NOTES, DIMENSIONS,** and **LABELS** that have been placed on top of, or very close to, other entities.

Some people like to keep different line types on different levels. With the use of attribute masking you can easily select any line type to be moved to a level all its own.

LEVEL MASKING

Level masking is also an aid in selection. It gives the user the ability to designate the level from which he wishes to select entities. Once a level mask has been set, the only entities that can be selected are those on the level being masked. The current level that is being masked can be found in the status window on the left of your display monitor, next to **MLEV**. It can also be accessed through the Immediate Mode command sequence **ALT-N**. The level on which you are masking can be changed at any time by pressing **ALT-N** or selecting **MLEV** from the Status Window. If no level masking is desired, the **MLEV** should be set to zero. This allows free selection from all the displayed levels.

There are 256 available levels. If the mask level is set to a positive number between 1 and 256, only that level number may be selected from, regardless of how many other levels are displayed. If the mask level is set to a negative number, the selection is possible from all displayed levels except the negative number.

 POSSIBLE USE

As with type masking, level masking is very helpful with large, complicated drawings with multiple entities on multiple levels. Often, when using different levels, entities are stacked on top of one another. If the selection of an entity on a specific level is desired, it can be achieved by setting a level mask. If all of the hidden lines in a particular drawing were placed on one level and the lines were very close to entities on other levels (Figure 12-3), a level mask could be used to aid in the selection and manipulation of the hidden lines.

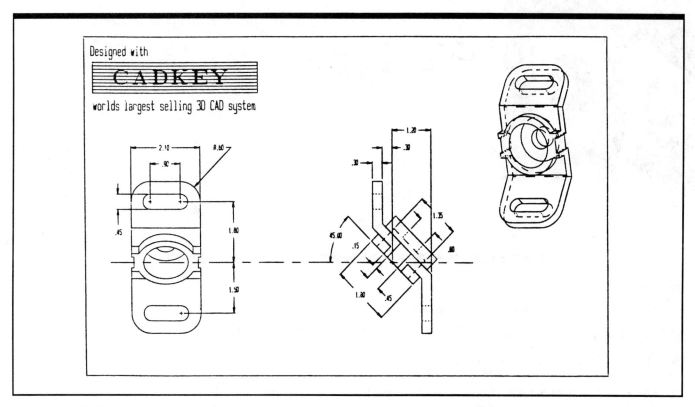

Figure 12-3 Bracket with hidden lines on a separate level (Courtesy of Cadkey, Inc.)

REVIEW QUESTIONS

1. What is masking?
2. When would masking be used?
3. What are three types of masks that can be set?
4. If a level mask were set for level 2, would you be able to select entities on level 1?
5. If you set the **EX TYPE** mask for **GEOM-ARC/CIR** and **POINT,** would you be able to select a circle?
6. If you set the **BY TYPE** mask for **DETAIL-NOTE,** would you be able to select a **NOTE?**

CHAPTER 13

File Types

OBJECTIVES

After completing this chapter, you will be able to:

- Create, load, and list part files.
- Create, retrieve, and list part files.
- Create, execute, and list CADL files.
- Create a plot file.
- Create and read a DXF file.
- Create a macro file.
- Create and retrieve an IGES file.
- Create and retrieve a DWG file (version 6).

CADKEY utilizes several different types of files in its file management system. A part can be stored in seven different file formats: **PART, PATTERN, PLOT, CADL, IGES, DXF**, and **DWG**. Two other types of files are **MACRO** and **CDE**. Each of these files has a specific

```
D:\CK35\PTN dir b:

 Volume in drive B has no label
 Directory of   B:\

CARB       PRT    102781    7-22-86     4:48a
CAR        PRT     30240    7-15-86     4:39a
GLASTRON   PRT     63894   10-07-86     8:34a
TAM2       PRT     19504    7-29-86    10:23a
FORMR2     PRT     49828    1-01-80     1:49a
TEST       PTN       580    4-29-89    12:33p
BRACKET    PRT     19762    7-23-86     2:32a
DOVE5      PTN      4856    5-06-89    12:26p
SOUTH      PTN      3708    4-08-90    12:20p
DOVE2      PTN      6154    5-06-89    12:25p
DOVE7      PTN      4856    5-06-89    12:25p
DOVE1      PTN      4856    5-06-89    12:24p
        12 File(s)       45056 bytes free

D:\CK35\PTN
```

Figure 13-1 The directory of a floppy disk containing different file types, as identified by their extensions

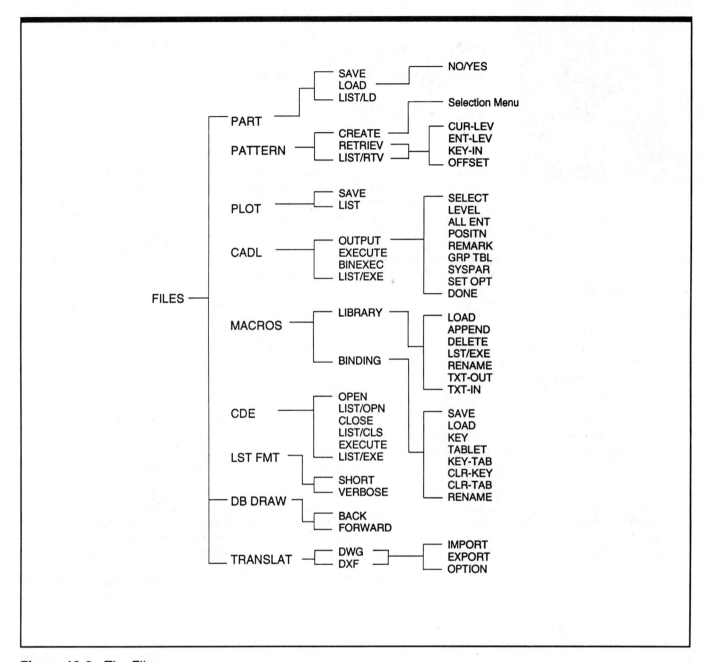

Figure 13-2 The Files menu

purpose and is used for this purpose only. Figure 13-1 is a directory of a floppy disk containing different types of files. You can tell what type of file each is by its extension. Figure 13-2 summarizes the file types.

PART FILES

A part file stores the complete data base for a part or drawing. It contains all of the information, such as modal parameters, views, and geometric data, that is used to extract the information needed

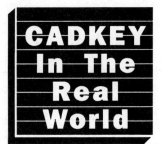

CADKEY In The Real World

Industrial Fan Manufacturer Creates Agricultural Vacuum Cleaner with CADKEY

What can you do if you are a farmer who needs to control insects in your food crops, but you don't want to use insecticides because of their potentially harmful effects on the people and animals that consume the food you grow? One answer is the Bug-Buster™, designed and manufactured by IAP, Inc. of Phillips, Wisconsin. The letters IAP stand for Industrial Air Products. IAP has manufactured custom-designed industrial fans for 16 years, and they have been using CADKEY to design their fans ever since Version 1.5. In 1989, one of their customers, a lettuce farmer, led IAP into an entirely new application of their technology: using the intake suction of fans, mounted on a frame attached to a tractor, to remove harmful insects from plants, so that they can do no more harm to the plants. The idea of the Bug-Buster had arrived.

IAP's Bug-Buster operates on the principle of a vacuum cleaner. As the fan hoods pass over the plants, the insects in and around the foliage are sucked into the fan. The centrifugal force of the fans kills the insects instantly and discharges them into the atmosphere. The Bug-Buster provides an environmentally safe alternative to chemical insecticides for agricultural and horticultural applications. Since the first Bug-Buster's success with lettuce crops, IAP's agricultural customers from California to England have applied the Bug-Buster to strawberries, potatoes, alfalfa, bell peppers, yellow squash, grapes, snap beans, carrots, asparagus, cauliflower, cabbage, celery, tomatoes, onions, broccoli, and flowers (particularly chrysanthemums and azaleas).

Custom Design

"Different plants have different characteristics, and they also have different types of insects that cause problems," said Ralph Mallwitz, engineering manager at IAP. "So, to an extent, we have to custom design fans and hoods for different crop types. For example, strawberries are delicate and grow close to the ground, and the problem insects are lightly attached to outer layers of foliage. We had to customize our hood design and the suction of the fan so that the farmer could pick off the bugs without picking the strawberries. On the other hand, potato plants have an entirely different foliage canopy, with larger, more dense insects that require much greater suction capacity to capture."

"We designed our first Bug-Buster on CADKEY from scratch—fans, frame, hoods, everything," Ralph said. "The ability to model the frame and fans in 3-D and to define our own 3-D views of the model was critical because the clearances were critical. We had to look down the frame from very particular views to see whether the fans would overlap the frame anywhere. Shortly thereafter, we designed another model of Bug-Buster as a self-propelled crop vacuum, with a wide wheelbase to support the weight and the vibration of its two fans."

"We like the ability to change colors in CADKEY, and to mask geometrics by color," Ralph said. "We now have written standards at IAP so that different parts are designed in specific colors. For example, housings are green and bearing pedestals are blue. Dimensions are light purple, and notes are light blue. That way, more than one engineer can work on a design. Also, by masking on

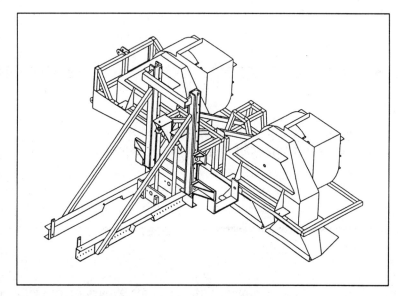

Two-fan Bug-Buster II ™

—Continued, next page

color, we can copy an existing design for a particular part of a fan into a new file so that it can be customized into a new design rapidly. We also use the snap and grid function a lot, especially for placing dimensions a specified distance from the geometry. It keeps the geometry clean and readable. IAP's drawings impress our customers very strongly. We are one of the few fan manufacturers that give customers customized drawings instead of standard data sheets."

"Getting back to the Bug-Buster," Ralph said, "each fan on a typical Bug-Buster can remove the insects from two rows of plants. Now, we make units ranging from one to four fans depending upon the customer's need. We make the Bug-Buster's frame to fit, as a front-mounted accessory unit to the customer's tractor, any type of tractor."

Adapted with permission from **3-D WORLD**, *v. 6, no. 3, 1992.*

to create a tool path in the CAM side of CAD/CAM. A part file can contain numerous pattern files but it cannot be merged with another part file. Each part file works independently of all other part files. When you list your files, all of the part files will have the extension (a descriptor added to the end of a filename) of **.prt**. CADKEY 6 will allow the use of any or no extension. The default pathname (the sub-directory path taken to save or load a file) for a part file as it is found in the configuration program is **\prt**. To access files in this sub-directory from DOS, you must type the pathname in front of the filename. CADKEY 6 has a dialog box which allows the user to access any logical drive and sub-directory. For a classroom situation it is recommended that you re-configure CADKEY, using the CONFIG program, so that the default pathname is **a:**. This way students' work will be saved automatically to the floppy disk instead of to the hard disk. The database associated with a part file is double precision to 16 places.

PATTERN FILES

When symbols that are used repeatedly in many parts are created, they are stored as pattern files. A pattern file is a part-independent file that contains geometric entities, notes, dimensions, labels, etc. These can be retrieved, rotated, scaled, and positioned within a part or drawing. One of the major advantages of CAD is being able to create a symbol or pattern once and store it away for future use. Then whenever the pattern is needed, it can be retrieved without having to be recreated. Many third-party vendors have developed whole libraries of symbols which can be purchased rather than creating them yourself. An example of a possible pattern is a standard-size nut or bolt that is used many times in many parts or drawings. Pattern files use the extension **.ptn** after their filename. The default path for pattern files is **\ptn** unless you have changed it to something else in the configuration program.

PLOT FILES

If you wish to plot off-line using CADKEY's PLOTFAST program, your drawing must be stored as a plot file. This codes the file to work only with the PLOTFAST program and cannot be retrieved any other way. For this reason it is recommended that you store the file as a part first. If you are plotting on-line using the **CONTROL, PLOT** function you do not have to create a plot file. In a file list the plot files will have the extension **.plt**. The default pathname for a plot file is **\plt** unless it has been changed in the configuration program. **RPLOT**—the CADKEY offline printing utility— also uses plot files. Some third-party utilities can convert a plot file into a **CADL** file.

CADL FILES

CADL stands for CADKEY Advanced Design Language. A CADL file is a readable ASCII text file. A CADL file allows you to print out the data base information for entities within a part. Values within the printout can be changed to manipulate the entities in the part. CADL files can also be used to interface with various third-party programs, such as numerical control (NC) or finite element analysis (FEA). This is done by making a CADL file of the part. Third-party application programs and customization programs can also be written using CADL. This is the true value of CADL. These files have a maximum precision of 10 decimal places and a minimum of 4 places. The default pathname for a CADL file is **\cdl**. A CADL file can be identified by its extension **.cdl** found after the filename. A CADL file can be compiled to run more quickly in a binary format. A compiled CADL would have a **.cdx** extension.

DXF FILES

DXF stands for Drawing Exchange Format. This standard was developed by AutoDesk Inc. of Sausalito, CA, as a way of transferring files between programs. Through the use of a DXF file you can transfer drawings that you may have created on such systems as AutoCAD or VersaCAD to CADKEY. A DXF file can be written or read with an external translator shipped with CAD-KEY 5. In CADKEY 6 the FILES-TRANSLAT option will read and write DXF files internally. A DXF file uses the extension **.dxf** after the filename.

MACRO FILES

A *macro* is a file that contains a set of commands which are frequently used or lengthy. Rather than having to execute each command individually, the macro file will execute all of the commands saved in the macro. A macro can be created simply by invoking the Immediate Mode command **CTRL-J**, which begins recording every command and keystroke you make until you again press **CTRL-J** to stop recording. The macro is stored in a macro library of related macros. The library is limited to four hundred macros, but you can create an unlimited number of macro libraries. You might have one library for mechanical macros and another library for architectural macros or some other application. You can bind (assign) each macro to keys on the keyboard or a region on a table.

IGES FILES

IGES stands for Initial Graphics Exchange Specification, which is the standard used to transfer 3D data. CADKEY has written a translator that changes the format of the CADKEY data base and creates an IGES file that can be transferred bi-directionally with other CAD/CAM systems that support IGES, such as Computervision and Unigraphics. The IGES Translator operates independent of CADKEY.

CDE FILES

CDE is an acronym for CADKEY Dynamic Extension. CDEs are miniapplications that run inside CADKEY to increase functionality or give you special features, such as special drafting functions or a set of templates for creating flowcharts. A CDE can contain many different functions that you can execute individually. The CDE option lets you use CDEs, either those you or your company created, CDEs from third-party developers, or CDEs from CADKEY. These files run at the same speed as CADKEY.

PART FILE OPTIONS

PART FILE OPTION: SAVE

The **SAVE** option allows you to store the part or drawing that you are creating as a part file. Once the part has been saved it can then be retrieved or loaded for future use or editing. The **SAVE** option can be accessed by using the Immediate Mode command **CTRL-F** for file save. It can also be accessed by using the Escape Code **F5-F1-F1 (FILES-PART-SAVE)**.

You are prompted to enter a filename. If there is a filename in parentheses, this is the default or current filename. It can be accepted by pressing **RETURN**. To store the part under a new name, type the disk drive specifier (**A:**, **B:**, or **C:**) and pathname to save to a disk drive and/or path other than the default setting, then enter the new filename. If you should enter a filename that already exists, you will be prompted:

File already exists, replace it?

The **NO/YES** menu will be displayed. You then have the options of replacing and updating your current file or saving it under a new name.

In many classroom situations it is preferred to have students save their work on floppy disks that they are responsible for, rather than cluttering up the hard disk. To do this, type **a:** before the filenames each time or change the path for your files in the configuration file so that whenever you save a file it will go directly to the **A:** drive.

 POSSIBLE USE

It is highly recommended that, as you are creating a part, you save that part at various stages (for example, every 15 minutes or just before you make a major transformation) so that if there should be a loss of power you would not have to start over from the beginning. This can be done quickly and easily by pressing the Immediate Mode command **CTRL-F**. CADKEY can be configured to automatically save drawings at a user-selected time interval.

The **SAVE** function saves your input to a storage device. The first file you save should be the "blank" file, which contains all of the status options (**VIEW, NOTE HEIGHT**, etc.) toggled to the options with which you most commonly start creating. The file does not contain any entities.

PART FILE OPTION: LOAD

The **LOAD** option allows you to recall a part file that has been previously stored. Generally, when you first initialize CADKEY, you are prompted to enter a filename. You can load an existing file, create a new file, or simply press the **ESC** key to begin working in CADKEY.

If you are already into a part file, you may wish to load another part file. Only one part file can be active at a time. The Escape Code for loading a new part file is **F5-F1-F2 (FILES-PART-LOAD)**. When this code is executed you will be prompted to save or replace the current part before loading the new part. You will respond using the **NO/YES** menu. You are then prompted to enter the filename of the new part. The filename, including pathname, is limited to forty characters. It is best to limit your character use to letters and numbers even though some special characters can be used in filenames. If a disk drive specifier (**A:**, **B:** or **C:**) was used when the part was saved, then it must also be used to load the part. The computer will search the appropriate disk, retrieve and display the new part.

 POSSIBLE USE

Many times, when you first begin working with CADKEY, you will just experiment with a series of different functions and then delete what you have created. All of the functions that you have deleted are still in memory and can be un-deleted. If a global un-delete were used it would bring back everything that you had ever created in that part file. To avoid a mess like this it is recommended that you load a blank file, that contains no entities, that you have previously stored. The storing of a blank file is described above. The blank file will give you a clean, empty file in which to begin creating entities.

PART FILE OPTION: LST/LD (List/Load)

The **LST/LD** option allows you to get a listing of all of the part files on a specific directory path. From this list, the desired file can be selected and loaded. This option can be accessed by the Escape Code **F5-F1-F3 (FILES-PART-LST/LD)**.

When the **LST/LD** option is selected you are prompted to enter a directory pathname. The default pathname on the prompt line will be the pathname under which you have configured your files to be saved. Press **RETURN** to accept the default directory. To specify a new path, use the backspace key to remove the default pathname, then type in the new pathname followed by **RETURN**.

At this point a list of the part files will appear on the screen in a pop-up menu. Highlight the page indicator to review additional pages. After reviewing the list and deciding which file to load, cursor-select the part from the pop-up menu. This is done by moving the screen cursor so that the file you wish to load is displayed in reverse video, then pressing the mouse button. Before the new part file is displayed you will have the option of saving the current part first. The **NO/YES** menu will be displayed for you to choose from.

In CADKEY 6 the **FILES-PART-LIST/LD** option is displayed as a dialog box (Figure 13-3). The dialog box contains a large amount of information about the file. This makes the management of your files much easier. The **File Spec:** displays the currently selected file. **Directory:** lists the current pathname from which the files are being listed. Directories allow the user to change the disk drive being read. The [..] option allows the user to move back through the sub-directories one at a time. **Sort** allows the user to define the files listing order by various options, such as Name, Extension, Date, Size, Ascending, or Descending. The Date, Time, Size, and Read/Write status of the current files is also listed in the dialog box. The file management options allow the user to Copy, Move, Delete, or Rename a file from the dialog box in CADKEY. Once you have selected the desired options, click on **OK** or press **Enter** from the keyboard. If you quickly double click the mouse button while highlighting the desired file, it will automatically be loaded after you respond **NO** or **YES** to the prompt "Save Current Part First."

 ## POSSIBLE USE

The **LST/LD** option is particularly helpful when you do not remember exactly what abbreviation you used for a filename or simply do not remember a filename. By using the list/load option you can review the available files and select and load the one that you desire, using the screen cursor.

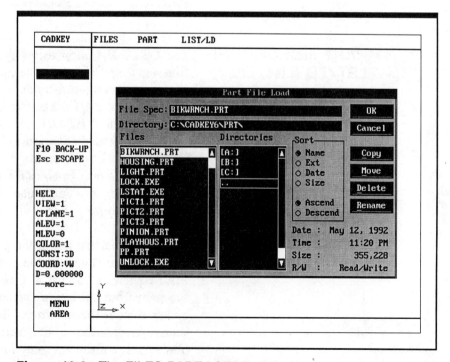

Figure 13-3 The FILES-PART-LST/LD dialog box

PATTERN FILE OPTIONS

PATTERN FILE OPTION: CREATE

The **CREATE** option allows you to save a pattern that can be retrieved and positioned as needed in a part file. The Escape Code for accessing this option is **F5-F2-F1 (FILES-PATTERN-CREATE)**.

After you select **CREATE**, the Selection Menu appears to aid you in your selection of entities that will be used in the pattern. Once the entities are selected, the Position Menu appears. You will be prompted to indicate the base position of the pattern. The base position is the reference point that is used to position the pattern when it is retrieved in the future. Finally, you will be prompted to enter the pattern filename. As with the part files, you can use a disk drive specifier and pathname if desired. If you enter a filename that has already been used you will have the option to replace it by choosing from the **NO/YES** menu.

 POSSIBLE USE

If you use a particular symbol or pattern over and over, you need to create it only once. In the future, whenever the pattern is needed it can simply be retrieved. An example is the electronic symbols used for electrical schematic drawing. A complete symbol library could be created by storing each electronic symbol as a pattern file. Then, every time a resistor or diode is needed in a drawing, it could be retrieved.

Another use for pattern files is in layout drawings that contain multiple views of an object. In order to display the multiple views of a 3D object each view must be saved as a separate pattern file. Then all of the patterns are retrieved and positioned within the border pattern. When creating a pattern of a view, the Position Menu option **ENDENT** is recommended for indicating base position. By using the **ENDENT** option and selecting an entity in the lower left corner, you will have selected the lower left corner exactly for accurate positioning of the pattern when it is retrieved (Figure 13-4).

PATTERN FILE OPTION: RETRIEV (Retrieve)

The **RETRIEV** option allows you to recall, display, and position a pattern file that has been previously stored. This function can be executed using the Escape Code **F5-F2-F2 (FILES-PATTERN-RETRIEV)**.

When this option is executed you are prompted to enter the filename of an existing pattern. The filename does not need to have the extension **.ptn** on it. When the pattern file is located you are then prompted as to whether you wish to group the pattern. Once again, the **NO/YES** menu is displayed for you to choose from.

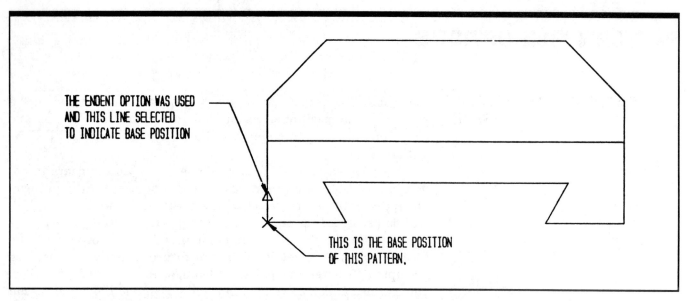

THE ENDENT OPTION WAS USED
AND THIS LINE SELECTED
TO INDICATE BASE POSITION

THIS IS THE BASE POSITION
OF THIS PATTERN.

Figure 13-4   Base position of a pattern file indicated using the ENDENT option from the Position Menu

If **YES** is selected you will be prompted to enter a group name. The default name in parentheses is the same as the pattern filename. You can accept this name or type in another group name. A quick check is made to see if the group already exists. If so, you are prompted: **Group already exists....create subgroup #nn?** (nn is the next available subgroup number in the group). The **NO/YES** menu is displayed for you to choose from. If **YES** is selected, the system proceeds by displaying the following menu options:

CUR-LEV (Current Active Level)—all entities in the pattern are assigned to the current active level.

ENT-LEV (Entity Level)—each entity in the pattern file retains its original level assignment from when the pattern was created.

KEY-IN—allows the user to assign the pattern to a specific level as it is retrieved. You are prompted to key in the level to which all entities in the pattern will be assigned.

OFFSET—a new level number is assigned to each entity in the pattern file by adding a user-specified integer to each entity's existing level number.

After you choose an option, you are prompted to enter a scale factor. The scale of the pattern can be changed so that when the pattern is retrieved, it is displayed at a scale larger or smaller than the original scale under which it was saved. When you are prompted to enter a scale factor you can accept the default scale factor by pressing **RETURN** or type in the new scale factor desired.

The next prompt requests that the rotation angle be entered. This allows for the pattern to be rotated from the position it was saved in. The standard angle convention is in effect for entering this value. Positive angles move in a counterclockwise direction from a 3 o'clock origin and negative angles move in a clockwise direction from the same origin. Enter the new value and press **RETURN**, or press **RETURN** to accept the default value.

After the angle is entered, the Position Menu is displayed. You can use any of these options to assist you in positioning the pattern within the displayed part. When the position for the pattern is selected, the pattern is then displayed.

You are then prompted for another base position for the pattern file. You can retrieve this pattern file as many times as you wish. If grouping was previously selected, the next copy of the pattern that is retrieved will automatically have the next available subgroup number assigned to it. Press **F10 BACKUP** if another group name is desired. If you do not wish to retrieve the pattern file again, press **ESCAPE** to return to the Main Menu.

 POSSIBLE USE

The possible uses that were discussed for creating a pattern file also apply to retrieving a file. The ability to scale and rotate makes this option valuable.

PATTERN FILE OPTION: LST/RTV (List/Retrieve)

The **LST/RTV** option allows the user to display a list of all of the available pattern files in a specific directory. From this list, the desired pattern file can be cursor-selected for retrieval. This function can be executed using the Escape Code **F5-F2-F3** (**FILES-PATTERN-LST/RTV**).

You will be prompted to enter the pattern directory path you wish to list. The pathname on the prompt line **\ptn** is the default pathname that CADKEY has been configured to use when saving pattern files. To accept the default path press **RETURN**. If you wish to list another directory path, use the backspace key to remove the default pathname, then enter the directory pathname and press **RETURN**.

You will see a listing of the pattern files in the specified directory in the pop-up menu on the screen. If you are not at the end of the list, you can cursor-select the page indicator to review the next page.

To select the file you wish to retrieve, move the screen cursor until the filename that you wish to select is highlighted in reverse video. Press the mouse button to make the selection.

The pattern will be retrieved and you are prompted: **Group pattern entities (YES)?** The **NO/YES** menu is displayed for you to choose from. Grouping combines all of the entities in the pattern into one entity.

When **YES** is accepted you are prompted to enter a group name. The default group name is the same as the pattern name. Press **RETURN** to accept the default name or enter a new name. A quick check is made to see if the group already exists; if so, you are prompted to create a subgroup. If **YES** is selected, the system proceeds by displaying the following menu options:

> **CUR-LEV (Current Active Level)**—all entities in the pattern are assigned to the current active level.
>
> **ENT-LEV (Entity Level)**—each entity in the pattern file retains its original level assignment from when the pattern was created.
>
> **KEY-IN**—allows the user to assign the pattern to a specific level as it is retrieved. You are prompted to key in the level to which all entities in the pattern will be assigned.
>
> **OFFSET**—a new level number is assigned to each entity in the pattern file by adding a user-specified integer to each entity's existing level number.

Choose one of the four menu options. Next, you are prompted to enter the scale factor for the pattern or press **RETURN** to accept the default scale. If you have the **AUTO UP** switch set to **YES** (see **SET** function), the dimensions will be automatically updated by the scale factor entered. This is followed by a prompt to enter the rotation angle value or press **RETURN** to accept the default value.

The final step is to indicate the base position of the pattern within the part. The Position Menu is displayed to aid in positioning the pattern. Once the base position is selected, the pattern is displayed within the current part.

In CADKEY 6 the **FILES-PATTERN-LIST/RTV** option is displayed as a dialog box (Figure 13-5). The dialog box contains a large amount of information about the file. This makes the management of your files much easier. The **File Spec:** displays the currently selected file. **Directory:** lists the current pathname from which the **Files** are being listed. **Directories** allows the user to change the disk drive which is being read or with the [..] option allows the user to move back through the sub-directories one at a time. **Sort** allows the user to define the files listing order by various options, such as Name, Extension, Date, Size, Ascending or Descending. The Date, Time, Size, and Read/Write status of the current files is also listed in the dialog box. The file management options allow the user to Copy, Move, Delete, or Rename a file from the dialog box in CAD-KEY. Once you have selected the desired options click on **OK** or press **Enter** from the keyboard. If you quickly double click the mouse button while highlighting the desired file it will automatically be retrieved.

 ## POSSIBLE USE

One possible use for for the **LST/RTV** option is when you are unsure of a pattern name. The list can be reviewed to aid in selecting the correct pattern.

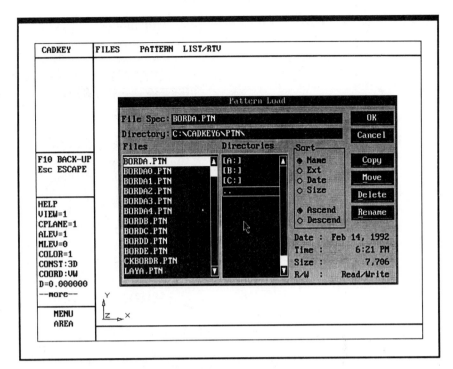

Figure 13-5   The FILES-PATTERN-LIST/RTV dialog box

PLOT FILE OPTIONS

PLOT FILE OPTION: SAVE

The **SAVE** option allows you to store the currently displayed portion of a part or drawing as a plot file to be used for off-line plotting. To save a part as a plot file you can use the Escape Code **F5-F3-F1** (**FILES-PLOT-SAVE**).

You are prompted to enter a plot filename and press **RETURN**. The plot filename can be the same as the part filename because they will have different extensions. A plot file uses the extension **.plt**. You can save the plot file by **PEN#**, using the pen numbers you assigned in the drawing, or by **COLOR**, using the colors assigned in the configuration program. The Selection Menu is then displayed to assist you in selecting the entities to be included in the plot file. The plot file is automatically created.

WARNING: if a plot file already exists by the same name it will be automatically replaced by the current part.

 POSSIBLE USE

The PLOTFAST program, which will plot the plot files, can be run independently of CADKEY. PLOTFAST only requires 128K of memory to be run. Therefore, you can use a computer of lesser expense (e.g., an IBM PC) to run your plotter as a dedicated plotter station. This way you are not tying up a workstation for plotting. You can simply save a plot file on floppy disk and carry it over to the plotter station. Plot files can also be plotted to disk in HPGL format to be imported into other application software, such as desktop publishing.

PLOT FILE OPTION: LIST

The plot file option **LIST** allows you to list the existing plot files in a specified directory path. The **LIST** option can be accessed using the Escape Code **F5-F3-F2** (**FILES-PLOT-LIST**).

You are prompted to enter the directory pathname where you stored your plot files. The displayed default directory path is the same as your configuration path. Enter a new directory pathname if you wish, or accept the default path by pressing **RETURN**.

A listing of the files in the specified directory is displayed in a pop-up menu on the screen. If all of the files are not displayed, select the **PAGE** indicator to review the other plot files.

In CADKEY 6 the **FILES-PLOT-LIST** option is displayed as a dialog box (Figure 13-6). The dialog box contains a large amount of information about the file. This makes the management of your files much easier. The **File Spec:** displays the currently selected file. **Directory:** lists the current pathname from which the **Files** are being listed. **Directories** allows the user to change the disk drive which is being read or with the [..] option allows the user to move back through the sub-directories one at a time. **Sort** allows the user to define the files listing order by various options, such as Name, Extension, Date, Size, Ascending or Descending. The Date, Time, Size, and Read/Write status of the current files is also listed in the dialog box. The file management options allow the user to Copy, Move,

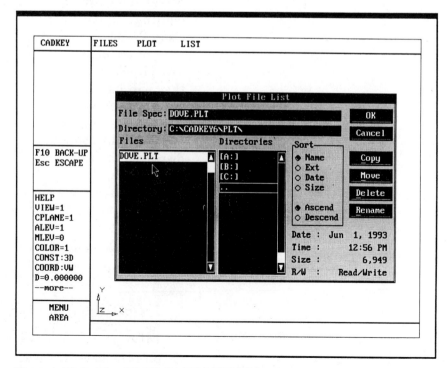

Figure 13-6 The FILES-PLOT-LIST dialog box

Delete, or Rename a file from the dialog box in CADKEY. Once you have selected the desired options click on **OK** or press **Enter** from the keyboard.

 POSSIBLE USE

When a plot file is saved it will automatically replace any plot file with the same name. To be sure that you are not incorrectly replacing a plot file that already exists, do a **PLOT-LIST** first.

CADL FILE OPTIONS

CADL FILE OPTION: OUTPUT

The CADL option **OUTPUT** writes entities, remarks, and the system group table to a CADL file. Selected entity types, attributes such as line type, line width, color, level number and text font are transferred to an ASCII file for use outside of CADKEY. The Escape Code to access the **OUTPUT** option quickly is **F5-F4-F1 (FILES-CADL-OUTPUT)**.

You are prompted to enter a filename for the CADL output file you are about to create. A new menu is displayed with nine CADL output options:

> **SELECT**
> **LEVEL**
> **ALL ENT**
> **POSITN**
> **REMARK**
> **GRP TBL**
> **SYS PAR**
> **SET OPT**
> **DONE**

SELECT—allows you to use the Selection Menu to select the entities that you wish to place in a CADL file.

LEVEL—allows you to output all of the entities on a specific level to a CADL file.

ALL ENT (ALL ENTITIES)—selects all entities such as **POINTS, ARCS, SPLINES, LINES,** and **NOTES** in the current data base and uses them to create a CADL file, regardless of whether or not they are displayed.

POSITN (POSITION)—displays the nine Position Menu options.

REMARK—allows you to enter one or more text strings which are written to your CADL file as remarks (**REM**).

GRP TBL (GROUP TABLE)—allows you to use the system group table to create a CADL file in the form of group primitives.

SYS PAR (SYSTEM PARAMETERS)—allows you to use the part header to the currently open CADL file. The CADL header is a combination of statements that describe the current system parameters.

SET OPT (SET OPTIONS)—allows you to set the output options before the CADL file is created. **DEC PRE (DECIMAL PRECISION)** sets the number of places to the right of the decimal point. **ATT OUT (ATTRIBUTES OUT)** sets the attributes for the CADL output file. **SPL REP (SPLINE REPRESENTATION)** sets the coefficient or node representation for splines. **DIM INFO (DIMENSION INFORMATION)** dimension text, reference and witness lines.

DONE—allows you to exit the CADL option menu after making your selections.

 POSSIBLE USE

A CADL output file is one method of transferring 3D data from a part file to a third party numerical control program or finite element analysis program. Figure 13-7 is an example of a CADL output file for Figure 13-8.

CADL FILE OPTION: EXECUTE

The CADL option **EXECUTE** reads in a CADL file and stores data primitives as entities. It also executes CADL commands and program instructions. You can access the CADL **EXECUTE** option

```
LINE      1.5000, 1.5000, 0.0000, 2.5000, 1.5000, 0.0000, 1, 1, 1, 0, 0, 1
LINE      2.5000, 1.5000, 0.0000, 3.0000, 2.0000, 0.0000, 1, 1, 1, 0, 0, 1
LINE      3.0000, 2.0000, 0.0000, 4.0000, 2.0000, 0.0000, 1, 1, 1, 0, 0, 1
LINE      4.0000, 2.0000, 0.0000, 4.5000, 1.5000, 0.0000, 1, 1, 1, 0, 0, 1
LINE      4.5000, 1.5000, 0.0000, 5.0000, 1.5000, 0.0000, 1, 1, 1, 0, 0, 1
LINE      5.0000, 1.5000, 0.0000, 6.0000, 2.5000, 0.0000, 1, 1, 1, 0, 0, 1
LINE      6.0000, 2.5000, 0.0000, 6.0000, 3.5000, 0.0000, 1, 1, 1, 0, 0, 1
LINE      6.0000, 3.5000, 0.0000, 4.0000, 3.5000, 0.0000, 1, 1, 1, 0, 0, 1
LINE      4.0000, 3.5000, 0.0000, 4.0000, 3.0000, 0.0000, 1, 1, 1, 0, 0, 1
LINE      4.0000, 3.0000, 0.0000, 3.2500, 3.0000, 0.0000, 1, 1, 1, 0, 0, 1
LINE      3.2500, 3.0000, 0.0000, 3.2500, 3.5000, 0.0000, 1, 1, 1, 0, 0, 1
LINE      3.2500, 3.5000, 0.0000, 2.7500, 3.5000, 0.0000, 1, 1, 1, 0, 0, 1
LINE      2.7500, 3.5000, 0.0000, 1.5000, 1.5000, 0.0000, 1, 1, 1, 0, 0, 1
CIRCLE    5.0000, 2.5000, 0.0000, 0.5000, 1, 1, 1, 1, 0, 0, 1
TEXT      3.2500, 4.8750, '1/4" CORK GASKET
', 0.0000, 0.3000, 0.5000, 1, 1, 1, 0, 0, 0, 1
```

Figure 13-7 A CADL output file in which values can be manipulated and executed

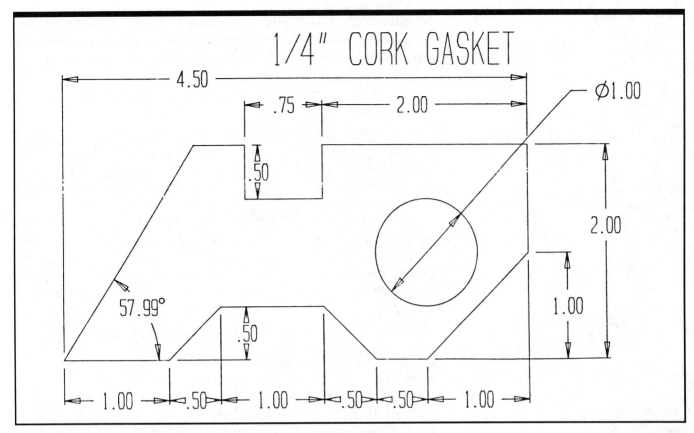

Figure 13-8 The result of the CADL file when executed

quickly by using the Escape Code **F5-F4-F2** (**FILES-CADL-EXECUTE**).

You are prompted to enter the CADL filename that you wish to execute and press **RETURN**. The CADL data primitives are read in and stored in the data base as entities, and the CADL command and program instructions are executed as required. If program statements or commands have been added to the program they will also be displayed.

 ## POSSIBLE USE

If you wished to modify an existing part it could be done through CADL. The teach files are executed as compiled CADL files.

CADL FILE OPTION: BINEXEC (Binary Execute)

This option reads in and executes previously compiled CADL files, stores data primitives as entities, and executes CADL commands. A compiled CADL file will have a **.cdx** extension. To execute it, simply enter the filename and press **RETURN**. The CADL data primitives are read in and stored in the data base as entities and the CADL commands are executed as required. In CADKEY 6 simply use the **LIST/EXE** option.

CADL FILE OPTION: LIST/EXE

This option allows you to list and execute existing CADL files in an ASCII-format from a pop-up menu. The cursor is used to highlight and select the desired file. The CADL data primitives are read in and stored in the data base as entities and the CADL commands are executed as required.

In CADKEY 6 the **FILES-CADL-LIST/EXE** option is displayed as a dialog box (Figure 13-9). The dialog box contains a large amount of information about the file. This makes the management of your files much easier. The **File Spec:** displays the currently selected file. **Directory:** lists the current pathname from which the **Files** are being listed. **Directories** allows the user to change the disk drive which is being read or with the [..] option allows the user to move back through the sub-directories one at a time. **Sort** allows the user to define the files listing order by various options, such as Name, Extension, Date, Size, Ascending or Descending. The Date, Time, Size, and Read/Write status of the current files is also listed in the dialog box. The file management options allow the user to Copy, Move, Delete, or Rename a file from the dialog box in CADKEY. Once you have selected the desired options click on **OK** or press **Enter** from the keyboard. If you quickly double click the mouse button while highlighting the desired file it will automatically be executed.

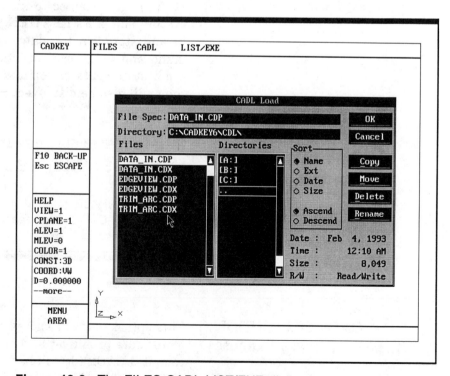

Figure 13-9   The FILES-CADL-LIST/EXE dialog box

DXF FILE OPTIONS

Included with CADKEY is a separate DXF translator which runs independently of CADKEY. This DXF translator can be used to convert CADKEY files to DXF files or DXF files to CADKEY files.

 POSSIBLE USE

If you have drawing files from a 2D CAD system you can transfer them into CADKEY and enhance them. Work that was done with the other system is not dead material; it can be used with CADKEY. To transfer an AutoCAD file into CAD-KEY you must first convert the file to the DXF format by using the AutoCAD command **DXFOUT**. Once the file is in the DXF format it can then be loaded into CADKEY.

A part saved in the DXF format can be used to transfer geometric data from a CADKEY part file to a Numerical Control package to define the tool path that will be followed when the part is machined. This option could also be used to transfer a part file to a Finite Element Analysis package to test the structural properties of a part design.

MACRO FILE OPTIONS

Macro files contain a set of CADKEY menu commands that can be executed with a single keystroke rather than many keystrokes or selections. Macros use four Immediate Mode commands:

CTRL-J—turns the **RECORD** function on and off, starting and ending the creation of a macro.

CTRL-K—turns the **PAUSE** mode on and off, allowing for user input in the macro.

CTRL-O—allows you to enter a **PROMPT**, overriding the standard prompt.

CTRL-E—allows you to execute a macro by entering the macro name.

MACRO FILE OPTION: LIBRARY

Macro files are stored in a macro library. Each macro library can store up to four hundred different macros. A unique library containing a set of macros can be developed for specific applications such as mechanical, detail drafting, or electrical.

The seven options available under **LIBRARY** are:

       LOAD
       APPEND

DELETE
LST/EXE
RENAME
TXT-OUT
TXT-IN

LOAD—A library must be defined or opened before you can create a macro. To open a macro library, use the Escape Code **FILES-MACRO-LIBRARY-LOAD**. You will then be prompted to enter the name of the library to load and press **RETURN**.

NOTE: You can also set the automatic file loading options when you configure CADKEY to load a specific macro library when you boot up CADKEY.

APPEND—allows you to combine macros from different libraries by adding the contents of a specific library to the library that is currently active. If the two libraries contain macros with the same name, the following menu will appear:

REPLACE—replaces the macro in the active library with the macro of the same name being read in.
REN-NEW—allows the macro being read in to be renamed.
REN-OLD—allows the macro in the active library to be renamed.
OMIT—ignores or omits the new macro being read in.

DELETE—Allows you to erase or remove a macro from the active library. You must also update the binding file which has been temporarily modified by using the **FILES-MACROS-BINDING-SAVE** command sequence.

LST/EXE—displays a list of the macros in the active library from which the desired macro can be cursor-selected and executed. If the list is longer than one page, press **RETURN** to see the next page and **F10** to see the previous page. This command is a good way to execute macros if you do not have a digitizer with a template overlay or if you are using a mouse. In CADKEY 6 the **FILES-MACROS-LIBRARY-LST/EXE** option is displayed as a dialog box (Figure 13-10). The dialog box contains a list of available macro files. Once you have selected the desired macro file, click on **OK** or press **Enter** from the keyboard. If you quickly double click the mouse button while highlighting the desired file it will automatically be executed.

RENAME—changes the name of an existing macro in the currently active macro library. Enter the current name of the macro and press **RETURN**. Then enter the new name for the macro and press **RETURN**. If a macro by that name already exists, your options will be:

REPLACE—writes the macro over the old one.
REN-NEW—you enter a different name for the new macro.

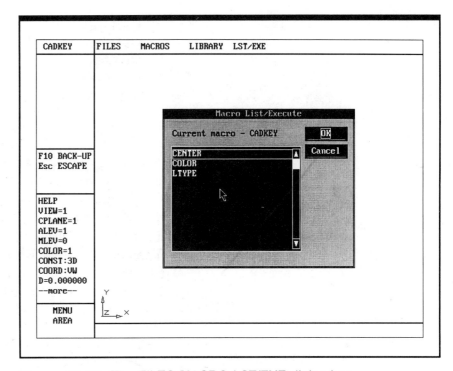

Figure 13-10   The FILES-MACRO-LST/EXE dialog box

REN-OLD—you enter a new name for the old macro.
OMIT—exit without renaming either macro.

TXT-OUT—Outputs a macro from the active library as an ASCII text file, which can then be edited using any word processor or line editor. You are prompted to enter the name of the macro and then the name of the text file, which will automatically have the extension **.txt**. The text file is stored in the default macro sub-directory. You can then exit CAD-KEY to DOS or use **CONTROL-SYSCMD-SHELL** and load this file into your line editor or word processor to review and make changes.

TXT-IN—used to bring the edited ASCII file back into CAD-KEY as a macro. Simply enter the name of the text file and press **RETURN** and it will become an active macro once again.

MACRO FILE OPTION: BINDING

BINDING allows you to assign or bind existing macros to keys on the keyboard and/or to column/row positions on a digitizing tablet. The options available under **BINDING** are:

> **SAV ASCII**
> **SAV BINRY**
> **LOAD**
> **KEY**
> **TABLET**
> **KEY-TAB**

CLR-KEY
CLR-TAB
RENAME

SAV ASCII—stores or updates the binding file in an ASCII format that you create using the other options in this command. The binding file contains the key and tablet assignments for executing the macros you have created. Enter a filename and press **RETURN**.

SAV BINRY—stores or updates the binding file in a binary format that you create using the other options in this command. The binding file contains the key and tablet assignments for executing the macros you have created. Enter a filename and press **RETURN**.

LOAD—used to retrieve a previously created binding file. If you change macro libraries you can load a new binding file to activate the key and tablet assignments for the active library.

KEY—allows you to assign a macro to a specific key or combination of keys except for **F1-F10**, **DEL**, **BACKSPACE**, **ENTER**, and the keys found on the numeric keypad. Enter the name of the macro and press **RETURN**, then press the key to which you wish to bind the macro. Repeat these steps for each macro you wish to bind to a key and then use the **SAVE** option to create a binding file after all of the macros have been bound.

TABLET—assigns a macro to a specific column and row location on a tablet. Enter the name of the macro you wish to bind and press **RETURN**. With your cursor control device (puck or stylus) select the column and row location to bind the macro. In the future, when you select this location the macro will be executed. Again, the save option must be used to create a binding file for the locations you have defined. If you do not save a binding file, the locations will be lost.

KEY-TAB—assigns a key from the keyboard to a tablet location. Press the key that you wish to assign and then select the column and row location on the tablet.

CLR-KEY—clears a macro from a previously assigned key. Press the key to clear and the assignment will be removed.

CLR-TAB—clears a macro from a previously assigned tablet column and row location. Select the location on the tablet and it will be cleared. If a macro is not assigned to that location, an error message will prompt you to press **RETURN**.

RENAME—changes the name of an existing macro in the currently active macro library. Enter the current name of the macro and press **RETURN**. Then enter the new name for the macro and press **RETURN**. If a macro by that name already exists, your options will be:

> **REPLACE**—writes the macro over the old one.
> **REN-NEW**—you enter a different name for the new macro.

REN-OLD—you enter a new name for the old macro.
OMIT—exit without renaming either macro.

Always remember: to create a binding file with the changes you have made, you must choose the **SAVE** option before escaping this function.

 POSSIBLE USE

You could create a macro that would draw a horizontal line through the end of another entity. This could be used when you project height between views in a 2D orthographic drawing. Begin by opening the macro library with the command sequence **F5-F5-F1-F1** (**FILES-MACRO-LIBRARY-LOAD**), entering a library name (e.g., CADKEY) and pressing **RETURN**. Next, press the Immediate Mode command **CTRL-J**, which will prompt you for a macro filename; in this case use **HORIZTL**. CADKEY will now start recording all of your keystrokes to create a macro. The word **RECORD** will be displayed above the Menu Window. Begin each macro by pressing the **ESC** key four times. This clears any previous functions that might still be active and starts the macro from the Main Menu. For this macro select **CREATE-LINE-HRZ/VRT-HORIZTL-ENDENT**. To end the macro press **CTRL-J**, and the **RECORD** sign will be turned off.

To test the macro, press the Immediate Mode command **CTRL-E**. When prompted for the name of the macro **HORIZTL** should be the default name, so press **RETURN** and the macro should **PLAY**. If not, use the **TXT-OUT** function and any line editor or word processor to compare your macro file to the one below.

ASCII File	Description
HORIZTL	**Macro Name**
ROOT	**Escape to Main Menu**
ROOT	
ACCEPT	
ROOT	
ROOT	
MENU1	**CREATE**
MENU1	**LINE**
MENU5	**HRZ/VRT**
MENU1	**HORIZTL**
MENU3	**ENDENT**

At this point you have created and executed a macro. To easily execute the macro in the future you can bind or associate the macro to a tablet location or to the keyboard. For this example you will bind it to the keyboard. To start this process press **ESC-F5-F5-F2-F3** (**ESCAPE-FILES-MACROS-BINDING-KEY**). You are prompted, **Enter the macro name:**

HORIZTL and press **RETURN**. You will be prompted to press the keys to which you wish to bind the macro. Press **CTRL-F1**. You can then continue to bind other macros to various keys and tablet locations. When done, press **F10 BACKUP and F1 SAVE**. You are then prompted: **Enter the name of the binding file to save** (i.e., CADKEY). *Do not attempt to save after each macro binding. Wait to save until all macros have been bound.* The binding file contains the key and tablet location for many different macros and can only be edited using a text editor.

CDE FILE OPTIONS

The CDE option lets you load and use CDEs, either those you or your company created, CDEs from third-party developers, or CDEs from CADKEY. CDEs are miniapplications that run inside CADKEY to increase functionality or give you special features, such as special drafting functions or a set of templates for creating flowcharts. A CDE can contain many different functions that you can execute individually.

CADKEY counts the CDEs you open, so if you open the same CDE twice in one work session you must close it twice. Make sure you set the default name or names of the CDE or CDEs you want CADKEY to open on when you start CADKEY, and the default path name for the CDE directory. The default pathname is **\cde**; the default filenames extension is **.cde**. The default CDE is the last CDE you worked with in the current work session. When you exit CADKEY, the default CDE returns to the default set in the configuration program. You can have multiple CDEs open at the same time.

CDE has six menu options:

> **OPEN**
> **LIST/OPN**
> **CLOSE**
> **LIST/CLS**
> **EXECUTE**
> **LIST/EXE**

CDE FILE OPTION: OPEN

OPEN loads a CDE module. The CDE module you open overlays any previously opened functions that have the same name. To make the functions inside the CDE available for you to work with, choose **EXECUTE** or **LIST/EXE** after you choose **OPEN**.

CADKEY will prompt you for the name of the CDE you want to open. Enter the name of the CDE and press **RETURN**. You will see a dialog box displaying the status of opening the CDE. CADKEY then opens the CDE.

CDE FILE OPTION: LIST/OPN

LIST/OPN displays a list of all CDEs found in the default CDE directory. To make the functions inside the CDE available for you to work with, choose **EXECUTE** or **LIST/EXE** after you choose **LIST/OPN**.

When you turn on the **VERBOSE** option when working with CDEs, the list box will show the name, size, and the date of creation of the CDE. If you cannot remember the name of the CDE, this information can help jog your memory.

If you try to open a CDE module with corrupt data in it or one that was not built correctly, you will see an **UNDEFINED SYMBOLS** error message. Choose again from the list and click on **OK** or press **RETURN**.

To use the **LIST/OPN** option enter the CDE path name and press **RETURN**. A list box will appear, displaying the names of all CDEs stored in the directory and the page number of the list. Choose the CDE you want to open. Click on **OK** or press **RETURN**. CADKEY then opens the CDE. In CADKEY 6 a dialog box will appear for this option.

CDE FILE OPTION: CLOSE

CLOSE closes a CDE and marks its functions for removal. Simply enter the name of the CDE you want to close and press **RETURN** to close a CDE. If you also used a CADL program to open the CDE module, the CDE module stays open until you close the CADL program, too.

CDE FILE OPTION: LIST/CLS

The **LIST/CLS** option displays a list box of all CDE modules currently open. Choose the CDE you want to close from the list box. Click on **OK** or press **RETURN** to close the CDE. If you used the **FILES-LST FMT-VERBOSE** option, the list shows the name of the function and a description from its definition file.

CDE FILE OPTION: EXECUTE

The **EXECUTE** option starts a function from a opened CDE module. Enter the name of the function you want to execute and press **RETURN**.

CDE FILE OPTION: LIST/EXE

The **LIST/EXE** option displays a list of all currently available CDE functions that you can execute. You can execute a function by picking it from the list. If you used the **FILES-LST FMT-VERBOSE** option, the list shows the name of the function and a description from its definition file. Choose the CDE function from the list. Click on **OK** or press **RETURN** to start the function.

In CADKEY 6 the **FILES-CDE-LIST/EXE** option is displayed as a dialog box (Figure 13-11). The dialog box contains a list of available CDE files. Once you have selected the desired CDE file click on **OK** or press **Enter** from the keyboard. If you quickly double click the mouse button while highlighting the desired file it will automatically be executed.

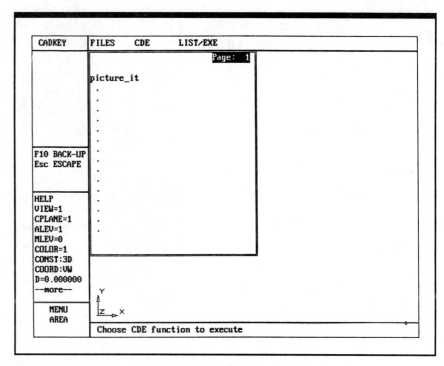

Figure 13-11   The FILES-CDE-LIST/EXE dialog box

LST FMT FILE OPTION

This option controls the format of the filenames displayed during listing operations in a pull-down menu format. The **SHORT** option lists only the filenames in a designated directory. The **VERBOSE** option lists the filename, size, date, and time of day the file was stored in the directory. This option no longer is needed in CADKEY 6 because of the addition of dialog boxes for all listing options.

DB DRAW FILE OPTION

This option controls the order in which entities are selected and redrawn from the data base. When set to **BACK**, the part will be redrawn from the last entity created to the first entity. The **FORWARD** option redraws the entities in the order in which they were created. The **DB DRAW** option can be used for selecting entities for deletion or editing when one is on top of the other. **DB DRAW** will allow you to pick either one of them.

TRANSLATE FILE OPTION

The **FILES-TRANSLAT** (Translate) option allows CADKEY 6 users to read and write files in AutoCAD's main formats, **DWG** and **DXF**. The options under both include **IMPORT, EXPORT**, and **OPTION**.

The **IMPORT** option in CADKEY 6 is displayed as a dialog box (Figure 13-12). The dialog box contains a large amount of information about the file. This makes the management of your **DWG** and **DXF** files much easier. The **File Spec:** displays the currently selected file. **Directory:** lists the current pathname from which the files are being listed. **Directories** allows the user to change the disk drive which is being read or with the [..] option allows the user to move back through the sub-directories one at a time. **Sort** allows the user to define the files listing order by various options, such as Name, Extension, Date, Size, Ascending or Descending. The Date, Time, Size, and Read/Write status of the current files is also listed in the dialog box. The file management options allow the user to Copy, Move, Delete, or Rename a file from the dialog box in CADKEY. Once you have selected the desired options, click on **OK** or press **Enter** from the keyboard. If you quickly double click the mouse button while highlighting the desired filed it will automatically be loaded.

The **EXPORT** option in CADKEY 6 is displayed as a dialog box (Figure 13-13). The dialog box contains information about the file. This makes the management of your **DWG** and **DXF** files much easier. **File Name** is where the name of the DWG or DXF file that you

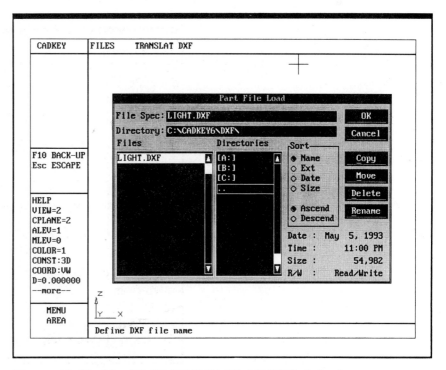

Figure 13-12 The FILES-TRANSLAT-IMPORT dialog box

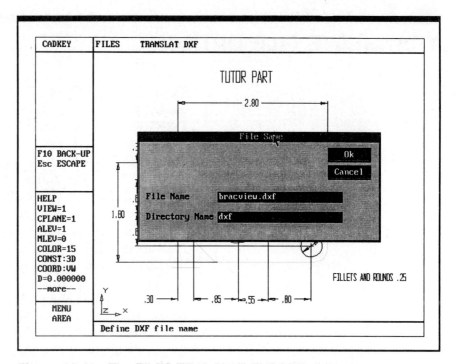

Figure 13-13 The FILES-TRANSLAT-EXPORT dialog box

wish to create is entered. **Directory** lists the current pathname to which the files are being saved.

The **OPTIONS** dialog box (Figure 13-14) allows the user to define the format of the **DWG** and **DXF** files when they are imported or exported. Import optons include linetype, Font, Entity Format and Levels. Export options include the AutoCAD Version, Font Name, Chord Height Tolerance and External Reference.

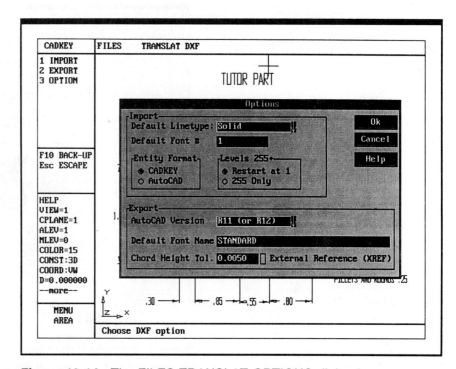

Figure 13-14 The FILES-TRANSLAT-OPTIONS dialog box

DELETING AND COPYING FILES

To copy or delete a file you must be at the system prompt **C:** or **A:**. There are two ways to get there. The first way is to exit CAD-KEY, type **cd**, and press **RETURN** to return to the **C:** or **A:**. The second way is to use the **SYSCMD** (DOS COMMAND) to get to the system prompt while you are still in CADKEY.

To try this enter CADKEY and create a few line and arc entities.

Press the Immediate Mode command **CTRL-F** to save the file. Use the name **jet** for the part name.

Use **SYSCMD** to get to the system prompt **c:\cadkey**. Use the Escape Code **F7-F8-F4 (CONTROL-SYSCMD-SHELL)**.

After leaving the drawing screen you will get a system prompt **C:\cadkey**. Type **cd** and press **RETURN** to enter the parent directory **C:**. Depending on how your system is set up, the system prompt will either be **C** or **C:**. At this point you can execute any DOS command.

To copy a file from the hard disk drive **c:** to the floppy disk drive **a:** type from the **C:** prompt **copy\cadkey\prt\jet.prt a:** and press **RETURN**. The bold text identifies: copy\drive specifier\pathname\filename.extension\drive specifier.

To make sure that the file has been copied to your floppy disk, from the **C:** prompt type **dir a:** and press **RETURN**.

To copy a file from a floppy disk **a:** to your hard disk drive **c:** type from the system prompt **C:\ copy a:jet1.ptn c:\cadkey\ptn**. Make sure to use the correct extension and path for the type of file you are working with.

To erase a file from the hard disk drive **c:** type from the **C:\ erase\cadkey\prt\jet.prt** and press **RETURN**.

To erase a file from a floppy disk **a:** type from the **C:** prompt **erase a: jet.prt** and press **RETURN**. The bold type identifies: erase drive specifier\pathname\filename.extension.

To return to CADKEY type **cd\cadkey** and press **RETURN**. Then type **exit**, press **RETURN**, and you will be back in CADKEY.

Listing and renaming files can be done through the use of the Files function in CADKEY. When a file is displayed on the screen it can be renamed during the saving procedure.

In CADKEY 6 this can be done in the file listing options.

 POSSIBLE USE

The following is an exercise to practice saving, copying, and erasing a file.

Step 1: Initialize CADKEY and place a floppy disk in drive **a:**. Create a few line and arc entities.

Step 2: Save the file by invoking the Immediate Mode command **CTRL-F,** entering the pathname and filename **c:\cadkey\prt\test,** and pressing **RETURN.**

Step 3: Use the Escape Code **ESC-F7-F8-F4 (ESCAPE-CONTROL-SYSCMD-SHELL)** to get to DOS.

Step 4: Type **cd** and press **RETURN** to exit the CADKEY directory.

Step 5: Type the copy command at the **C:\ copy\cadkey\prt\jet.prt a:** and press **RETURN** to copy the file from the hard disk to the floppy disk.

Step 6: Type the erase command at the **C:\ erase\cadkey\prt\test.prt** and press **RETURN** to erase the file from the hard disk drive.

Step 7: To return to CADKEY type **cd\cadkey** and press **RETURN.** Then type **exit** and press **RETURN** and you will be returned to CADKEY.

Step 8: To check to see that the file is now on the floppy disk execute the Escape Code **ESC-F5-F1-F3 FILES-PART-LST/LD** and enter **a:** as the pathname to list the part files on the floppy disk.

REVIEW QUESTIONS

1. What is the difference between a part file and a pattern file?
2. What is the extension that is automatically attached for each of these files?

> **PART**
> **PATTERN**
> **PLOT**
> **DXF**
> **CADL**
> **MACRO**

3. What is the Immediate Mode command for saving a part file?
4. How often should a part file be saved?
5. What is the advantage of the **LST/LD** and **LST/RTV** options over the **LOAD** and **RETRIEVE** options?
6. When you are indicating a base position, how can you accurately position the base point?
7. What can a **CADL** file be used for?
8. What is the Immediate Mode command for creating and ending a macro?
9. What is a macro?
10. Identify three ways that a macro can be executed.

Display Options

OBJECTIVES

After completing this chapter, you will be able to:

- Zoom in or out of an object on screen.
- Use the **PAN** option to change the displayed view.
- Add, remove, move, and mask entities on different levels.
- Use the **ACTIVE** and **LIST** level options.
- Control the grid display, increments, and alignment.
- Control the snap display, increments, and alignment.
- Redraw the screen.
- Identify the eight predefined views and view numbers.
- Create and save new views of a part.
- Add, remove, move, resize, and use the **AUTOSET** icon menu for viewports.
- Control the display of the construction and display view axes.
- Control the display of cursor tracking.

CADKEY's display options change the viewing area, set an on-screen grid to facilitate drawing, and control the level options. Many of these functions are accessed through the Immediate Mode commands. Using the **ZOOM** option temporarily enlarges, shrinks, fills the screen, doubles, or halves the size of the drawing currently displayed on the screen. CADKEY also maintains a rolling stack of the last three windows that were used. With the **LEVEL** option, details can be placed on different levels. The **GRID** and **SNAP** options can be used together to produce a drawing. Figure 14-1 shows the **DISPLAY** menu structure.

ZOOM OPTIONS

CADKEY's **ZOOM** option changes the window or viewing area on the drawing. It is useful when working on small details such as fillets, erasing entities that are close to other entities, or

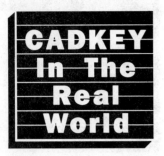

Sailing New Waters with CADKEY

The technology of wooden boat design and boat building had changed remarkably little until the late nineteenth century. Within the last forty years, however, the introduction of synthetic resins has created significant advances in the manufacture of high-performance boats. Now, with the help of CADKEY, one craftsman in Maine is aiming at new, even more significant changes in how boats are conceived and built.

Timothy Allen, president of T.N. Allen Company in Blue Hill, Maine, has a long history of manually drafting the essential technical drawings and handcrafting the wooden cabinetry found in better boats. At first glance, you might think that traditional boat building would not lend itself easily to computer-aided design and drafting, and that most traditional craftsmen would be skeptical about relying on a computer. However, an acquaintance who teaches yacht design at the Maine Maritime Academy introduced Tim to CADKEY.

New Light on Design Process

"He was using CADKEY for some basic boat designs," Tim said. "I was intrigued by how easy it is to create interchangeable modules with CAD software, and I recognized that I could save some money with computerized design tools." Tim liked what he saw well enough to call Cadkey, Inc. He spoke with Chris Williams, who happens to be an amateur boatbuilder.

"We were looking for actual applications to include in our next trade show, when Tim contacted us out of the blue," Chris said. "I'm currently restoring a 1946 raised-deck wooden boat, so Tim and I immediately recognized that we could help each other."

Tim concurs. "I've received some outstanding assistance from Cadkey. Chris Williams and Danielle Cote (Cadkey's Trade Show Events Manager) in particular have been very supportive and really helped push this project along. While I'm working primarily with boats constructed in wood, this new technology applies to any type of composite boat building, especially wood, metal, and fiberglass."

However, there's more to building boats than saving time and money. With CADKEY, Tim plans to build boat-interior components that will be better integrated structurally with the hull. That requires the complex 3D visualizing capabilities at the heart of CADKEY.

Fluid dynamics tax the capabilities of standard CAD programs, so boat designers rely on programs such as FairLine™ from AeroHydro Corporation and Prolines™ from Vacanti Yacht. "FairLine and Prolines are specialized boat-design programs that mathematically define the hull. From my perspective, though, what's interesting about it is that I can dump the mathematics that define the actual lines of the hull into CADKEY to generate the necessary technical drawings. And after the external components of the boat are complete, I can use the same models to build the interior and smaller details of the boat."

At some point, Tim would like to explore the possibility of using CADKEY to design the hull from scratch, without the FairLine or Prolines programs. "But right now," he says, "I'm just excited by the ways that CADKEY can help build better boats more efficiently. After I'm more proficient with CADKEY

A sample of a significant advance in the technology of building boats: Decolite® composite panel bent at 60-degree and 90-degree angles, with two-inch outside radii, epoxied with teak veneer using the vacuum-bagging technique, produced by Tim Allen.

—Continued, next page

CADKEY In The Real World—Continued

in automating the essentials of my craft, I can explore some of the more sophisticated aspects of the program."

Areas Where CADKEY Can Help

Among the areas where CADKEY could help are:

- Using CADKEY SOLIDS to calculate hull displacement, weight, and center of gravity to indicate how the hull will react to changes in design;

- Using CADKEY SURFACES to create a data base for finite element analysis with third-party programs; and
- Using CADKEY RENDER for the most exciting part of the design process: letting customers visualize precisely what a boat will look like.

"With CADKEY, I have the luxury of knowing that the sophistication is there when I'll need it. But right now, I don't have to worry about learning that to get my work done. And, I can add other software packages later that work well together," Tim said.

Adapted with permission from **3-D WORLD**, *v. 5, no. 2, Mar./Apr., 1991.*

---

detailing large drawings. Remember that when the window on a drawing is changed, only the display of the part is changed, not the data base. A display change is a temporary alteration which can be made at any time. However, any changes to the data base of the drawing, such as deleting a line or adding entities, will become part of the data base and remain after the viewing area is changed back. To actually change the scale of the part and the data base, use the **X-FORM SCALE** option described in Chapter 17.

Figure 14-1 shows the six different **ZOOM** options available. All six can be accessed through the Immediate Mode commands. These six options provide an unlimited number of choices for changing the viewing area of a drawing. You can window in to the smallest detail and make it fill the whole screen or window out and make a house floor plan look like a postage stamp. Remember, these are temporary alterations that do not have any effect on the data base of the drawing. The **BACK-1** option returns the part back to its previous size on the screen.

ZOOM OPTION: WINDOW

The **WINDOW** option defines the area of the drawing to be displayed on the screen through the use of a rubberbox. This rubberbox is stretched around the part of the drawing that you would like to display on the screen. Areas outside of this rubberbox are not displayed on screen. This option is accessed through the Immediate Mode command **ALT-W**. Figure 14-2 shows two views of a drawing that have filled up the screen. At times working on small details may be difficult using the current window, so the **WINDOW** option is used to enlarge the area to be used.

ZOOM OPTION: BACK-1

After having finished with the window (Figure 14-3), you may want to return to the previous window (Figure 14-2). This is accomplished

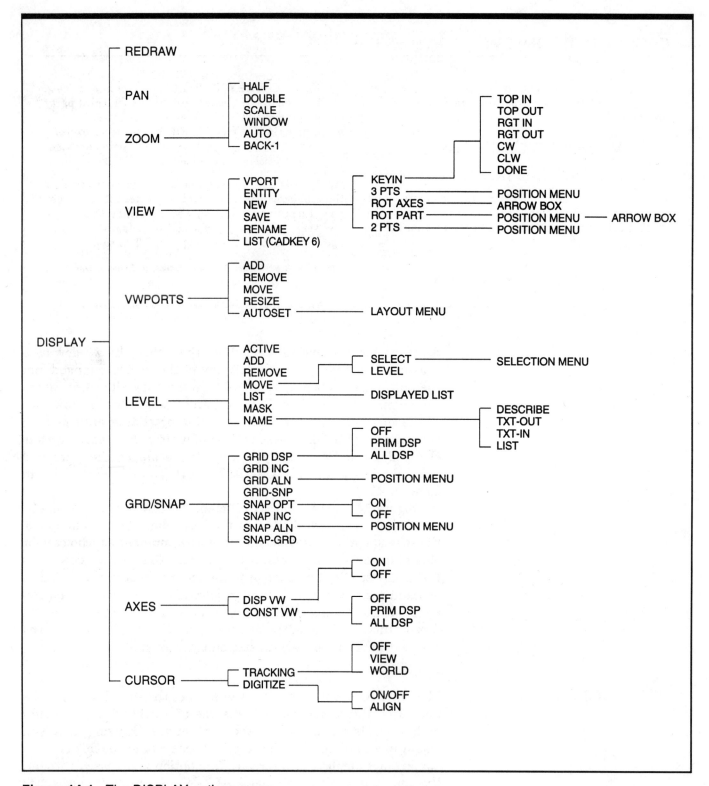

Figure 14-1 The DISPLAY options menu

using the **BACK-1** option, the Immediate Mode command **ALT-B**, or the Escape Code **ESC-F6-F3-F6** (**DETAIL-ZOOM-BACK-1**). The **BACK-1** option returns the display to the last window.

POSSIBLE USE

The following steps demonstrate the **WINDOW** option used to change the viewing area.

1. Invoke the Escape Code sequence **ESC-F6-F3-F4 (DISPLAY-ZOOM-WINDOW)**.

2. A prompt reads: **Cursor indicate 1st corner of new window.** Move the cursor to a position on screen that will be one of the corners for the new window. Pick this position by pressing the mouse button.

3. A second prompt reads: **Cursor indicate 2nd corner of new window.** A rubberbox is now functioning to assist in defining the new window. Move the cursor horizontally and vertically to stretch the rubberbox around the area. When the desired area for enlargement is surrounded by the rubberbox pick the point by pressing the mouse button. The window and its contents are immediately expanded to fill the screen (Figure 14-3).

4. Detail work can now be performed on the part. Press **ESC** to exit this option and return to the Main Menu. It is possible to enlarge the drawing even further if the need arises by using the **WINDOW** option on the current window.

Now, shrink the drawing back to its original window with the **BACK-1** option.

1. Use the Immediate Mode command **ALT-B**, or the Escape Code **ESC-F6-F3-F6 (DETAIL-ZOOM-BACK-1)**.

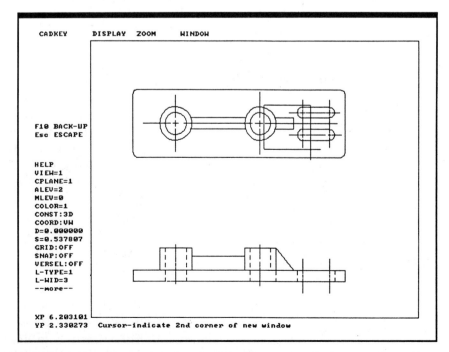

Figure 14-2 The area located within the square in the top view will be enlarged on screen using the WINDOW option.

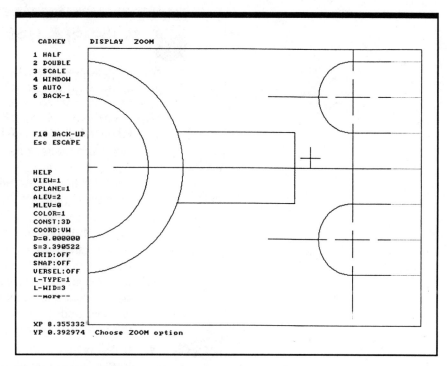

Figure 14-3 The new displayed window

2. The system automatically redraws the screen to the window immediately preceding the current display. In this example, the screen returned from Figure 14-3 to Figure 14-2. Press **ESC** to exit this option.

ZOOM OPTION: SCALE

The **SCALE** option allows the user to enlarge or shrink a part to any size that is input using this command. This option does not change the actual size of the part, only the display of the drawing. After the scale factor is input, use the cursor to define a new center position for the new window. For example, Figure 14-2 can be shrunk to half size to add a right side view. In this case, .50 is the scale factor to use. To enlarge a drawing, use 2.0 for doubling the size, 4.0 for quadrupling, and so forth. Scale factors larger than 1 will enlarge a drawing and scale factors less than 1 shrink drawings. After the scale factor is input cursor-indicate the new center or press **RETURN** to automatically center the part.

Use the Status Window to determine the current scale value of the window. Notice that in Figure 14-2, the scale is 0.53665. This value also appears in the Prompt Line when using the **SCALE** option. This command is accessed by using the Immediate Mode command **ALT-S**, or the Escape Code sequence **ESC-F6-F3-F3** (**DETAIL-ZOOM-SCALE**).

POSSIBLE USE

There are times when it is necessary to view a drawing at different scales to get an idea how the layout might look before plotting or simply to determine where to place details on the drawing. In this example Figure 14-4 is going to be shrunk so that a right side view can be added to the drawing. The following steps demonstrate the **ZOOM-SCALE** option.

1. Use the Immediate Mode command **ALT-S** or the Escape Code **ESC-F6-F3-F3 (DETAIL-ZOOM-SCALE)**.

2. A prompt reads: **Enter new scale factor (current= 0.536646)=** Notice that this scale is the same as displayed in the Status Window. Key in **.30** for the new scale factor and press **RETURN**.

3. A prompt reads: **Cursor indicate new center in viewport.** Move the cursor to a new center located slightly to the right of the current center of the display and press the cursor button. A new screen is immediately displayed (Figure 14-5).

4. Press **ESC** to exit this function.

ZOOM OPTION: HALF

The **HALF** option reduces the part display scale to one-half of the current scale. To use this option choose the Escape Code sequence **ESC-F6-F3-F1 (DISPLAY-ZOOM-HALF)** or the Immediate

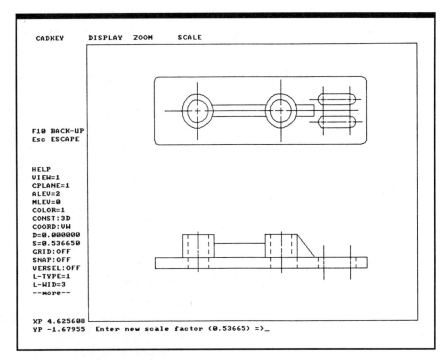

Figure 14-4 Part to be shrunk using the SCALE option

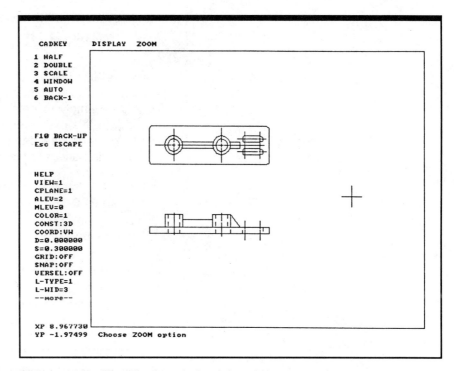

Figure 14-5 The displayed view shrunk to .30 scale

Mode command **ALT-H**. As soon as the option is chosen, the drawing is reduced to one-half size.

 POSSIBLE USE

The **HALF** option can be used in place of the **SCALE** option in the previous example to make room for a right side view. The following steps show how this option might be used.

1. Press **ALT-B** to return to the part display (Figure 14-4).
2. Use the Immediate Mode command **ALT-H** or the Escape Code sequence **ESC-F6-F3-F1**. The part is immediately redrawn on screen one-half its original size (Figure 14-6). Notice that the Status Window scale reads 0.26832, which is one-half of 0.536646.
3. Press **ESC** to exit this option.

ZOOM OPTION: DOUBLE

As you might have guessed, the **DOUBLE** option doubles the current size of the part display. This option is assigned the Immediate Mode command **ALT-D** and the Escape Code sequence **ESC-F6-F3-F2**. This option is immediately executed after it is selected. If the **DOUBLE** option were to be used on Figure 14-6 it would double back to its original size (Figure 14-7).

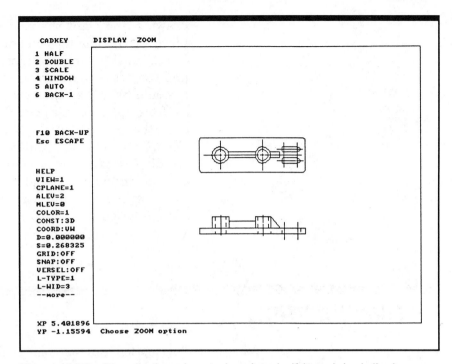

Figure 14-6   The displayed part shrunk to half its original display.

ZOOM OPTION: AUTO

AUTO automatically scales the entire working part to fit on screen at the largest possible size. Entities not currently displayed, such as those on a non-displayed level, are not considered part of the working part. For example, suppose a drawing had been completed so the border lines and title blocks were added to the drawing. Afterwards you decide to make some changes to the

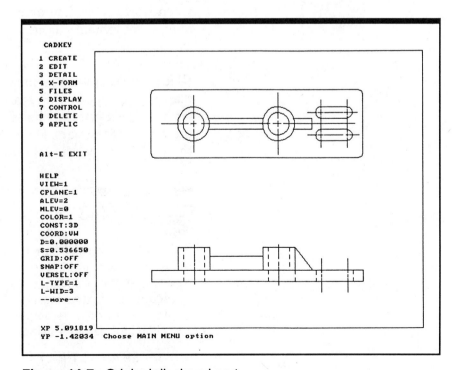

Figure 14-7   Original displayed part

drawing and want the part to automatically fill the screen. You place the border line and title block onto a non-displayed level. Those entities are not considered part of the working drawing and the actual part drawing would appear on screen larger than it would have, had the border and title block been displayed.

This option is assigned the Immediate Mode command **ALT-A** and the Escape Code sequence **ESC-F6-F3-F5 (DISPLAY-ZOOM-AUTO)**. The working part is immediately scaled to fill the display after the **AUTO** option is selected. Press **ESC** to exit this option. As an example, Figure 14-6 is taken from its half-size fit on screen to a full screen display (Figure 14-7) by using the **AUTO** option.

THE PAN OPTION

This option can be compared to looking through a pair of field glasses at some distant object on the horizon. As you look through the field glasses you might decide to change your field of sight by moving the glasses slightly to the right. When you do this you are changing your field of sight by panning across the horizon. Think of the **PAN** option used with CADKEY in the same way.

The **PAN** option redefines the center of the display relative to the part. This option scans across the working part on screen. It looks as if the Drawing Window is shifted with respect to the part. It can be used whether the whole working part is displayed or is at half, double or some other scale. It is particularly useful when working on fine details in a close zoom window because it allows movement around the screen without changing the current zoom scale. **PAN** is selected using the Immediate Mode command **ALT-P**, or the Escape Code sequence **ESC-F6-F2 (DISPLAY-PAN)**. The new center position for the working display is then input through cursor position. When the cursor is moved, a box attaches itself to the cursor. The cursor represents the center of the part. Move the cursor to a new position and pick the point. The part is repositioned on screen.

 POSSIBLE USE

The **PAN** option can be used to view different parts of a zoomed drawing in order to add detail. It can also be used to change the working display so that other features can be added to a drawing. In this example, Figure 14-7 will be panned across the screen. The following steps describe how this might be done using CADKEY.

1. Select **ALT-P** or **ESC-F6-F2 (DISPLAY-PAN)**. A prompt reads: **Cursor indicate new window position**.

2. Move the cursor to the right center of the screen and press the mouse button. The display changes (Figure 14-8), so that a right-side view can be added to the drawing. Select **ESC** to exit this function.

THE LEVEL OPTIONS

The **LEVEL** options manipulate, move, delete, name, list or recall levels of a drawing. Levels separate different entities that are part of a drawing. The **LEVEL** function controls which newly created entities are to be assigned through the **ACTIVE** option. A list of the 256 levels can also be displayed on screen to change their display status. Remember that level 256 is never displayed even if you have it turned on or active. It is reserved for those entities which are chosen not to be displayed. Entities can also be moved from one level to another with this display function. Separating entities by level allows the user to display different entities and to plot entities by level.

LEVEL OPTION: ACTIVE

ACTIVE allows the choosing of levels to which newly created entities are assigned and is changed before the entities are to be drawn.

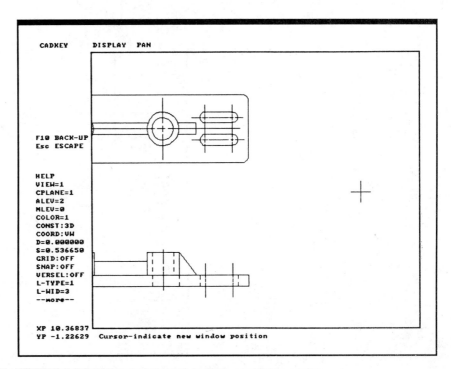

Figure 14-8 View of the part after panning

If the active level is not already displayed, it is added to the system's display list of levels. The current active level is displayed in the Status Window as **ALEV:1**. The number 1 represents the current active level as being level 1. This also is the default setting used by CADKEY when the system is turned on. Only one level can be active at a time. This option is accessed using the Immediate Mode command **CTRL-L** or the Escape Code **ESC-F6-F6-F1**.

 POSSIBLE USE

One possible use of the **ACTIVE** option is to assign different entities to different levels so they can be separated for plotting. For example, Figure 14-9 shows an orthographic drawing of a part that has had some entities separated by level. All the hidden lines are placed on level 2, which is not active at this time and is not displayed on screen. The following steps demonstrate how the **ACTIVE** option is used to change the active level.

1. Notice that on Page One of the Status Window, the current active level is **1 (ALEV:1)**. Press **CTRL-L** or **ESC-F6-F6-F1 (DISPLAY-LEVEL-ACTIVE)**. You are prompted: **Enter new level number from 1-256 (current=1) =**. Press the number **2** key and **RETURN**.

2. Status Window **ALEV:2** is displayed. Every entity drawn from this point on will be assigned to level 2 unless the active level is again changed. Entering **ESC-F6-F6-F2**

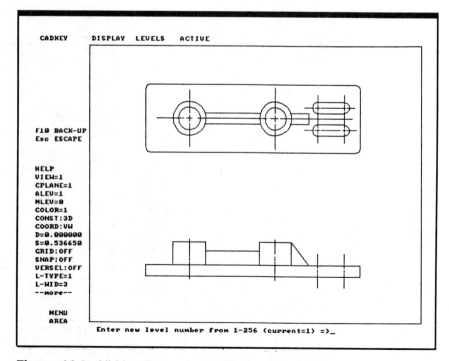

Figure 14-9 Hidden lines are not displayed.

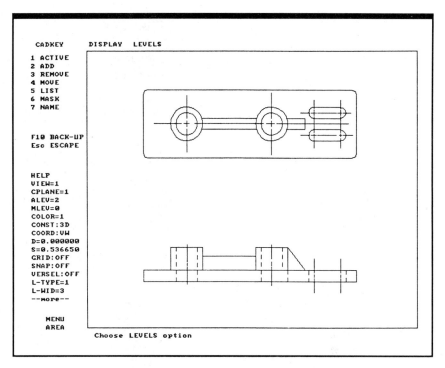

Figure 14-10   Hidden lines are displayed after the ADD option is used.

(DISPLAY-LEVEL-ADD) then entering the number **2** will display the hidden lines (Figure 14-10).

3. Another prompt reads: **Enter level descriptor:**. A name using up to thirty characters can be used to further describe this level. For this example, enter **dimensions**.

4. By selecting **ESC-F6-F6-F7-F4 (DISPLAY-LEVEL-NAME-LIST)**, a list of level numbers and names will be displayed on screen. This list will show the level numbers, active and currently displayed levels (marked with asterisks), the number of entities on each level, and the level name (Figures 14-11 and 14-12).

5. The active level can be changed by selecting **F1 ACTIVE**, highlighting the new active level with the cursor, and pressing the mouse button. Levels can be turned off or on by selecting **F2 DISPLAY**, highlighting the level to be turned on or off, and pressing the mouse button. The name of the level can be changed by selecting **F3 DESCRIBE**, highlighting the level to be named or changed, pressing the mouse button, and entering the name at the Prompt Line.

LEVEL OPTIONS:
ADD AND REMOVE

The **ADD** and **REMOVE** options turn levels on or off for viewing purposes within the current display. Do not confuse this option with the **DELETE-LEVEL** option, which deletes all the entities on that level. The **ADD** and **REMOVE** options can be compared

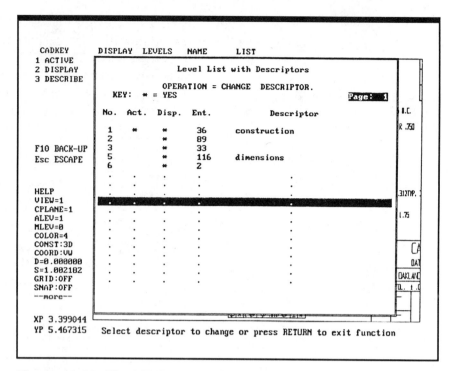

Figure 14-11 The LEVEL-LIST menu for CADKEY 5

to turning a light switch on or off. They only temporarily turn off those entities found on designated levels. These options are extremely valuable when working with complicated drawings because they allow you to view only those entities that are important to the current task. Turning off entities by level will

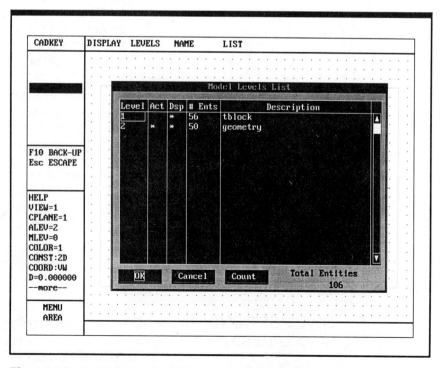

Figure 14-12 DISPLAY-LEVELS-NAME-LIST (CADKEY 6)

increase the redraw time and the time it takes the system to search entities for various tasks.

ADD designates the levels that are to appear in the current display. Any number of levels can be visible at one time. **REMOVE** designates those levels that you do not want to have displayed. For either option more than one level may be designated by separating the level numbers with commas (1,4,5,6), by using a hyphen or dash (1-4, 5-7) to identify a series of levels, or by using a combination of these methods (1,2,3,4-6,7-10).

POSSIBLE USE

The **REMOVE** option temporarily turns off entities such as dimensions and cross-hatching. In Figure 14-13, the part displayed has crosshatching. The level having the crosshatching could be turned off using the **REMOVE** option. The following steps show how the **REMOVE** option turns off entities.

1. Select **ESC-F6-F6-F3 (DISPLAY-LEVELS-REMOVE).** A prompt reads: **Enter level(s) to be removed**. Press the number **2** key. The number 2 is added to the prompt line.
2. Press **RETURN** to remove the display features on Level 2 (Figure 14-14).

The **ADD** option can be used to add levels that are to appear in the current display. Crosshatching has been placed on level 2 of the drawing in Figure 14-14. However, level 2 has previously

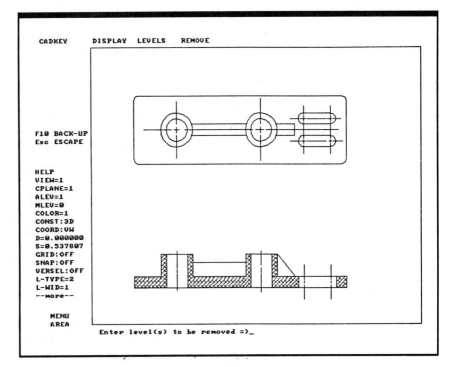

Figure 14-13 Using the LEVELS-REMOVE option to turn off hatching

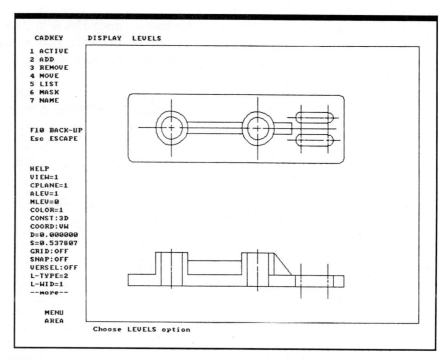

Figure 14-14 The hatching removed from the drawing

been removed. The steps shown below will describe how the **ADD** option is used to turn level 2 on and view the cross-hatching.

1. Select **ESC-F6-F6-F2 (DISPLAY-LEVELS-ADD)**. A prompt reads: **Enter level(s) to be added =**
2. Enter the number **2** and **RETURN**. The crosshatching re-appears (Figure 14-13).

LEVEL OPTION: MOVE

The **MOVE** option moves entities to another level. For example, if all notes had been drawn on active level 1 and now you wanted all the notes separated from the other drawing entities on level 1, they could be moved to another level using **MOVE**.

There are two methods that can be used to move entities to different levels: **SELECT** and **LEVEL**. The **SELECT** option will display the Selection Menu so you can select entities to move using **SINGLE, CHAIN, WINDOW, POLYGON, GROUP, PLANE, ALL DSP**. All of these options are described in Chapter 11.

The **LEVEL** option is used to identify all entities that are assigned to a level so they can be moved to another level. This option is accessed by using the Escape Code. **ESC-F6-F6-F4-F2 (DISPLAY-LEVEL-MOVE-LEVEL)**. A prompt will follow: **Move entities from level =**. The level number is then input followed by **RETURN**. A second prompt will be displayed: **To**

level =. Input the level that the entities are to be moved to and press **RETURN**. The screen is automatically updated. Press **ESC** to exit the function.

 POSSIBLE USE

Entities are moved to different levels to facilitate deleting, plotting, viewing, redrawing, and editing. As an example, a single entity will be moved to another level (Figure 14-15).

1. Press **ESC-F6-F6-F4 (DISPLAY-LEVEL-MOVE)**. The menu displays two options, **F1 SELECT** and **F2 LEVEL**.

2. Press **F1 SELECT**. A prompt reads: **Move entities to level =**. Enter number **4** to move the entity to level 4. The number **4** appears in the prompt. This number can be changed by using **BACKSPACE** and typing in a new number.

3. Press **RETURN**. The Selection Menu will appear on screen. Only a single item will be moved, so select **F1 SINGLE**.

4. A prompt reads: **Select entity 1 (press RETURN when done)**. Pick one of the entities from Figure 14-14. A small triangular marker is displayed over the picked entity, or the entity is highlighted, depending on the setting in the configuration program. Notice that the entity is removed from the drawing if level number **4** is not a visible level

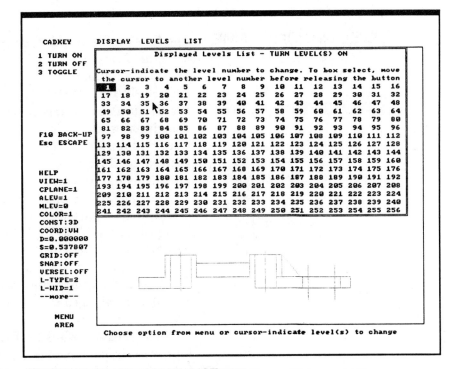

Figure 14-15   The LEVEL-LIST screen.

after **RETURN** is pressed. Use the **ADD** option to turn the level on.

5. Press **ESC** to exit this function.

All the other options available under the Selection Menu are used in the same way as described in Chapter 11. For example, to move entities to a new level using the **WINDOW** option pick the first position for the window and the second position for the window. Any entity enclosed by the window is moved to the specified level.

LEVEL OPTION: LIST

The **LIST (ESC-F6-F6-F5)** option allows the user to generate a list of levels on the screen and change the current display status. When this option is selected a pop-up menu is shown and level numbers 1-256 are displayed on screen. A prompt will ask: **Choose option from menu or cursor-indicate level(s) to change** (Figure 14-15).

Active levels are displayed in inverse video. The cursor is used to turn levels on or off by positioning the cursor over the level number and pressing the mouse button. If a level is turned **OFF** the inverse video is removed from around the level number. If the level is turned **ON** the inverse video is drawn around the level number.

After all changes have been made the **RETURN** or **ESC** key will exit this function. The screen will also be redisplayed with the image which was removed to display the level numbers. This command temporarily disables the Immediate Mode commands.

LEVEL OPTION: NAME

The **NAME** option has four sub-menu items used to describe, list, save, and retrieve level descriptors. The **DESCRIBE** option attaches a name to a level. Enter **ESC-F6-F6-F7-F1 (DISPLAY-LEVEL-NAME-DESCRIBE)**, and a prompt will ask: **Enter level number (current=1)**. Enter the level number for the descriptor, such as level 4. The next prompt asks: **Enter level descriptor:**. Enter the name of the descriptor you want. You are limited to thirty-one characters in a name. After entering the name this descriptor is attached to the name (Figure 14-16).

The **TXT-OUT** option saves the current list of descriptors as a text file to be used in a word processor or to use with the **TXT-IN** option. The **TXT-IN** option will read in level descriptors and assign them to the current drawing. This option is especially useful if you have a standard method of separating entities. The **LIST** option displays the levels and descriptors on screen (Figure 14-15), and changes the active, displayed, and descriptor for each level.

In CADKEY 6 the **LEVEL-NAME-LIST** option is displayed as a dialog box as seen in Figure 14-17. The **COUNT** option displays entity types by level.

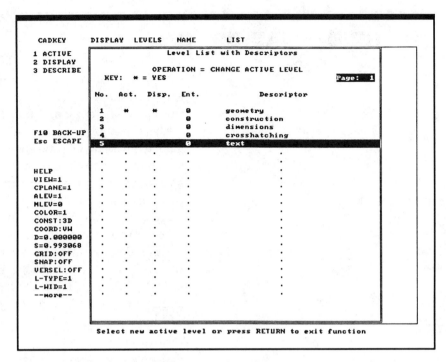

Figure 14-16 The NAME-LIST option displays the current level settings. (CADKEY 5)

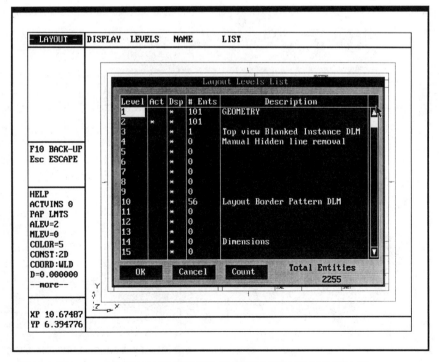

Figure 14-17 DISPLAY-LEVELS-NAME-LIST (CADKEY 6)

THE GRID/SNAP OPTION

The **GRID/SNAP** option allows the user to draw with greater ease for some tasks. A grid is a series of small dots of equal or unequal X,Y spacing that facilitates some drawing tasks with CAD. Entities, patterns, objects, and other features can use the grid as a point of reference. For example, lines can be drawn on screen by snapping to a series of grid points. The grid can be turned on or off using the Immediate Mode command **CTRL-G** or by using the Escape Code **ESC-F6-F7 (DISPLAY-GRID/SNAP)**. Page One of the Status Window can be used to view the current state of the grid or to toggle between its on and off states.

The default value used when the grid is turned on is .50 inches on the X and Y axes. This spacing can be changed by using the **GRID-SNP** or **GRID INC** options. The grid can also be set to align with an object already drawn on screen using the **GRID ALN** option.

GRID/SNAP OPTION:
GRID DSP (Grid Display)

Selecting **GRID DSP** offers three options: **PRIM DISP, ALL DSP**, and **OFF**. The primary display (**PRIM DSP**) option turns the grid on only in the primary viewport. The all display (**ALL DSP**) option turns the grid on in all the viewports (Figures 14-18 and 14-19). The Immediate Mode command **CTRL-G** can be used to toggle the grid on and off, or the Escape Code **ESC-F6-F7-F1 (DISPLAY-GRD/SNAP-GRID DSP)** can be used to turn it on. After the grid is turned on, the grid will immediately be displayed. However, if the current viewing scale makes the grid

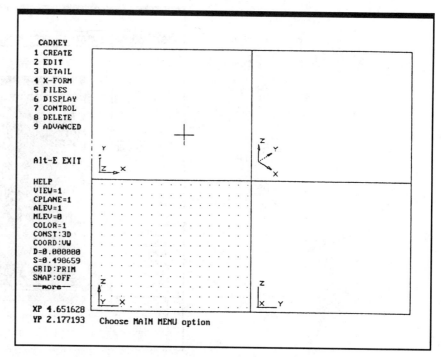

Figure 14-18   Grid turned on in the primary viewport

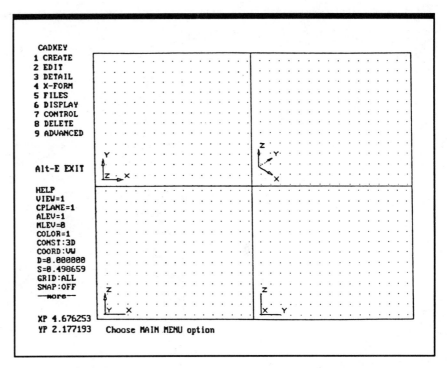

Figure 14-19   Grid turned on in all the viewports

appear too close together to be useful, the grid is automatically disabled. The grid will automatically be re-enabled when the scale will allow a reasonable number of grid points on screen. The grid display can be stopped before the screen is filled by pressing the **ESC** key. **OFF, CTRL-G,** or the Status Window Page One turn the grid off.

GRID/SNAP OPTION: GRID INC (Grid Increment)

This option sets the X,Y spacing of the grid. Different values can be assigned to each grid so they are rectangular and not square. You are limited to .00005 spacing or greater, up to a total of 10,000 units. This option is accessed by pressing **ESC-F6-F7-F2 (DIS-PLAY-GRD/SNAP-GRID INC)**. A prompt is displayed: **Enter X grid increment (0.500000) =**. Enter a value such as **.25** and press **RETURN**. Another prompt asks: **Enter Y grid increment (0.500000) =**. Enter the same value or **.50** followed by **RE-TURN**, or simply press **RETURN** to accept the current value of .500000. The new grid is displayed on screen. Figure 14-20 shows a grid with different X and Y values.

GRID/SNAP OPTION: GRID-SNP (Grid-Snap)

This option controls the grid spacing by the size of the snap spacing. For example, if the current snap spacing is set at .25″ and the grid is set at .50″, the grid can be set to the snap spacing using **GRID-SNP** much more quickly than using **GRID INC**. The sequence of commands to use for this option is **ESC-F6-F7-F4 (DISPLAY-GRD/SNAP, GRID-SNP)**. The screen is updated immediately.

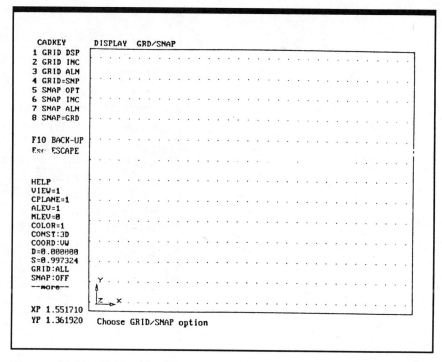

Figure 14-20   Grid with different X and Y values

GRID/SNAP OPTION: GRID ALN (Grid Align)

This option aligns a grid with an entity on screen, such as the end-point of a line or the center of a circle. This option is selected by pressing **ESC-F6-F7-F3 (DISPLAY-GRD/SNAP-GRID ALN)**. The Position Menu is displayed and a prompt requests that a new grid alignment point be indicated.

 POSSIBLE USE

This option is useful when an object has been drawn without the use of a grid and the user would like to use a grid to place dimensions on the part. If you were to just place a grid on screen it might not align with the part because the part was not drawn using the grid. By using the grid align function it is possible to have the grid aligned with the object. The following steps show how this might be done, using Figure 14-21 as an example.

1. Using the **GRID INC** option set the grid spacing to .25″ horizontally and .25″ vertically.
2. Press **ESC-F6-F7-F3 (DISPLAY-GRD/SNAP-GRID ALN)**.
3. The Position Menu is displayed. Select **F3 ENDENT** so the end of the line will be the position to which the grid will be aligned. For this example pick the lower right corner of the object using the **SINGLE** option.

THE SNAP OPTIONS

The **SNAP** options control the movement of the cursor. Without the snap feature, the cursor is free to move to any position on the screen. However, there are times when you may want to have some measure of control over the movement of the cursor. For example, the object shown in Figure 14-21 could have been drawn using the **SNAP** function. The **SNAP** option can be turned on or off using one of three methods. It can be toggled on and off using the Immediate Mode command **CTRL-X**. The Status Window can turn it on or off, and it can be used to determine its present status. For example, if the snap function is off, the Status Window will show: **SNAP:OFF**. The third method that can be used to turn **SNAP** on or off is **ESC-F6-F7-F5**.

The default value used for the snap function is .50″, which is the same as the default value for the grid function. Moving the cursor around the screen after having turned the snap function on will cause the cursor to snap or jump in .50″ increments around the screen. It is not possible to move between the .50″ increments unless the snap is turned off or the spacing is changed. The spacing can very quickly be changed by using the **PgUp** and **PgDn** keys located on the right side of the keyboard on most computers. The **PgUp** key will cause the snap increment to become double the current snap increment. So the default increment of .50 will change to 1.00 with one press of the key, 2.00 after two presses of the key, and so forth. The **PgDn** key will change the current snap increment by one-half. So the default increment of .50 will become .25 with one press of the key, .125 with two presses of the key, and so forth.

The snap spacing can be changed by using the **SNAP INC** option (**ESC-F6-F7-F6**), which sets the X,Y spacing by assigning a value to each axis. The snap spacing can also be changed by

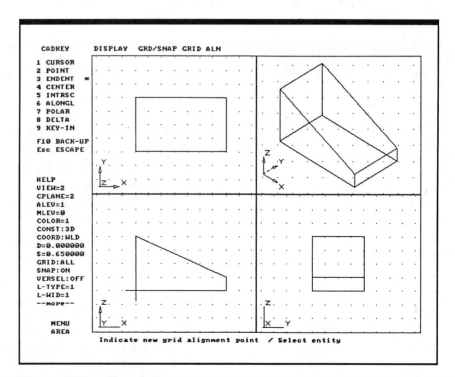

Figure 14-21   Screen display after the grid is added

using the **SNAP-GRD** option (**ESC-F6-F7-F8**), which automatically sets the snap spacing to the current grid. The **SNAP ALN** option (**ESC-F6-F7-F7**) aligns the snap with an indicated position on screen. With this option, the Position Menu is displayed so that the point for snap alignment can be easily defined.

THE REDRAW OPTION

REDRAW cleans up a drawing when markers and other screen symbols have to be removed from the screen. It eliminates broken lines that result from creating and editing entities. The most effective method to use with this option is the Immediate Mode command **CTRL-R**. This will automatically cause the screen to be refreshed. If multiple viewports are displayed, the viewport to be redrawn must be cursor-selected, or press **RETURN** to redraw all the viewports. While the display is being redrawn, a message in the prompt area will read: **Redrawing display...press ESC to abort**. The **CTRL-BREAK** keys or **ESC** can be used at any time to stop the redraw. The other method that can be used for **REDRAW** is **ESC-F6-F1** (**DISPLAY-REDRAW**). This sequence will automatically cause a redraw of the screen if there is only one viewport.

THE VIEW OPTIONS

The **VIEW** options allow the user to manipulate a 3D wireframe part and view it from virtually any angle. When more than one viewport is displayed, you must specify the viewport. From the wireframe model, the six principal orthographic views can be created. Each orthographic view has a view number assigned, as shown below. In addition, there are two pictorial views that have assigned view numbers: Isometric and Axonometric.

VIEW ASSIGNED NUMBER
Top 1
Front 2
Back 3
Bottom 4
Right 5
Left 6
Isometric 7
Axonometric 8

As new views are created, they are added to the list and assigned the next available view number. After a 3D model of a part has been drawn, it is possible to automatically extract the eight assigned views shown above by simply choosing the view number. Selecting the **AUTOSET** viewport for four views displays views 1, 2, 5, and 7 (Figure 14-22). Selecting **VIEW** or the Immediate Mode command **ALT-V** prompts you to enter another view number; the display will change to the new assigned view.

Views other than those listed are defined using one of the **VIEW** options. The **NEW** option has five sub-menu items: **KEYIN, 3 PTS, ROT AXES, ROT PART,** and **2PTS**, which create different viewpoints of the part. The **ENTITY** option defines a new view by selecting a planar entity, such as a circle or arc, conics, or 2D splines. The **VIEWPORT** option changes the primary viewport's assigned view.

The following steps demonstrate how to change a viewport to one of the eight assigned views.

1. Select **DISPLAY-VIEW (ESC-F6-F4)**, press **ALT-V**, or select **VIEW** from the first page of the Status Window.

2. A prompt reads: **Cursor select viewport**. Pick the isometric viewport.

3. A prompt reads: **Enter view number or choose menu option (1-8) =**.

4. Enter the number **8** and press **RETURN**. The isometric viewport changes to the axonometric view. Enter **ALT-A** to automatically scale the view to fit in the viewport (Figure 14-23).

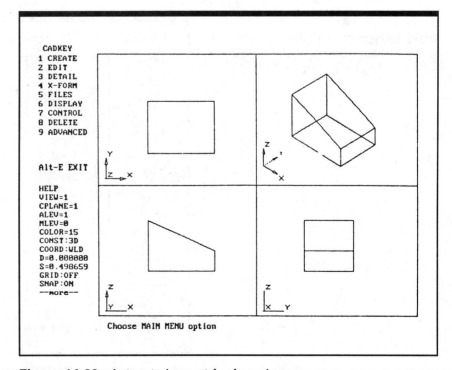

Figure 14-22 Autoset viewport for four views

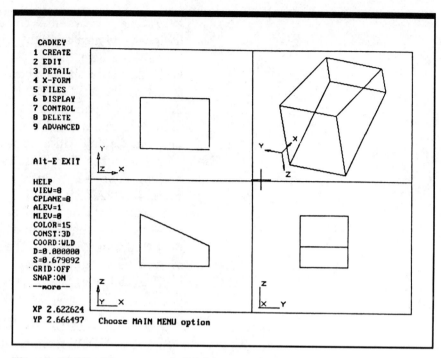

Figure 14-23 The upper-right viewport changed to view number 8

VIEW OPTION: ENTITY   The **ENTITY** option selects an existing planar entity, such as an arc, conic, 2D spline, or line, to create a new view. Figures 14-24 and 14-25 demonstrate creating a new view. Enter **ESC-F6-F4-F2 (DISPLAY-VIEW-ENTITY)**. A prompt requests that an

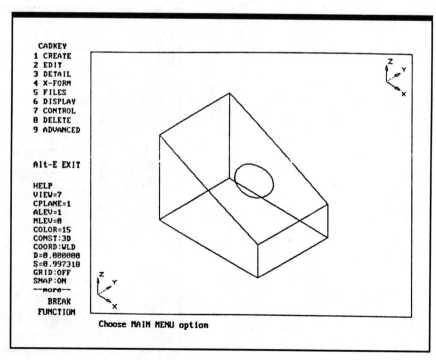

Figure 14-24 The ENTITY option creates a new view.

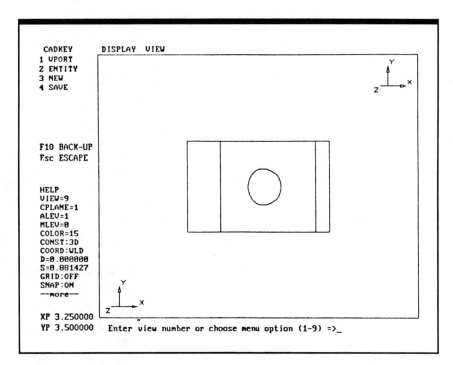

Figure 14-25 New view number 9 created by picking the circle

entity be selected. Pick the circle located on the inclined surface in Figure 14-24. A new view is displayed on screen (Figure 14-25).

VIEW OPTION: VPORT

The **VPORT** option assigns an existing view in a multiple view-port screen to the primary viewport. It is a way of changing the primary viewport from its default to a user-specified one. Enter **ESC-F6-F4-F1 (DISPLAY-VIEW-VPORT)**. A prompt requests that the primary viewport be selected with the cursor. The selected viewport automatically takes on the new display view.

VIEW OPTION: NEW

This option creates a new view either by specifying three points to define a plane using the **3 PTS** option, by using the **KEYIN** option to enter an angle to create a new view, or by specifying a rotation of the axes or the part to create a new view using the **ROT AXES** or **ROT PART** options. Whenever a new view is created using one of these options it is assigned the next available number in the view list and it will be displayed in the Status Window after the **VIEW-SAVE** option is selected. By defining a viewpoint, the user can view any object from any direction.

The 3 PTS Option

With this option any three points can be defined, which will create a unique viewing direction. After this option is selected, the Position Menu is displayed to more accurately define the three points. If the points that you have selected create a unique view, a temporary view is created and assigned the number -1. You must **SAVE** the view with **ALT-V SAVE** or else you will have to define it again.

POSSIBLE USE

The **3 PTS** option is useful when adding details to an inclined or oblique surface. The following steps show how **3 PTS** creates a view perpendicular to the inclined surface of the part shown in Figure 14-25.

1. Select **ESC-F6-F4-F3-F2 (DISPLAY-VIEW-NEW-3 PTS)**. The Position Menu and a prompt are displayed: **Indicate 1st pt on X axis/Cursor-indicate position**
2. From the Position Menu select **F3 ENDENT**.
3. Move the cursor to one of the lines on the inclined surface and pick it. This point is marked with an "**X**" (Figure 14-26).
4. A second prompt reads: **Indicate 2nd pt on X axis/Cursor-indicate position.**
5. Move the cursor to the other endpoint of the line and pick it. This point is marked with a small "**X**" (Figure 14-26).
6. A third prompt reads: **Indicate direction for Y axis/Select ENDENT.**
7. Pick a line on the inclined surface that is perpendicular to the first line chosen, as shown by the "**X**" in Figure 14-24. A new view is displayed and added to the view list (Figure 14-27).

The KEYIN Option   **KEYIN** defines a new view by assigning a series of rotations to the currently displayed part. After the **KEYIN** option is selected a menu is displayed showing six different methods of rotating the

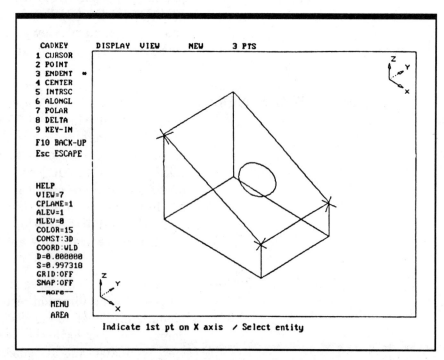

Figure 14-26 The 3 PTS option creates a new view.

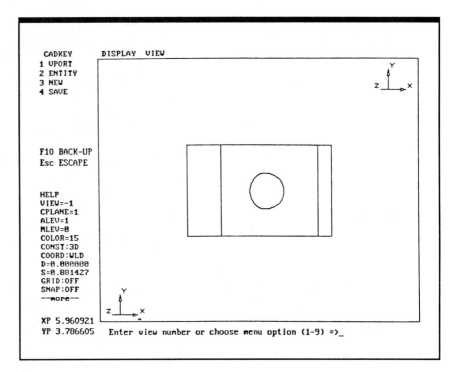

Figure 14-27   A view perpendicular to the inclined plane using the 3 PTS option

part: **F1 TOP IN, F2 TOP OUT, F3 RIGHT IN, F4 RIGHT OUT, F5 CW (clockwise),** and **F6 CCW (counterclockwise).** A seventh option, **F7 DONE,** indicates to the system that you have completed the number of rotations for the new view. The **IN** options rotate the view into the screen. The **OUT** options rotate the part out of the screen. A rotation order can be specified by using the displayed menu options. For each option chosen a rotation angle must be input. This is done by entering a positive value for the angle and pressing **RETURN**. The part is rotated about either the X, Y or Z axis or any combination of them to define a new view.

 POSSIBLE USE

The uses for this option are the same as described under the **3 PTS** option. The **KEYIN** option is just another way of creating new views that you may find more useful for some views than using the **3 PTS** option. The following steps show how a view might be created using the **KEYIN** option in Figure 14-27.

1. Select **ESC-F6-F4-F3-F1 (DISPLAY-VIEW-NEW-KEYIN).** A prompt reads: **Choose rotation direction(s).** A menu of choices is displayed as described earlier.
2. Select **F1 TOP IN.** A prompt reads: **Enter rotation angle =.**
3. Enter the number **45** and press **RETURN**.

4. The same menu as in Step One is displayed. At this point you could specify another rotation angle using one of the other menu items, such as **F4 RGT OUT** or **F5 CW**. By doing this, a series of rotations can be connected to specify a new view. In this example one rotation angle is sufficient, so menu item **F7 DONE** is selected.

5. If necessary, select **ALT-A** to scale the drawing to the screen. This new view will be displayed (Figure 14-28).

6. This new view is not automatically added to the display list. To add the new view to the list enter **F4 SAVE**. A message appears **"Saving view"** and the view is assigned the next available view number.

The ROT AXES (Rotate Axes) Option

The **ROT AXES** option creates a new viewing direction by changing the axes. Selecting this option displays the axes of the selected viewport in the menu area along with arrows indicating the direction that the axes can be moved to create a new view (Figure 14-29). Select the arrow direction to define the new view by moving the cursor onto the arrow in the menu area and picking it. Holding down on the pick button of the input device causes continuous rotation of the axes. After the axes are in the desired position for the new view, press **RETURN**. The active viewport is redisplayed with the new view rotated into the position defined by the new axes. New views are created by selecting one arrow key in the menu area or by combining the selection of many arrow keys. Figure 14-30 shows how the part shown in Figure 14-29 is changed by picking the right arrow key four times

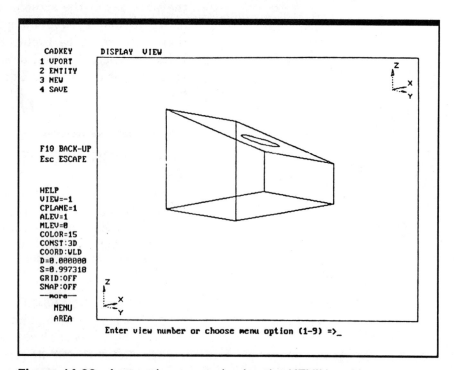

Figure 14-28 A new view created using the KEYIN option

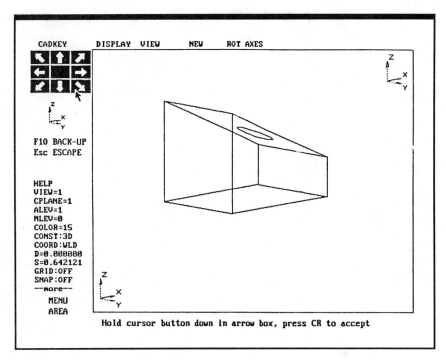

Figure 14-29   The ROT AXES option

and the down arrow key four times. To add the new view to the
list of views, use **VIEW-SAVE**.

**The ROT PART
(Rotate Part) Option**

ROT PART creates a new view of a part by dynamically rotating
the view in the active viewport. After selecting **ROT PART** you
are prompted to select the active viewport (if more than one is

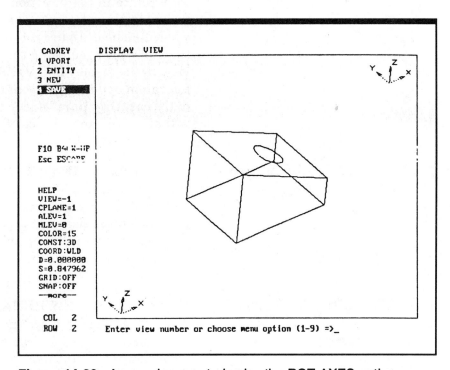

Figure 14-30   A new view created using the ROT AXES option

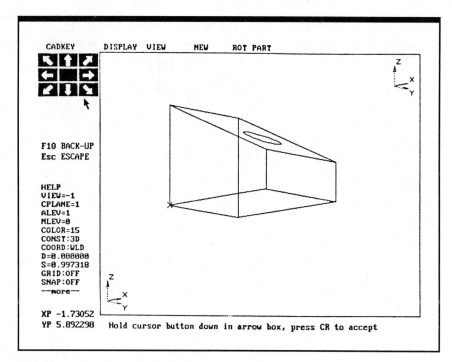

Figure 14-31   The ROT PART menu

displayed), then cursor-indicate the center of the part for rotation. The Position Menu is available to assist in picking the desired center point. After the point is selected, the rotation arrow icon is displayed in the menu area (Figure 14-31). To change the view move the cursor over one of the arrow icons and pick it. Holding down the mouse button will cause the view to dynamically rotate in the direction of the arrow. Pick other arrow icons to rotate the part into the desired view and press **RETURN** when finished. Figure 14-32 shows a series of views of a part that were changed using the **ROT PART** option. To add the new view to the display list, **VIEW-SAVE** must be executed. The configuration option "**MAX # of entities to display while dynamically rotating part**" under the **GRAPHICS** option in **CONFIG** will affect the appearance of the rotated part.

VIEW OPTION: SAVE

The **DISPLAY-VIEW-SAVE** option allows the user to save a system view or auxiliary view which they have defined. Commonly it is a surface where a construction plane has been required to place a circle in its true size and shape. The view will have the view number -1 assigned to it until it has been saved. When the view is saved the next available view number is assigned.

VIEW OPTION: RENAME

The **DISPLAY-VIEW-RENAME** option allows the user to rename or assign a name to a system view which has been defined for the model on which they are currently working. The advantage of naming a view is that the name better describes the view and is easier to remember than a number. To rename a view use the rename

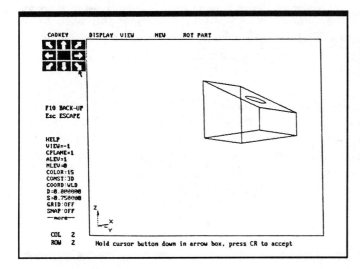

Figure 14-32a

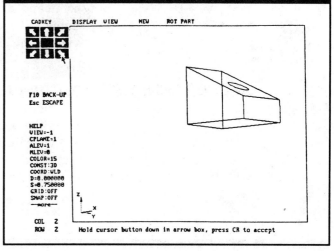

Figure 14-32b

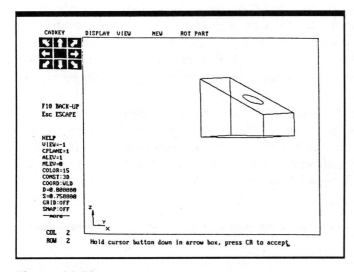

Figure 14-32c

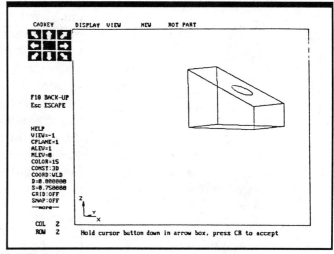

Figure 14-32d

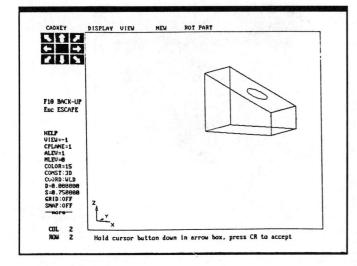

Figure 14-32e

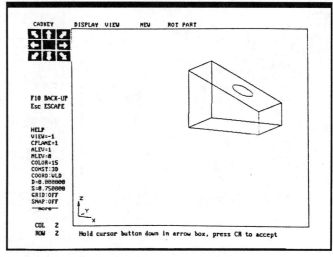

Figure 14-32f

command and then enter the number of the view you wish to rename followed by **Enter**. You will then be prompted to enter the view name. Type the name and press **Enter**.

VIEW OPTION: LIST

The **DISPLAY-VIEW-LIST** option allows the user to see a list of the views which have been defined for the model on which they are currently working. The advantage of the list is that it displays both the view number and its name in a dialog box (Figure 14-33). To select a view double click on the view name. After the view has been changed Autoscale the view by pressing **ALT-A**.

THE VIEWPORTS OPTIONS

The viewports options allow the user to divide the screen into different viewing areas, which is helpful when creating 3D models. Normally, a user would use a single viewport and change to the desired view of a model by entering **ALT-V** or selecting **DISPLAY-VIEW**. With viewports, more than one view of the model can be viewed at once. Figure 14-34 shows a 3D model displayed in four viewports arranged orthographically, with the exception of the upper right viewport, which is view 7 (isometric). View 1 (top view) is in the upper left viewport, view 2 (front view) is in the lower left viewport, and view 5 (right side view) is in the lower right viewport. The configuration program determines the

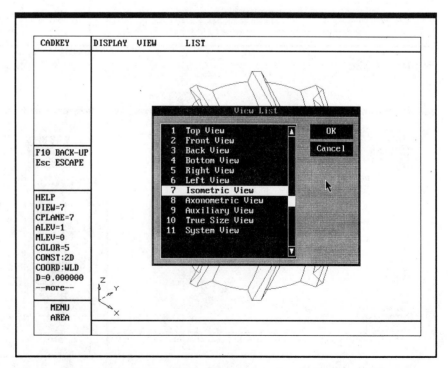

Figure 14-33 DISPLAY-VIEW LIST dialog box for CADKEY 6

default setting for the viewports displayed when CADKEY is booted up.

The viewports option is controlled by entering **DISPLAY-VWPORTS (ESC-F6-F5)**. The **VWPORTS** options are used to **ADD, REMOVE, RESIZE, MOVE,** or automatically set (**AUTOSET**) the viewports on screen. When more than one viewport is displayed on screen it affects many other CADKEY options. For example, if you turn the grid on when multiple viewports are displayed, you must choose to display the grid in all the viewports or only in the primary one. Most of the display options, such as zoom, pan, and auto-scale, also are affected by multiple viewport displays. The primary viewport is usually the view in which you are doing most of the construction. It is the active viewport in which all operations are assigned unless you specify otherwise.

VIEWPORTS OPTION: AUTOSET

The **AUTOSET** option displays one of eight predefined viewports that the CADKEY program displays in the menu area (Figure 14-35). Selecting **DISPLAY-VWPORT-AUTOSET** automatically displays the icon menu in the menu area. A prompt reads: **Cursor-select viewport layout from menu**. To change the viewport display, cursor-pick one of the eight icons in the menu area. The new viewport configuration is displayed on screen. A second prompt reads: **Cursor-indicate primary viewport**. Move the cursor into the desired viewport and click on it.

VIEWPORT OPTIONS: REMOVE

The **REMOVE** option deletes an existing viewport from the screen. If the primary viewport is removed, a new one must be

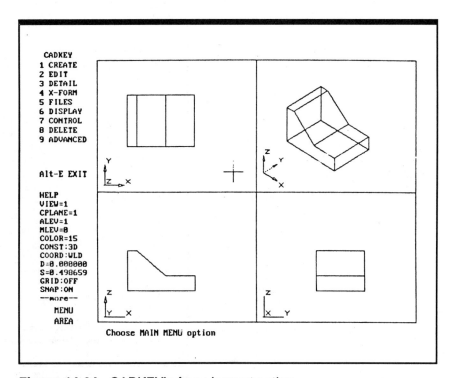

Figure 14-34 CADKEY's four-viewport option

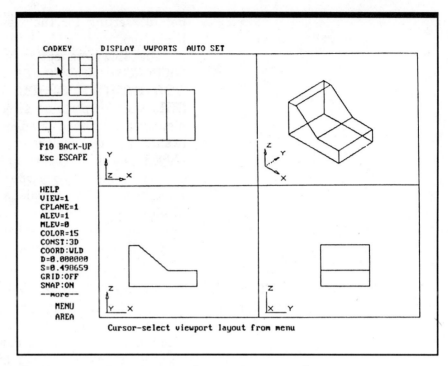

Figure 14-35   The viewport displayed in the upper left

cursor-selected. Figures 14-36 and 14-37 demonstrate the **RE-MOVE** option.

Select **DISPLAY-VWPORTS-REMOVE**. A prompt reads: **Cursor-indicate viewport to remove**. The viewport is automatically deleted after it is picked (Figure 14-37).

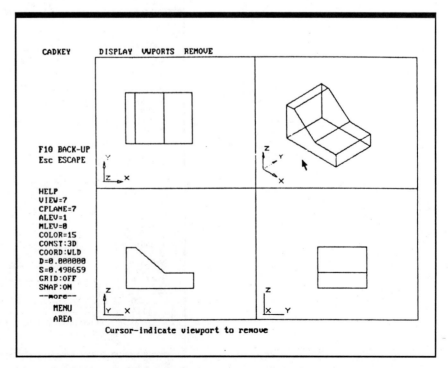

Figure 14-36   Pointing at the viewport to be removed

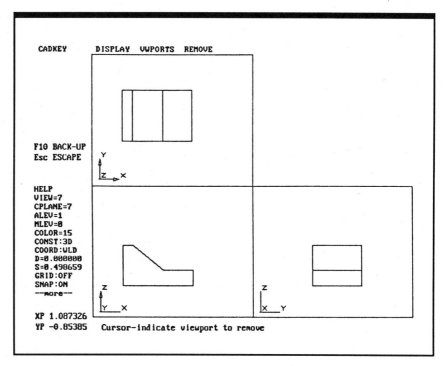

Figure 14-37 Display after selecting the viewport to be removed

VIEWPORT OPTIONS: RESIZE

The **RESIZE** option changes the size of a selected viewport. Before a viewport can be resized, remove the viewports not needed. Overlapping viewports cannot be created. To resize the lower right viewport in Figure 14-37 enter **DISPLAY-VWPORTS-RESIZE**. A prompt reads: **Cursor-indicate viewport near corner to move**. Pick a point near the upper right corner of the viewport. A second prompt reads: **Cursor-indicate new position for viewport corner**. Move the cursor to the new location near the upper right of the screen. The new viewport is displayed on screen (Figure 14-38).

VIEWPORT OPTIONS: MOVE

The **MOVE** option places an existing viewport anywhere within the drawing window. Before a viewport is moved there must be a place to move it to, so use the **REMOVE** option to create a place for the viewport. Figure 14-39 shows front and isometric views of a part. The right half of the screen does not have a viewport. To demonstrate the **MOVE** option, the isometric view will be moved to the right side of the screen.

Enter **DISPLAY-VWPORTS-MOVE**. A prompt reads: **Cursor-indicate viewport to move**. Click on the isometric viewport. Another prompt reads: **Cursor-indicate new viewport position**. An outline of the viewport moves when the cursor is moved to assist in positioning the new location. Move the cursor and pick the new point. The new viewport is automatically displayed (Figure 14-40).

VIEWPORT OPTIONS: ADD

The **ADD** option creates a new viewport on the display. This option is used if you would like to create a custom-designed

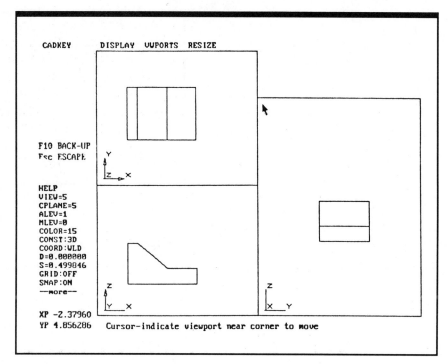

Figure 14-38   The RESIZE option used to enlarge the profile viewport

viewport configuration. Before the **ADD** option can be used, an existing viewport must be removed or resized to make room for the new viewport. The maximum number of viewports allowed on screen is determined by the number set in the configuration program. Overlapping viewports are not allowed.

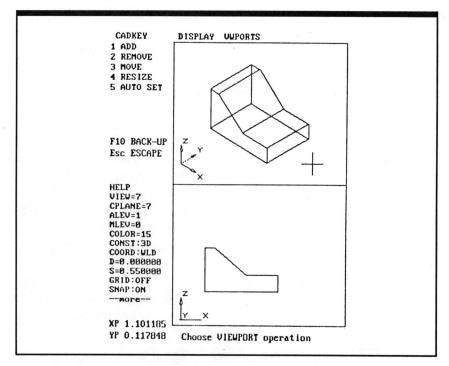

Figure 14-39   A two-viewport display with an empty area for moving

Figure 14-40 demonstrates the **ADD** option. Selecting **ADD** displays a prompt: **Cursor-indicate lower left corner of viewport**. Pick a point on screen to define the lower left corner of the new viewport. Another prompt reads: **Cursor-indicate upper right corner of viewport**. Moving the cursor causes a rectangle to drag as an aid in defining the new viewport. Position the cursor and pick the point. The new viewport will automatically be displayed and assigned view number 1 (Figure 14-41). To change the view number enter **ALT-V**.

THE AXES OPTIONS

The AXES options control the display of the axes gnomon which identifies the direction of the world or construction view axes relative to the Drawing Window. Two type of axes gnomons can be displayed: construction view and displayed view. The axes can be turned on in all viewports or only in the primary viewport. The default setting is to have the displayed view axes on. Each leg of the axes is identified with a letter for the X, Y, and Z axes and shown with a solid or hidden line. A hidden line represents the axis that is pointing away from the user (Figure 14-42). Having the axes displayed on screen is very useful when creating 3D models.

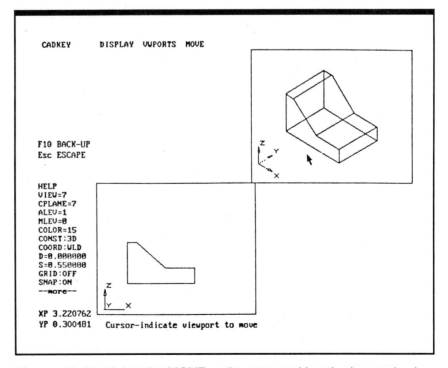

Figure 14-40 Using the MOVE option to reposition the isometric view

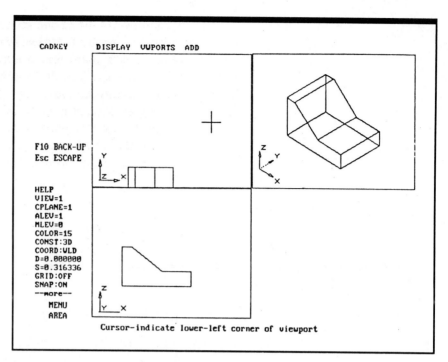

Figure 14-41   Using the ADD option to create a new viewport

AXES OPTION: CONST VW (Construction View)

The construction view appears only when a construction plane is active. Construction planes are defined using the **CONTROL CONST (Construction)** command. The construction icon is displayed in the upper right-hand corner of the viewport and shows the positions of the X, Y, and Z axes in the construction plane.

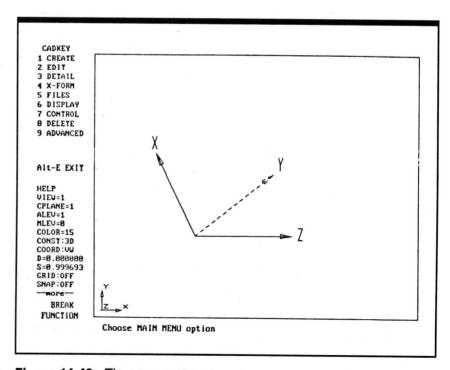

Figure 14-42   The axes gnomon

Figure 14-43 shows the construction axes in the upper right-hand corner of each viewport for the construction plane. In Figure 14-43, the inclined surface is the construction plane. Because the construction plane is parallel to the inclined plane, the grid is not shown in the front view. The construction view axes option is turned on by entering **DISPLAY-AXES-CONST VW**. A prompt reads: **Choose option for construction view axes display (current=off)**. There are three options: **OFF, PRIM DSP** (on in the primary display only), and **ALL DSP** (on in all the viewports).

AXES OPTIONS: DISP VW (Displayed View)

The displayed view option turns the world coordinate axes on or off. This axes option is displayed in the lower left-hand corner of the viewports and its default is to be on. To turn the displayed view axes on, select **DISPLAY-AXES-DISP VW**. A prompt reads: **Choose option for display view axes display (current=on)**. There are two options: **ON** and **OFF**. Figure 14-44 shows a part with both the display and construction axes on.

CURSOR TRACKING

An X, Y coordinate read-out of the current cursor location in 3D or view coordinate space is possible with CADKEY. The coordinate

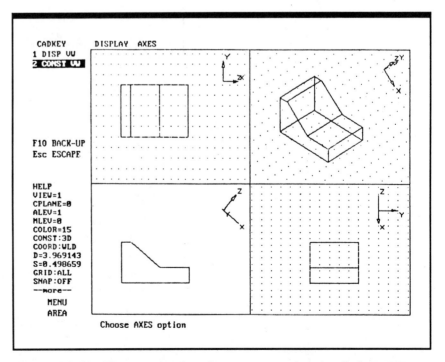

Figure 14-43   The construction view axes turned on in all viewports

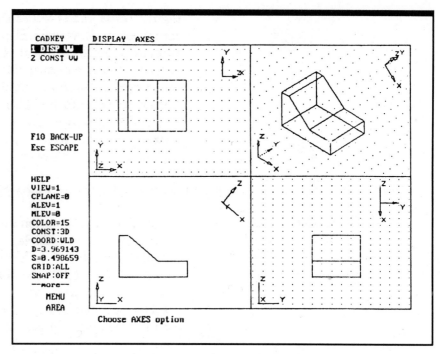

Figure 14-44 The display and construction view axes on in all viewports

read-out is located in the lower left corner of the screen below the Status Window. Cursor tracking is turned on from its off default setting by entering **DISPLAY-CURSOR-TRACKING** or **CTRL-T**. A prompt reads: **Choose option for tracking coordinate display (current=OFF).** There are three cursor tracking options:

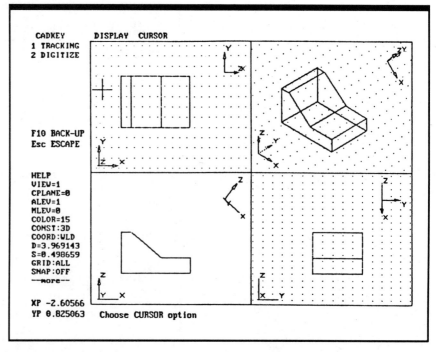

Figure 14-45 View coordinate cursor tracking located in the lower left

OFF, VIEW, and WORLD. The VIEW option will turn cursor tracking on that will display in the lower left corner of the screen, with the cursor position in view coordinates on the current plane (Figure 14-45). The WORLD option displays the cursor position in world coordinates on the current construction plane (Figure 14-44).

REVIEW QUESTIONS

1. What is the name of the option that will automatically return to the previous window?
2. What will happen to the scale of the screen image if the number **2** is input when using the ZOOM-SCALE option?
3. How many levels are available with CADKEY? Which level can never be displayed?
4. Which LEVEL option would one use to temporarily turn off entities on a specified level(s)?
5. How are active levels identified when listing levels on screen using the LEVEL-LIST option?
6. What is the default grid setting used with CADKEY?
7. Name the Immediate Mode command used to turn the grid on and off.
8. What is the default snap setting used with CADKEY?
9. Name the Immediate Mode command used to turn the snap on and off.
10. List the eight predefined views and view numbers.
11. Sketch the eight autoset viewports.
12. List the different methods that can be used to create a new view.
13. Describe the difference between the displayed and construction view axes gnomons.

CHAPTER 15

Editing Your Design

OBJECTIVES

After completing this chapter, you will be able to:

- Trim or extend lines, arcs, and circles.
- Divide lines, arcs, and circles.
- Break entities.
- Recall the last entity deleted.
- Recall erased entities by level or all.
- Move entities within a box.
- Create, edit, delete, and list groups.
- Automatically segment entities.
- Edit splines.
- Edit entities by defining a plane.
- Define a reference plane.

THE EDIT FUNCTION

When using many of the precision entity creation techniques in CADKEY, such as creating a line parallel to another line, you often end up with a line that is longer than needed. To remedy this situation you have to edit your design or part. Many different options allow you to edit the entities that have been created. **TRIM/EXT** cuts off excess portions of entities or extends them until they intersect with other entities. **BREAK** breaks a single entity into two or three separate entities. **RECALL** calls back entities that have been deleted. **BX MOVE** moves entities within the part and associatively updates all the affected entities. **GROUP** combines several entities into a single unit. **AUTOSEG** creates segments on top of existing entities. All of these options and the options found on the sub-menus provide you with a tremendous editing

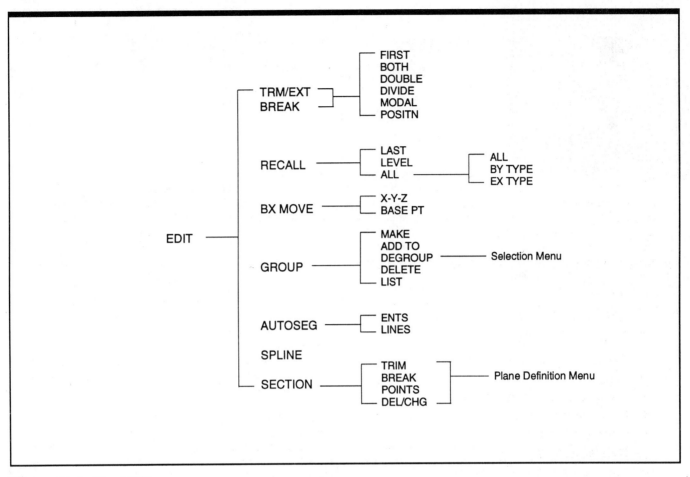

Figure 15-1   The EDIT menu

capability to be used to refine and accurately create a part of the highest quality. The precision of CADKEY's editing options makes them a pleasure to work with. Figure 15-1 shows the hierarchy of commands within **EDIT**.

THE TRM/EXT (Trim/Extend) OPTION

The six **TRM/EXT** options are:

> **FIRST**
> **BOTH**
> **DOUBLE**
> **DIVIDE**
> **MODAL**
> **POSITN** (POSITION)

Each option offers a unique method of trimming entities (see Reference Chart 15-1).

Interlock Uses CADKEY to Design Locks and the Dies to Make Them!

In an economy that has experienced one of its lowest ebbs since the Great Depression, Interlock Hardware Developments, Ltd., of Auckland, New Zealand, has been swimming successfully against the tide. Interlock designs and manufactures hardware fittings for doors and windows using CADKEY® and CADKEY SOLIDS™. Brett May and John Paterson founded the company in 1984 with some new ideas about door and window locks, which they patented worldwide in 1985. Their customers already span the globe, and the number of Interlock's employees has swelled to 110.

Lock Design

Interlock features left-hand and right-hand models of its locking hardware. The locks are manufactured from die-cast zinc alloy. Some window locks feature a nylon flap that extends and retracts as the window is locked or unlocked, to prevent damage to the window sash.

"Our designers do everything using CADKEY in 3-D, right from the initial concept on the computer screen," said Brett May. Interlock's door and window hardware include many interchangeable parts for which they have created a library of pattern files. "After building a wire-frame model of a lock," Brett continued, "we generate a solid model of the lock in CADKEY SOLIDS. This solid model serves two purposes. We use the solid model first to check for interference. Then, the solid model also creates an image that we use as an illustration in our catalogue. It lets the customer *see* the product without having to build a prototype."

Die Design

Interlock's tool designers have created a standard set of die blocks as pattern files in CADKEY. Using the volume of the solid model of the lock generated by CADKEY SOLIDS, they calculate the cavity volume for the die that will be used to make the lock. Then, they copy the pattern files of the die blocks into CADKEY part files and modify them to create a new die. The tool designers also design the placement of water lines and cooling channels. "We attempt to maximize the cooling efficiency of the die," Brett added, "so that we can run the die fast. That reduces the cost per part."

Interlock does not manufacture its own die sets. An outside manufacturer produces the tools for them. "Die casting involves two parts, the fixed half and the movable half," Brett said. "You clamp the movable half to the fixed half and inject the molten metal into the cavity through a nozzle connected via a gooseneck tube to a vat of molten zinc alloy. The molten metal is then forced into the gooseneck tube by a hydraulic piston."

Profitable Productivity

"A prospective customer in the United Kingdom had received a proposed product design from us," Brett said, "but they required an adjustment to a sculptured handle which they sent to us by fax. We modified the design in CADKEY and the next day we faxed them a print produced on a laser printer. The next morning we received a fax from them with a firm order for 120,000 locks."

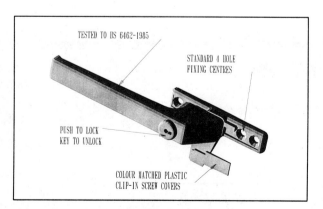

TESTED TO BS 6462-1985

STANDARD 4 HOLE
FIXING CENTRES

PUSH TO LOCK
KEY TO UNLOCK

COLOUR MATCHED PLASTIC
CLIP-IN SCREW COVERS

One of Interlock's left-handed window fasteners.

Adapted with permission from **3-D World**, *v. 5, no. 1, Jan./Feb., 1991.*

THE EDIT TRIM/EXTEND AND BREAK OPTIONS

FIRST   BOTH   DOUBLE   DIVIDE   MODAL   POSITION

NUMBERS INDICATE ORDER OF SELECTION
DASHED LINES INDICATE PORTION OF LINE TRIMMED OR BROKEN AWAY

Reference Chart 15-1   Selection orders for TRIM and BREAK options

The **TRM/EXT** function is view dependent. The 2D intersection is computed and the proper trimming occurs according to the entities selected. In cases where both entities do not intersect, one or both of the entities will be projected to the intersection point. Each **TRM/EXT** option will remain active until you press **ESCAPE**, allowing you to select additional entities. The only **TRM/EXT** option that will trim a circle is the **DOUBLE** option.

UNDO   The **EDIT** commands that have the **UNDO (ALT-I)** option available are **TRM/EXT** and **BREAK**. **UNDO** appears in the Break Option Window when it is available with these commands. When it is selected the entities which were just trimmed, extended or broken will be restored to the way they were prior to using the command.

TRM/EXT OPTION: FIRST   The **FIRST** option allows you to trim or extend the first entity selected with reference to another existing entity. The Escape Code for quick access to this function is **ESC-F2-F1-F1 (EDIT-TRM/EXT-FIRST)**.

You are prompted to select the entity to trim. The selection point on this entity must be the portion of the entity that you wish to

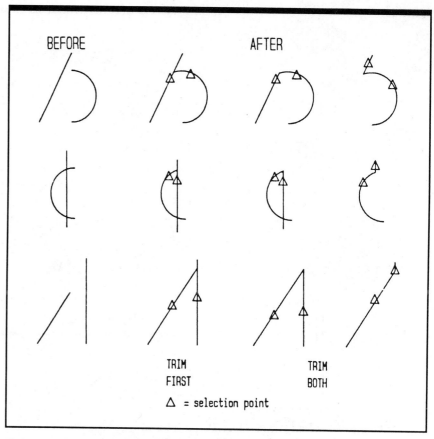

BEFORE   AFTER

TRIM   TRIM
FIRST   BOTH

 = selection point

Reference Chart 15-2 Examples of the TRM/EXT options FIRST and BOTH

remain after trimming. You are then prompted to select the trimming entity. This entity serves as a reference for trimming the first entity.

Once the reference entity is selected, the first entity is trimmed to the entities' intersection point. If the entities do not intersect they will be extended until they intersect (see Reference Chart 15-2).

POSSIBLE USE

The **FIRST** option for trimming and extending entities is the one most often used to edit a part. In Figure 15-2 it could be used a total of eight times to trim the excess portion of the lines. First, the line that you wish to trim is selected on the portion that you wish to remain after trimming. Second, the line that you select is the trimming entity to which the first line is trimmed. To keep the selection order correct it helps to say to yourself (that's right, CAD sometimes makes you talk to yourself), "I want to trim this (first) entity with reference to this (second) entity." Select each entity as you say this. Another method is to remember to pick the "keeper" first; then pick the "cutter."

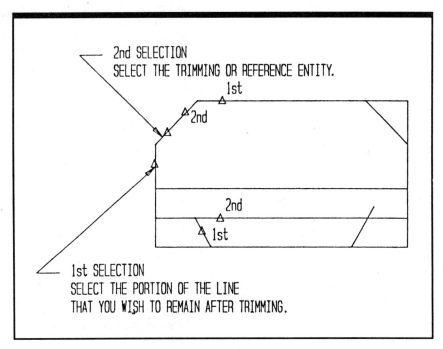

Figure 15-2   Using the TRM/EXT option FIRST to trim the first line selected with reference to the second line selected

TRM/EXT OPTION: BOTH

The **BOTH** option allows you to trim or extend two intersecting entities with reference to each other. The Escape Code for this option is **ESC-F2-F1-F2 (EDIT-TRM/EXT-BOTH)**.

You are prompted to select the first entity to trim or extend. Remember to select the portion that you wish to remain after trimming. You are then prompted to select the second entity to trim or extend. Once both entities are selected they are trimmed or extended until they just intersect without any overlap.

QTRIM (Quick Trim)

The **QTRIM** option is found on the Status Window. It must be turned ON to work with the **EDIT-TRM/EXT-BOTH** option. The **BOTH** option is the only option it works with. Rather than having to select both of the entities which you would like to trim or extend, you simply select inside of the corner intersection you wish to have trimmed. The lines or arcs will be trimmed or extended to their intersection point.

 POSSIBLE USE

In the example in Figure 15-3, the two lines that were trimmed could have been trimmed using the **FIRST** option, but the **BOTH** method allows you trim both lines at the same time with reference to each other. The **BOTH** method is more efficient and saves you time. This method can be used to trim or extend two lines with reference to each other at their intersection point.

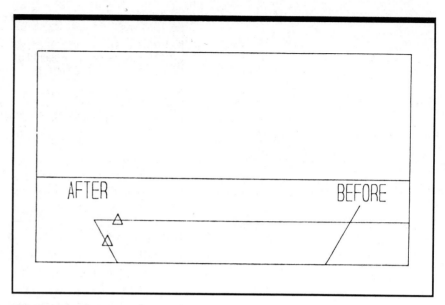

Figure 15-3 Both lines selected are trimmed or extended to their intersection point.

TRM/EXT OPTION: DOUBLE

The **DOUBLE** option allows you to trim or extend an entity with reference to two other entities, which results in the entity being trimmed twice. The Escape Code for this option is **ESC-F2-F1-F3 (EDIT-TRM/EXT-DOUBLE)**.

You are prompted to select the entity to trim or extend. Select the portion that you wish to remain after trimming. You are then prompted to select the first trimming entity and then the second trimming entity. The entities that you select must intersect. Once all

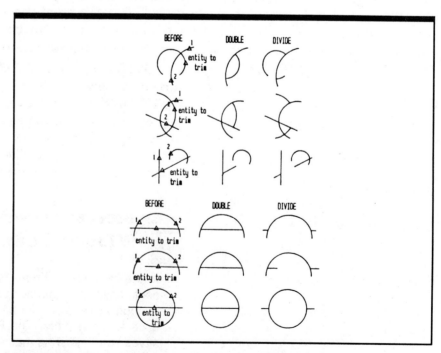

Reference Chart 15-3 Examples of the TRM/EXT options DOUBLE and DIVIDE

of the entities are selected, the first entity selected is trimmed or extended with reference to the other two entities (see Reference Chart 15-3).

POSSIBLE USE

A common use for the **DOUBLE** option is to trim circles in an isometric or rotated view. This is because the **DOUBLE** option is the only option that can be used to trim circles. The **DOUBLE** option can also be used to trim or extend a line with reference to two other lines. In this case only the center portion of the line will remain after trimming. The example in Figure 15-4 shows a circle being trimmed in the isometric view number 7.

TRM/EXT OPTION: DIVIDE

The **DIVIDE** option divides an entity into two entities by removing the middle portion of the entity. The Escape Code to quickly access this option is **ESC-F2-F1-F4** (**EDIT-TRM/EXT-DIVIDE**).

You are prompted to select the entity that you wish to divide. **NOTE:** select the portion of the entity that you want removed in this case. This is the opposite of the previous options.

Then you are prompted to select the first and second trimming entities into which you wish the first entity to be divided. At this point, the entity is divided into two entities and the middle portion of the entity is removed.

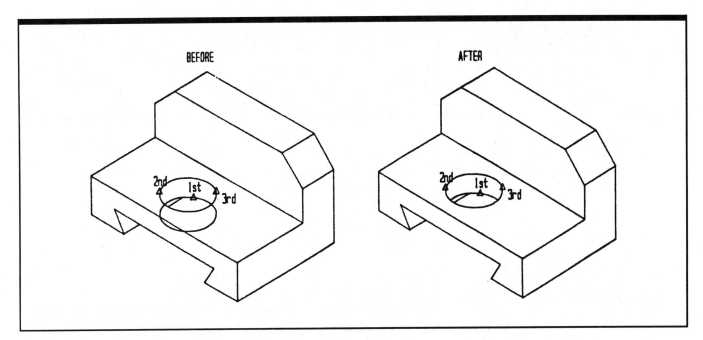

Figure 15-4 A circle trimmed with the DOUBLE option

POSSIBLE USE

In Figure 15-5 the base line in this view must be divided to illustrate that the dovetail goes all the way through the object. To divide this line, the first selection point is the portion to be removed and the second and third selections are the trimming and reference entities.

TRM/EXT OPTION: MODAL

The **MODAL** option allows you to trim or extend several entities with reference to one trimming entity. The Escape Code for this option is **ESC-F2-F1-F5 (EDIT-TRM/EXT-MODAL)**.

You are prompted to select the trimming or reference entity first in this option. This is the entity to which all of the other entities will be trimmed or extended. Then you are prompted to select the entities that you wish to trim or extend. Remember to select the portion of the entity that you wish to remain. The entities are trimmed as soon as they are selected.

 POSSIBLE USE

The **MODAL** option is great when you have multiple entities that you wish to trim with reference to one trimming entity. In the example in Figure 15-6, the trimming entity is the first selection. Every entity that is selected after this is either trimmed or extended to the trimming entity.

TRM/EXT OPTION: POSITN (Position)

The **POSITN** option allows the user to trim or extend an entity to any location, which can be specified using the Position Menu. The Escape Code for this option is **ESC-F2-F1-F6 (EDIT-TRIM/EXT-POSITN)**.

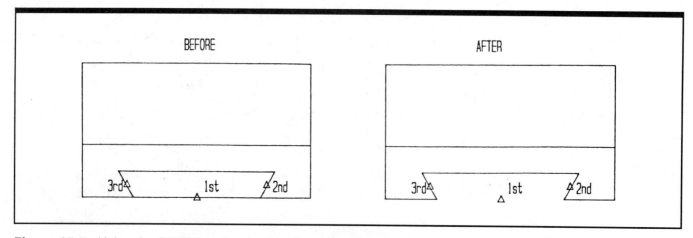

Figure 15-5 Using the DIVIDE option to trim the midsection of a line

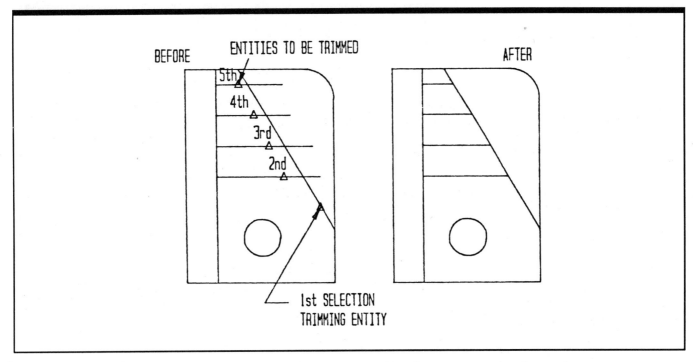

Figure 15-6 The MODAL option trims and extends entities with reference to a trimming entity.

You are prompted to select the entity to trim. Select the entity on the portion that you wish to remain after trimming. You are then prompted for the position to which to trim. The Position Menu is displayed and any of the nine options can be used, including **CURSOR**, to determine the location where the entity will be trimmed.

 POSSIBLE USE

Many times when you are placing center lines on a detailed drawing they need to be trimmed or extended beyond the circle. This can be done with the **POSITN** option without having to create lines to trim or extend (Figure 15-7).

THE BREAK OPTION

BREAK breaks an existing entity into two or three separate entities. When an entity is broken, the portion of the entity to break that is selected will retain its original attributes. The other side or portion of the entity that is broken away will assume the current attributes as they are set in the Status

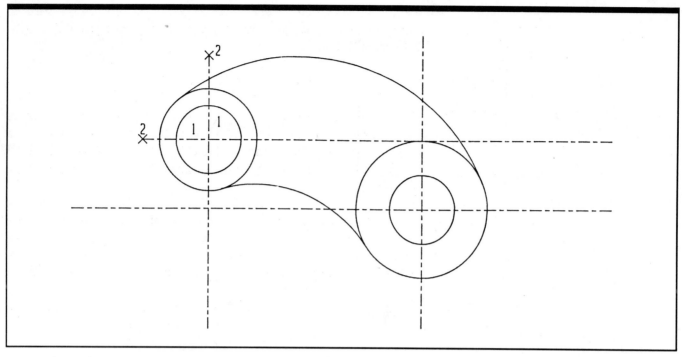

Figure 15-7 Placing center lines with the POSITN option without creating lines to trim or extend

Window. The only break option that will work on circles is the **DOUBLE** option (see Reference Chart 15-4). After the entity has been broken, either half can be deleted without affecting the other half.

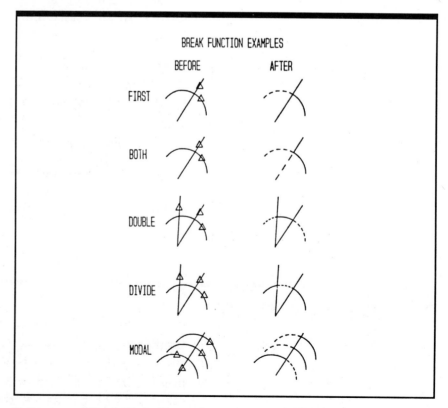

Reference Chart 15-4 Examples using the BREAK option

You will remain in the **BREAK** function and can break as many entities as needed. When you have finished breaking entities press **ESCAPE** to exit.

BREAK OPTION: FIRST

FIRST breaks an entity (excluding circles) into two separate entities. The break point is determined by the intersection point of two entities. The same is true for the **TRM/EXT-FIRST** option. The Escape Code for this option is **ESC-F2-F2-F1** (**EDIT-BREAK-FIRST**).

You are prompted to select the entity to break. The portion of the entity that is selected will retain its original attributes. The other portion of the entity that is broken will assume the current attributes for line type, color, etc.

Next, you are prompted to select the breaking or intersecting entity near the intersection point which becomes the break point for the entity to be broken. Once the breaking entity is selected, the break is made. If you are using the same attributes for both sides of the break, you will not see the break, but it's there!

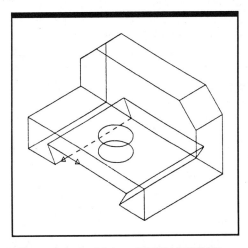

Figure 15-8 Using BREAK-FIRST to break one line into two

 ## POSSIBLE USE

The **BREAK-FIRST** option works similarly to the **TRM/EXT-FIRST** option, except that instead of deleting the portion of the line that was not selected, that portion is assigned the current attributes for line type, etc. One use for this option occurs if you are creating an isometric view from a wireframe model and, for clarity, you leave the hidden lines in the drawing. By changing the current line type to hidden, you can now break a line with reference to a breaking entity. The portion selected first will retain its attributes and the other portion will be assigned the current attributes, as in Figure 15-8. The second entity selected is the breaking entity.

BREAK OPTION: BOTH

The **BOTH** option allows you to break two entities (excluding circles) which intersect into two or three separate entities. The Escape Code for this option is **ESC-F2-F2-F2** (**EDIT-BREAK-BOTH**).

When this option is chosen you are prompted to select the first entity to break. Then you are requested to select the second entity to break. Both of the entities selected are broken at their intersection point. The portion of the entity that is selected will retain its original attributes. The other portion that is broken will assume the current attributes for line type, color, pen number, and line width from the intersection on.

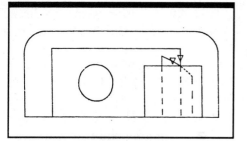

Figure 15-9 BREAK-BOTH used to break a line and arc with reference to their intersection point

POSSIBLE USE

In Figure 15-9, two intersecting entities were broken with reference to each other. This option can be used any time you wish to break two entities with reference to their intersection points.

BREAK OPTION: DOUBLE

The **DOUBLE** option allows you to break an entity (including circles) into three separate entities. The break points are the intersection points of the entity with two other entities. The Escape Code to access this option quickly is **ESC-F2-F2-F3** (**EDIT-BREAK-DOUBLE**).

This option begins with a prompt to select the entity to break. Select the portion of the entity that will retain its original attributes. You are then prompted to select the first breaking entity near the intersection point, and then the second breaking entity the same way. The same entity can be selected twice if it intersects the entity to be broken in two different places. The two outer portions of the broken entity assume the current attributes.

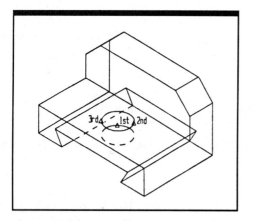

Figure 15-10 Breaking a circle with the BREAK-DOUBLE option

POSSIBLE USE

As in the **TRM/EXT-DOUBLE** option, the **BREAK-DOUBLE** option is the only **BREAK** option that will work with circles. In the example in Figure 15-10, the lower portion of the drill hole was broken to indicate that it was hidden. That circle is now two separate arc entities that can be deleted separately if desired.

BREAK OPTION: DIVIDE

The **DIVIDE** option works just the opposite of the **DOUBLE** option. It allows you to break an entity into three separate entities, but this time the entity selected to break retains the original attributes on the two outer portions of the entity, and the middle portion assumes the current attributes. The Escape Code for this option is **ESC-F2-F2-F4** (**EDIT-BREAK-DIVIDE**).

You are prompted to select the entity to break. Select the middle portion which will assume the current attributes. You are then prompted to select the first breaking entity and then the second breaking entity. Select both of these entities near their intersection point. The entity is broken and redrawn as three separate entities. Any of these entities can now be singly selected for any reason needed.

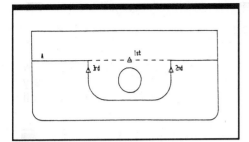

Figure 15-11 A line broken into three lines with the BREAK-DIVIDE option

POSSIBLE USE

The **BREAK-DIVIDE** option is a good way to break a continuous line into a hidden line when part of the line is hidden by another surface. In Figure 15-11, the line to be broken was selected first and the breaking entities were selected second and third.

BREAK OPTION: MODAL

The **MODAL** option selects one breaking entity that will serve as a reference entity for breaking several entities which intersect with the reference entity. The Escape Code for this option is **ESC-F2-F2-F5** (**EDIT-BREAK-MODAL**).

You are prompted to select the breaking entity. You are then prompted to select the entity to break. The portion of the entity that is selected will retain its original attributes. The portion on the other side of the intersection point will assume the current attributes. You can continue to select and break entities which intersect with the breaking entity. When you are done press **ESCAPE** to exit the function.

POSSIBLE USE

The **BREAK-MODAL** option was used in Figure 15-12 to break a series of four lines with reference to the breaking entity, which was the first entity selected. This allows you to draw a single line and then break it into hidden lines as needed.

BREAK OPTION: POSITN (Position)

The **POSITN** option breaks an entity to any location that can be specified using the Position Menu. The Escape Code for this option is **ESC-F2-F2-F6** (**EDIT-BREAK-POSITN**).

You are prompted to select the entity to break. Select the entity on the portion that will retain its current attributes after the break has occurred. You are then prompted for the position to which to break. The Position Menu is displayed and any of the nine options, including **CURSOR**, can be used to determine the location where the entity will be broken.

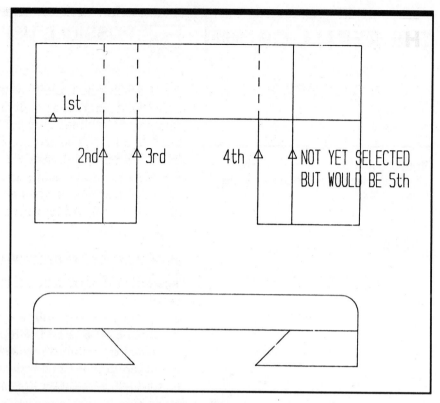

Figure 15-12 Using the BREAK-MODAL option to break several lines with reference to one breaking entity

POSSIBLE USE

Sometimes, when constructing a model, you need to break a portion of a line off. The **POSITN** option can do this without you having to create intersecting lines (Figure 15-13).

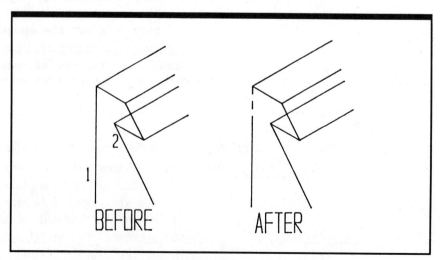

Figure 15-13 Breaking a portion of a line with the POSITN option without creating intersecting lines

THE RECALL OPTION

RECALL OPTION: LAST

The **LAST** option allows you to recall or undelete the last 100 entities that have been deleted. It will redisplay the deleted entities one at a time. This option will not recall entities that have been trimmed. The Escape Code for this option is **ESC-F2-F3-F1 (EDIT-RECALL-LAST)**. You can bypass the menu and use the Immediate Mode command **CTRL-U** to invoke this function. When this option is selected, a search of the data base locates the last entity deleted and recalls it onto the screen.

 POSSIBLE USE

In Figure 15-14, the horizontal line on the top of the dovetail was deleted incorrectly. Because it was the last entity deleted it can be recalled with **RECALL-LAST**. Use this option anytime you wish to recall an entity you have mistakenly deleted.

RECALL OPTION: LEVEL

The **LEVEL** option recalls an entire level that has been deleted and displays it on the screen in the current part. This option counters the **DELETE-LEVEL** option. The Escape Code for this option is **ESC-F2-F3-F2 (EDIT-RECALL-LEVEL)**.

 POSSIBLE USE

In a large complicated part you may decide to recall a whole level if the level that you deleted is not the one that you had intended to delete. Use the **RECALL-LEVEL** option to recall and display the level that was previously deleted (Figure 15-15).

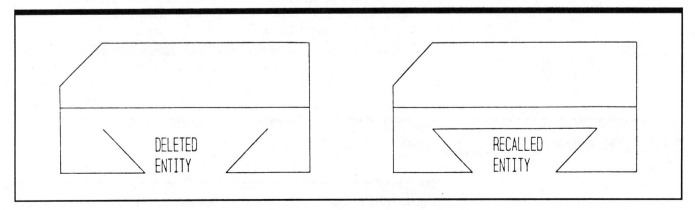

Figure 15-14   Recalling an entity that was deleted using the RECALL-LAST option

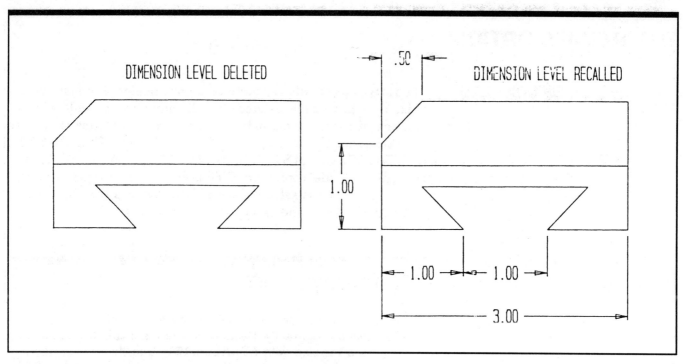

Figure 15-15 The dimension is recalled after being deleted.

RECALL OPTION: ALL

The **ALL** option recalls all of the entities that have been deleted in the entire file. Everything that was previously deleted is displayed on the screen. You have the option to use the Masking Menu **BY TYPE** or **EX TYPE**, allowing you to recall only certain entity types, which may be an advantage. In some cases it may be much more than you ever expected. The Escape Code for this option is **ESC-F2-F3-F3** (**EDIT-RECALL-ALL**).

 POSSIBLE USE

The **RECALL-ALL** option is a last resort option because when you use **RECALL-ALL** everything that you have created, and deleted, will be recalled. If you have been working in the same file but have deleted **ALL** several times, even those things will be recalled (Figure 15-16).

THE BX MOVE (Box Move) OPTION

BX MOVE moves the entities selected within the rubberbox using two different methods for indicating the new position. Entities that are entirely inside the box will be moved intact. Entities

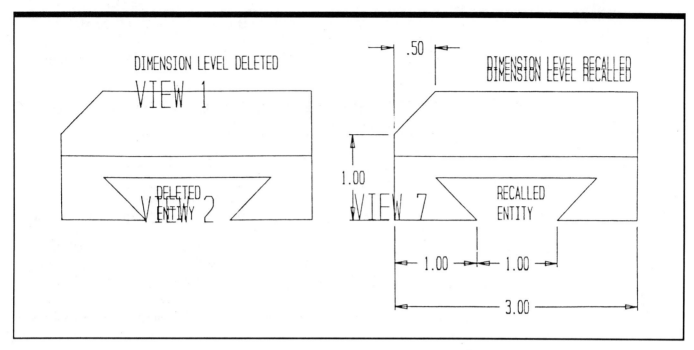

Figure 15-16   An example of the RECALL-ALL option redisplaying everything that had been deleted during the current work session

that are partially in the box will be updated automatically. Lines and polygons will be extended or compressed and dimensions will be updated. The Escape Code for this option is **ESC-F2-F4** (**EDIT-BX MOVE**).

You are prompted to select the first corner of the window or rubberbox and then to select the second corner. Position the box so that all of the entities that you desire to move are completely contained within the box. Once the entities have been selected a new menu with two options, **X-Y-Z** and **BASE PT**, is displayed.

BX MOVE OPTION: X-Y-Z (Delta View Coordinates)

X-Y-Z moves the entities using the delta view coordinate system. This allows you to move the entities relative to their current position. You are prompted **dx=** for the X value, **dy=** for the Y value, and **dz=** for the Z value. A negative X value will move the entities to the left and a negative Y value moves the entities down from their current position. The positive Z value moves the entities on the depth toward the viewer. The values **dx= –1.5 dy= 2** would move the entities one-and-a-half units to the left and two units up.

BX MOVE OPTION: BASE PT (Base Point)

BASE PT specifies a base position, which is used to indicate the new base position for the entities in the window. Once the new base point is indicated, all of the entities are moved to the new position. You can continue to indicate base positions until you have the position desired.

The **BASE PT** option displays the Position Menu. Using one of the locations found on the Position Menu, select a base position from which to move the selected entities. Again, the Position Menu is displayed and you are prompted to indicate a new base position to which the entities in the window will be moved.

 POSSIBLE USE

Often in a family of parts one or two dimensions change, such as the overall length of the part or the depth of the round end slot. Rather than constructing the geometry from scratch for the new part, the original part can be updated, using the **BX MOVE** option, to reflect the new dimensions and be saved under a new name. In Figure 15-17, the round end slot was moved –.25 on the Y axis. The round nose feature was moved .75 on the X axis. In both cases the lines attached to the arcs were stretched and the dimensions updated.

THE GROUP OPTION

GROUP combines or links many entities together so that they can be treated as a single unit. The group is assigned an eight-character name. When you are selecting a group you can use the cursor to select the group or you can use the group name to identify the group you wish to select.

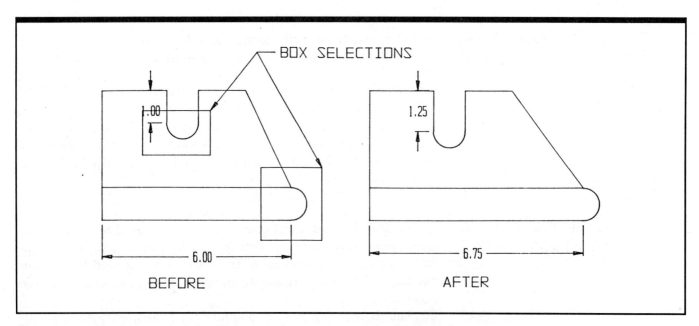

Figure 15-17   Using the BX MOVE-BASE PT option to move the text into proper position

Each group that is created can be divided into a maximum of 256 subgroups. The first time a group is created it is assigned subgroup number 1. Each time that group is used again in the part it is assigned the next available subgroup number in one-unit increments (i.e., 1, 2, 3 . . .). By using the **GROUP-LIST** option you can review the group table to find out how many subgroups exist for each group. This information can then be used for a bill of materials.

In most cases you will want to group such items as a polygon or a pattern, both of which are made up of several entities. By grouping a pattern or a polygon you are then able to treat the group as a single entity. If you wish to delete a group you only have to select one of the entities associated with the group and the whole group will be deleted. If there are subgroups associated with the group they will also be deleted.

GROUP OPTION: MAKE

The **MAKE** option allows you to create a group. The Escape Code for this option is **ESC-F2-F5-F1 (EDIT-GROUP-MAKE)**.

You are prompted to enter a group name of up to eight characters and press **RETURN**. If the group already exists you are asked if you wish to create a subgroup. The **NO/YES** menu is displayed for you to choose from. If **YES** is chosen, the Selection Menu is displayed to aid you in selecting the entities to be included in the group.

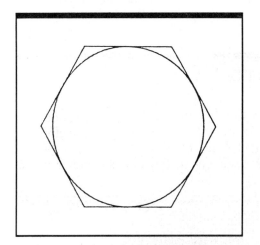

Figure 15-18 A group formed by associating six lines and a circle to represent a hex head bolt

 POSSIBLE USE

A common use for making a group is to associate several entities with one another as a group. In Figure 15-18, the six line entities which make up the hexagon and the circle have been grouped using the **MAKE** option to represent the pattern of a hex head bolt.

GROUP OPTION: ADD TO

The **ADD TO** option allows you to add additional entities to a group that already exists. The Escape Code for this option is **ESC-F2-F5-F2 (EDIT-GROUP-ADD TO)**.

You are prompted to enter the existing group name and press **RETURN**. If the group exists you are requested to enter a subgroup number between 1 and 256. If the group doesn't exist an error message is displayed.

Next, the Selection Menu is displayed for you to select the entities that you wish to add to the group. When all of the desired entities have been added to the group press **ESCAPE** to exit the function.

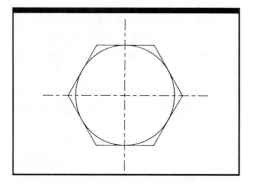

Figure 15-19 Using the ADD TO option to add center lines to the hex head bolt group

POSSIBLE USE

The **GROUP-ADD TO** option allows you to add other entities to an existing group. In Figure 15-19, the center lines have been added to the hex head bolt group.

GROUP OPTION: DEGROUP

The **DEGROUP** option allows you to disassociate or degroup selected entities contained in a group without removing the group and subgroup references. The Escape Code for this option is **ESC-F2-F5-F3** (**EDIT-GROUP-DEGROUP**).

You are prompted to select the entities that you wish to degroup. The Selection Menu is displayed to aid you in selecting the entities that you wish to degroup. Press **RETURN** to exit this function.

POSSIBLE USE

If you need to delete a single entity from a group you cannot simply select one entity to delete. Once one entity is selected the whole group is selected. To delete one entity you must degroup the entire group or disassociate the entities and then delete the desired entities.

GROUP OPTION: DELETE

The **DELETE** option allows you to delete a group from the group table without deleting its grouped entities contained in the subgroups. The Escape Code for this option is **ESC-F2-F5-F4** (**EDIT-GROUP-DELETE**).

You are prompted to enter the name of the group which you wish to delete and press **RETURN**. The group is then removed from the group table and its entities are degrouped.

POSSIBLE USE

GROUP-DELETE deletes a group from the group list without deleting the entities from the subgroups. This is helpful when you have a large part with many groups and subgroups, and one group must be deleted without affecting the other groups.

GROUP OPTION: LIST   The **LIST** option allows you to display and review the group table, which contains a listing of all of the groups that have been used in the current part file. The Escape Code for this option is **ESC-F2-F5-F5** (**EDIT-GROUP-LIST**).

The list is displayed on the screen. If there is another page to the list, you are prompted to press **RETURN**. The total number of groups in the current part is displayed at the bottom of the list. Press **RETURN** to return to the **GROUP** menu.

 ## POSSIBLE USE

GROUP-LIST will give you a complete list of all of the groups contained in a part. From this list you can determine the total number of groups that have been used and the total number of a specific group that has been used. This information can then be used to generate a bill of materials. Figure 15-20 is an example of a group list. Information about the groups can be obtained from this table.

THE AUTOSEG (Auto Segment) OPTIONS

AUTOSEG OPTION: **ENTS** option creates a specified number of new entity segments
ENTS (Entities)   on top of an existing entity. The segments which are created are

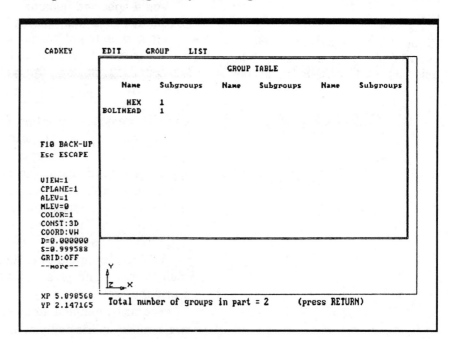

Figure 15-20   A group list identifies the groups by name and lists how many subgroups have been created.

the same type of entity as the reference entity. If an arc were used as the reference entity, the segments created would be arcs. The Escape Code for this option is **ESC-F2-F6-F1** (**EDIT-AUTO-SEG-ENTS**).

You are prompted to enter the number of segments that you wish to create on top of the entity and press **RETURN**. The number of segments must be between 1 and 99.

You are prompted to select the entity to segment. If an arc or circle entity is selected you are prompted to enter the start angle for where the segmentation should begin. The start angle of the arc or circle is the default value in parentheses. Press **RETURN** to accept this value. You are then prompted to enter the end angle for the segmentation. Again, the default value is the end value of the arc. If you wish to segment only part of the arc enter the start and end angles for the portion of the arc to be segmented.

Temporary markers are displayed on the reference entity to represent where the entity will be segmented. You are prompted to accept the segmentation. The **NO/YES** menu is displayed for you to choose from. Choose **F2 YES** to accept or **F1 NO** to reselect.

Next, you are given the option to group the entities. The **NO/YES** menu is displayed. If **YES** is selected you are prompted to enter a group name and press **RETURN**. The entity segments are then created and are assigned the current attributes for line type, color, etc.

Figure 15-21 A circle that has twelve arc segments created on top using the AUTOSEG-ENTS option

POSSIBLE USE

This option can be used prior to transforming a circle to give you a specified number of join lines to simulate the cylindrical surface of a drill hole. A new set of twelve arc segments is created on top of the circle. When they are transformed, a join line is created at each of the segment points (Figure 15-21).

AUTOSEG OPTION: LINES

LINES creates a specified number of new line entities of equal lengths on a referenced entity. The only difference between this option and the **ENTS** option is that all of the segments created are lines regardless of the type of entity used for reference. If a circle were used and five segments were requested, five equally spaced line segments would be inscribed in the circle, appearing as a pentagon. The Escape Code for this option is **ESC-F2-F6-F2** (**EDIT-AUTOSEG-LINES**).

You are prompted to enter the number of line segments that you wish to create and press **RETURN**. The number must be between 1 and 99.

You are prompted to select the entity to segment. If an arc or circle entity is selected you are prompted to enter the start angle for where the segmentation should begin. The start angle of the arc or circle is the default value in parentheses. Press **RETURN** to accept this value. You are then prompted to enter the end angle for the segmentation. Again, the default value is the end value of the arc. If you wish to segment only part of the arc enter the start and end angles for the portion of the arc to be segmented.

A series of temporary markers is displayed on the reference entity to represent where the entity will be segmented. You are prompted to accept the segmentation. The **NO/YES** menu is displayed for you to choose from. Choose **F2 YES** to accept or **F1 NO** to reselect.

Next, you are given the option to group the entities. The **NO/YES** menu is displayed. If **F2 YES** is selected you are prompted to enter a group name and press **RETURN**. The entity segments are then created and are assigned the current attributes for line type, color, etc.

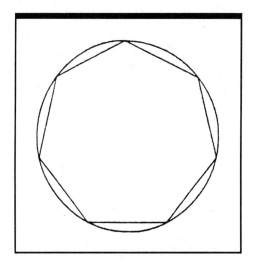

Figure 15-22 A circle that has seven line segments created using the segment points as end points

 POSSIBLE USE

The **AUTOSEG-LINE** and **ENTS** options are used primarily in the generation of a mesh for Finite Element Analysis (FEA). In this option, line segments are created on top of the entity that is segmented regardless of the shape of the entity (Figure 15-22).

THE SPLINE OPTION

When this option is selected you are prompted to select the spline that you wish to edit. The three ways in which a spline can be edited or modified are: **MOV NOD (move node)**, **CHG END (change end)**, and **2D/3D (convert 2D splines to 3D)**.

SPLINE OPTION: MOV NOD

The **MOV NOD** option allows you to move individual nodes on a spline, thus changing the overall shape of the spline without having to respecify all of the node points and end conditions. The Escape Code for this option is **ESC-F2-F7-F1 (EDIT-SPLINE-MOV NOD)**.

When this option is selected, temporary markers highlight the node locations on the existing spline to make it easier to select the nodes to be edited. You are prompted to select the node that you wish to move. Then the Position Menu is displayed to assist you in identifying the new location of the node point. Once the new node has been specified, the spline is regenerated passing through the new node.

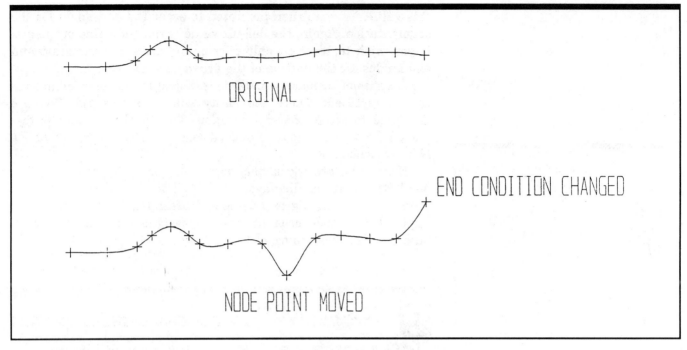

Figure 15-23   A spline that has had a node point and an end condition moved

---

![disk icon] **POSSIBLE USE**

If the location of a node along a spline has to be changed rather than deleted, you can simply move the nodes needed to make the required curve to recreate the spline (Figure 15-23).

SPLINE OPTION: CHG END
(Change End)

CHG END edits the end position of a cubic spline. The Escape Code for this option is **ESC-F2-F7-F2 (EDIT-SPLINE-CHG END)**.

You are prompted to select a spline near the endpoint that you wish to edit or modify. Once this selection has been made, the start/end condition menu that was used in the creation of the cubic spline is displayed. Choose the option for redefining the desired new end condition. After the appropriate data is entered, the spline will be redrawn using the new end condition.

![disk icon] **POSSIBLE USE**

Use **CHG END** to change the end condition on a spline without recreating the spline.

SPLINE OPTION: 2D/3D

The **2D/3D** option automatically converts a 2D spline to a 3D spline. The node points are freed so that they are no longer forced to lie within a specific 2D plane. The data base changes allow for 3D node coordinates. To execute this option, simply select the 2D spline that you wish to convert and the conversion is done automatically. The Escape Code for this option is **ESC-F2-F7-F3 (EDIT-SPLINE-2D/3D)**.

 POSSIBLE USE

Splines fit a curve through several node points for the smoothest curve possible. In the case of aerodynamics, the spline may need to be edited to reduce resistance on a particular surface. This can be done by moving a node point or changing the start or end condition of the spline. If a spline has been drawn as a 2D entity it can be transformed or converted into a 3D entity.

THE SECTION OPTION

The **EDIT-SECTION** option can be used to identify a cross-section or intersection of specific entities through a defined plane (Figure 15-24). There are five ways to identify the plane for the

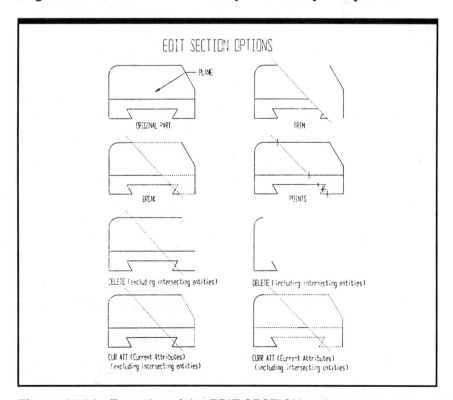

Figure 15-24 Examples of the EDIT-SECTION options

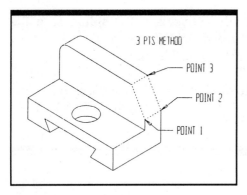

Figure 15-25 A plane defined by selecting three points

section and four types of sections which can be created. The four section methods are:

TRIM—will trim selected entities with reference to a plane.

BREAK—will break selected entities with reference to a plane.

POINTS—creates points at the intersection of selected entities and a reference plane.

DEL/CHG—deletes or changes the attributes of selected entities.

PLANE DEFINITION

Once a plane has been defined a temporary icon representing the plane will be displayed so that you can see the plane clearly. The icon will be displayed in the same color as your cursor. The five ways to define or identify a reference plane are:

3 PTS—defines a plane using three positions in space. Use Position Menu options other than **CURSOR** (Figure 15-25).

PT/LINE—defines a plane using a point or position in space and a line (Figure 15-26).

2 LINES—defines a plane using two lines and their 3D intersection or projected intersection (Figure 15-27).

VW/DEPTH—defines a plane by specifying a view number and a specific depth in that view. The depth can be specified by entering a value or by indicating a position using the Position Menu (Figure 15-28).

ENTITY—defines a plane by selecting a line, arc, conic, or 2D spline (Figure 15-29).

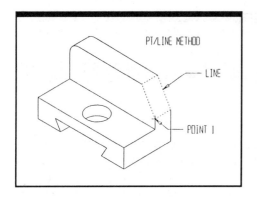

Figure 15-26 A plane defined by selecting a point and a line

SECTION OPTION: TRIM

This option will trim selected entities with reference to a plane. The Escape Code for this option is **ESC-F2-F9-F1** (**EDIT-SECTION-TRIM**).

To start, the Selection Menu is displayed and you are prompted to select the entities that you wish to trim. Then choose a method for defining a plane. Once the plane has been defined you must choose (with the aid of the Position Menu) the side of the plane on which you wish the entities to remain displayed. CADKEY then identifies the intersection of the selected entities and the plane and trims them appropriately.

 POSSIBLE USE

This option is very helpful when you have projected entities beyond a surface in the **X-FORM** option. Using **EDIT-SEC-TION-TRIM** you can trim all of the lines to a selected plane or surface.

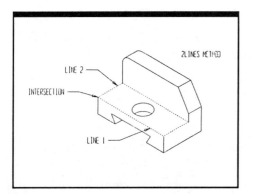

Figure 15-27 A plane defined by selecting two lines

SECTION OPTION: BREAK

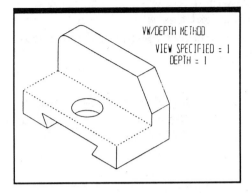

Figure 15-28 A plane defined by identifying a view and depth

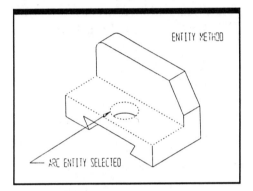

Figure 15-29 A plane defined by selecting an entity within the plane

This option will break selected entities with reference to a plane. The Escape Code for this option is **ESC-F2-F9-F2** (**EDIT-SEC-TION-BREAK**).

To start, the Selection Menu is displayed and you are prompted to select the entities that you wish to break. Then choose a method for defining a plane. Once the plane has been defined you must choose (with the aid of the Position Menu) the side of the plane on which you wish the entities to retain the original attributes. CAD-KEY then identifies the intersection of the selected entities and the plane and breaks them appropriately.

📀 POSSIBLE USE

If you needed to change the line type of all of the entities on one side of a plane, you could break the entities with reference to the plane and they would take on the current line type attributes.

SECTION OPTION: POINTS

This option creates points at the intersection of selected entities and a reference plane. The Escape Code for this option is **ESC-F2-F9-F3** (**EDIT-SECTION-POINTS**).

To start, the Selection Menu is displayed and you are prompted to select the entities on which you wish to create points. Then choose a method for defining a plane. Once the plane has been defined you must respond **NO/YES** to reject or accept the plane. When the plane is accepted, point entities are automatically generated at the points of intersection.

SECTION OPTION: DEL/CHG

This option deletes or changes the attributes of selected entities. The Escape Code for this option is **ESC-F2-F9-F4** (**EDIT-SEC-TION-DEL/CHG**). Choose **DELETE** to erase entities, or **CUR ATT** to change attributes.

THE DELETE OPTION

The **DELETE** option allows you to delete selected entities which intersect a defined plane. When the **NO/YES** menu is displayed, choose **F1 NO** to exclude intersecting entities from deletion and **F2 YES** to delete them. The Selection Menu is then displayed and you are prompted to select the entities that you wish to delete. Then choose a method for defining a plane. Once the plane has been defined you must choose (with the aid of the Position Menu) the side of the plane on which you wish to retain the entities. CADKEY then deletes those entities which lie on the other side of the plane and those that intersect the plane if you choose **F2 YES** from the **NO/YES** menu.

THE CUR ATT
(Current Attribute) Option

CUR ATT changes the attributes (color, line type, pen number, line width, and level) of entities which lie on a selected side or intersect a defined plane. When the **NO/YES** menu is displayed, choose **F1 NO** to exclude intersecting entities from the change and **F2 YES** to change them. The Selection Menu is then displayed and you are prompted to select the entities that you wish to change. Then choose a method for defining a plane. Once the plane has been defined you must choose (with the aid of the Position Menu) the side of the plane on which you wish to retain the original attributes. CADKEY then changes the attributes of those entities which lie on the other side of the plane and those that intersect the plane if you chose **F2 YES**. If **F1 NO** is selected, the intersecting entities remain unchanged.

REVIEW QUESTIONS

1. Which **TRM/EXT** option can be used to trim a circle?
2. What will happen when you select two lines to **TRM/EXT** which do not intersect?
3. When using the **TRM/EXT-FIRST** option, do you select the line to be trimmed first or second?
4. What happens when you **BREAK** an entity?
5. What is the Immediate Mode command which will allow you to bring back entities that were deleted?
6. What happens when the function **EDIT-RECALL-ALL** is used?
7. When would you use the **BX MOVE** option?
8. If you wanted to move a note over one inch, which **BX MOVE** option would you use?
9. What is a **GROUP**?
10. What is an advantage of a **GROUP**?
11. What is the difference between the **AUTOSEG** options **ENTS** and **LINES**?
12. What is the purpose of **EDIT-SECTION**?

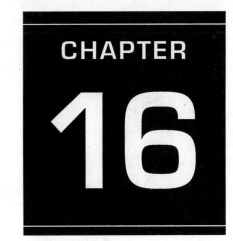

Detailing Your Design

OBJECTIVES

After completing this chapter, you will be able to:

- Use the **SET** function to control the output of the detailing options.
- Change the text options used for detailing.
- Set tolerancing values for dimensions.
- Change the dimensioning units.
- Place linear, radial, angular, diametral, and ordinate dimensions on a drawing.
- Place notes on a drawing.
- Place labels on a drawing.
- Create arrow and witness lines.
- Use the **CHANGE** function to change the existing text and dimensions.
- Use the **UPDATE** function to change attributes of dimenions.
- Create sectional views with CADKEY.

After a design is created, detailing is one of the most important elements in completing an engineering drawing. Detailing a drawing includes dimensioning, adding notes or text, placing pointers and labels, and adding cross-hatching to sectional views. CADKEY offers the user a very powerful sequence of commands that are used to detail drawings. CADKEY also gives the user many different options as far as the changing of the default values of the entities used for detailing. For example, you are not limited to one lettering style but can choose from six different ones or define your own using the font program. Dimensions can be in inches, feet, yards, millimeters, meters, centimeters, or user defined. The size and style of dimension text can also be changed. These and many more options are described, as well as the steps to follow to place details on a drawing. Figures 16-1 and 16-2 show the menu commands covered in this chapter.

THE SET FUNCTION

CADKEY 6 uses a dialog box to set the various detail attributes for dimensioning (Figure 16-3). To change one of the options, simply click on it and a list of selections will appear. The detail attributes

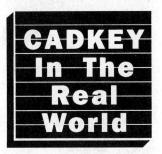

NCGA Honors High-Tech Headache Researcher

The National Computer Graphics Association presented its second annual award for innovative use of microcomputer-based CAD technology to Ruth Ann Fraser for her unique application of CAD software, designed for mechanical and reverse engineering, in the study of cluster headaches. The NCGA made the presentation on April 23, 1991, during the **NCGA '91 Conference and Exposition** in Chicago, Illinois.

Ruth Ann, a registered nurse and graduate student at Carleton University, Ottawa, Ontario, Canada, had been doing research into the dimensions of headache pain at Ottawa Civic Hospital in 1988, and at the Headache Research Foundation of Faulkner Hospital, Boston, Massachusetts in 1989-1990. During her research, Ruth Ann obtained the collaboration of Wence Daks of CAD WIRE, a CADKEY dealer in Markham, Ontario, and of Gary Magoon and

Ujjwell Trivedi of Cadkey, Inc. She used a combination of Cadkey's products, CADDInspector™/ Copy CAD™, CADKEY®, CADKEY SOLIDS™, and Perceptor® to create a reliable technique by which sufferers of cluster headaches can report the locations of their pain in three dimensions. Pain has not only height and width (area), but also depth (volume). (See **High-Tech Headache Research, 3-D WORLD**, May/June 1990, page 1.)

Ruth Ann created an anatomically correct, three-dimensional CAD model of the human head which Santin Engineering of Beverly, Massachusetts, manufactured into a physical model, assembled from precisely machined slices of ⅜-inch-thick, transparent acrylic. The slices are assembled as a physical model of the head by sliding each slice down two dowels mounted into a base, through precisely drilled holes, one slice at a time, from the neck to the top of the head, similar to a child's ring-stacking toy.

A cluster headache patient can help medical personnel to study his/her pain by drawing its location(s), level by level on the acrylic slices, in erasable marker. Using the Perceptor's stylus, a technician creates a digital tracing of the patient's 2-D drawings on each respective level. The accumulated headache data from all the levels, provided through the Perceptor, creates a three-dimensional representation of the pain as a part file in CADKEY. Using CADKEY SOLIDS on this part file allows the medical personnel to calculate the mass properties of the pain, specifically its volume and its centroid. The centroid serves as the focal point of each pain region that the patient has described.

Ruth Ann's research into headache pain is the foundation of her thesis for a Master of Arts degree in Psychology at Carleton University.

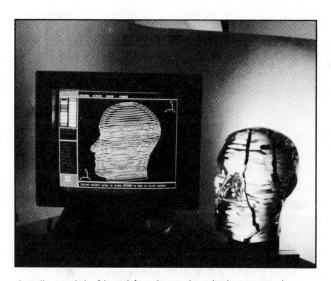

Acrylic model of head for cluster-headache research.

*Adapted with permission from **3-D WORLD**, v. 5, no. 4, July/Aug., 1991.*

---

are divided into four categories: Dimension, Tolerance (Figure 16-4), Text (Figure 16-5), or Lead/Wit (Figure 16-6).

A new option allows the user to save the settings as a template file which can be recalled and used whenever needed. The template command buttons are **Save** (saves the template file to disk), **Load** (ability to load, copy, move, delete or rename a template file), **Reset** (restores the values to the previous settings), and **Default**

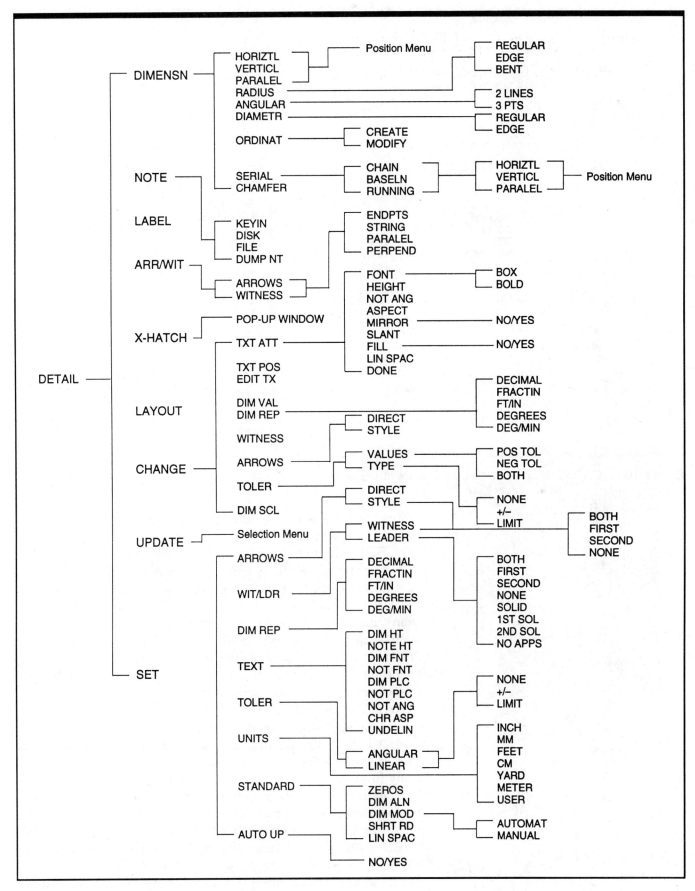

Figure 16-1   The DETAIL menu (CADKEY 5)

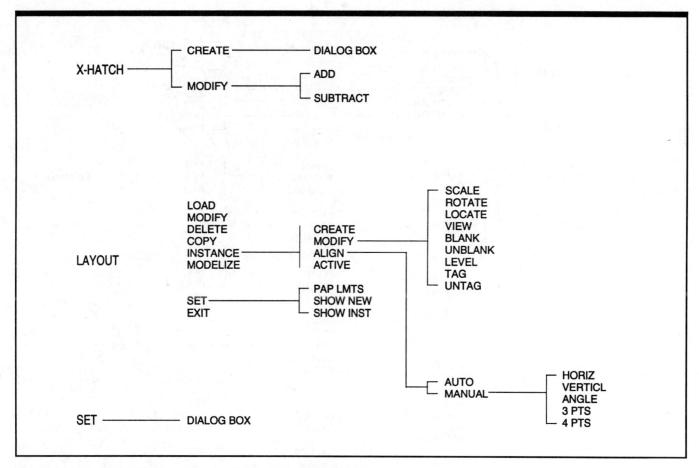

Figure 16-2   CADKEY 6 new or revised options in DETAIL

(restores the values to the settings which were active when the current part was loaded).

The **SET** function is a series of nine options that control the output of various detailing options. The **ARROWS** option controls the direction that arrows are placed for dimensioning. **WIT/LDR** (Witness Leader) controls the location of witness (extension) lines on dimensions. **DIM REP** (Dimension Representation) controls the dimensions in decimals, feet and inches, or fractions in the dimension figure. **TEXT** controls the height, lettering font, alignment position for text placement, angle, and the aspect ratio of the text used in dimensions, notes, and labels. The **TOLER** (Tolerance) option allows control of angular and linear tolerancing limits for dimensions. **STANDRD** (Standard) controls whether the trailing or leading zero in a dimension is displayed and whether a dimension is displayed parallel or horizontally in an aligned dimension. The **UNITS** option controls the units used in the dimensioning. **AUTO UP** controls whether displayed values of selected dimensions are automatically updated when retrieving a pattern or using an **X-FORM** option. Finally, the **VER DIM** (Verify Dimension) option controls the displaying of a prompt to speed up the dimensioning task.

These **SET** options offer the user a wide range of latitude when detailing. Learning to control these options and use them to

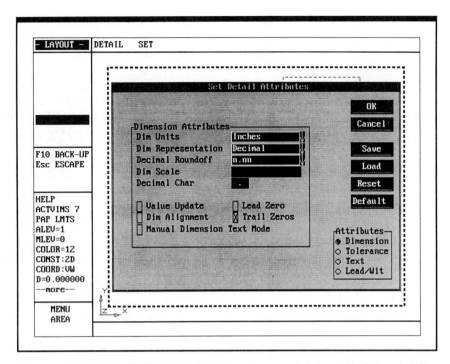

Figure 16-3 The DETAIL-SET dialog box containing all of the detail attributes

full advantage greatly enhances the appearance of most drawings. The first part of this chapter explains how the **SET** function controls the detailing process. From this knowledge you will be better prepared to detail a drawing to your specifications using the full power and many options offered by CADKEY.

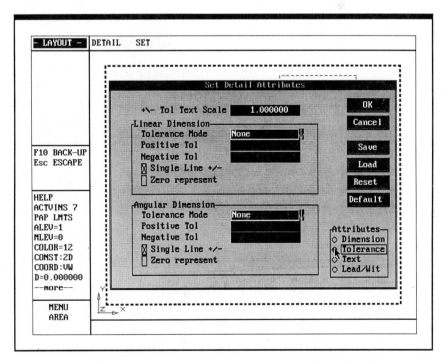

Figure 16-4 The DETAIL-SET dialog box containing all of the tolerance attributes

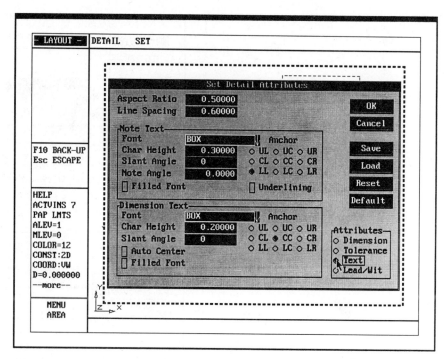

Figure 16-5 The DETAIL-SET dialog box containing all of the text attributes

SET OPTION: ARROWS

This option controls the direction in which arrows are positioned relative to the extension lines on a linear or angular dimension or to the geometry on a circle or arc. The default position for this option is **IN**. CADKEY offers three different methods of controlling

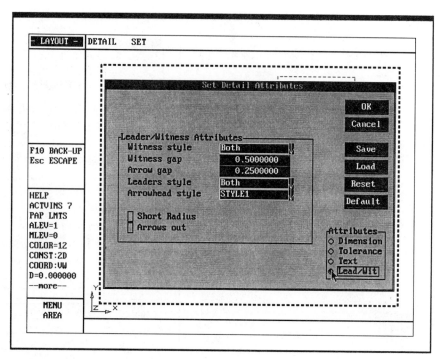

Figure 16-6 The DETAIL-SET dialog box containing all of the Lead/Wit attributes

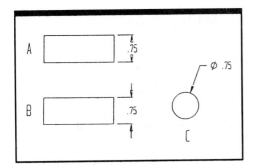

Figure 16-7   Locations of arrows for dimensions

the arrows. One is through the menu commands **DETAIL-SET-ARROW**. Another method is to use the Immediate Mode command **CTRL-A**. The current status of the arrow option is displayed in the Status Menu under **ARROW:** and it can be toggled between **IN** and **OUT** by moving the cursor onto the menu command and pressing the mouse button. **IN** means that the arrows are located on the same side of the extension lines (witness) or part geometry as the dimension figure. **OUT** places the arrows outside of the extension (witness) lines or part geometry as the dimension figure (Figure 16-7). The **STYLE** option provides two styles of arrows: **STYLE1** (open) and **STYLE2** (filled). **STYLE1** is the default setting, as shown in Figures 16-7A and B, and **STYLE2** is shown in Figure 16-7C.

POSSIBLE USE

The **ARROW** in or out feature is an extremely valuable aid when dimensioning small details. Figure 16-7A shows the conventional and the default methods used to dimension a detail when there is enough room to have the dimension figure on the same side of the extension (witness) line as the arrows. However, sometimes the detail to be dimensioned is small and the arrows must be placed outside of the extension (witness) lines or part geometry (Figure 16-7B). The following steps show how this option is changed using the **ARROW** option.

1. Use the commands **ESC-F3-F9-F1-F1** (**DETAIL-SET-ARROWS-DIRECT**).

2. A prompt reads:

    Choose arrowhead direction (current=IN).

    A menu is displayed showing the **IN, OUT** options. The default is **IN**; to change to **OUT**, press **F2 OUT**. The Status Window **ARR:** changes to **ARR:OUT**.

3. Anything dimensioned after the **OUT** option is selected is dimensioned as shown in Figures 16-7B and C.

SET OPTION: WIT/LDR (Witness/Leader)

WIT/LDR controls the display of extension (witness) and dimension or leader lines of dimensions. The **WITNESS** option is selected from the menu, through the Status Window (**WIT**) or Immediate Mode command **CTRL-B**. The **LEADER** option is selected from the menu or from the Status Window (**LDR**).

Witness

This option controls which of the two extension (witness) lines are displayed for linear or angular dimensions. CADKEY uses the term "witness" in place of extension line, which is the more commonly accepted term used in drafting and design. The default is for both witness lines to be displayed (see Figure 16-8A). This option is changed using the Immediate Mode command **CTRL-B**. The following steps

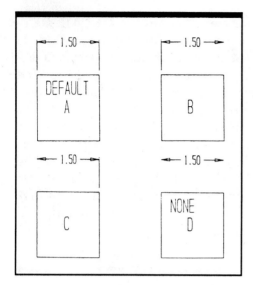

Figure 16-8 Location of witness
lines using the SET-WITNESS option

demonstrate how this option is used to control the placement of witness lines.

1. Select **ESC-F3-F9-F2-F1** **(DETAIL-SET-WIT/LDR-WITNESS)**.

2. A prompt reads:

    Choose witness lines display option (current=BOTH).

    A menu is displayed listing the options for this command: **BOTH, FIRST, SECOND, NONE.**

3. Select **F2 FIRST**. This option causes a witness line to be drawn at the first point picked for the location of the dimension. The second point picked does not have a witness line (Figure 16-8B).

The steps used for the **SECOND** and **NONE** options are the same as those described above, except for the **SECOND** option. With this option, the second point picked will have a witness line and the first will not. Examples of both the **SECOND** and **NONE** options are shown in Figures 16-8C and 16-8D, respectively. Once the **WITNESS** option is changed it will remain that way until it is changed again.

Leader   The **LEADER** function controls the display of leader lines in **ANGULAR** and **LINEAR** dimensions, which are more commonly called "dimension lines" by drafters and designers. Sometimes it is more convenient to leave one or more dimension (leader) lines off a drawing because of space limitations. There are eight methods of controlling leader lines: **BOTH, FIRST, SECOND, NONE, SOLID, 1ST SOL, 2ND SOL, NO ARRS.** The default setting is to display **BOTH** leader lines (Figure 16-9A).

The **FIRST** option displays only one of the leader lines. The leader line that is displayed is the first point picked when locating the dimension (Figure 16-9B). The **SECOND** option will display the leader line for the second dimension point only (Figure 16-9C). **NONE** is used when you do not want any leader lines to be displayed. This option is used quite often when there is only room for the dimension text but not for dimension (leader) lines or arrowheads (Figure 16-9D). An alternative to using the **NONE** option would be to use the **ARROWS-OUT** option described earlier in this chapter. **SOLID** is used for architectural dimensions that require no breaks in the dimension line (Figure 16-9E). **1ST SOLID** displays an arrow at the first point selected when placing the dimension (Figure 16-9F). **2ND SOLID** displays an arrow at the second point selected (Figure 16-9G). **NO ARRS** displays a dimension without arrows (Figure 16-9H).

The **LEADER** function is used by selecting **ESC-F3-F9-F2-F2** **(DETAIL-SET-WID/LDR-LEADER)**. A menu is displayed having all eight options. Select the desired option and place the dimension(s). The system will remain set at the selected **LEADER** options until it is changed again.

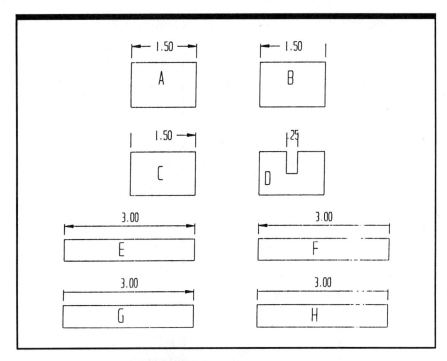

Figure 16-9   The SET-LEADER options

SET OPTION: DIM REP
(Dimension Representation)

DECIMAL   The dimension representation-decimal option controls the decimal precision of dimension values in the tolerancing of a part or the decimal precision that a part should be held to when manufacturing. CADKEY uses the standard two-place decimal precision as the default value for dimensions. However, CADKEY also allows the value to change from no decimals to six places of decimal accuracy, or to one ten-millionth of an inch. When this option is chosen it is displayed as a menu on the screen as follows:

   n. (full units) (Figure 16-10A)
   n.n (tenths) (Figure 16-10B)
   n.nn (hundredths) (Figure 16-10C)
   n.nnn (thousandths) (Figure 16-10D)
   n.nnnn (ten-thousandths) (Figure 16-10E)
   n.nnnnn (hundred-thousandths) (Figure 16-10F)
   n.nnnnnn (millionths) (Figure 16-10G)

After the decimal precision is changed, it will remain that way.

 POSSIBLE USE

The **DIM REP** option is used whenever the decimal precision of a part has to be controlled. For example, the decimal precision necessary to manufacture a commercial airline jet engine is much greater than that necessary to manufacture a bicycle.

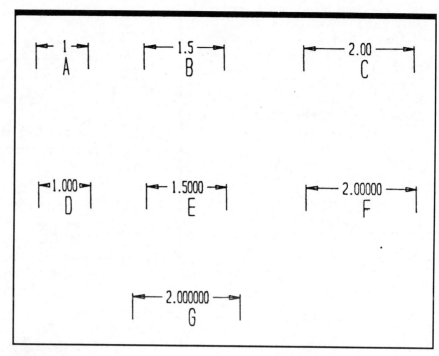

Figure 16-10 Some of the decimal precisions available

The decimal precision may change from part to part for certain types of assemblies. For these reasons it is necessary to be able to control the decimal precision of your dimension values. Figure 16-10 shows the various decimal precision values that can be used with CADKEY. The following steps show how to change the decimal precision from the default (.nn) to another value.

1. Select **ESC-F3-F9-F3-F1 (DETAIL-SET-DIM REP-DECIMAL)**.
2. A prompt reads:

    Choose decimal precision (current=n.nn).

    A menu is displayed listing the options from **n.** to **.nnnnnn**.
3. Choose **F1 N**. This changes the decimal precision from n.nn the default) to n., which will display only full units (Figure 16-10A).
4. Press **ESC** to exit this function.

The steps necessary to change to the other decimal precision options are the same as those described above.

FRACTIN (Fraction)   The **FRACTIN** option controls the fractional precision of dimensions. To assign the fractional precision select: **ESC-F3-F9-F3-F2 (DETAIL-SET-DIM REP-FRACTIN)**. A prompt displays: **Choose round off factor (current = 1/2)**. The screen menu displays:

1/2
1/4
1/8
1/16
1/32
1/64

Make the selection from the menu to change the fractional precision. All calculated dimensions from this point will be to the nearest fractional value set.

**DEGREES and DEG/MIN
(Degrees and Minutes)**

The **DEGREES** option sets angular dimension text to display in degrees when using the **ANGULAR** dimension option (Figure 16-11A). The **DEG/MIN** option sets angular dimension text to display in degrees, minutes, and seconds when using the **ANGULAR** dimension option (Figure 16-11B).

FT/IN (Feet and Inches)

The **FT/IN** (feet and inches) option sets feet and inches and the rounding value to dimension values. Select **ESC-F3-F9-F3-F3 (DETAIL-SET-DIM REP-FT/IN).** A prompt reads: **Choose round off factor (current = 1/2).** Select the round off factor from the same six options displayed with **DIM REP-FRACTIN.** All calculated dimensions from this point will be in feet and inches to the nearest fractional value chosen.

 POSSIBLE USE

One use of the **FT/IN** option is for architectural drawings. The standard method of dimensioning architectural drawings is to use feet and inches displayed above the dimension line. By

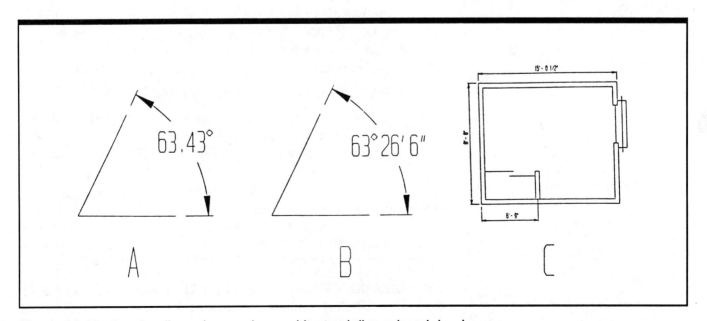

Figure 16-11 Angular dimensions and an architectural dimensioned drawing

setting the **DIM REP** option to **FT/IN** it is possible to create architectural dimensions to standards (Figure 16-11C).

SET OPTION: TEXT

This function provides the user with nine different options for controlling the text used in notes, labels, and dimensions. These options control the text height, slant, width, font, alignment, and angle. The default values for note and label text and dimension text are listed below.

	NOTE/LABEL	*DIMENSIONS*
HEIGHT	.3 English mode 5mm 5mm Metric mode	.3 English mode 5mm 5mm Metric mode
FONT	Box	Box
ALIGNMENT	Left, Bottom	Center, Center
CHARACTER ASPECT	.5	.5
NOTE ANGLE	0.0	0.0
NOTE SLANT	0	0
NO FILL	0	0
UNDERLINE	None	None

The following paragraphs describe how to change these values for various functions.

DIM HT (Dimension Height)

This option changes the height of the dimension text from the default value of .3. The smallest value that can be entered using **DIM HT** is .0005. Any new value input will be the height assigned to dimension text from this point on. Figure 16-12 shows a dimension using the default value of .3 and a new value of .125. The following steps show how to change the text dimension height.

1. Press **ESC-F3-F9-F4-F1** (**DETAIL-SET-TEXT-DIM HT**). The **DH** Status Window parameter also could be used to change the dimension height.
2. A prompt reads:

   Enter dimension text height (current=0.3) =.

   Enter the new value of **.125** and press **RETURN**. This becomes the value assigned to dimension text, as shown in Figure 16-12 and displayed in the **DH** Status Window parameter.

NOTE HT (Note Height)

This option changes the default value for note height from .3 to any value above .0005. All text entered after it has been changed takes on the new value until it is changed or the system is re-booted. Figure 16-12 also shows the default text height and a smaller value assigned using the **NOTE HT** option as explained below.

1. Select **ESC-F3-F9-F4-F2** (**DETAIL-SET-TEXT-NOT HT**). The **NH** Status Window parameter could also be used to change the note height.

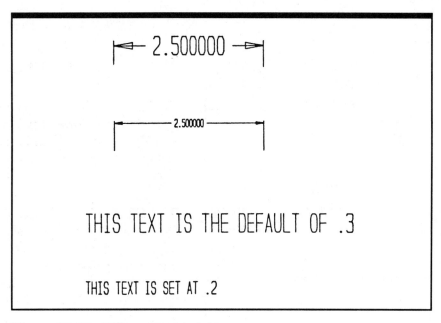

Figure 16-12 Different text heights

2. A prompt reads:

    Enter note text height (current=0.3) =.

    Enter **.125** and press **RETURN**. This becomes the value used for all notes and labels placed on the drawing, as shown in Figure 16-12 and displayed in the NH Status Window parameter.

DIM FNT (Dimension Font)

This option specifies one of the many lettering fonts (styles) for use with dimension text. The default style is **BOX** and is the type that is recommended for most applications because of faster redraw and plotting speed. It is highly recommended that fonts other than the **BOX** style be changed after the drawing is completed using the **CHANGE** and/or **UPDATE** functions that are described later in this chapter. The following steps demonstrate how to change the dimension text.

1. Press **ESC-F3-F9-F4-F3** (**DETAIL-SET-TEXT-DIM FNT**) or select **DF** (Dimension Font) from the Status Window.

2. A prompt reads:

    Choose dimension text style (current=BOX).

    A menu is displayed listing the two styles available as **BOX** and **BOLD**. Select the desired dimension text style and press **RETURN**.

3. Another prompt reads:

    Enter text slant angle=.

    Enter an angle between −31 degrees and +31 degrees.

4. A prompt reads:

    Fill Font?

A **YES/NO** menu is displayed. Enter **F1 NO** or **F2 YES**.

Figure 16-13 shows dimension text created using **BOLD**, 10 degrees slant, and filled. Press **ESC** to exit this function.

NOT FNT (Note Font)

NOT FNT is basically the same option as the **DIM FNT** except it controls the type of font used with **NOTES** and **LABELS**. The default font is **BOX** and any new notes or labels will take on that style until it is changed again. The same styles used for dimensions are available for notes and labels.

DIM PLC (Dimension Placement)

DIM PLC controls where the dimension text is placed relative to the point picked when placing the dimension text. The default value for this option is **CENTER, CENTER**. This means that wherever the point is picked for the placement of the dimension text, the text is centered left to right and centered top to bottom relative to the picked point (Figure 16-14E). The picked point is represented by the small "X" on the dimension text.

This option increases the ease with which dimension text placement is controlled. When this option is selected, three options are displayed: **HORIZTL, VERTICL** (horizontal and vertical) and **AUTOCEN**. Nine different alignment points can be assigned using the **LEFT, CENTER, RIGHT** options which are displayed after **HORIZTL** is selected and **BOTTOM, CENTER, TOP** after **VERTICL** is selected. All of these are shown in Figure 16-14 with the picked points represented by the small "X" located on or near the dimension text.

The following steps show how the **DIM PLC** option is used to change the alignment points for dimension text.

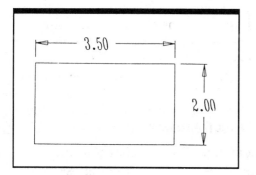

Figure 16-13

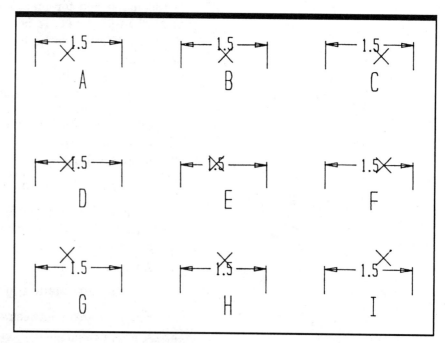

Figure 16-14   The picked points relative to the placement of dimension text on screen

1. Select **ESC-F3-F9-F4-F5 (DETAIL-SET-TEXT-DIM PLC).**

2. A menu is displayed having the **HORIZTL** and **VERTICL** options. Select **F1 HORZNTL.**

3. A prompt is displayed:

   Choose horizontal alignment (current=CENTER).

   Another menu is displayed having the **LEFT, CENTER, RIGHT** options. Choose **F1 LEFT.** Dimension text is displayed relative to the picked point (Figure 16-14D).

The default mode when placing a dimension is to choose the text placement with the input device. However, CADKEY gives the user the option of having the dimension text automatically centered between the extension lines on linear and angular dimensions. If **ARROW-IN** is selected with automatic centering, the text for radius dimensions is centered between the tip of the arrow on the leader line and the center of the circle, and the text for diameter dimensioning is placed at the center of the arc or circle.

To automatically center dimensioning text, select **ESC-F3-F9-F4-F5-F3 (DETAIL-SET-TEXT-DIM PLC-AUTO CEN).** A prompt reads: **Automatically center dimension text? (current=NO).** To change the setting from its default, select **F2 YES** from the menu.

NOT PLC (Note Placement)

NOT PLC is similar to the **DIM PLC** option. It is used to control the placement of **LABELS** and **TEXT** relative to the picked point used for locating the text on the drawing. The default alignment for notes and labels is bottom, left. Again, there are two options: **HORIZTL** and **VERTICL** with a total of nine possible alignment points. Figure 16-15 shows the different alignment points relative to the digitized points used when locating the text on screen. The following steps show how this option is used.

1. Press **ESC-F3-F9-F4-F6 (DETAIL-SET-TEXT-NOT PLC).**

2. A new menu is displayed having the **HORIZTL** and **VERTICL** options. Select **F2 VERTICL.**

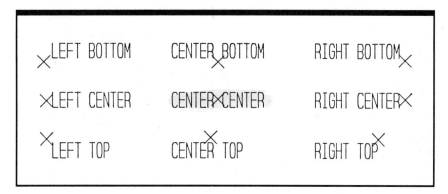

Figure 16-15 The picked points relative to the placement of the note and label text on screen

3. A prompt reads:

    Choose vertical alignment (current=BOTTOM).

    A menu is displayed listing the options: **BOTTOM, CEN-TER, TOP**. Select **F2 CENTER**. Labels and notes are displayed relative to the picked point to place the text (Figure 16-15). This setting is used on all labels and notes until it is changed again.

4. Press **ESC** to exit this option.

NOT ANG (Note Angle)

Figure 16-16 An example of the different angles that text can be placed with CADKEY

NOT ANG controls the angle that the line of text in a **NOTE** or **LABEL** is drawn on the screen. The default angle for all text is **0** degrees. The text is rotated about the current alignment point of the text. For example, if the current alignment point is center-center, the note is rotated about that point when it is placed on screen. Figure 16-16 shows text that is rotated at different angles using the bottom-left alignment point. The following steps show how this option is used to change the text angle.

1. Press **ESC-F3-F9-F4-F7** (**DETAIL-SET-TEXT-NOT ANG**).
2. A prompt reads:

    Enter new note angle (current=0).

    Enter the angle desired using the angle convention (right hand) used for CADKEY as described in Chapter 3.

3. Press **ESC** to exit this option.

CHR ASP (Character Aspect)

DEFAULT ASPECT RATIO .5

ASPECT RATIO .2

ASPECT RATIO .7

ASPECT RATIO 1

Figure 16-17 Default aspect ratio and other aspect ratios for comparison

CHR ASP (Character Aspect ratio-width to height) controls the width of text characters relative to the height. The default value is .5, which means that the width of the text characters is one-half their height. The limits for this option are greater than or equal to .01 and less than or equal to 100. Figure 16-17 shows the default value and three other aspect ratios for comparison. The following steps show how the aspect ratio is changed.

1. Press **ESC-F3-F9-F4-F8** (**DETAIL-SET-TEXT-CHR ASP**).
2. A prompt reads:

    Enter new character aspect ratio (current=0.5).

    Enter the number desired and press **RETURN**. This aspect ratio is used on all text until it is changed again using the **CHR ASP** option.

3. Press **ESC** to exit this option.

UNDELIN (Underlining)

Note and label text can be automatically underlined using the **UNDERLINE** option. To create text with underlining, select

ESC-F3-F9-F4-F9 (DETAIL-SET-TEXT-UNDELIN). A prompt reads: **Underline note and label text? (current=NO).** Select **F2 YES** to automatically underline text (Figure 16-18).

SET OPTION: TOLER (Tolerance)

Figure 16-18 Underlined text

This option sets the tolerancing that CADKEY automatically applies to dimensions when it is set to on. The default for the system is to have the tolerance off. To turn the tolerance on you must use the **TOLER** option and select the type of tolerance text to be displayed. CADKEY gives you a number of options in selecting limits and how the tolerance value is displayed. This section explains how this function is used.

CADKEY divides tolerancing into two major types: **ANGULAR** and **LINEAR. ANGULAR** affects only the angular dimension values and has default values of +.05 and −.05. **LINEAR** affects only linear dimensions, such as horizontal, vertical, parallel, radial, and diametrical, and has a default value of +.01 and −.01. When either of these two functions is selected, there are three possible options: **NONE, +/-,** and **LIMIT.** These options control how the tolerance is displayed on screen.

The number of decimal places assigned to a tolerance depend on the number of decimals assigned using the **DECIMAL** function described earlier in this chapter. For example, the default decimal value is set at .nn (hundredths) when using inches. Therefore, any tolerance is limited to the default value which is in hundredths. If three-place decimal precision is required for tolerances, then the **DECIMAL** precision must also be changed to .nnn or three-place precision. Figure 16-19 shows various tolerancing options assigned with the **TOLER** function.

LINEAR-NONE

The **NONE** option turns tolerancing values off after they are turned on. As a default, CADKEY has the tolerancing values turned off.

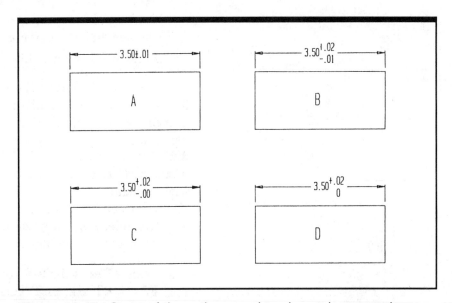

Figure 16-19 Some of the options used to place tolerance values

The **NONE** option is used only after tolerances have been turned on using the **ANGULAR** or **LINEAR** commands and tolerance values are no longer desired. **ESC-F3-F9-F5-F1-F1** or **ESC-F3-F9-F5-F2-F1** (**DETAIL-SET-TOLER-ANGULAR** or **LINEAR-NONE**) will turn either angular or linear tolerances off.

LINEAR-+/–

The **+/–** option sets a plus-minus expression after the displayed dimension. There are five options available under this function: **POS TOL, NEG TOL, 0 REP, DBL UP, TOL SIZ**. Each of the options is explained and an example is demonstrated. Simply selecting the **+/–** option under **LINEAR** allows tolerances to be displayed using the default values (Figure 16-19A).

POS TOL assigns a positive value to the tolerance. With this option you can change the default positive value of +.01 to a value of your specifications. Remember that the decimal precision that you use is controlled by the **DECIMAL** command and not the **POS TOL** option. If you want to change from two-place decimal precision to three-place precision with the **POS TOL** option, you must first change the decimal precision from two to three using **DETAIL, SET, DIM REP, DECIMAL**. Figure 16-19B shows a dimension with a positive tolerance of .02 which was changed using the **POS TOL** option. The steps used to change this value are shown below.

1. Press **ESC-F3-F9-F5-F1-F2-F1** or **ESC-F3-F9-F5-F2-F2-F1** (**DETAIL-SET-TOLER-ANGULAR** or **LINEAR-+/– POS TOL**).

2. A prompt reads:

   Enter positive tolerance (current=0.01) =.

   Enter the desired positive value, such as **.02**, and press **RETURN**. Any linear dimensions placed on a drawing display a positive value of .02 for a tolerance (Figure 16-19B). The same steps are used to assign a positive tolerance value for angular dimensions, except that **ANGULAR** is chosen in the sequence of commands instead of **LINEAR**.

3. Press **ESC** to exit this option.

NEG TOL changes the negative tolerance value from the default of .01 to any negative value within the system limits (.n to .nnnnnn). The default decimal precision of .nn must be changed using the **DECIMAL** option before a negative tolerance value of different precision is used. When the negative tolerance value is input with the **NEG TOL** option, the value must be preceded with a negative (–) sign. The value is not accepted if the negative sign is not entered. Figure 16-19C shows a dimension that had a negative tolerance value of .00 input using the **NEG TOL** option. The following steps show how this option is used.

1. Select **ESC-F3-F9-F5-F1-F2-F2** or **ESC-F3-F9-F5-F2-F2-F2** (**DETAIL-SET-TOLER-ANGULAR** or **LINEAR-+/– NEG TOL**).

2. A prompt reads:

    Enter negative tolerance (current=–0.01) =.

    Enter **–.00** and press **RETURN**. Be sure to enter the negative value before the number. Figure 16-19C is an example of a dimension using the negative tolerance assigned with this option.

3. Press **ESC** to exit this option.

0 REP is a toggle switch that controls the display of the + or – sign in front of zero tolerance values. The default setting for this option is to display the sign before tolerance values of zero. For example, Figure 16-19C shows a negative tolerance value of –.00. The negative value could be turned off and displayed, as shown in Figure 16-19D, by using the **0 REP** option. The use of this option is shown using the following steps.

1. Select **ESC-F3-F9-F5-F1-F2-F3** or **ESC-F3-F9-F5-F2-F2-F3** (**DETAIL-SET-TOLER-ANGULAR** or **LINEAR-+/–0 REP**).

2. A prompt reads:

    Display 0 tolerance without decimal or sign? (current=NO).

    A **NO/YES** menu is displayed. To remove the sign before zero tolerance values press **F2 YES**.

3. Press **ESC** to exit this option.

DBL UP (Double Up) is a toggle switch that shows both the positive and negative tolerance values on the same line of text. The default setting is **ON**, which displays both values in one line if they are equal. Both the positive and negative values must be equal for this option to work. For example, 2.50 with a tolerance of +.05 and –.05 is displayed as 2.50 ±.05 when this option is **ON**. Another example is shown in Figure 16-20A. However, if the tolerance value of 2.50 is +.05 and –.03, this option will not work because both tolerance values must be equal. The following steps show how to use this option.

1. Press **ESC-F3-F9-F5-F1-F2-F4** or **ESC-F3-F9-F5-F2-F2-F4** (**DETAIL-SET-TOLER-ANGULAR** or **LINEAR-+/– DBL UP**).

2. A prompt reads:

    Display+/– tols as one value if possible? (current=YES).

    A **NO/YES** menu will also be displayed. Press **F2 YES** or press **RETURN**. Dimensions are displayed as shown in Figure 16-20A.

3. Now change the option to **NO** by pressing **F1 NO**. Dimensions are displayed as shown in Figure 16-20B.

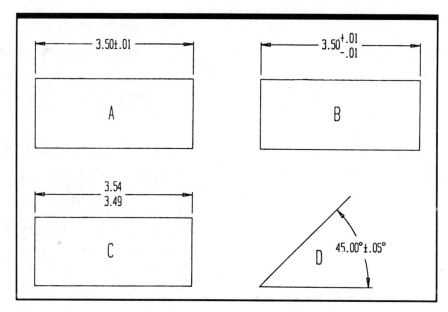

Figure 16-20   Other examples of toleranced dimensions

LINEAR-LIMIT   This option displays the upper and lower limits of a dimension (Figure 16-20C). In that example, the basic size is 3.50 +.04 and −.01 but it is displayed with the upper limit value above the lower limit value. The upper limit value is computed as the basic size plus the positive tolerance value. The lower limit value is computed as the basic size minus the negative tolerance value. The system will do these calculations automatically when the **LIMIT** option is on. When this option is chosen, two parameters can be set: **POS TOL** and **NEG TOL**. Choose the tolerance value to set and enter the desired value. If both values are to be set, then **POS TOL** and **NEG TOL** must be set one at a time. The following steps demonstrate how the **LIMIT** option is used.

1. Select **ESC-F3-F9-F5-F1-F3** or **ESC-F3-F9-F5-F2-F3** (**DETAIL-SET-TOLER-ANGULAR** or **LINEAR-LIMIT**).

2. A menu is displayed having two options: **POS TOL** and **NEG TOL**. Select the desired tolerance to change. In this example the positive tolerance value is to be changed so **F1 POS TOL** is selected.

3. A prompt reads:

   Enter positive tolerance (current=0.010000) =.

   Enter the value desired, in this example .04, followed by **RETURN**. Linear dimensions are displayed (Figure 16-20C).

4. Press **ESC** to exit this option.

ANGULAR   The **ANGULAR** option affects only those dimensions that are used for angular measurements. All the options described above will work identically whether the **ANGULAR** or **LINEAR** option is chosen. The **ANGULAR** option will work the same as the

LINEAR option for these commands: **NONE, +/−, LIMIT, POS TOL, NEG TOL, 0 REP, DBL UP**. The default tolerance values for the **ANGULAR** option are +.05 and −.05. Figure 16-20D shows an angular dimension using the assigned default values.

SET OPTION: STANDRD (Standard)

The **STANDRD** function controls whether leading or trailing zeros are displayed with a dimension and determines if the dimension figure for parallel dimensions is horizontal or aligned with the dimension line. These standards vary for ANSI or ISO. There are five options offered with the **STANDRD** function: **ZEROS, DIM MOD, SHRT RD, LIN SPAC** and **DIM ALN**.

ZEROS

ZEROS controls the display of leading or trailing zeros through two options: **LEAD** and **TRAIL**. The **LEAD** option controls the display of leading zeros in a dimension figure. There are two conditions offered: **NO** and **YES**. Figure 16-21A is an example of the default condition, which is without leading zeros. Figure 16-21B is an example of the ISO condition, which will display leading zeros. The following steps show how this option is used to display leading zeros.

1. Press **ESC-F3-F9-F6-F1-F1 (DETAIL-SET-STANDRD-ZEROS-LEAD)**.

2. A prompt reads:

   Display leading zero before decimal point? (current= NO).

   A **NO/YES** menu is displayed. Choose **F2 YES**. All dimensions are displayed with leading zeros (Figure 16-21B).

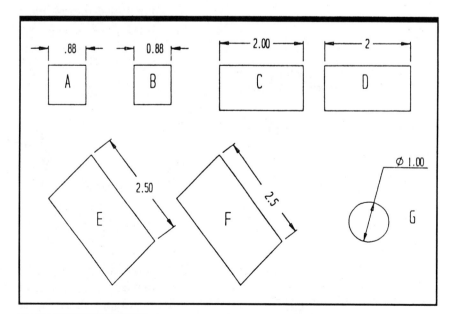

Figure 16-21 Various options assigned to dimension text

The **TRAIL** option controls whether trail zeros on dimension text are displayed. For example, if the **TRAIL** option was in the **ON** state, a dimension of 2 would be displayed as 2.00 (Figure 16-21C). This same dimension would be displayed as 2 without trailing zeros (Figure 16-21D), if the **TRAIL** option was in the **OFF** state. The default is to have **TRAIL** in the **ON** state. The following steps show how this option is used to turn the trailing zeros **OFF**.

1. Press **ESC-F3-F9-F6-F1-F2** (**DETAIL-SET-STANDRD-ZEROS-TRAIL**).

2. A **NO/YES** menu is displayed and a prompt:

   Display trailing zeros? (current=YES).

3. Select **F1 NO**. All dimensions added to the drawing are displayed without trailing zeros (Figure 16-21D).

4. Press **ESC** to exit this function.

DIM ALN (Dimension Align)

DIM ALN (Dimension Align) controls the text for parallel dimensions. The default for this option is for the dimension text to be placed horizontally and not aligned with the dimension line. Figure 16-21E shows the default setting and Figure 16-21F shows the aligned setting. The following steps show how this option is used to change from the default setting (horizontal) to the aligned setting.

1. Press **ESC-F3-F9-F6-F2** (**DETAIL-SET-STANDRD-DIM ALN**).

2. A prompt reads:

   Align dimension text (current=NO)?

   A menu appears with two options: **NO** and **YES**. Select **F2 YES** to align the dimensions with the dimension line for parallel dimensions.

3. Press **ESC** to exit this option.

SHRT RD (Shorten Radius/Diameter)

This option, used with solid leader lines, will shorten the leader line and display below the dimension text (Figure 16-21G).

DIM MOD (Dimension Mode)

DIM MOD allows control between automatic and manual creation of dimension text. **DIM MOD** is always in the automatic mode when CADKEY is booted. To change to the manual mode enter **ESC-F3-F9-F6-F3** (**DETAIL-SET-STANDRD-DIM MOD**). The **DIM MOD** menu has two options: **AUTOMAT** (automatic) and **MANUAL**. To change to manual entry of dimension text enter **F2 MANUAL**. In the manual mode every dimension that is placed on screen using the **DETAIL-DIMENSN** option must have the dimension text entered from the keyboard.

When creating a dimension in manual mode, indicate the first and second positions for the dimension. Instead of indicating the text position a prompt reads: **Enter text:**. The text string to appear in the dimension is entered at the prompt. Pick the position for the text and the dimension and text are displayed (Figure 16-22).

When creating a text string for dimensions you may have to create the diameter symbol, degree symbol, or the plus/minus symbol. Most IBM compatibles will use the numeric entries listed below with the NUM LOCK on and entered from the number pad on the keyboard to create the desired symbol.

DIAMETER SYMBOL

1. Press the **ALT** key and hold down.
2. Press the **1** key on the numeric keyboard.
3. Press the **7** key on the numeric keyboard.
4. Press the **4** key on the numeric keyboard.
5. Release the **ALT** key. (This creates the first half of the diameter symbol.)
6. Press the **ALT** key and hold down.
7. Press the **1** key on the numeric keyboard.
8. Press the **7** key on the numeric keyboard.
9. Press the **5** key on the numeric keyboard.
10. Release the **ALT** key. (This creates the second half of the diameter symbol.)

DEGREE SYMBOL

1. Press the **ALT** key and hold down.
2. Press the **2** key on the numeric keyboard.
3. Press the **4** key on the numeric keyboard.
4. Press the **8** key on the numeric keyboard.
5. Release the **ALT** key.

PLUS/MINUS SYMBOL

1. Press the **ALT** key and hold down.
2. Press the **2** key on the numeric keyboard.
3. Press the **4** key on the numeric keyboard.
4. Press the **1** key on the numeric keyboard.
5. Release the **ALT** key.

LIN SPAC (Line Spacing)

When placing multiple lines of notes with CADKEY, the default space used between lines of text is set for a ratio of .6 (Figure 16-23). This ratio is a factor of the current text height. For example, if your text height was 1″, the spacing used between the lines of text would be .6″. To change the ratio setting, select **ESC-F3-F9-F6-F5 (DETAIL-SET-STANDRD-LIN SPAC)**. **LIN SPAC** (line spacing) is the set option used to change the distance between lines of text. A prompt reads: **Enter note text line spacing factor (current=.6)**. Enter a new spacing ratio and press **RETURN**. Figure 16-23 shows the difference in spacing between the default ratio of .6 and 1.

SET OPTION: UNITS

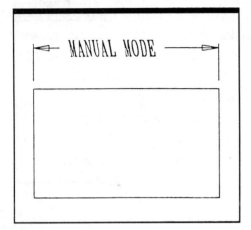

Figure 16-22 A dimension created using the manual mode

The **UNITS** option controls the units that are used when dimensioning a drawing. These units can be different from the units used to construct the drawing. If CADKEY is configured for the English system of units, the default unit is 1.0 inch. If the unit is changed before the drawing is dimensioned, all dimensions are scaled to the new units from the units used to create the part. For example, if the drawing is created using the default setting of inches and is changed to feet, all dimension values are divided by 12 inches. So a measurement of 9 inches is dimensioned as .75. The units are displayed without identification symbols, such as an apostrophe (') for feet or mm for millimeters.

There are seven different options listed under the **UNITS** command. Six are standard units and the seventh, **USER**, is used to enter your own scale factor. Thus CADKEY offers an unlimited variety of units that can be used with a drawing. All **UNITS** options are applied to the scale of dimensions placed after the unit is changed. Any dimensions already displayed are not affected by the **UNITS** option. Figure 16-24 shows a horizontal dimension created using all seven **UNITS** options. The following steps show how the **UNITS** option is used.

1. Select **ESC-F3-F9-F7** (**DETAIL-SET-UNITS**).

2. A prompt reads:

   Choose dimension scaling units (current=INCH).

   A menu also appears with the seven **UNITS** options: **INCH, MM, FEET, CM, YARD, METER, USER**. Select the desired units and dimension the drawing.

   The steps are the same for all **UNITS** in the menu except **F7 USER**. With this command, the prompt reads: **Enter dimension scale factor (current=1)=**. At this point, the user would input the desired scale and press **RETURN**. For the **USER** example shown in Figure 16-24, .5 is input making the 8.50 full-scale dimension read as one-half or 4.25.

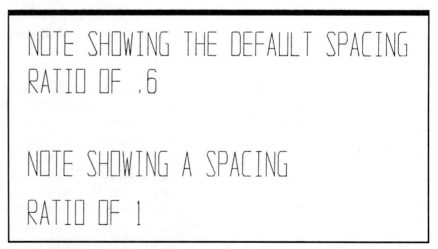

Figure 16-23 Changing spacing of text lines

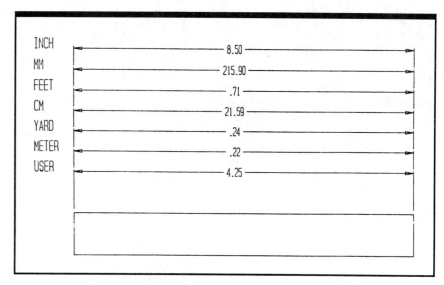

Figure 16-24   Different units created by CADKEY

**SET OPTION: AUTO UP
(Automatic Update)**

AUTO UP controls whether CADKEY automatically updates the displayed values of dimensions when using the **PATTERN-RE-TRIEVE** or **X-FORM** options. The default setting is set to **F1 NO**, which means that the scale factor applied when retrieving a pattern or using the **X-FORM** option is not applied to the displayed dimension value. Setting **AUTO UP** to **F2 YES** causes retrieved pattern files or **X-FORM** scaling functions to apply the new dimensions except for angular and manually entered dimensions.

 POSSIBLE USE

As experience is gained with CADKEY you will find the **SET** function is used quite often to control the lettering style, height, decimal precision, tolerances, change dimension characters, and so forth. The dimensioned drawing shown in Figure 16-25 was detailed using many of the options listed under the **SET** function. Many of the default values were changed to create the dimensions, notes, and labels. By using these different options available with CADKEY, the general appearance of a drawing will greatly improve. The following steps show how the **SET** function was used to create the details on the drawing in Figure 16-25.

1. Select **ESC-F3-F9-F4-F1** to set the dimension height (**DIM HT**).
2. Change the value from .3 to **.2** and press **RETURN**.
3. Select **ESC-F3-F9-F4-F2** to set the note height (**NOTE HT**).
4. Change the value from .3 to **.25** and press **RETURN**.
5. Select **ESC-F3-F9-F4-F3-F2** and change the dimension font (**DIM FNT**) to **BOLD**. Enter a slant angle of **0** degrees and **NO** for fill font.

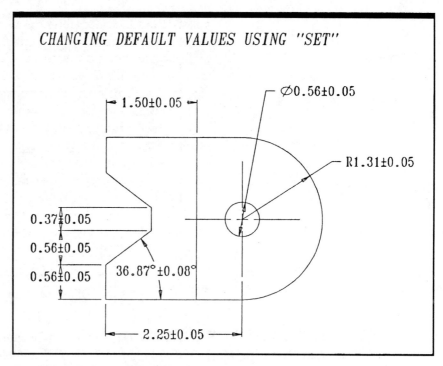

Figure 16-25 An example of how the SET function can be used to control the output of dimensions and notes

6. Select **ESC-F3-F9-F4-F4-F4-F2** and change the note font (**NOT FNT**) to **BOLD**. Enter a slant angle of **−10** degrees and **NO** for fill font.

7. Select **ESC-F3-F9-F4-F8** and change the aspect ratio (**CHR ASP**) of the characters to **.75.**

8. Select **ESC-F3-F9-F5-F1-F2-F1** and **F2, F4** and change the **ANGULAR** tolerance value to **+.080 −.080** and **DBL UP**.

9. Select **ESC-F3-F9-F5-F2-F2-F1** and **F2, F4** and change the **LINEAR** tolerance value to **+.05 −.05** and **DBL UP, YES.**

10. Select **ESC-F3-F9-F6-F1-F1-F2** to turn the lead **ZEROS** option on.

11. Select **ESC-F3-F9-F7-F7** to change the units to **USER.** Input **.75** and press **RETURN.**

12. Dimension the object and notice the changes to the notes, labels and dimensions from the default settings.

THE DIMENSN (Dimension) FUNCTION

The **DIMENSN** function places horizontal (**HORIZTL**), vertical (**VERTICL**), parallel (**PARALEL**), radial (**RADIUS**), angular (**ANGULAR**), diametral (**DIAMETR**), ordinal (**ORDINATE**), and **SERIAL** dimensions on a drawing. With CADKEY, horizontal, vertical, and angular dimensions are placed on a drawing by picking the two points to be dimensioned and picking the location for the dimension text. CADKEY automatically draws the dimension (witness) line, extension (leader) lines, arrows, and text. Radial and diametral dimensions are placed on a drawing by picking the entity and picking the location of the label. The system prompts for the direction that the leader should be related to the label. The system then places the leader, arrow, and label on the drawing. For diametral dimensions, the phi symbol is placed before the dimension according to ANSI standards.

The default values associated with dimensioning include: .3 dimension text height, box character font, center-center label alignment, arrows in, both witness lines, two-place decimal precision, no tolerance values, no lead zeros, inch or mm units depending on how CADKEY is configured, both leaders, and the dimension verification turned on. The **DIM MOD** option is set to automatic when CADKEY is booted so the dimension is calculated. If **DIM MOD** is set to manual, the on-line text editor is automatically activated and you are prompted to enter the dimension text. **DIM MOD** and the text editor (**EDIT TX**) options are explained later in this chapter.

By using the **SET** function reviewed earlier in this chapter, you can control the dimension text height and font using the **TEXT** option, the placement of the witness line using the **WITNESS** option, and the placement of the leader lines using the **LEADERS** option. You have a variety of methods for placing dimensions as well as a number of different character styles, heights, slants, and aspect ratios.

To erase a single dimension, use **DELETE-SINGLE** or **CTRL-Q**. The dimension text is picked to erase a dimension. Dimensions are considered to be one entity, so the whole dimension is erased including the text, arrows, dimension and extension lines by picking the text position. The **CHANGE** function changes some of the characteristics of a dimension on a drawing, such as text font and height, and arrow in or out. The **CHANGE** option is explained later in this chapter.

When placing dimensions it may be more convenient to display a **GRID** or use the **SNAP** option. These two options control the distance between dimensions so that they are uniformly placed on the drawing.

This part of the chapter guides you through the placement of dimensions. Each step is explained and an example given for the specific task. Figure 16-26 is used as the drawing to be dimensioned.

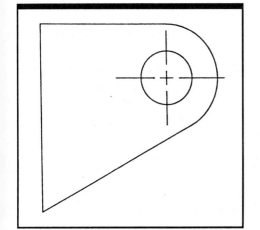

Figure 16-26 The drawing to be dimensioned

DIMENSN OPTION: HORIZTL (Horizontal)

The **HORIZTL** option places horizontal dimensions onto a drawing. The two points to be dimensioned are picked using the Position Menu for accuracy. The position for the text is then picked and the dimension is automatically placed by CADKEY. Use the **SET** function to change any default values before dimensions are placed on a drawing. Place a grid on screen for uniform placement of

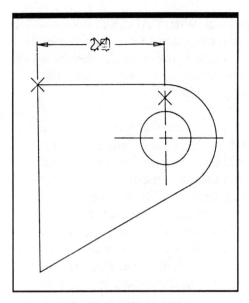

Figure 16-27 Picked points to place a horizontal dimension

dimensions. Refer to Figure 16-27 and the following steps to place horizontal dimensions. The picked points are represented by an "**X**" on the drawings.

1. Select **ESC-F3-F1-F1 (DETAIL-DIMENSN-HORIZTL)**.

2. The Position Menu is displayed defaulted to **ENDENT** along with a prompt:

    Indicate 1st position.

    Select the proper item from the displayed Position Menu. For this example **F3 ENDENT** is selected. Pick the endpoint of the entity to be dimensioned. A small triangular marker or X will identify the entity selected.

3. A prompt is displayed along with the Position Menu. The prompt reads:

    Indicate 2nd position.

    ENDENT is used again so the second entity is picked and the system places a small marker on the line.

4. The system now prompts:

    Indicate dimension text position.

    Pick the location for the dimension text. The dimension is displayed along with **ALT-I UNDO** in the Break Menu area.

    If the position for the dimension is acceptable, you can place another dimension. If the dimension is not placed properly, select **ALT-I UNDO**. The dimension is deleted from the screen and you will be prompted to indicate the dimension text position again.

DIMENSN OPTION: VERTICL (Vertical)

The **VERTICL** option places vertical dimensions onto a drawing. The steps to follow are exactly the same as for placing horizontal dimensions. To change default settings use the **SET** function. Refer to Figure 16-28 and the following steps to place a vertical dimension.

1. Select **ESC-F3-F1-F2 (DETAIL-DIMENSN-VERTICL)**.

2. The Position Menu is displayed along with a prompt:

    Indicate 1st position.

    Select the proper item from the displayed Position Menu. For this example **F3 ENDENT** is selected. Pick the endpoint of the entity to be dimensioned. A small triangular marker or X identifies the entity selected.

3. A prompt is displayed along with the Position Menu. The prompt reads:

    Indicate 2nd position.

    ENDENT is used again, so the second entity is picked and the system places a small marker on the line.

4. The system now prompts:

    Indicate dimension text position.

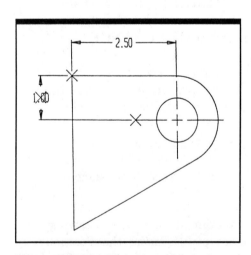

Figure 16-28 Picked positions for vertical dimensions

Pick the location for the dimension text. If the dimension is not placed properly, select **ALT-I UNDO**. The dimension is deleted from the screen and you are prompted to indicate the dimension text position again.

DIMENSN OPTION: PARALEL (Parallel)

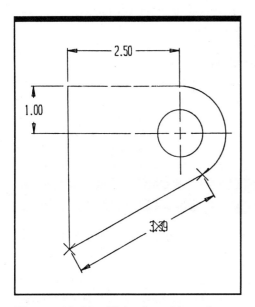

Figure 16-29   Picked positions for parallel dimension

PARALEL places a parallel dimension between two specified points. The most obvious use is to dimension lines which are not horizontal or vertical. With this option all the default characteristics and procedures for placing the dimension are the same as for horizontal and vertical dimensions. The **SET** function is used to change the default settings. The dimension text is placed horizontally for the dimension. If for some reason you would like the dimension text placed parallel or aligned with the dimension line, the **DIM ALN** option under the **SET** function is used to change it. Figure 16-29 shows a parallel dimension placed on a drawing and the following steps show the procedures to follow.

1. Select **ESC-F3-F1-F3 (DETAIL-DIMENSN-PARALEL)**
2. The Position Menu is displayed along with a prompt:

   Indicate 1st position.

   Select the proper item from the displayed Position Menu. For this example **F3 ENDENT** is selected. Pick the endpoint of the entity to be dimensioned. A small triangular marker or X identifies the entity selected.

3. A prompt is displayed along with the Position Menu. The prompt reads:

   Indicate 2nd position.

   ENDENT is used again, so the second entity is picked and the system places a small marker on the line.

4. The system now prompts:

   Indicate dimension text position.

   Pick the location for the dimension text. If the dimension is not placed properly, select **ALT-I UNDO**. The dimension is deleted from the screen and you are prompted to indicate the dimension text position again.

DIMENSN OPTION: RADIUS

The **RADIUS** option automatically dimensions the radius of a circle or an arc. The arc must be parallel to the current viewing plane or seen true size and not as an ellipse. Arcs are dimensioned by picking the arc to be dimensioned followed by picking the position for the label. A prompt asks for the leader side before the dimension is placed on the drawing. There are two options: **LEFT** or **RIGHT** (Figure 16-30). CADKEY automatically places the arrows, dimension line, and label on the drawing after the leader direction is chosen. The dimension label is preceded by the letter "R" for radius. The default settings are the same as for all other dimensions and are changed using the **SET** command.

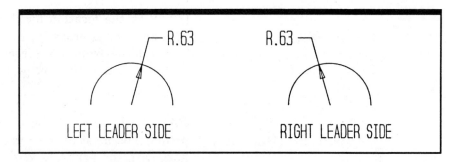

Figure 16-30 Examples of the LEFT and RIGHT leader side options

When selecting the **RADIUS** command, there are three options: **REGULAR, EDGE** or **BENT**. The **REGULAR** option dimensions the radius of a circle or arc in the same view you are dimensioning (Figure 16-31). The **EDGE** option dimensions an arc or circle using one of three choices: **HORIZONTAL, VERTICAL,** or **PARALLEL** (Figure 16-31). If the arc or circle is not shown in the current view the **EDGE** option dimensions it as shown in Figure 16-32. The **BENT** option dimensions an arc or circle whose center is not on screen. The **BENT** option creates a zigzag line (Figure 16-33).

Refer to Figure 16-34 and the following steps to place radial dimensions on a drawing.

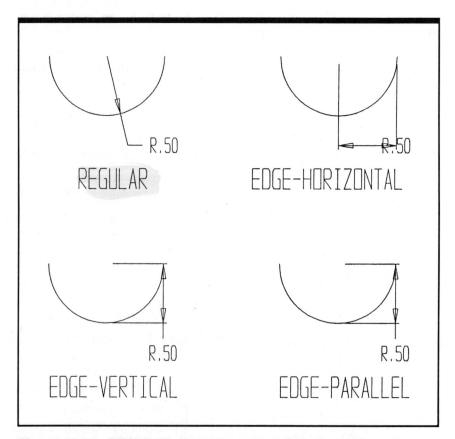

Figure 16-31 REGULAR dimensions the radius of a circle or arc in the same view you are dimensioning.

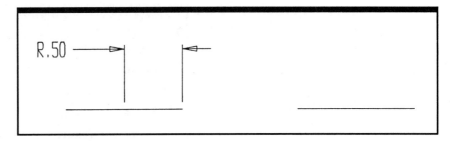

Figure 16-32 EDGE will dimension the arc or circle if it is not shown in the current view.

1. Select **ESC-F3-F1-F4-F1** (**DETAIL-DIMENSN-RADIUS-REGULAR**).

2. A prompt reads:

    Select arc to be dimensioned.

    Pick the arc or circle to be dimensioned. A small marker is displayed on the entity.

3. A prompt reads:

    Indicate dimension text position.

    Pick the position for the dimension label.

4. A prompt reads:

    Choose leader side.

    A **LEFT/RIGHT** menu is displayed. For this example **F1 LEFT** is selected.

5. The dimension is displayed as shown in Figure 16-34.

DIMENSN OPTION: ANGULAR

The **ANGULAR** option automatically dimensions the angle between two non-parallel lines using **2 LINES** or **3 POINTS** options. The displayed portions of the lines do not have to actually

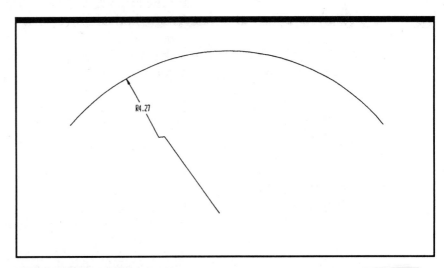

Figure 16-33 BENT dimensions an arc or circle whose center is not on screen

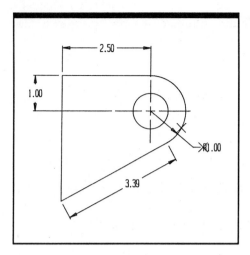

Figure 16-34 Picked positions to place an arc dimension

intersect. Default settings are the same as for other dimensions and the **SET** function is used to make changes. Figure 16-35 and the following steps show how the **ANGULAR** option is used to dimension non-parallel lines.

1. Select **F3-F1-F5-F1** (**DETAIL-DIMENSN-ANGULAR-2 LINES**).

2. A prompt reads:

   Select first line.

   Pick one of the non-parallel lines to be dimensioned. A marker is placed on the line.

3. A prompt reads:

   Select 2nd line.

   Pick the other line making up the angle. Temporary lines are drawn through the two lines to make angle selection easier.

4. A prompt reads:

   Select angle to be dimensioned.

   Pick the angle to dimension.

5. A prompt reads:

   Which angle do you wish to measure? (current= INNER)

   Select **F1 INNER**, then pick the text position.

DIMENSN OPTION: DIAMETR (Diameter)

The **DIAMETR** option automatically dimensions the diameter of an arc or circle using the **REGULAR** or **EDGE** options. The arc or circle must lie parallel to the current view to use the **REGULAR** option. This is accomplished by picking the arc or circle and picking the leader line. The phi diameter symbol specified by ANSI standards is automatically displayed in front of the dimension number. The default values for **DIAMETR** are the same as for the other dimensions. Figure 16-36 illustrates some of the different methods of displaying a diametral dimension by controlling the **ARROWS-IN/OUT** option and whether the position picked for the label is inside or outside the circle or arc. Figure 16-36A shows the **ARROWS-IN** and the label picked outside the circumference of the circle. Figure 16-36B has the **ARROWS-OUT** and digitized label position outside the circle. Figure 16-36C has the **ARROWS-IN** and the label position picked inside the circumference of the circle. Figure 16-36D has the **ARROWS-OUT** and the label position picked inside the circle.

Figure 16-37 and the following steps show how to use the **DIAMETR** option.

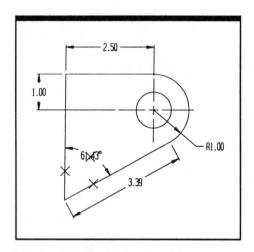

Figure 16-35 Picked positions to place angular dimensions

1. Select **ESC-F3-F1-F6-F1** (**DETAIL-DIMENSN-DIAMETR-REGULAR**).

2. A prompt reads:

   Select arc to be dimensioned.

Pick the arc or circle to be dimensioned. A small marker is displayed on the entity.

3. A prompt reads:

Indicate dimension text position.

Pick the position for the dimension label.

4. A prompt reads:

Choose leader side.

A **LEFT/RIGHT** menu is displayed. For this example **F1 LEFT** is selected.

5. The dimension is displayed, as shown in Figure 16-37.

DIMENSN OPTION: ORDINAT (Ordinal Dimensioning)

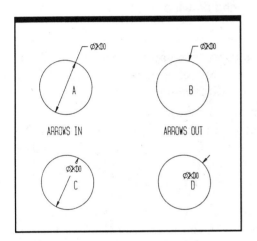

Figure 16-36 Examples of the ARROWS IN and ARROWS OUT options used for diametral dimensions

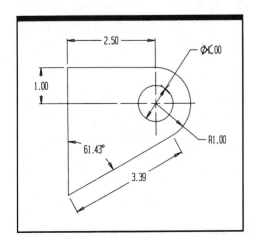

Figure 16-37 Picked positions to place diametral dimensions

Ordinal or datum dimensioning uses a datum surface from which all dimensions are measured. The datum surface would have a dimension value equal to zero. Ordinate dimensions have extension lines and dimension text but no arrows or dimension lines. **ORDINAT** is the CADKEY command used to **CREATE** and **MODIFY** ordinate dimensions.

The **CREATE** option is used to place a string of ordinate dimensions on the horizontal, vertical, or parallel axis in an **ALIGNED** or **NON-ALN** (non-aligned) mode. Figure 16-38 and the following steps demonstrate the **ORDINAT** function.

1. Turn the grid and snap on by entering **CTRL-G** and **CTRL-X**. Press **ESC-F3-F1-F7** (**DETAIL-DIMENSN-ORDINAT**).
2. **CREATE** and **MODIFY** options are listed on the screen menu. Select **CREATE** by entering **F1**.
3. **ALIGNED** and **NON-ALN** modes are listed in the menu. Select **F1 ALIGNED**.
4. Three ordinal dimensioning options are listed in the menu: **HORIZONTL**, **VERTICL**, and **PARALEL**. Select **F1 HORIZONTL** to place horizontal ordinal dimensions.
5. The Position Menu is displayed and a prompt:

Indicate base position.

Select **F3 ENDENT** and pick the upper right corner of the part, shown in Figure 16-38. This point is the datum point for the horizontal dimensions.

6. A prompt reads:

Indicate dimension text position.

Pick a point about 1/2 inch directly above the upper right corner of the part.

7. A prompt reads:

Indicate position (RETURN when done).

CADKEY is requesting you to pick the points on the part that are to be dimensioned. As each point is picked, the dimension is automatically placed in alignment with the first dimension

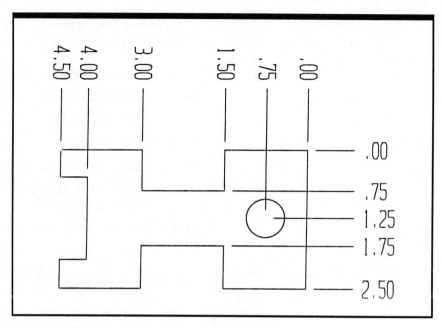

Figure 16-38   Ordinate dimensioned drawing

because the **ALIGNED** option was selected in step 3. If an error is made in the placement of a dimension, enter **F10 BACKUP** to erase the last dimension entered.

8. After all the points are dimensioned press the **RETURN** key. CADKEY returns to step 5, to choose the base position. To finish dimensioning this part, enter **F10-F2 (BACKUP-VERTICL)** and repeat steps 5 through 8 to place the vertical dimensions on the drawing (Figure 16-38).

The **NON-ALN** mode creates ordinate dimensions that are individually placed on the drawing by the user. The **PARALEL** option places ordinal dimensions parallel to a selected entity. The **MODIFY** option provides four methods of changing existing ordinal dimensions. **ADD TO** adds more ordinal dimensions to an existing string. **MOVE** moves any existing ordinal dimension except the base ordinal. **ALIGN** aligns ordinal dimensions that were created in the **NON-ALN** mode. **REMOVE** erases ordinal dimensions by picking the lower left corner of the text.

DIMENSN OPTION: SERIAL

SERIAL dimensioning is a technique used to place a series or chain of dimensions. There are three options to the **SERIAL** command: **CHAIN, BASELN,** and **RUNNING.** These options are used to create a series or chain of dimensions on a drawing using one of three options: **HORIZTL, VERTICL,** or **PARALEL.**

BASELN (Baseline)

BASELN creates a series of dimensions that have one reference point or baseline. The dimensions are stacked so that larger dimensions are outside of smaller dimensions (Figure 16-39). The following steps describe how to create serial dimensions using the **BASELN** option.

1. Select **ESC-F3-F1-F8-F2** (**DETAIL-DIMENSION-SERIAL-BASELN**).

2. A prompt reads:

   Choose dimension axis option.

   There are three options: **HORIZTL, VERTICL,** and **PARALEL.** Select **F1 HORIZTL** to create a horizontal chain of dimensions.

3. A prompt reads:

   Indicate base position.

   Select the upper-left corner of the figure labeled as point **1** using **ENDENT** from the Position Menu.

4. Another prompt reads:

   Indicate position (RETURN when done).

   This prompt is requesting you to select the second point for the first dimension. Select the point labeled **2** in Figure 16-39.

5. Repeat step 4 until all the points have been selected. For Figure 16-39, points **3-5** were selected. Press **RETURN** after all the points are selected.

6. A prompt reads:

   Indicate height of witness lines.

   Pick a point .50″ above the object.

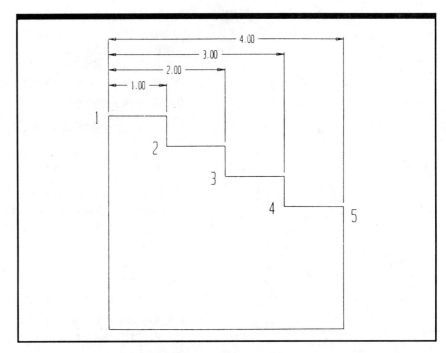

Figure 16-39 Using SERIAL to chain dimensions

7. The first extension lines are displayed on screen along with a prompt that reads:

Indicate dimension text position.

Pick a point to center the first dimension between the extension lines (Figure 16-39).

8. Select **ALT-I** to erase the dimension text placement.

9. The second set of extension lines are drawn on screen and a prompt reads:

Indicate dimension text position.

Pick a point that centers the dimension text between the given extension lines.

10. The third set of extension lines are drawn on screen and a prompt reads:

Indicate dimension text position.

Pick a point that centers the dimension text between the given extension lines.

11. The fourth set of extension lines are drawn on screen and a prompt reads:

Indicate dimension text position.

Pick a point that centers the dimension text between the given extension lines. All of the dimensions have been placed on screen (Figure 16-39). Press **ESC** to exit this function.

The **VERTICL** and **PARALEL** options work the same as **HORIZTL**.

CHAIN　　The following steps describe how to create **CHAIN** dimensions on a drawing.

1. Select **ESC-F3-F1-F8-F1-F2** (**DETAIL-DIMENSN-SERIAL-CHAIN-VERTCL**).

2. A prompt reads:

Indicate position (RETURN when done).

Select the lower-left corner of the part indicated by the number **1** in Figure 16-40.

3. Repeat step 2 until all the corners have been selected. For this example pick the corners labeled **2-5** and press **RETURN** after the last point is selected.

4. A prompt reads:

Indicate height of witness lines.

Pick a point .50″ to the right of the object.

5. The first extension lines are displayed on screen along with a prompt that reads:

Indicate dimension text position.

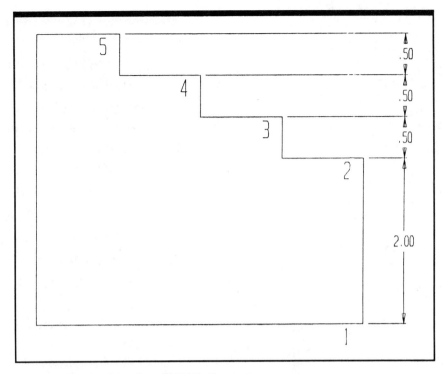

Figure 16-40   Drawing CHAIN dimensions

Pick a point to center the first dimension between the extension lines, as shown in Figure 16-40.

6.   Use **ALT-I** to erase the last dimension placed.

7.   The second set of extension lines are drawn on screen and a prompt reads:

Indicate dimension text position.

Pick a point that centers the dimension text between the given extension lines.

8.   The third set of extension lines are drawn on screen and a prompt reads:

Indicate dimension text position.

Pick a point that centers the dimension text between the given extension lines.

9.   The fourth set of extension lines are drawn on screen and a prompt reads:

Indicate dimension text position.

Pick a point that centers the dimension text between the given extension lines. All of the dimensions have been placed on screen (Figure 16-40). Press **ESC** to exit this function.

The **HORIZNTL** and **PARALEL** options work the same as **VERTICL.**

RUNNING   The **RUNNING** dimension option places the dimension text at the end of the extension lines and at the same angle as the extension line (Figure 16-41). The following steps describe how to create **RUNNING** dimensions on a drawing.

1. Select **ESC-F3-F1-F8-F3-F1 (DETAIL-DIMENSN-SERIAL-RUNNING-HORIZTL)**.

2. A prompt reads:

   Indicate base position.

   Select the upper-left corner of the figure labeled as point **1** using **ENDENT** from the Position Menu.

3. A prompt reads:

   Indicate height of witness lines.

   Pick a point .50″ above the object.

4. Another prompt reads:

   Indicate position (RETURN when done).

   This prompt is requesting you to select the second point for the first dimension. Select the point labeled **2** in Figure 16-41.

5. Repeat step 4 until all the points have been selected. For Figure 16-41 points **3-5** were selected. All of the dimensions have been placed on screen (Figure 16-41). Press **ESC** to exit this function.

The **VERTICL** and **PARALEL** options work the same as **HORIZTL**.

DIMENSN OPTION: CHAMFER

The **CHAMFER** option automatically dimensions chamfers like those shown in Figure 16-42. There are two options: **C** and **X**

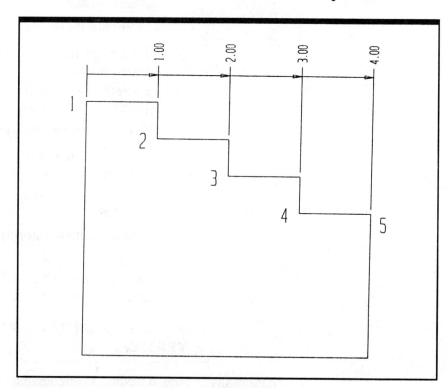

Figure 16-41 Drawing RUNNING dimensions

45DEG. The **C** option adds a C as a prefix to the linear chamfer dimension (Figure 16-43). The **X 45DEG** option adds X 45° as the chamfer dimension's suffix (Figure 16-43). The following steps show how to dimension using the **CHAMFER** option.

1. Press **ESC-F3-F1-F9 (DETAIL-DIMENSN-CHAMFER).**
2. Choose **F2 X 45 DEG.**
3. A prompt reads:

    Indicate the chamfer line.

    Pick the upper-left chamfered corner of the drawing in Figure 16-42.
4. A prompt reads:

    Indicate the text position.

    The Position Menu is displayed to assist in the selection point. For this example use the **CURSOR** option to select the point outside of the boundaries of the object. After the point is selected the dimension is automatically placed on the drawing showing the linear dimension with X 45° appended (Figure 16-43). The **C** option works the same way as the **X 45DEG** option.

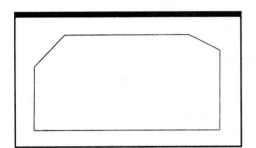

Figure 16-42   Using the CHAMFER option

THE NOTE FUNCTION

The **NOTE** function adds text to a drawing by typing the note from the keyboard or from text read in from a disk file. The default values or notes include: text height of .3, box note font, bottom left note alignment, .5 character aspect ratio. These characteristics can be changed using the various **SET-TEXT** commands described earlier in this chapter.

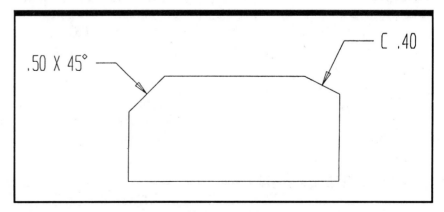

Figure 16-43   The C option adds a C as a prefix to the linear chamfer dimension.

NOTE OPTION: KEYIN

This option places notes or lines of text on a drawing using characters from the computer keyboard. The maximum note length using keyboard input is 1024 characters. However, only one line of text is displayed in the prompt area. Lines longer than 128 characters will be truncated. Underscores and other special characters using keyboard input are not allowed. Any changes in the text height, aspect ratio, and so forth must be made before the text is entered.

Text is entered with the keyboard one line at a time. All corrections and editing must occur before the **RETURN** key is pressed. After the **RETURN** key is pressed, the second line of text is entered. When the last line of text is typed in, press **F6 SAVETX** twice to terminate text input. The text is positioned on screen by moving the cursor to the point where the line of text will begin and picking that point or typing in X,Y view coordinates. Remember that the default alignment point for **NOTE** is bottom left, so the note is drawn on screen slightly above and to the right of the point picked. To erase a single note, use the **DELETE-SINGLE** option or **CTRL-Q** and digitize the alignment point, which is near the bottom of the first character in the note if the default setting for note alignment was used.

Figure 16-44 and the following steps show how the **KEYIN** option is used to locate text on a drawing.

1. Press **ESC-F3-F2-F1 (DETAIL-NOTE-KEYIN)**.

2. A prompt reads:

   Enter text:

   and the Edit Text Menu is displayed. Refer to **EDIT TX** later in this chapter for explanations of these options.

   Enter the first line of text using the keyboard. Press **RETURN** once if another line of text is to be input or press **F6 SAVETX** when finished. For this example type in **NOTE:** and press **RETURN** once.

3. A prompt reads:

   Enter text:

   Enter the second line of text and press **RETURN** once if another line of text is to be input or twice if you are finished. For this example, type in **.125 Brass FAO** and press **F6 SAVETX**.

4. A prompt reads:

   Indicate text position.

   Pick the start position for the text. The text is displayed on screen, as shown in Figure 16-44. Press **ESC** to exit this function.

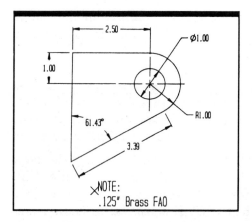

Figure 16-44 Notes can be added to the drawing using the KEYIN or DISK option.

NOTE OPTION: DISK

The **DISK** option is used to read ASCII files from a disk which has prepared text. For example, a word processing program or text editor such as WordStar or Edlin could be used to develop a text file that could be used on a CADKEY drawing for a note or a paragraph for a technical manual. You are limited to 1024 characters per file to be read in at one time, and 256 characters is the maximum allowed in any single line.

To place a text file on screen, the pathname of the location of the file is input along with the filename and extension (if applicable). For example, in **A:NOTE.DOC**, the letter **A:** is for disk drive A, **NOTE** is the filename, and **.DOC** is the filename extension. If a pathname is not specified, the current directory (**\cadkey**) is assumed. After the file is found, cursor-indicate the position for the text or key-in X,Y view coordinate values.

Figure 16-45 and the following steps will show how to use the **DISK** option.

1. Select **ESC-F3-F2-F2 (DETAIL-NOTE-DISK)**.

2. A prompt reads:

    Enter disk note filename:.

    Type in the pathname, filename, and extension and press **RETURN**. For this example **A:EXAMPLE** is entered followed by **RETURN**.

3. If the file is not found a prompt reads:

    Cannot find file . . . press RETURN.

    You will then be given the opportunity to input another name or press **ESC** to exit this function. If the file is found, a prompt reads:

    Indicate text position.

    Pick the location for the text or input X,Y view coordinates. The text is displayed, as shown in Figure 16-44, on the lower right portion of the drawing using current **SET** values for font, alignment, height, and so forth.

NOTE OPTION: FILE   **NOTE FILE** is very similar to **NOTE DISK** in operation and placement on a drawing. The major difference between the two is

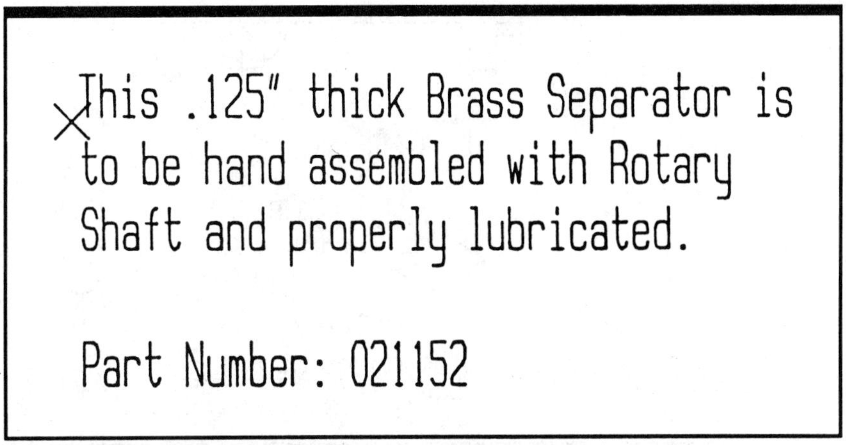

This .125" thick Brass Separator is
to be hand assembled with Rotary
Shaft and properly lubricated.

Part Number: 021152

Figure 16-45 Notes can be added using the KEYIN or DISK option.

that the note text placed on a drawing using the **NOTE FILE** option is not stored in the part file program. This means that CAD-KEY must have access to the file-based note when saving a part, plotting, printing, or drawing.

NOTE OPTION: DUMP NT
(Dump Note)

DUMP NT (Dump Note) is used to output text from existing notes created with the **KEYIN** or **DISK** option to an ASCII text file.

THE LABEL FUNCTION

LABEL creates a label with a leader and arrow. The maximum label length is 128 characters. The default values or the current values assigned using **SET-TEXT** are used when the **LABEL** is placed. With this option, type in the label and press **RETURN** after each line of text. Press **F6 SAVETX** after all the text is input. The text position is then picked or located using X,Y view coordinates. The label's leader direction is then specified as being located to the **RIGHT** or **LEFT** of the text. The position for the tip of the arrow using the Position Menu is then specified using cursor selection or coordinate values.

Figure 16-46 and the following steps show how this command is used to place a **LABEL** on a drawing.

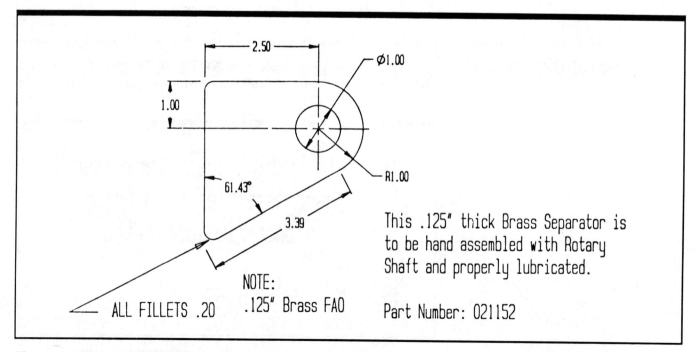

Figure 16-46 Label added to the drawing

1. Press **ESC-F3-F3 (DETAIL-LABEL)**.

2. A prompt reads:

   Enter number of arrowheads for label (1)=>.

   Press **RETURN** for one arrowhead.

3. A prompt reads:

   Enter text:

   Enter the first line of text using the keyboard and press **RE-TURN** once if another line is to be input or twice if it is the only line of text. For this example, type in **ALL FILLETS .20** and press **F6 SAVETX**.

4. A prompt reads:

   Indicate text position.

   Pick the desired location. The text is displayed on screen.

5. A prompt reads:

   Cursor-indicate leader side.

   Pick a point to the left or right of the displayed text.

6. A prompt reads:

   Indicate arrow tip position.

   A rubberband line extends from the point picked in step 4 to the current cursor position. The Position Menu will also be displayed. Select the Position Menu option and pick the arrow tip location. For this example, **CURSOR** is used to pick the round shown in Figure 16-46. The leaders, arrows, and label are displayed as shown in Figure 16-46.

THE ARR/WIT (Arrows and Witness Lines) FUNCTION

ARR/WIT creates arrows and witness (dimension) lines as separate entities from the dimension function. It gives users the capability of creating their own dimensions using separate text, arrow and witness line entities. **ARR/WIT** options can be created using four different options: endpoints, string, parallel, and perpendicular. The size of the arrow is drawn according to the dimension height assigned in the **SET-TEXT** function.

ARR/WIT OPTION: ARROWS

ENDPNTS (Endpoints)

The **ENDPNTS** option places arrows by using the Position Menu to define the location of the start and endpoints. This option works just like the **LINE-ENDPNTS** option used to draw separate lines.

Figure 16-47 and the following steps demonstrate placing arrows on a drawing.

1. Select **ESC-F3-F4-F1-F1 (DETAIL-ARR/WIT-ARROWS-ENDPTS)**.

2. The Position Menu is displayed along with a prompt that reads:

    Indicate start point.

    Select the position to pick from the Position Menu. Move the cursor to the position and press the mouse button.

3. A rubberband line is connected to the point picked to assist in locating the second point. A prompt reads:

    Indicate end point.

    The Position Menu is displayed. Pick the desired position. This second point is where the arrow tip is located (Figure 16-47A).

STRING   The **STRING** option will create a string of arrows by connecting the endpoint of one arrow to the starting point of another. This option works just like the **LINE-STRING** option used to place a connected series of lines. A Position Menu is displayed to help in the location of the start and endpoints of the arrow. Figures 16-47B, C, and D show examples of how this option is used and the following steps demonstrate the drawing of arrows using the **STRING** option.

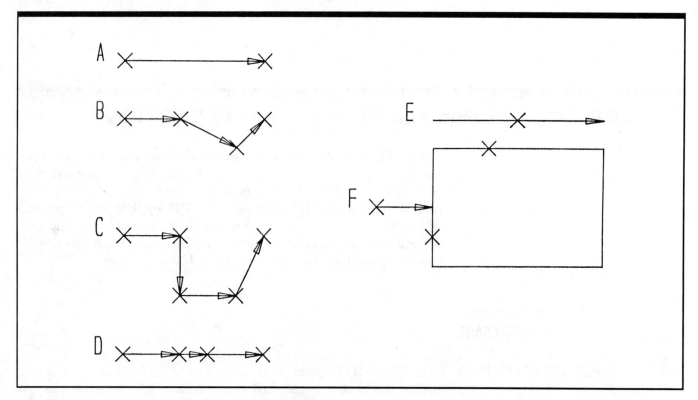

Figure 16-47   Examples of the ARROWS options

1. Select **ESC-F3-F4-F1-F2 (DETAIL-ARR/WIT-ARROWS-STRING).**

2. The Position Menu is displayed along with a prompt that reads:

   Indicate start point.

   Select the position to pick from the Position Menu. Move the cursor to the position and press the mouse button.

3. A rubberband line is connected to the point picked to assist in locating the second point. A prompt reads:

   Indicate end point.

   The Position Menu is displayed. Pick the desired position. This second point will be where the arrow is located.

4. Repeat step 3 to make a string of arrows (Figures 16-47B, C, and D).

PARALEL (Parallel)   The **PARALEL** option creates arrows parallel to a previously created line through a specified point. Figure 16-47E and the following steps show how this option is used.

1. Select **ESC-F3-F4-F1-F3 (DETAIL-ARR/WIT-ARROWS-PARALEL).**

2. A prompt reads:

   Select reference line.

   Pick the line to which the arrow is to be drawn parallel. A small marker is displayed on the line.

3. A prompt reads:

   Indicate position.

   Move the cursor to the position for the arrow and select it. An arrow and a line of the same length and parallel to the reference line are drawn (Figure 16-47E).

PERPEND (Perpendicular)   **PERPEND** draws an arrow perpendicular to a previously created line. Figure 16-47F and the following steps show how to use this option.

1. Select **ESC-F3-F4-F1-F4 (DETAIL-ARR/WIT-ARROWS-PERPEND).**

2. A prompt reads:

   Select reference line.

   Pick the line to which the arrow is to be drawn perpendicular. A small marker is displayed on the line.

3. A prompt reads:

   Indicate position.

Move the cursor to the position for the arrow and pick it. An arrow and a line are drawn perpendicular to the reference line (Figure 16-47F).

ARR/WIT OPTION: WITNESS

ENDPTS (Endpoints)   This option locates witness lines by picking the endpoints. Figure 16-48A and the following steps show how this option is used.

1. Select **ESC-F3-F4-F2-F1 (DETAIL-ARR/WIT-WITNESS-ENDPTS)**.

2. The Position Menu is displayed along with a prompt that reads:

    Indicate start point.

    Select the position to pick from the Position Menu. Move the cursor to the position and press the mouse button.

3. A rubberband line is connected to the point picked to assist in locating the second point. A prompt reads:

    Indicate end point.

    The Position Menu is displayed. Pick the desired position and the witness line is displayed (Figure 16-48A).

STRING   This option places a string of witness lines similar to placing a string of lines using the **LINE** function. Figure 16-48B and the following steps show how this option is used.

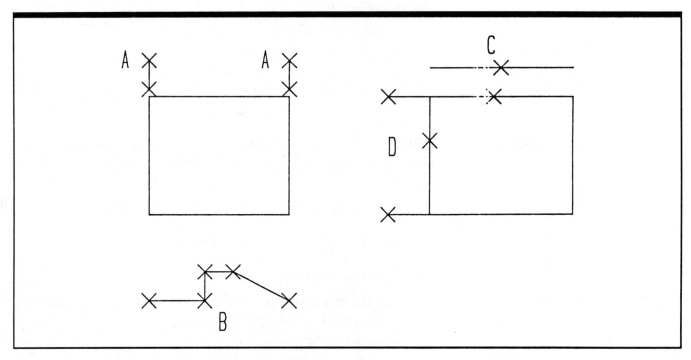

Figure 16-48   Examples of the WITNESS options

1. Select **ESC-F3-F4-F2-F2 (DETAIL-ARR/WIT-WITNESS-STRING)**.

2. The Position Menu is displayed along with a prompt that reads:

   Indicate start point.

   Select the position to pick from the Position Menu. Move the cursor to the position and press the mouse button.

3. A rubberband line is connected to the point picked to assist in locating the second point. A prompt reads:

   Indicate end point.

   The Position Menu will also be displayed. Pick the desired position.

4. Repeat step 3 to make a string of witness lines (Figure 16-48B).

PARALEL (Parallel)   This option places witness lines parallel to a previously placed line. Figure 16-48C and the following steps show how this option is used.

1. Select **ESC-F3-F4-F2-F3 (DETAIL-ARR/WIT-WITNESS-PARALEL)**.

2. A prompt reads:

   Select reference line.

   Pick the line to which the witness line is to be drawn parallel. A small marker is displayed on the line.

3. A prompt reads:

   Indicate point.

   Move the cursor to the position for the witness line and pick it. A witness line of the same length and parallel to the reference line is drawn (Figure 16-48C).

PERPEND (Perpendicular)   This option draws a witness line perpendicular to a previously drawn line. Figure 16-48D and the following steps show how this option is used.

1. Select **ESC-F3-F4-F2-F4 (DETAIL-ARR/WIT-WITNESS-PERPEND)**.

2. A prompt reads:

   Select reference line.

   Pick the line to which the witness line is to be drawn perpendicular. A small marker is displayed on the line.

3. A prompt reads:

   Indicate point.

   Move the cursor to the position for the end of the witness line and digitize. A line is drawn perpendicular to the reference line (Figure 16-48D).

POSSIBLE USE

One possible use for the **ARROW** and **WITNESS** line options is to place cutting plane lines on a drawing. The **ARROW** option is used to place an arrow indicating the direction of sight for the sectional view. The **WITNESS** option is used to create the cutting plane. Figure 16-49 and the following steps show how to create a cutting plane line.

1. The first step would be to change the line width from 1 to 3 because a cutting plane line is a thick line. This is done by pressing **ALT-Y**. A prompt reads:

    Select new line width or press RETURN to key in line width.

    Pick line width **3** from the pop-up menu. Display a grid to assist in locating the endpoints for the arrows. Change the line type to phantom using **L-TYPE** in the Status Window.

2. Window-in to an area of the screen to facilitate the drawing of the cutting plane line by pressing **ALT-W**. Pick two points defining the new window.

3. Select **ESC-F3-F4-F2-F3** (**DETAIL-ARR/WIT-WITNESS-PARALEL**). A prompt reads:

    Select reference line.

4. Pick the vertical center line of the drawing. A second prompt reads:

    Indicate point.

    The Position Menu is displayed. Select **F3 ENDENT** and pick the center line. This causes the new witness line to be drawn on top of the center with a line thickness of 3. Now the arrows are added to the drawing.

5. Press **ESC-F3-F4-F1-F1** (**DETAIL-ARR/WIT-ARROW-ENDPTS**). A prompt is displayed:

    Indicate start point.

    The Position Menu is displayed. Select **F3 ENDENT** to position the arrow line at the end of the center line. Pick the end of the center line.

6. A rubberband line is attached to the end of the center line. A prompt is displayed:

    Indicate end point.

    Move the cursor horizontally and to the left of the end of the center line and pick it. The arrow is drawn (Figure 16-49).

7. Repeat step 6 to locate the second arrow on the opposite end of the cutting plane line.

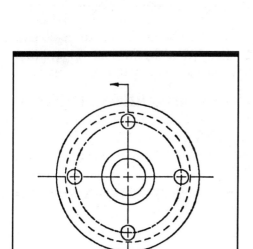

Figure 16-49 Arrows added to create a cutting plane line

THE CHANGE FUNCTION

The **CHANGE** function is used to change the parameters assigned to existing entities. Some of the parameters that can be changed are: text attributes, dimension values, decimal precision, tolerances, and dimension scale. With the **CHANGE** function, the attributes of the entity are immediately updated after it is changed. By using the **CHANGE** function it is possible to make changes or corrections without deleting the entity.

CHANGE OPTION: TXT ATT (Text Attributes)

TXT ATT changes the attributes assigned to text. You can change the font, text height, note angle, mirror, slant, fill, line spacing, and the aspect ratio of existing notes, labels, and dimensions. After the attributes have been changed, the Selection Menu is used to identify the text to be changed. Figure 16-50 and the following steps show how this option is used to change text attributes.

1. Select **ESC-F3-F7-F1 (DETAIL-CHANGE-TEXT ATT)**.
2. A menu is displayed with eight different attributes of text that can be changed. Select **F1 FONT**.
3. A menu is displayed with the two different fonts available. Choose **F2 BOLD**.
4. To change the text height select **F2 HEIGHT**. A prompt is displayed:

   Enter new text height =.

   Input the numeric value desired, such as **.5**, and press **RETURN**.
5. The note angle is changed by using the **F3 NOT ANG** option. For this example, the angle is not changed.
6. To change the aspect ratio for the text choose **F4 ASPECT**. A prompt reads:

   Enter new character aspect ratio =.

   For this example, **1** is input followed by **RETURN**.
7. Select **F5 MIRROR**. A prompt reads:

   Mirror note text?

   A **NO/YES** menu appears. Select **F2 YES** to mirror the text string.

NOTE BEFORE CHANGES

NOTE AFTER CHANGES

Figure 16-50   Results of the CHANGE TEXT option

8. Select **F6 SLANT**. A prompt reads:

    Enter text slant angle =.

    Enter an angle of **15** degrees.

9. Select **F7 FILL**. A prompt reads:

    Fill font?

    A **NO/YES** menu appears. The **FILL** option will fill the out-
    line of the text letter. Enter **F2 YES**.

10. After all necessary changes have been made select **F9 DONE**.

11. The Selection Menu is now displayed. This option is used to
    complete the **TXT ATT** menu sequence and make the
    changes to the text as specified. Select **F1 SINGLE** and
    pick the line of text. In this example, the line of text is
    changed from the default values of .3 height, box font, and
    .5 aspect ratio to the values listed in the steps and shown in
    Figure 16-50.

CHANGE OPTION: TXT POS (Text Position)

The **TXT POS** option moves the position of the text in an existing dimension, note or label. With this option the text is picked
and a box cursor is displayed the same size as the text. This cursor is then moved to the new text position. The space bar or
mouse button is then pressed and the text is moved to the new
position. Figures 16-51, 16-52, and 16-53 and the following steps
explain the operation of this option.

1. Select **ESC-F3-F7-F2 (DETAIL-CHANGE-TXT POS)**.

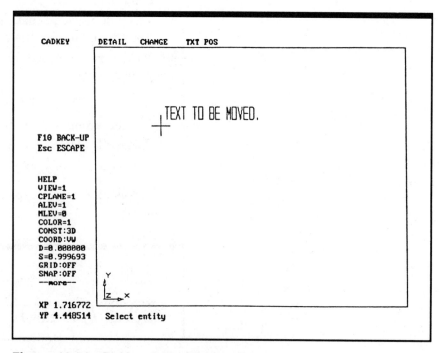

Figure 16-51 Picking the text to be moved

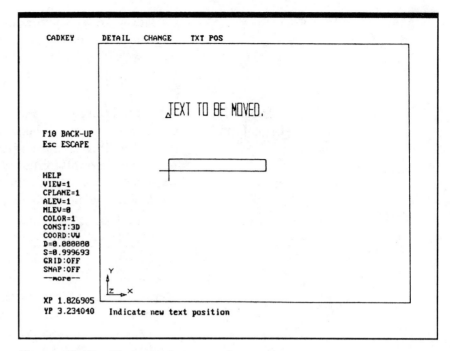

Figure 16-52 The text placement box

2. You are prompted:

   Select entity.

   Pick the note to be moved.

3. A box the size of the selected text replaces the normal plus (+) cursor (Figure 16-52). A prompt reads:

   Indicate new text position.

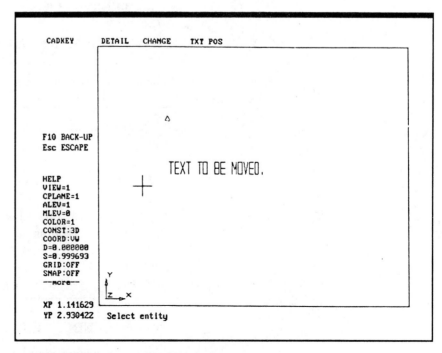

Figure 16-53 Text after being moved

Move the box cursor to the new position and pick that point. The text is deleted from its original position and moved to the new position (Figure 16-53).

CHANGE OPTION: EDIT TX (Edit Text)

EDIT TX is used to start the on-line text editor that modifies text in dimensions, labels, and keyed-in and disk-based notes. The text editor is automatically started when creating dimensions with manually created text, keyed-in notes, and labels. To edit existing text, select the text string with the cursor. The text is edited using the keyboard and these special options:

KEY	FUNCTION
Home	Moves the cursor to the start of the current line.
Up Arrow	Moves the cursor up one line.
Down Arrow	Moves the cursor down one line.
Right Arrow	Moves the cursor one character to the right.
Left Arrow	Moves the cursor one character to the left.
End	Moves the cursor to the end of the current line.
Page Up	Moves the cursor to the last character of the first line.
Page Down	Moves the cursor to the last character of the last line.
Back Space	Deletes the character to the left of the current cursor position.
Del	Deletes the character at the current cursor position.
Ins	Toggles between insert and overwrite mode.
Enter	

Insert Mode—saves all text up to the current cursor position and creates a new line containing the text to the right of the cursor.

Overwrite Mode—saves all text up to the current cursor position and advances the cursor to the end of the next line.

CADKEY's text editor also displays seven menu options:

OPTION	FUNCTION
DEL LIN	Deletes the current line and displays the preceding line.
JOIN LN	Joins the current line with the following line.
INSERT	Inserts a blank line above the current line.
APPEND	Inserts a blank line below the current line.
DEL END	Deletes all characters from current cursor position to end of the line.
SAVE TX	Saves the edited text.
ABORT	Exits the text editor without changing the text.

To use the **EDIT TEXT** option follow the steps below.

1. Select **ESC-F3-F7-F3 (DETAIL-CHANGE-EDIT TX)**.

TEXT STRING TO BE EDITED.

Figure 16-54   Editing a text string

TEXT STRING TO BE EDITED. (AFTER EDITING)

Figure 16-55   Text added to the string

2.   A prompt reads:

Select text for editing.

Pick the text string (Figure 16-54).

3.   Another prompt reads:

Enter text: TEXT STRING TO BE EDITED.

The text string selected appears in the Prompt Line. Any of the editing features listed earlier or the options shown in the menu area can be used to edit the text. Select **F2 JOIN LN.**

4.   Add the text **(AFTER EDITING)** to the text string and press **RETURN** twice. The new text string is shown on screen (Figure 16-55).

CHANGE OPTION: DIM VAL (Dimension Value)

The **DIM VAL** option changes an existing dimension value. The dimension is picked and the true size of the dimension is displayed in a prompt ignoring the scale or previous dimension value changes. To accept that dimension press **RETURN**. To change that value, input the new value to be displayed and press **RETURN**. The dimension is updated to the new value input. The dimension value is changed but not the entity. In other words, if the dimension is 4 inches and it is changed to 8 inches using the **DIM VAL** option, the actual length of the line does not change; it remains four inches long. This option gives the user the opportunity to assign any value to a dimension. Figures 16-56 and 16-57 and the following steps show how this option is used.

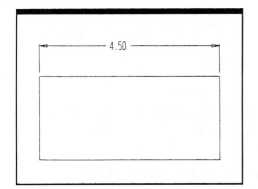

Figure 16-56   Dimension value to be changed

1.   Select **ESC-F3-F7-F4 (DETAIL-CHANGE-DIM VAL)**. A prompt reads:

Select dimension.

Pick the dimension shown in Figure 16-56.

2.   Another prompt reads:

Enter new dimension value (actual=4.5) =.

Input the new value, such as **9**, and press **RETURN**. The new value is displayed (Figure 16-57).

CHANGE OPTION: DIM REP (Dimension Representation)

This function changes the decimal, fractional, or ft/in (feet and inches) precision of an existing dimension. There are seven decimal options available, six round-off fraction factors, and six round-off factors for ft/in, as described earlier in this chapter under **SET-DIM REP**. After the precision is changed, the Selection Menu is used to assist in selecting those dimensions to be changed. With this menu it is possible to change a single dimension, a group of dimensions, all, by level, in a window, and so forth. Figures 16-57 and 16-58 and the following steps show how the decimal option is used.

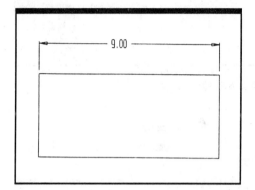

Figure 16-57 The changed dimension value

1. Select **ESC-F3-F7-F5-F1** (**DETAIL-CHANGE-DIM REP-DECIMAL**) to change the decimal precision.

2. A prompt reads:

   Choose decimal precision.

   A menu is displayed with the seven decimal options. For this example **F5 N.NNNN** is chosen.

3. The Selection Menu is displayed. Select **F1 SINGLE**.

4. A prompt reads:

   Select entity 1 (press RETURN when done).

   Pick the single dimension in Figure 16-57 and press **RETURN**. It is changed to four decimal places (Figure 16-58).

CHANGE OPTION: WITNESS

This option changes the witness line display on an existing dimension. The default for dimensions is to have two witness lines. With this option it is possible to change the displayed witness lines to: **BOTH, FIRST, SECOND,** and **NONE**. The dimension is picked and the prompt displays its current value, such as **BOTH** or **NONE**, for example. The new witness line option is chosen, then the dimension is re-displayed using the current witness value. See Figure 16-58 and the following steps to use this option.

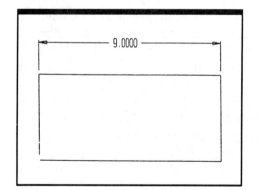

Figure 16-58 The changed dimension representation

1. Select **ESC-F3-F7-F6** (**DETAIL-CHANGE-WITNESS**).

2. A prompt reads:

   Select dimension.

   Select the dimension to be changed by moving the cursor over the dimension text in Figure 16-58 and picking it.

3. A prompt reads:

   Choose new witness line display option (current=BOTH).

   A menu is displayed with four options: **BOTH, FIRST, SECOND,** and **NONE**. For this example choose **F4 NONE**.

4. The dimension is displayed (Figure 16-59).

CHANGE OPTION: ARROWS

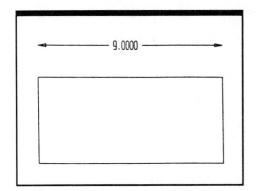

Figure 16-59 Suppressing the extension lines

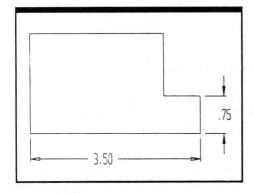

Figure 16-60

CHANGE OPTION: TOLER (Tolerance)

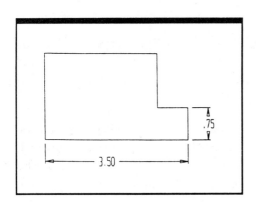

Figure 16-61 Vertical dimension arrows changed

This option is used to change the arrow direction (**DIRECT**) to in/out placement or the **STYLE** on existing dimensions. To use this option pick the dimension and select **IN** or **OUT** as the new direction or **STYLES 1, 2, 3,** or **4.** The dimension is re-displayed with the new arrow direction or style assigned. Figures 16-60 and 16-61 and the following steps show how to change the arrow direction and style.

1. Press **ESC-F3-F7-F7-F1** (**DETAIL-CHANGE-ARROWS-DIRECT**).

2. A prompt reads:

   Select dimension.

   Pick the dimension to be changed by moving the cursor over the dimension and pressing the mouse button.

3. A prompt reads:

   Choose new arrow direction (current=IN).

   Select **F2 OUT.** The dimension is redrawn with the new arrow direction.

4. Select **F10 BACKUP** once to return to the **DIRECT/ STYLE** menu.

5. Select **F2 STYLE.**

6. A prompt reads:

   Choose new arrowhead style.

   Select **F2 STYLE2.**

7. The Selection Menu is displayed. Select **F1 SINGLE** and pick the vertical dimension and press **RETURN.** The dimension is changed, as shown in Figure 16-61.

Figure 16-62 shows the four arrowhead styles.

The **TOLER** function changes the tolerance values and type assigned to an existing dimension. There are two options available: **VALUES** and **TYPE.** The **VALUES** option is used to change the **POS TOL** (positive tolerance), **NEG TOL** (negative tolerance), or **BOTH** tolerance values on a current dimension. After the tolerance value is changed, the Selection Menu is displayed to offer the user many different ways of selecting those dimensions to be changed. The **TYPE** option is used to change the type of tolerancing used in a selected dimension. This option has three choices: **NONE,** which turns all tolerance values off; **+/−,** which is used to display a dimension with a plus-minus tolerance; and **LIMIT,** to display the high limit over the low limit. Figures 16-63 and 16-64 and the following steps show how this **CHANGE-TOLER** option is used.

1. Select **ESC-F3-F7-F8** (**DETAIL-CHANGE-TOLER**).

2. Select **F1 VALUES.**

3. A menu is displayed with **POS TOL, NEG TOL, BOTH**. Choose **F3 BOTH**. A prompt reads:

    Enter new positive tolerance =.

    Enter **.03** and return. A second prompt reads:

    Enter negative tolerance =.

    Enter a value of **–.03** and return.

4. The Selection Menu is displayed. Choose **F7 ALL DSP**, then **F1 ALL**. All the dimensions are changed to the new values input (Figure 16-64).

CHANGE OPTION: DIM SCL (Dimension Scale)

The **DIM SCL** option is used to change a dimension scale by assigning a scale factor. The dimension scale can be changed on single dimensions, groups of dimensions, all, in a window, and so forth using the Selection Menu. Tolerance values and angular dimensions are not changed when using this option. The factor chosen is applied to the original dimensions, ignoring any other changes. Figures 16-64 and 16-65 and the following steps show how this option is used.

1. Select **ESC-F3-F7-F9** (**DETAIL-CHANGE-DIM SCL**).
2. A prompt reads:

    Enter new dimension scale factor =.

    Enter the new scale, such as **.5**, and press **RETURN**.

3. The Selection Menu is displayed so that the dimensions to be changed can be identified. For this example, choose **F7 ALL DSP** so that all dimensions are changed.

4. Another menu is displayed in the menu area. Choose **F1 ALL**. All the dimension text is changed from the full-scale value (Figure 16-64) to .5 (Figure 16-65).

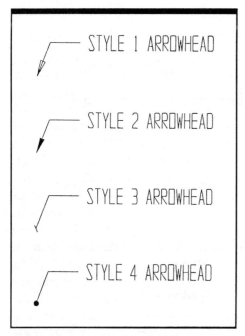

Figure 16-62   Four arrowhead styles

THE UPDATE FUNCTION

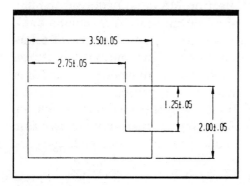

Figure 16-63

This function updates certain attributes of dimensions to the current value in effect as set by the **DETAIL-SET** function. **NOTE** and **LABEL** entities that are affected by the **UPDATE** function are height and font. Dimension entities affected by the **UPDATE** function are height, font, decimal precision, dimension scale, trailing and leading zeros, tolerance mode, and tolerance value. The Selection Menu defines those entities that are to be updated. Figures 16-65 and 16-66 and the following steps show how the **UPDATE** function is used to change dimension units.

1. The first step is to change the note or dimension values, using the **SET** function. For this example, the dimension units

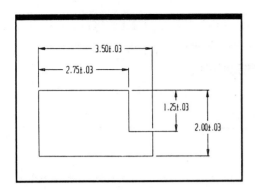

Figure 16-64 Tolerance values changed from .05

are changed from inches to millimeters by selecting **ESC-F3-F9-F7-F2 (DETAIL-SET-UNITS-MM)**.

2. Select **ESC-F3-F8 (DETAIL-UPDATE)**.

3. The Selection Menu is displayed to select the desired entities to be updated. From the list of options choose **F7 ALL DSP**.

4. A masking menu is displayed. Choose **F1 ALL**. Figure 16-65 is re-displayed with the dimension text updated to mm (Figure 16-66).

CROSS-HATCHING

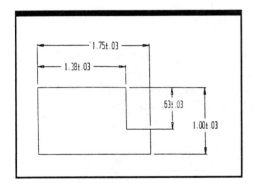

Figure 16-65 Values changed to half size

Cross-hatching represents section lines in a sectional view drawing. With CADKEY, individual lines do not have to be drawn to fill an area, as with hand tools. Instead, CADKEY offers the user eighteen ANSI standard styles of line patterns which can be spaced and placed at any angle specified by the user. Areas to be cross-hatched are identified using the Selection Menu options described in Chapter 11. After the area is defined, the specified cross-hatching pattern is placed on the drawing. Refer to the tutorial section located in Chapter 6 for additional practice in creating sectional drawings with CADKEY.

Only **LINE, ARC, SPLINE** and **CIRCLE** entities can be cross-hatched. Only closed boundaries can be cross-hatched using the **X-HATCH** function. In some cases you may have to break entities using the **TRIM** option to make a closed boundary area before the **X-HATCH** function will work properly. A cross-hatched area is considered one entity, not a number of separate lines. Therefore, to erase a cross-hatch area it is necessary to select one of the four extreme limits of the area. For example, to **DELETE-SINGLE** a cross-hatch area located in a square, the cursor must be placed near one of the corners of the square and the space bar or mouse button pressed. Be careful not to get too close to the corner of the square or one of the lines of the square may be selected. Cross-hatching is also view dependent. This means that the cross-hatching will only be displayed and plotted in the view of its creation. Placing cross-hatching on a separate layer is highly recommended to avoid difficulties in erasing or editing hatch patterns.

There are eighteen cross-hatching patterns available, as shown in Figure 16-67. For each of these patterns, the spacing of the line pattern and the angle can be specified. The default values of the patterns are .25″ spacing between lines, and a 45 degree angle.

As mentioned earlier, some areas are more difficult to cross-hatch than others. The right side view of Figure 16-68 is the type of

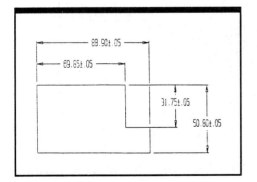

Figure 16-66 Dimension text updated to millimeters

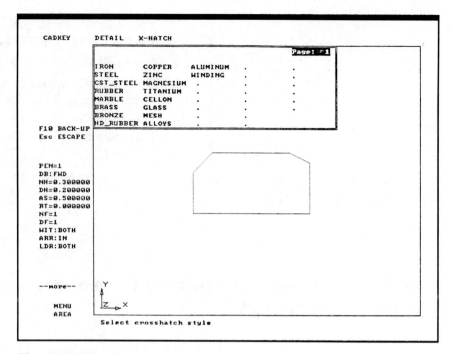

Figure 16-67   Cross-hatching patterns

drawing that is difficult to cross-hatch unless some precautions are taken. Trying to cross-hatch the view would probably result in an error message that reads: **Invalid data. . .press RETURN**. If this happens try using the **EDIT-TRIM** or **BREAK** option or the **DELETE** option to create separate areas to cross-hatch (Figure 16-69).

The separate areas are easily cross-hatched using the **X-HATCH** function and the **SINGLE, WINDOW,** or **CHAIN** options.

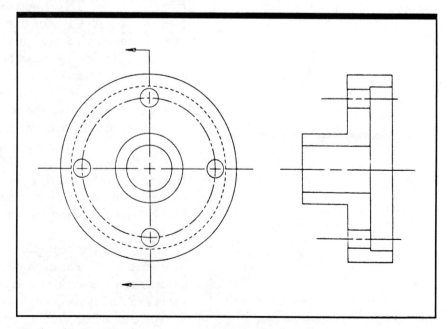

Figure 16-68   The right side view to be sectioned

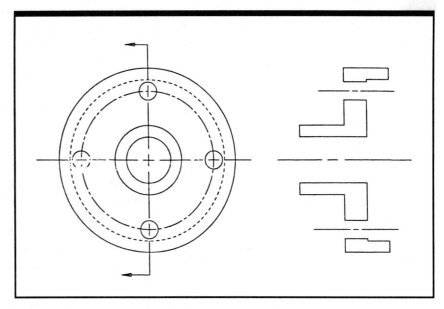

Figure 16-69   The EDIT TRIM option to separate areas

The missing lines are added to the drawing after the areas have been filled (Figure 16-70).

The placing of cross-hatching is the same no matter what type of pattern is selected. However, the Selection Menu will allow the user to choose a number of different methods to locate the cross-hatching in a specific area. The following examples show how to use the **X-HATCH** function for different hatch types and areas.

USING DEFAULT VALUES

Figure 16-71A and the following steps show how the **X-HATCH** function is used to place steel cross-hatching in a square area. Cursor picks are identified as an **"X"** in the figures.

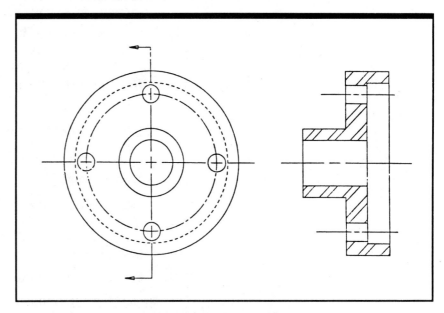

Figure 16-70   The completed section view

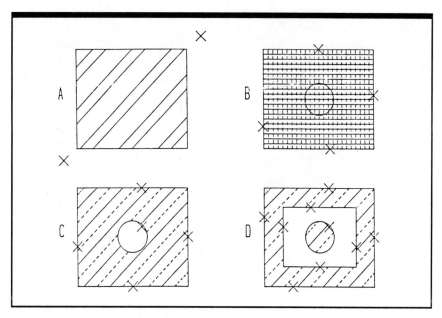

Figure 16-71 Various applications of X-HATCH

1. Select **ESC-F3-F5 (DETAIL-X-HATCH)**.

2. A pop-up window displays the eighteen hatch patterns available. Cursor-select the **STEEL** pattern (Figure 16-67).

3. A prompt reads:

   Enter angle for hatch lines (45.00000) =.

   This is the default angle of 45 degrees. Press **RETURN** to accept this value.

4. A prompt reads:

   Enter distance between hatch lines (0.250000) =.

   To accept the default value of .25 press **RETURN**.

5. The Selection Menu is displayed. For this example, **SINGLE**, **CHAIN, WINDOW**, or **ALL DSP** will work well. However, the **WINDOW** option is probably the most efficient. **F3 WINDOW** is selected from the menu.

6. A prompt reads:

   Indicate position for 1st selection window corner.

   Pick the lower left corner of the window.

7. A prompt reads:

   Indicate position for 2nd selection window corner.

   Pick the upper right corner of the window.

8. A prompt reads:

   Selection complete? (YES)

   Press **RETURN** for **YES**. The steel cross-hatching pattern automatically fills the square (Figure 16-71A).

CROSS-HATCHING ISLANDS

The following example demonstrates how to change the default values and introduces the concept of cross-hatching islands. Islands are entities found within other entities. For example, a circle drawn in the center of a square is considered an island. If the area within the circle is not to be cross-hatched then it must be identified in some manner. The following example illustrates what happens if the circle is not identified when defining an area to be cross-hatched.

1. Select **ESC-F3-F5** (**DETAIL-X-HATCH**).
2. Select **ALUMINUM** from the pop-up window.
3. A prompt reads:

   Enter angle for hatch lines (45) =.

   Input the number **0** for a zero degree angle and press **RETURN**.
4. A prompt reads:

   Enter distance between hatch lines (0.25) =.

   Input **.125** and press **RETURN**.
5. Select **F1 SINGLE** from the Selection Menu and pick each line of the square. After all four lines have been picked, press **RETURN**.
6. A prompt asks if the selection is complete. Press **RETURN** for **YES**. The area is filled (Figure 16-71B).

Notice that the area within the circle has also been filled. If the area within the circle is to be left without cross-hatching then it must be picked. This is demonstrated in the following steps and shown in Figure 16-71C.

1. Select **ESC-F3-F5** (**DETAIL-X-HATCH**).
2. Select **COPPER** from the pop-up window.
3. A prompt reads:

   Enter angle for hatch lines (45) =.

   Press **RETURN** to accept the 45 degree default value.
4. A prompt reads:

   Enter distance between hatch lines (0.25) =.

   Press **RETURN** to accept the .25 value.
5. Select **F1 SINGLE** from the Selection Menu. Pick each line of the square and the circle and press **RETURN**.
6. A prompt asks if the selection is complete; press **RETURN** for **YES**. The area is filled (Figure 16-71C). Notice that the area within the circle is not filled.

The following steps show how the **X-HATCH** function works with multiple islands. Refer to Figure 16-71D.

1. Select **ESC-F3-F5** (**DETAIL-X-HATCH**).
2. Select **COPPER** from the pop-up menu.
3. A prompt reads:

Enter angle for hatch lines (45) =.

Press **RETURN**.

4. A prompt reads:

    Enter distance between hatch lines (0.25) =.

    Press **RETURN**.

5. Select **F1 SINGLE** from the Selection Menu. Pick each line of the squares and the circle and press **RETURN**.

6. A prompt asks if the selection is complete. Press **RETURN** for **YES**. The area is filled (Figure 16-71D).

CADKEY 6 NEW FEATURES

CADKEY 6 has greatly enhanced the **X-HATCH** option. The 18 default cross-hatch patterns can be customized to make 32 patterns. In the status window, two new options, **HTCHINT** and **HTCH_UP**, enhance cross-hatching. Hatch Intersection (**HTCHINT**) allows the boundary entities to overlap. When this option is ON it will fill boundary entities to their intersections with other boundary intersections. This saves you the time of trimming or breaking all of the boundary entities as was done in the 2D section view tutorial in chapter 5 and Figure 16-69. Hatch Update (**HTCH_UP**) will adjust the cross-hatch lines when you modify or edit the boundary entities (Figure 16-72).

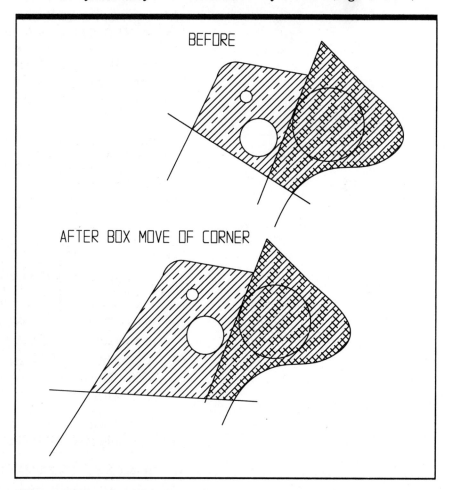

Figure 16-72 Hatch Update adjusts cross-hatching as boundaries change.

The **X-HATCH** option has been broken into two options—
CREATE and **MODIFY. CREATE** brings up a dialog box which
contains the 18 cross-hatch patterns as icons, angle and spacing information boxes, and the **Quick Hatch** option (Figure 16-73). The
Quick Hatch option allows the user to place the cursor inside an
area filled with cross-hatch lines and click. CADKEY 6 then starts a
chain selection process that highlights the entities which make up
the boundary. If there is a question as to which is the next entity in
the chain, two entities will be highlighted as dashed lines (Figure
16-74). Select the entity which defines part of the boundary area you
wish to fill. **MODIFY** is used to edit the existing entities which define the cross-hatch boundary or to create islands (Figure 16-75).
ADD adds entities to the existing cross-hatch boundary or island.
SUBTRACT subtracts entities from the existing cross-hatch boundary or island.

THE LAYOUT FUNCTION

CADKEY's Drawing Layout Mode (DLM) is a system mode that
automatically creates views of a part model without having to create
pattern files. The 3D model is converted to a 2D drawing in the layout format. Select **DETAIL-LAYOUT**, then enter a layout name,
drawing scale, and paper size to enter the Drawing Layout Mode.

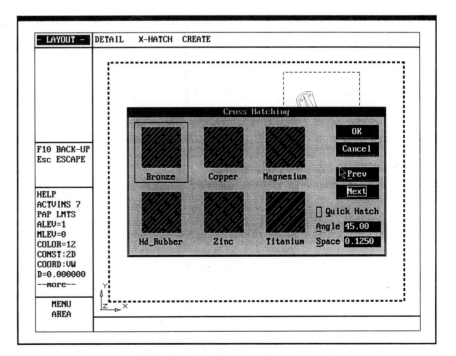

Figure 16-73 The DETAIL-X-HATCH dialog box with material,
angle, and space attributes

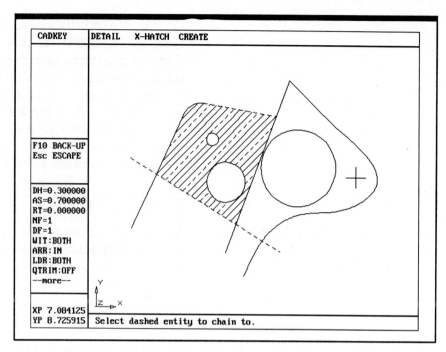

Figure 16-74   Quick Hatch chain selecting boundary entities

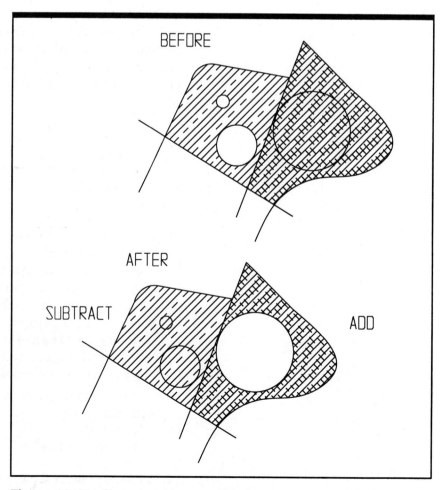

Figure 16-75   Cross-hatching modifications

Once in the Drawing Layout Mode CADKEY provides many features useful in the creation of multiview engineering drawings. Some of the more prominent features are the automatic alignment of views, scaling of views, and associativity.

The Drawing Layout Mode is meant to supplement the Model Mode of CADKEY. CADKEY defaults to the Model Mode at startup and is where three-dimensional modeling should occur. After the model is created, enter the Drawing Layout Mode and use it as a drafting tool to create 2D views of dimensions, notes, and labels.

ASSOCIATIVITY

Associativity is the relationship between the 3D model in Model Mode and the 2D drawing in Drawing Layout Mode. What this means is that any changes to the 3D model in Model Mode will automatically be reflected in Drawing Layout Mode. For example, if a hole is added to the 3D model in Model Mode that hole will be automatically represented in the drawing in Drawing Layout Mode. There is no associativity from Drawing Layout Mode to Model Mode. That is, if changes were made to a view in the Drawing Layout Mode no changes would be reflected in the 3D model in Model Mode.

Associativity also applies to dimensions placed on a 2D drawing in Drawing Layout Mode. The part can be changed in Model Mode and when you return to Drawing Layout Mode the dimension will automatically be updated to reflect the change made in Model Mode. For example, a 1″ dimensioned circle will automatically change to a 1/2″ dimensioned circle if the circle is changed to 1/2″ in Model Mode.

ENTERING DRAWING LAYOUT MODE

The three-dimensional part shown in Model Mode in Figure 16-76 will be used to create a multiview drawing on an A-size sheet in Drawing Layout Mode.

1. Enter **ESC-F3-F6** to access Drawing Layout Mode.

2. A pop-up window is displayed (Figure 16-77). This menu is used to name the layout, enter the drawing size, and assign a paper size. To change the drawing scale point at the drawing scale numbers and click the mouse button. A pull-down menu is displayed showing some of the pre-set scales (Figure 16-78). To display more scale choices point at the arrow buttons to the right of the displayed scales to page through the pre-set choices. For this example select the default setting of 1:2.

   The paper size can be changed in the same way as the drawing scale by clicking on the letter to reveal a pull-down menu (Figure 16-79). For this example use the default setting for the A-size sheet.

3. Enter the name for the drawing sheet by pointing and clicking in the name box, then entering a name.

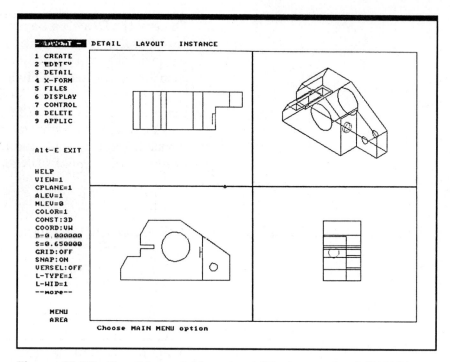

Figure 16-76   Use this model to extract 2D views using LAYOUT

4. Click on the **OK** button to enter the Drawing Layout Mode. The screen shown in Figure 16-80 is displayed. The thick dashed line surrounding the drawing screen represents the extents of the A-size drawing paper. Also displayed is the Main Layout Menu with seven options plus an exit

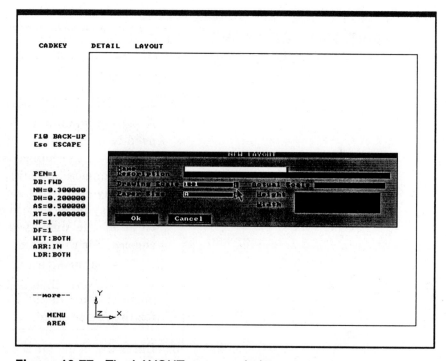

Figure 16-77   The LAYOUT pop-up window

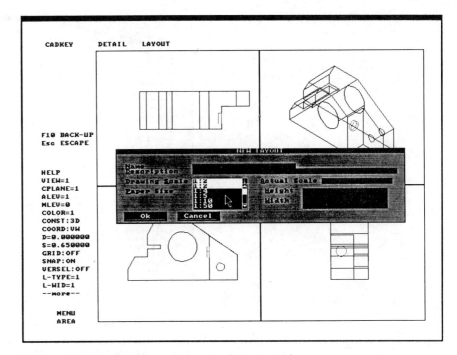

Figure 16-78   Pull-down menu to layout scale

function. To edit the 3D model you must exit Drawing Layout Mode using **F8 EXIT** from the Main Layout Menu.

INSTANCING THREE VIEWS OF THE 3D MODEL

Instancing is the term CADKEY uses to create a 2D view from a three view model. Select **F5 INSTANCE** to reveal four options: **CREATE, MODIFY, ALIGN**, and **ACTIVE**. **CREATE** is used

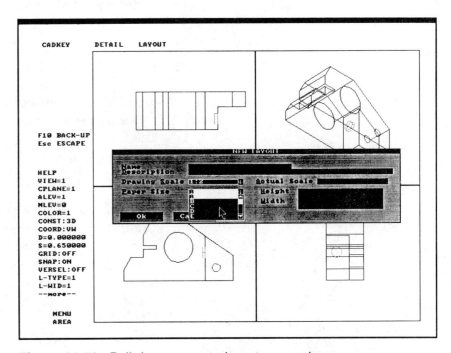

Figure 16-79   Pull-down menu to layout paper size

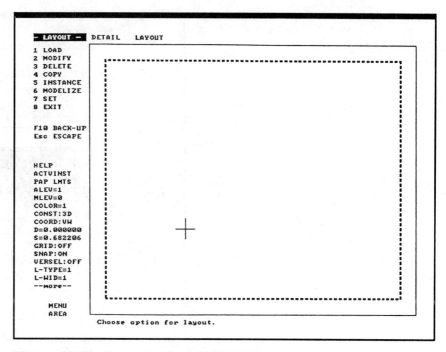

Figure 16-80 Paper border and LAYOUT options menu

to create and place a view of the model anywhere on the paper. The **MODIFY** option changes a drawing instance's scale, location. rotation angle, view, and can be used to change the level and to blank or unblank entities. **ALIGN** lines up drawing instances so that multiview drawings can be created to standard drawing practice. The **ACTIVE** option changes the active instance. This command can also be invoked by pressing **ALT-V** or by clicking on **ACTVINST** in the Status Window (Figure 16-80).

To create a three view drawing of the 3D model the following steps are used.

1. Select **F5 INSTANCE**.
2. Select **F1 CREATE**. A pop-up window is displayed which lists all the available views (Figure 16-81). The eight standard CADKEY views are listed along with any user defined views. To select a view point at it with the mouse and click. The front view is selected by pointing at it and clicking.
3. A prompt reads:

    Enter drawing instance rotation angle (0.0000)=>.

    No rotation is desired so **RETURN** is entered.
4. The instance box is centered about the cursor to indicate the approximate size of the view to be placed (Figure 16-82).
5. Move the box to the approximate position for the front view, then click. The view is displayed on screen along with a dotted line box (Figure 16-83). The view can be changed by using the **MODIFY** option.
6. Select **F1 CREATE** and select the top view from the pop-up window.

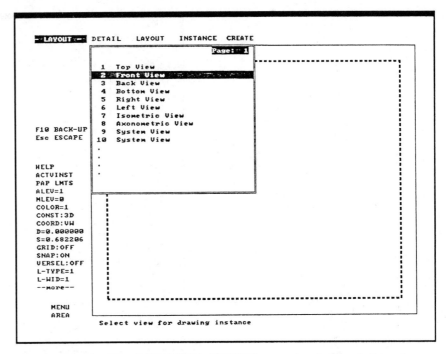

Figure 16-81 The INSTANCE CREATE pop-up window listing view options

7. A prompt reads:

   Enter drawing instance rotation angle (0.0000)=>.

   No rotation is desired, so press **RETURN**.

8. The instance box is displayed on screen representing the extents of the top view. Move the cursor to the approximate

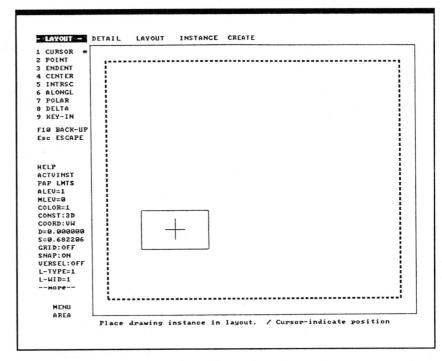

Figure 16-82 Placing the front view using the cursor

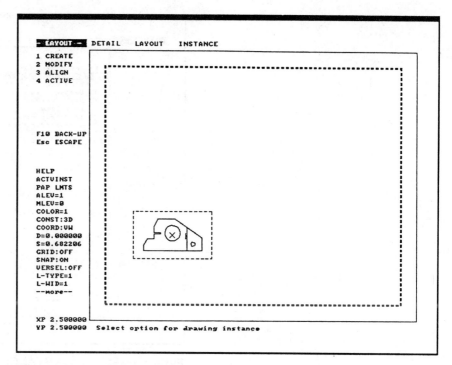

Figure 16-83 Placement of front view in the layout

location for the top view and click (Figure 16-84). Do not be too concerned about alignment with the front view as that can be taken care of later.

9. Select **F1 CREATE** and select the right side view from the pop-up window.

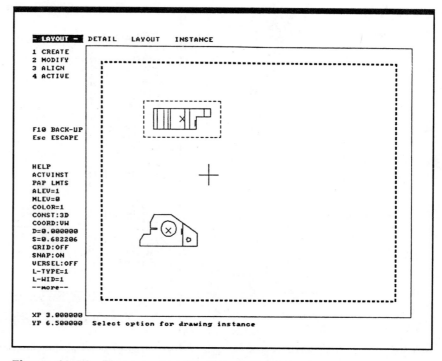

Figure 16-84 Placement of top view in the layout

10. A prompt reads:

    Enter drawing instance rotation angle (0.0000)=>.

    No rotation is desired, so press **RETURN**.

11. The instance box is displayed on screen. Move the cursor to the approximate location for the right side view then click. The right side view is drawn on screen (Figure 16-85).

12. Select **F1 CREATE** and select the isometric view from the pop-up window.

13. A prompt reads:

    Enter drawing instance rotation angle (0.000)=>.

    No rotation is desired, so press **RETURN**.

14. Move the instance box to the desired location for the isometric view and click. The isometric view is added to the drawing layout (Figure 16-86).

AUTOMATICALLY ALIGNING VIEWS

The front, top and right side views are not in alignment as required for engineering drawings. The automatic alignment option can be used to align the views as demonstrated in the following steps.

1. Select **F5-F3** (**INSTANCE-ALIGN**).

2. There are two options: automatic (**AUTO**) and **MANUAL**. The manual option aligns views relative to selected points on the drawing. The automatic option will align views to a selected view. For this example select **F1 AUTO**.

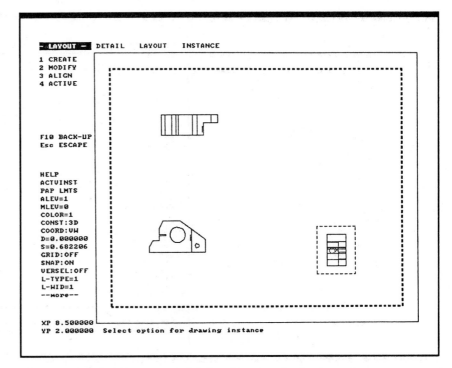

Figure 16-85 Placement of right-side view in the layout

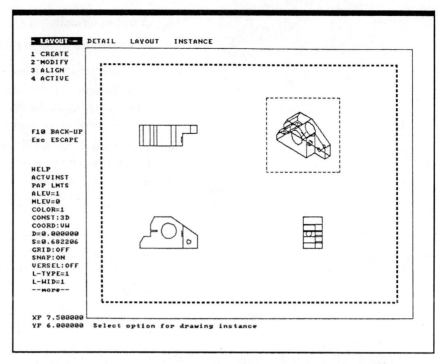

Figure 16-86 Adding the isometric view to the layout

3. A prompt reads:

   Select fixed drawing instance.

   Pick the front view, which will be used as the view to align the top and right side views. After selection a dotted instance box is drawn around the front view (Figure 16-87).

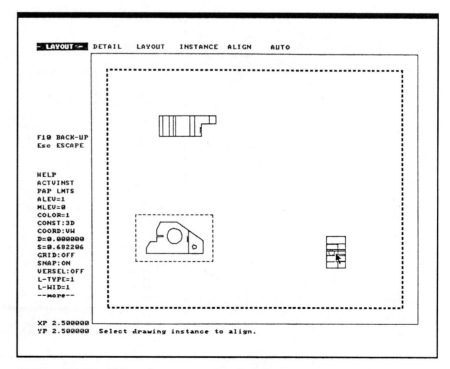

Figure 16-87 Using the automatic align feature

4. Another prompt reads:

   Select drawing instance to align.

   Point at and select the right side view as the first view to align with the front view. The view changes into an instance box after selection (Figure 16-88).

5. Moving the cursor will cause the instance box to move only horizontally in alignment with the front view. The box cannot be moved vertically on the drawing because it is locked onto the front view. Pick the location for the right side view and click. The right side view is displayed on screen (Figure 16-89).

6. A prompt reads:

   Select fixed drawing instance.

   Pick the front view again. After selection a dotted instance box is drawn around the front view.

7. A prompt reads:

   Select drawing instance to align.

   Point at and select the top view as the second view to align with the front view. The view changes into an instance box after selection (Figure 16-88).

8. Moving the cursor will cause the instance box to move only vertically in alignment with the front view. The box cannot be moved horizontally on the drawing because it is locked onto the front view. Pick the location for the top view and click. The top view is displayed on screen (Figure 16-89).

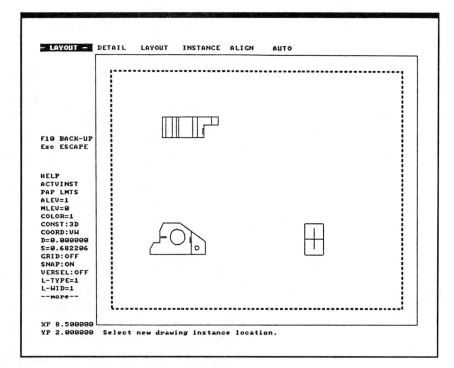

Figure 16-88 The right-side view is locked into horizontal alignment.

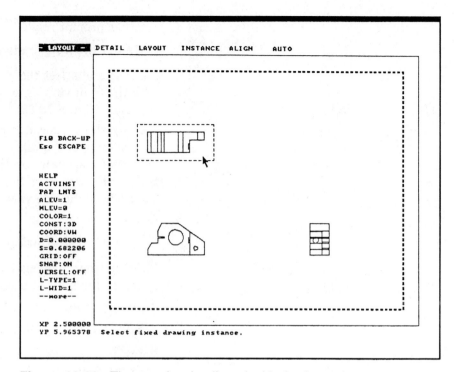

Figure 16-89   The top view is aligned with the front view.

CHANGING THE SCALE OF A VIEW

The **INSTANCE-MODIFY** option is used to change the scale of a view. The **MODIFY** option can also be used to rotate, locate or change a view, and to blank or unblank entities in a view. The following steps show how to increase the size of the isometric view.

1. Select **F5-F2-F1 (INSTANCE-MODIFY-SCALE)**.
2. A prompt reads:

    Select drawing instance to modify.

    Pick the isometric view by pointing at it and clicking.
3. After selection a dashed line box is drawn around the view and a prompt reads:

    Enter new scale (0.50000).

    For this example enter **.75** to increase the size of the isometric view (Figure 16-90).

OTHER LAYOUT COMMANDS

The **MODIFY** command changes the scale or paper size of the current layout. From the main layout menu select **F2 MODIFY**. A pop-up menu is displayed (Figure 16-91). Click on the paper size and change it to a B-size. The drawing layout is changed (Figure 16-92).

The **DELETE** command deletes any drawing layout assigned to the current part file. A part file can have more than one drawing

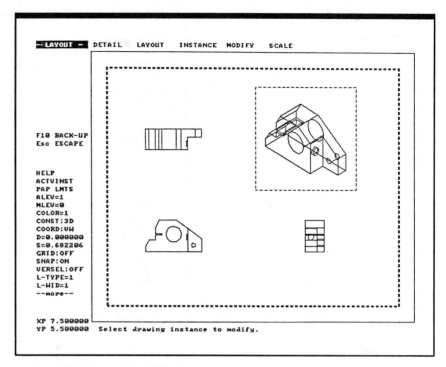

Figure 16-90   Isometric view after scaling modification

layout assigned to it. Selecting **F3 DELETE** displays the pop-up window shown in Figure 16-93. Clicking on the selected layout, then the **OK** button, will delete the layout from the part file.

The **COPY** command copies any existing drawing layouts. Selecting **F4 COPY** from the main layout menu displays the pop-up window shown in Figure 16-94. The **LOAD** command creates a new

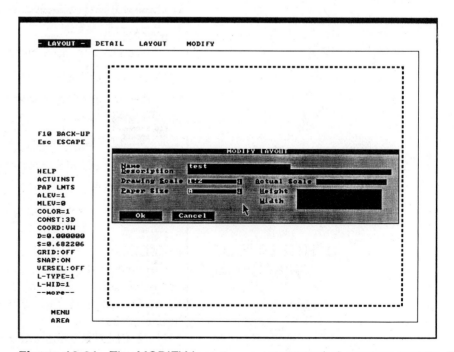

Figure 16-91   The MODIFY layout menu pop-up window

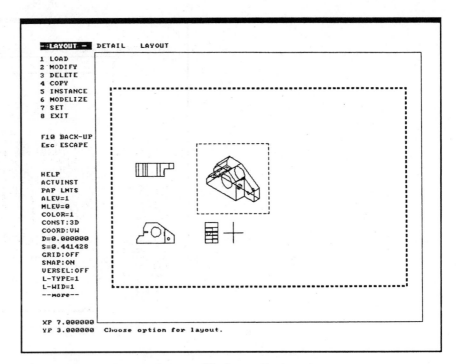

Figure 16-92 Paper size modified to a B-size sheet

layout or loads an existing one. Selecting **F1 LOAD** from the main layout menu displays the pop-up menu shown in Figure 16-95.

The **MODELIZE** command changes model entities to layout entities. Modelized entities can be edited without affecting the part file. Modelizing breaks the associativity with the part file.

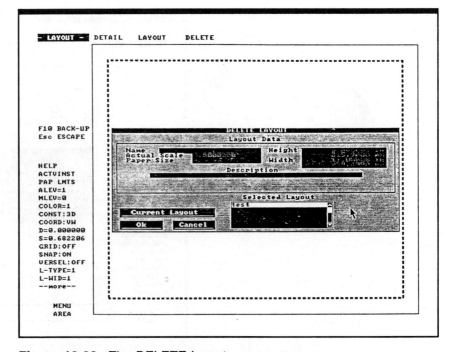

Figure 16-93 The DELETE layout pop-up menu

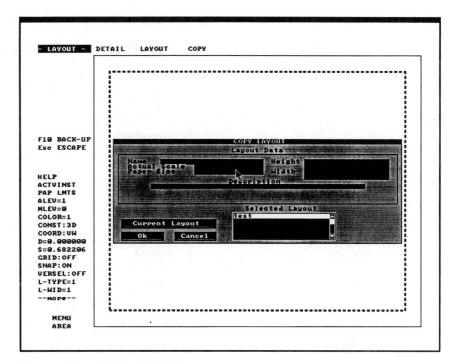

Figure 16-94   The COPY layout pop-up menu

After modelizing, no instance options can be used. Modelizing could be used to trim entities and for hidden line removal. The **MODEL** command has three options: **ENTITY, INSTANCE,** and **LAYOUT**. **ENTITY** selects entities one at a time. **INSTANCE** selects a whole instance with one pick. **LAYOUT** picks

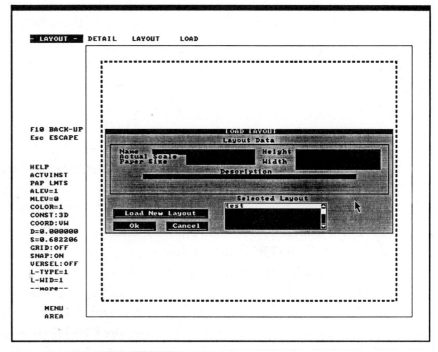

Figure 16-95   The LOAD layout pop-up menu

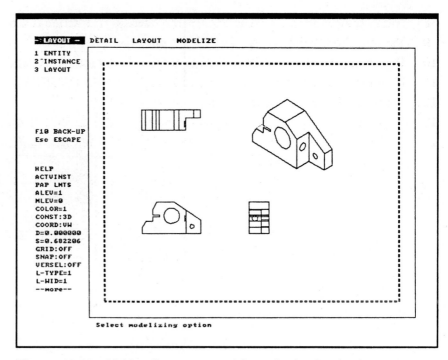

Figure 16-96 Hidden lines removed from the isometric view

the whole layout when modelizing. Figure 16-96 shows the isometric view with hidden lines removed after modelizing using the **MODELIZE-INSTANCE** option.

SET toggles the paper limits on and off using the **PAP LMTS** (paper limits) option. **SHOWNEW** is an on/off toggle that highlights any automatically updated dimensions and entities that were created since the last time you worked in layout. **SHOW INST** highlights all instance specific geometry in the current layout. **EXIT** exits the layout module and returns you to the model module.

REVIEW QUESTIONS

1. What is the default decimal inch precision used with CADKEY?
2. List the default **NOTE/LABEL** values.
3. List the default **DIMENSION** values.
4. What is the default position used for the alignment of **NOTES**?
5. What is the default position used for the alignment of **DIMENSION** text?
6. What are the six standard units listed under the **UNITS** option?
7. What is the pick position on a dimension for deletion?

8. Name the command used to change the decimal or fractional precision of dimensions.
9. Explain the function of **AUTO UP**.
10. How many points must be picked to place a horizontal or vertical dimension?
11. How many default hatch patterns are available with CADKEY?
12. Where is the hatch pattern selected when deleting?
13. Explain what associativity means as related to the **LAYOUT** command.
14. CADKEY defaults to which mode (model or layout) at start-up?

The X-FORM (Transform) Options

OBJECTIVES

After completing this chapter, you will be able to:

- Move entities on a drawing.
- Make single or multiple copies of entities.
- Copy and join entities (extrude).
- Create a mirror image of selected entities.
- Perform a 2D rotation of entities.
- Change the scale of a drawing.
- Project entities onto a specified plane.

The **X-FORM** function can move, copy and join entities. **NOTE**: for users of previous versions of CADKEY, **TRANS-R** has been renamed **DELTA** and **TRANS-A** has been renamed **OLD-NEW** in CADKEY Version 4. The eight **X-FORM** options include **DELTA, OLD-NEW, MIRROR, ROTATE, SCALE, PROJECT, HLX ROT** and **C-ARRAY** (Figure 17-1). These options are very useful in creating designs having similar characteristics, such as symmetrical parts, repetitive features such as gear teeth, oblique or perspective drawings, multiple copies, multiple scaled copies, and enlarging or shrinking details on a drawing. The primary use of the **X-FORM** function is in 3D wireframe modeling of parts.

THE UNDO FUNCTION

All of the **X-FORM** commands have the **UNDO (ALT-I)** option available. **UNDO** appears in the Break Option Window when a transformation has been completed. When it is selected the entities which were just moved, copied, or joined will be restored to the way they were prior to using the command. This is helpful if you accidentally copy entities on the wrong axis or move them to the wrong location. Simply select **UNDO** and your original geometry is reinstated.

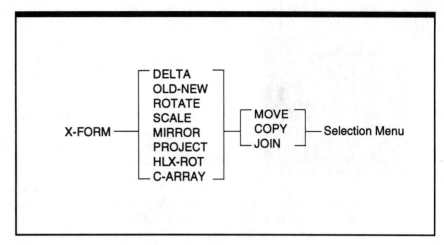

Figure 17-1   The X-FORM commands

THE MOVE/COPY/JOIN MENU

After selecting any of the eight **X-FORM** options a new menu gives the user three options: **MOVE, COPY, JOIN**. **MOVE** changes the position of selected entities or notes by moving those entities to a new position on screen without leaving an original copy (Figure 17-2A). **COPY** will make another copy of the selected entities or notes without changing the original entities. Each new copy is based upon the original entities selected. The number of copies is limited to the number of entities that CAD-KEY and the hardware can support (Figure 17-2B). **JOIN** makes single or multiple copies as described in the **COPY** option.

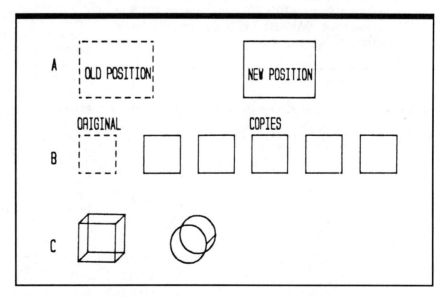

Figure 17-2   X-FORM, MOVE, COPY, and JOIN options

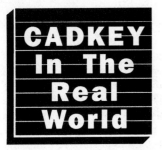

Progressive Die Design with CADKEY

By Calvin Sams

Tool and die shops have not been very progressive in the CAD/CAM marketplace until a few years ago. If a tool and die shop had CAD/CAM, they were really ahead of their time. Today is different. Without CAD or CAM, a shop cannot survive the demands of industry: shorter lead times, better quality of tools, and accurate drawings of tools.

Advantage of CAD

Some of the advantages of CAD in designing a tool or die include faster turn-around of designs, cleaner and more accurate drawings, the ability to make easier changes to the drawings when necessary, less chance of dimensional errors, and the ability to transfer data electronically.

Advantage of CADKEY

When first starting to lay out a part, it is a good idea to go ahead and draw the part in 3-D. Designing directly in 3-D is not yet the most common approach to die design. However, designers are finding that the ability to design in 3-D has distinct advantages.

Drawing in 3-D allows you, the designer, to see what the part will be. If possible, have the customer bring the drawings to you on disk. If the part was not originally designed in CADKEY, you can translate the geometry into CADKEY format through IGES or DXF translation. CADKEY has IGES and DXF translators that work really well.

After the part file exists in CADKEY, you can unfold the part, that is, make a flat pattern of the part, using a third-party program like Profold™ from Applied Production, Inc. This program allows a designer to fold or unfold parts with a complete user interface for radius compensation and angles.

The designer can take the flat-pattern layout and make the strip layout for the die. This is very easy to do in CADKEY because you know the length of the part. All that you need to do is to copy the geometry using the functions: **X-FORM-TRANS-R-COPY**. You select the geometry, specify the number of copies desired, and key in the x, y, z coordinates required. The coordinates in most cases will be the length of the part plus material thickness multiplied by 2, according to the formula: [LOP+(MT*2)]. This is the point where a designer's discretion enters the picture. In most cases, you will know what your customer expects.

An easy-to-use reference procedure, using drawing numbers with different letter extensions, can keep multiple drawing files in order. For example, use the number 1000-P as the drawing number/filename when referring to a part drawing and/or to a flat-pattern layout. Use 1000-S to refer to the strip layout. From this you can start to see how to place the cuts needed to achieve the part's geometry. After completing this, use **X-HATCH** to designate the cut areas.

Your next step is to decide on the best location for the strip carriers. These carriers actually carry and keep the strip together while the part progresses through the die. Then, you need to decide where to form the part in the die, and to **X-HATCH** the areas where the forming of the part will take place. With CADKEY's **EDIT** and **BREAK** functions and its various line types, it is very easy to show where all these actions take place in a die.

After the initial strip layout, review what you have done. It may be possible to merge stations together to shorten the length of the tool. Avoid unnecessary stations and secondary operations, if at all possible. After you have made any corrections that appear necessary, you need to show cross-sectional views of what happens in the form areas of the die. The form areas are very clearly visible when they have been designed in 3-D. 3-D allows the designer and the customer to see exactly what is expected to happen in each one of the stations.

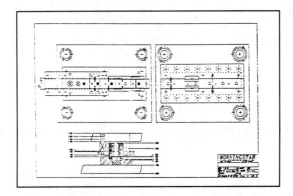

Plan views of progressive die.

—*Continued, next page*

CADKEY In The Real World—Continued

After you have completed the strip layout, the die plan is your next step. This process involves laying out the die-cutting sections and the die-forming sections. Then, if necessary, lay out the die chase. A die chase, or some form of holding all of the die's components in a solid section of steel, is not required for every die.

Your next decision is what die set to use. If you have the option of not needing to use a purchased die set, you can very easily design a custom die set. Many die-set manufacturers will make a customized set of dies for a customer, according to their drawings.

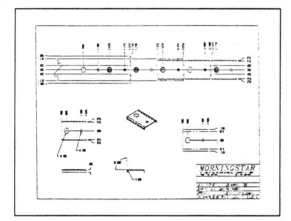

Strip layout of progressive die.

The punch layout is the next step in the process of die design. CADKEY makes this easy to do. Make a different level of the part file active, and use the functions **X-FORM, MIRROR,** and **COPY** to create the punch layout on that active level of the file. You can determine individual punch locations, or you can locate all of the punches with a chase. After you have determined the location of the punches, make a pattern file of the die plan and of the punch plan. Then, you can create a blank part file and insert the pattern file of the die plan and the punch plan. This keeps the size of your die's original part file very workable.

The next step is the stripper layout. The punch plan allows you to determine where to place the die springs and shoulder bolts.

After a designer has completed all of the plan drawings, it is up to the customer to approve the die design. Only then can the designer start to work on the detail drawings. Dimensioning in CADKEY takes 50% to 75% less time than other leading CAD systems, and does not detract from the accuracy of the drawings. This time savings allows you to be more competitive in costing designs.

Completion of the detail drawings means that the design is ready for release to the tool room for building the die. The toolmaker can build a better quality die from the drawings because the drawings are more accurate and cleaner than board-drawn designs.

Adapted with permission from 3-D World, v. 5, no. 1, Jan./Feb., 1991.

However, the endpoints of lines, arcs, circles, and points are connected with lines (Figure 17-2C). Circles are connected with a line segment at the 0 degree position. After the **COPY, JOIN,** or **MOVE** option is selected, the Selection Menu is displayed to provide different methods of selecting entities.

THE DELTA FUNCTION

The **DELTA** (Translate Relative) option translates or moves an object or entities from one location to another relative to its current position, by specifying X, Y, Z coordinate values for the new position. There are three options: **MOVE, JOIN,** and **COPY.**

DELTA OPTION: MOVE

Objects are moved from one position to another on the drawing using the **MOVE** command and inputting the coordinate values for the new position. This is different from using the **PAN** command, which changes only the viewing window. The **MOVE** option changes the position of entities relative to those entities which are not moved. For example, the **DELTA** option can be used to move an object whose lower left corner is at coordinate points 4,4,4 to coordinates 0,0,0. In the example shown in Figures 17-3 and 17-4, one view of the object is moved to a position over the other so that they are in alignment for an orthographic drawing. The following steps show how to use the **DELTA-MOVE** option to change the position of one of the views.

1. Select **ESC-F4-F1-F1 (X-FORM-DELTA-MOVE)**.
2. The Selection Menu appears. From this menu you are to select those entities to be moved. For this example, the figure on the left side of the screen in Figure 17-3 is moved to the right. **F3 WINDOW** is selected.
3. Pick the lower left and upper right corners of a window that will surround the view on the left.
4. A prompt reads:

    dXV=.

    Using view coordinate values input the X value needed to move the object to its new position. For this example, input the number **4** and press **RETURN**.
5. A new prompt reads:

    dXV=4   dYV=

    Using view coordinate values input the Y value needed to move the object to its new position. For this example, input **0** and press **RETURN** or simply press **RETURN** to assign zero.
6. The prompt now reads:

    dXV=4   dYV=0   dZV=.

Using view coordinate values input the Z value to which the object is to be moved. For this example input **0** and press **RETURN** or simply press **RETURN** to assign zero. The object is moved to the new coordinate values (Figure 17-4).

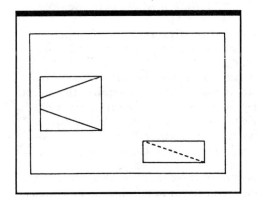

Figure 17-3 Drawing before MOVE

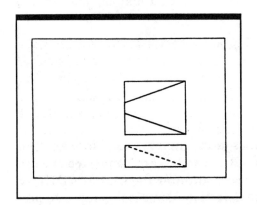

Figure 17-4 Drawing after MOVE is executed

DELTA OPTION: COPY

The **COPY** option is used to copy entities and position the copy(s) using X, Y, Z coordinate values. Figures 17-4 and 17-5 and the following steps show how this option is used.

1. Select **ESC-F4-F1-F2 (X-FORM-DELTA-COPY)**.
2. The Selection Menu appears. From this menu you are to select those entities to be moved. For this example, the figure on the right side of the screen in Figure 17-4 is copied to the left. Select **F3 WINDOW**.

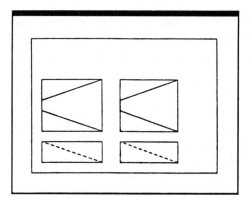

Figure 17-5 Drawing after COPY is executed

3. Pick the lower left and upper right corners of a window that will surround the view on the right.

4. A prompt reads:

   Enter number of copies (1) =.

   For this example only one copy is made, so simply press **RETURN**.

5. A prompt reads:

   dXV=.

   Using view coordinate values input the X value needed to move the object to its new position. For this example, input a negative **-4** and press **RETURN**.

6. A new prompt reads:

   dXV= -4   dYV=.

   Using view coordinate values input the Y value needed to move the object to its new position. For this example input **0** and press **RETURN** or simply press **RETURN** to assign zero.

7. The prompt now reads:

   dXV= -4   dYV=0   dZV=.

   Using view coordinate values input the Z value to which the object is to be moved. For this example, input **0** and press **RETURN** or simply press **RETURN** to assign zero. The object is moved to the new coordinate values (Figure 17-5).

DELTA OPTION: JOIN

The **JOIN** option joins the copy with the original entities with lines from endpoints of lines, arcs, and the start point of circles. This option, useful in creating a 3D model of a part, is referred to as *extrusion*.

Many times when creating a 3D model of a part, you will find it is most easily produced by drawing the front view of the object or the view that shows the most characteristic shape of the part. The third dimension can then be added using the **JOIN** option. For example, the object shown in Figure 17-6 is an isometric view of a two-dimensional object. A 3D wireframe model of the part could easily be produced using the **X-FORM-DELTA-JOIN** option. The following steps show how this is accomplished.

1. Select **ESC-F4-F1-F3 (X-FORM-DELTA-JOIN)**.

2. The Selection Menu appears. From this menu you are to select those entities to be moved. For this example, all the entities are joined. **F3 WINDOW** is used.

3. Pick the lower left and upper right corners of a window that will surround the entities you want.

4. A prompt reads:

   Number of copies (1).

   Press **RETURN**.

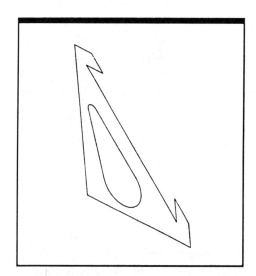

Figure 17-6 Drawing to be copied and joined

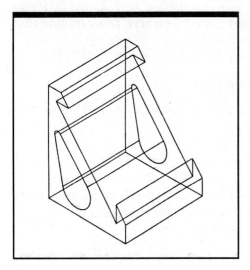

Figure 17-7 Drawing after using the JOIN option

Using world coordinate values input the X value needed to copy the object to its new position. For this example input **0** and press **RETURN** or simply press **RETURN**.

5. A new prompt reads:

    dX=0   dY=

    Using world coordinate values input the Y value needed to move the object to its new position. For this example, input the number **4** and press **RETURN**.

6. The prompt now reads:

    dX=0   dY=4   dZ=.

    Using world coordinate values input the Z value to which the object is to be moved. For this example, simply press **RETURN** to assign zero. The object is moved and joined to the new coordinate values (Figure 17-7).

THE OLD-NEW FUNCTION

The **OLD-NEW** option is used to **MOVE, COPY,** or **JOIN** entities by indicating first an old base position and then a new base position. The Position Menu locates the base position in a number of different ways, such as the end of an entity, center of an entity, or intersection of two entities. The **OLD-NEW** option is similar to the **DELTA** option. The only difference between the two options is the method used to locate the transformed entity. **DELTA** uses keyed-in X-Y-Z values while **OLD-NEW** uses Position Menu locations. After indicating the base position, you are prompted for the first direction. If you press **RETURN** to this prompt, the transformation will be an X-Y move. If you select a first point, this will define the X axis and the second direction will define the Y axis of the original entity plane. In the same way, the X-Y plane for the copy can be defined.

OLD-NEW OPTION: MOVE

The **MOVE** option moves entities from one position to another on the drawing. For this example, the arc and circle shown in Figure 17-8 is moved to a new depth using the **MOVE** option. The following steps show how to use the **OLD-NEW-MOVE** option.

1. Select **ESC-F4-F2-F1 (X-FORM-OLD-NEW-MOVE)**.

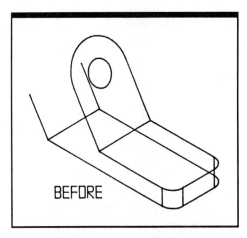

Figure 17-8   Part before MOVE

2. The Selection Menu appears on screen. For this example, use **F1 SINGLE**.
3. **SINGLE** select the arc and circle and press **RETURN**.
4. A prompt reads:

   Indicate base position.

   The Position Menu appears and is used to locate the base position for the part. The base position can be thought of as being the point or handle on the object used to locate the new position of the part. For this example, **F3 ENDENT** is used.
5. Pick the lower left corner of the arc. When prompted to indicate first direction desired, press **RETURN**.
6. A prompt reads:

   Indicate new base position.

7. With **ENDENT** selected from the Position Menu, move the cursor to the new position and pick by pressing the mouse button. The object is moved (Figure 17-9).
8. Press **UNDO** to restore the original position or **CONT** to continue selecting new positions.

OLD-NEW OPTION: COPY

The **COPY** option makes single or multiple copies of entities that are located using the Position Menu. Figures 17-9 and 17-10 and the following steps show how this option is used.

1. Select **ESC-F4-F2-F2 (X-FORM-OLD-NEW-COPY)**.
2. The Selection Menu appears on screen. For this example, use **F1 SINGLE**.
3. **SINGLE** select the arc and circle and press **RETURN**.
4. A prompt reads:

   Indicate base position.

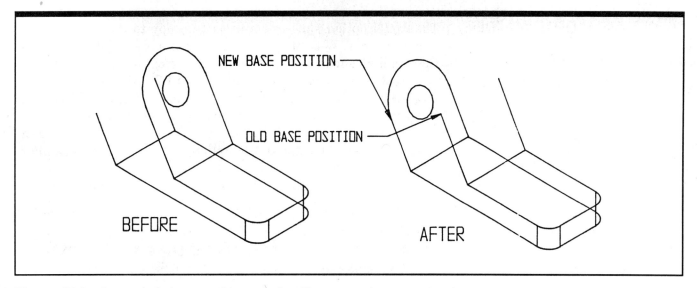

Figure 17-9   Arc and circle moved to end of entity

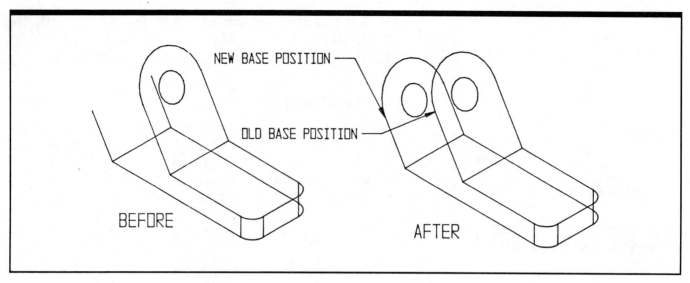

Figure 17-10 Arc and circle copied to end of entity

The Position Menu appears and is used to locate the base position for the part. The base position can be thought of as being the point or handle on the object used to locate the new position of the part. For this example use **F3 ENDENT**.

5. Pick the lower left corner of the arc. When prompted to indicate first direction desired, press **RETURN**.

6. A prompt reads:

   Indicate new base position.

   Select **F3 ENDENT** from the Position Menu.

7. Move the cursor to the new position and pick by pressing the mouse button. The object is copied to the new location (Figure 17-10).

8. Press **ESC** to exit this function.

OLD-NEW OPTION: JOIN

The **JOIN** option joins or connects the copy to the original entities with lines. With this option lines are added from the copy to the original entities from each endpoint of lines, arcs, and the start point of circles. As with the **DELTA-JOIN** option, the **OLD-NEW-JOIN** option is referred to as *extrusion* and is especially useful in creating three-dimensional wireframes.

You can transform the arc and circle in Figure 17-8 into a complete 3D model using the **JOIN** option. The following steps show how this was done.

1. Select **ESC-F4-F2-F3 (X-FORM-OLD-NEW-JOIN)**.

2. The Selection Menu appears on screen. For this example, use **F1 SINGLE**.

3. **SINGLE** select the arc and circle and press **RETURN**.

4. A prompt reads:

   Enter number of copies(1) =.

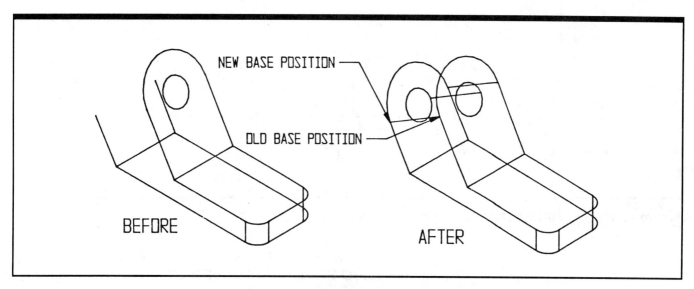

Figure 17-11   OLD-NEW join of the circle and arc using the end of the entity as the new base position

Input the number **1** and press **RETURN** or simply press **RETURN**.

5. A prompt reads:

   Indicate base position.

   The Position Menu appears and is used to locate the base position for the part. The base position can be thought of as being the point or handle on the object used to locate the new position of the part. For this example, **F3 ENDENT** is used.

6. Pick the lower left corner of the part. When prompted to indicate first direction desired, press **RETURN**.

7. A prompt reads:

   Indicate new base position.

   Select **F3 ENDENT** from the Position Menu.

8. Move the cursor to the corresponding line in the back plane of the object and press the mouse button. The object is copied and joined (Figure 17-11).

THE MIRROR FUNCTION

The **MIRROR** option is used to **MOVE, COPY,** or **JOIN** entities by producing a mirror image of the original. When an object is mirrored in the X-Y plane, view data are changed, but the Z depth is unchanged. Dimensions, notes, and labels are not mirrored with this option but can be using **DETAIL-CHANGE-TXT ATT-MIRROR,** as explained in Chapter 16. To mirror an object, two points are selected that define the line or axis about which

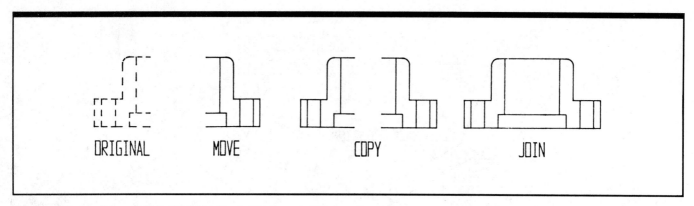

Figure 17-12 The MIRROR MOVE, COPY, and JOIN options

the entities are mirrored or reflected. The Escape Code for this option is **ESC-F4-F5 (X-FORM-MIRROR)**.

1. The **MOVE, COPY, JOIN** menu appears so you can select the appropriate option (Figure 17-12).

2. The Selection Menu will appear so you may identify the entities to rotate about the axis.

3. You are prompted to define the mirror plane. A menu with the following options appears:

1 PT H—One point is selected using the Position Menu options. The mirror image is created horizontal to the construction plane (Figure 17-13).

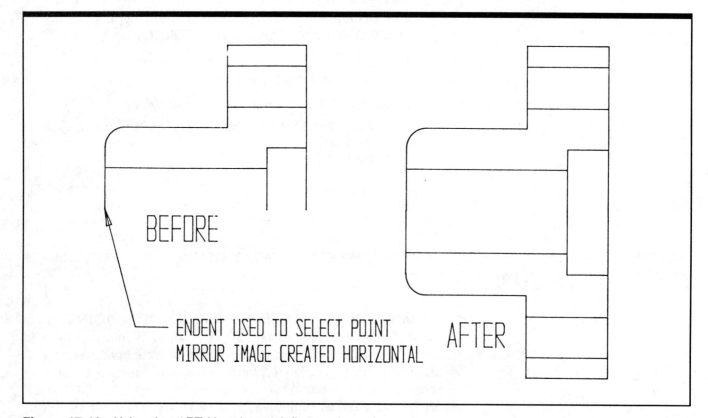

BEFORE

ENDENT USED TO SELECT POINT
MIRROR IMAGE CREATED HORIZONTAL

AFTER

Figure 17-13 Using the 1 PT H option to define a mirror plane

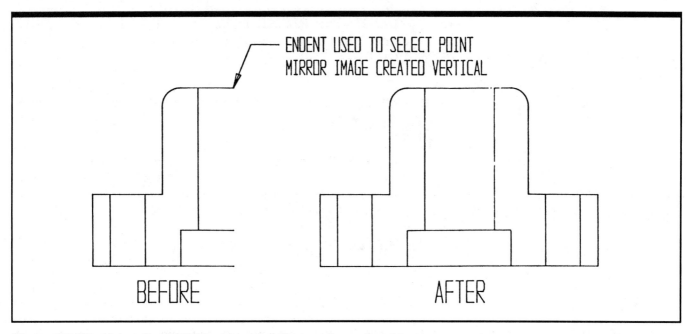

Figure 17-14   Using the 1 PT V option to define a mirror plane

1 PT V—One point is selected using the Position Menu options. The mirror image is created vertical to the construction plane (Figure 17-14).

2 PTS—Two points are selected using the Position Menu. A plane is defined perpendicular to the construction plane (Figure 17-15).

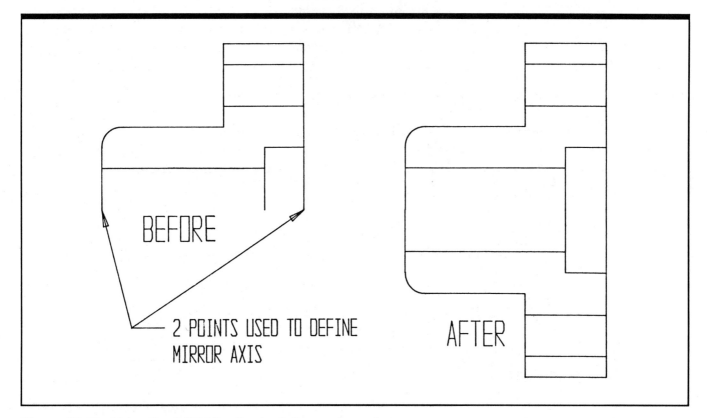

Figure 17-15   Using the 2 PT H option to define a mirror plane

PLANE—The **PLANE** menu is used to define the mirror plane, and the mirror image is copied to the plane (Figure 17-16).

 POSSIBLE USE

The **MIRROR** option can be used for any application that requires a reflected or mirror image. Generally, this applies to symmetric parts. The CADKEY user can create half of the part geometry and make a mirror copy of the other half, saving the time of creating the other half.

THE ROTATE FUNCTION

The **ROTATE** option is used to rotate selected entities about the axis pointing into the screen. For example, for view number 1 in world coordinates, the Z axis points into the screen and is the axis about which the object is rotated. Three options are available with this function: **MOVE, COPY,** and **JOIN**. These options are similar to those described for **OLD-NEW** and **DELTA**. The rotation function is very valuable when an object has to be rotated for clarity or descriptive geometry purposes, or to copy a number of objects about a center point.

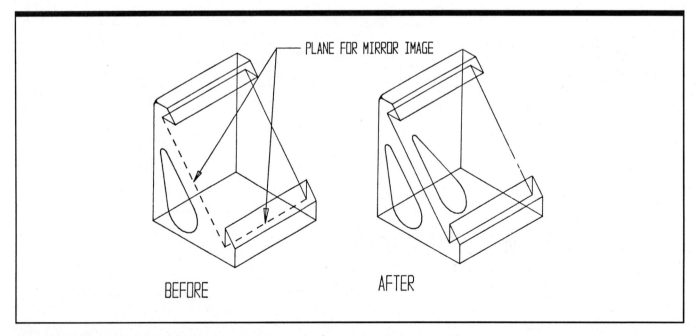

Figure 17-16 MIRROR-COPY of a round end slot from one plane to another

ROTATE OPTION: MOVE

This option moves and rotates a group of entities to a new position. Figures 17-17 and 17-18 and the following steps show how this option is used.

1. Select **ESC-F4-F3-F1 (X-FORM-ROTATE-MOVE)**.
2. The Selection Menu appears. Select **F3 WINDOW** and pick the lower left and upper right corners of a rectangle that encloses the object (Figure 17-17).
3. A prompt reads:

   Indicate first point on axis.

   The Position Menu is displayed. Select **F3 ENDENT**.
4. Pick the lower right corner of the part. When prompted to indicate second point on axis, press **RETURN**.
5. A prompt reads:

   Enter rotation angle (0)=.

   The rotation angle follows the right hand rule. For this example, input the number **45** to rotate the object 45 degrees, then press **RETURN**. The object is rotated (Figure 17-18).
6. Press **ESC** to exit this option.

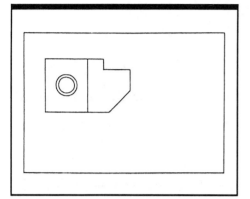

Figure 17-17 Drawing before rotation

ROTATE OPTION: COPY

The **COPY** option is used to rotate and copy entities at the same time. This option is useful when it is necessary to make multiple copies of a part that must be rotated and copied. For example, Figure 17-17 is one gear tooth that could be copied and rotated to make the finished spur gear. This is done by defining the rotation origin as the center of the spur gear, which is not currently displayed in Figure 17-17, determining the number of copies of the tooth, and entering the rotation angle. This is demonstrated in the following steps.

1. Select **ESC-F4-F3-F2 (X-FORM-ROTATE-COPY)**.
2. The Selection Menu is displayed. Choose **F1 SINGLE**, pick the five entities that are part of the gear tooth, and press **RETURN**.
3. A prompt reads:

   Enter the number of copies (1) =.

   For this example, **19** copies are needed, so that number is entered. Press **RETURN**.
4. A second prompt reads:

   Indicate first point on axis.

   To display the center of the spur gear press **ALT-A** to automatically display the full object on screen. The Position Menu will also be displayed. Select **F5 INTRSC** and pick the vertical and horizontal center lines. This makes the

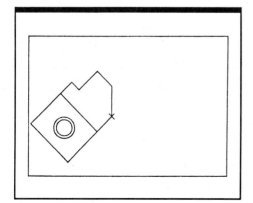

Figure 17-18 Drawing after a 45-degree rotation

rotation origin the center of the spur gear. When prompted to indicate second point on axis, press **RETURN**.

5. A prompt reads:

   Enter rotation angle (0) =.

   For this example, a gear tooth has to be located every 18 degrees. Input the number **18** and press **RETURN**. The gear tooth entities are transformed (Figure 17-19).

ROTATE OPTION: JOIN

With this option it is possible to copy, rotate, and join entities using one command. This option is very useful in developing 3D models of parts, especially symmetrical objects. Figure 17-20 demonstrates how this option can save the user hours of time by taking a single 2D image and producing a sophisticated 3D wireframe model.

Figure 17-20 is a six-sided object that has been displayed in view number 1. To create a 3D model it is necessary to change the view number from 1 to 2 because the rotation has to occur about the axis pointing into the screen. This is done by pressing **ALT-V**, entering the number **2** and pressing **RETURN**. This produces a view of the hexagon (Figure 17-21). The following steps show how to use the **JOIN** option.

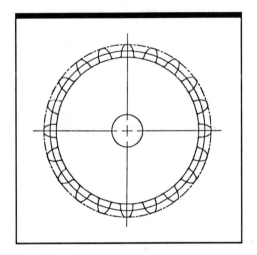

Figure 17-19 Completed gear profile

1. Select **ESC-F4-F3-F3 (X-FORM-ROTATE-JOIN)**.
2. The Selection Menu is displayed. Choose **F6 ALL DSP**, then **F1 ALL**. This automatically selects all the entities of the hexagon.
3. A prompt reads:

   Enter the number of copies (1) =.

   For this example, **18** copies are needed, so enter that number and press **RETURN**.
4. A second prompt reads:

   Indicate first point on axis.

   Pick a point 2 inches to the right of the part (Figure 17-21). This can be done by using **GRID** and **SNAP**. When prompted to indicate second point on axis, press **RETURN**.
5. A prompt reads:

   Enter rotation angle (0) =.

   For this example input the number **20** and press **RETURN**. The object will immediately be copied, joined and rotated (Figure 17-22).
6. To view the object in 3D select **ALT-V**, change the view to number **7** (isometric), and press **RETURN**. The object is displayed (Figure 17-23).

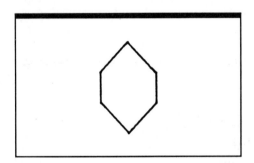

Figure 17-20 Drawing to be joined and rotated

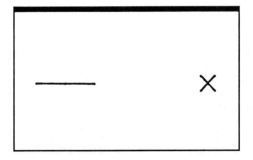

Figure 17-21 View #2 showing the pick point (X) for rotation

Carefully look at the object that was created and is illustrated in Figure 17-23. Consider how long it would take to create Figure 17-23 without the **X-FORM** options and you can easily see how powerful the **ROTATE-JOIN** option is.

THE SCALE FUNCTION

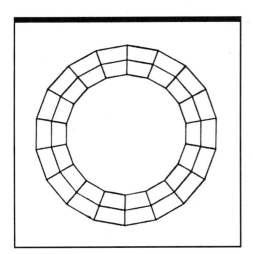

Figure 17-22   A torus is produced using ROTATE-JOIN.

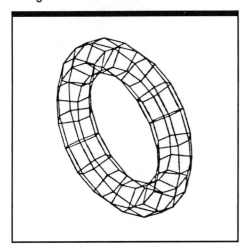

Figure 17-23   An isometric view of the torus

The **SCALE** option allows the user to increase or decrease the size of selected entities in directions defined by the user. The Escape Code for this option is **ESC-F4-F4 (X-FORM-SCALE)**. This command is not the same as using the **DISPLAY** options described in Chapter 14, which only change the current viewing window and not the data base. The **SCALE** option changes the data base so an object that was drawn four inches high and scaled to one half will now measure only two inches high. The **SCALE** function is used with the **COPY, MOVE**, and **JOIN** options. The **SCALE** function is very useful when creating enlarged views of small details on working drawings. With the **SET-AUTO** option set to **NO**, the displayed dimension values are not scaled.

1. The **MOVE, COPY, JOIN** menu appears so you can select the appropriate option.
2. The Selection Menu will appear so you can identify the entities to rotate about the axis.
3. The Position Menu is displayed to aid you in selecting the base position of the selected entities.
4. Next, a menu will appear giving you three options for defining the scaling direction. The options are:

FULL—scales selected entities in all directions at a scale factor defined by the user (Figure 17-24).

UNIDIR—(unidirectional) scales the selected geometry only in one direction (Figure 17-25). The direction is defined by one of the following options:

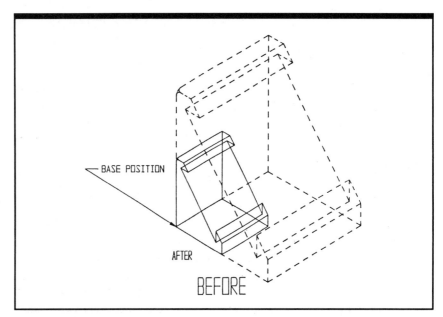

Figure 17-24   Using the FULL option, the part was scaled down equally on all axes, using a value of .5.

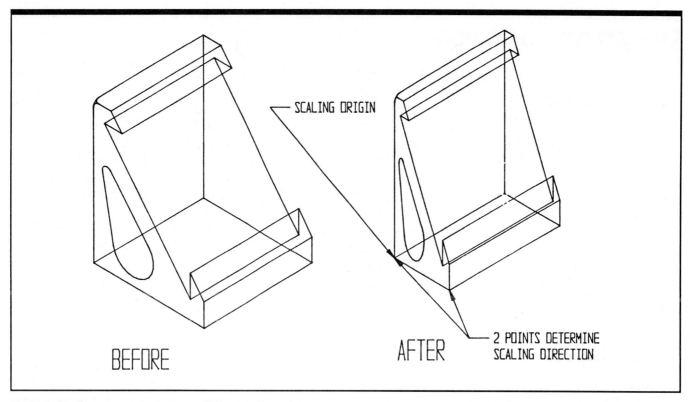

SCALING ORIGIN

2 POINTS DETERMINE
SCALING DIRECTION

BEFORE   AFTER

Figure 17-25   The part was scaled using a value of .5 on one axis using SCALE-UNIDIR.

2 PTS—defines a projection direction/axis between two locations selected with the aid of the Position Menu.

KEY-IN—defines a projection direction/axis between two X, Y, and Z locations.

VIEW—defines a projection direction/axis perpendicular/ normal to a view.

2D ENT—defines the projection direction/axis as perpendicular/normal to a 2D entity such as an arc, circle conic, or 2D spline.

AXES—similar to **UNIDIR** except the scaling is done three times, once on each of the X, Y, and Z axes as defined by the coordinate system in use under COORD:VW or WLD in the Status Window. A different value can be entered for each of the axes (Figure 17-26).

 ## POSSIBLE USE

The **SCALE** option can be used for any application that requires an entity's size to be changed. To make corrections or to redesign a part that requires dimensional changes can also be done with the **SCALE** option (Figures 17-27 and 17-28).

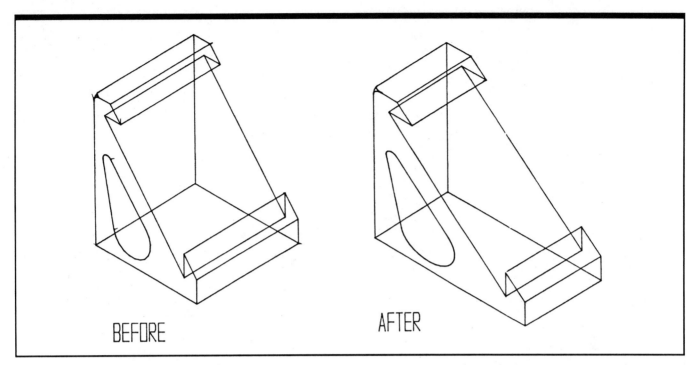

Figure 17-26   Using world coordinates, the part was scaled with the AXES option SX=1.5, SY=.75, SZ=1.

THE PROJECT FUNCTION

PROJECT transforms selected entities onto a specified projection plane and along a designated direction. **PROJECT** is used

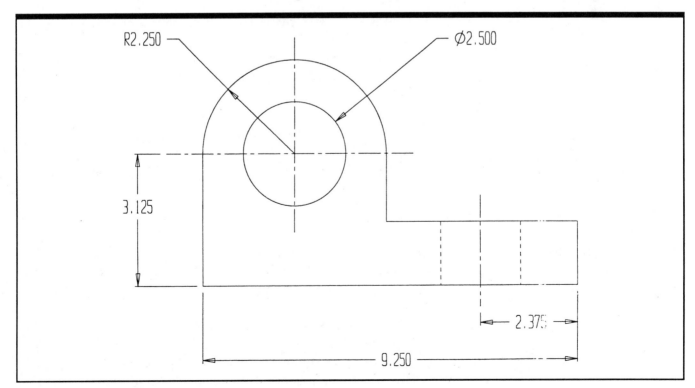

Figure 17-27   A part to be scaled down by a value of .5

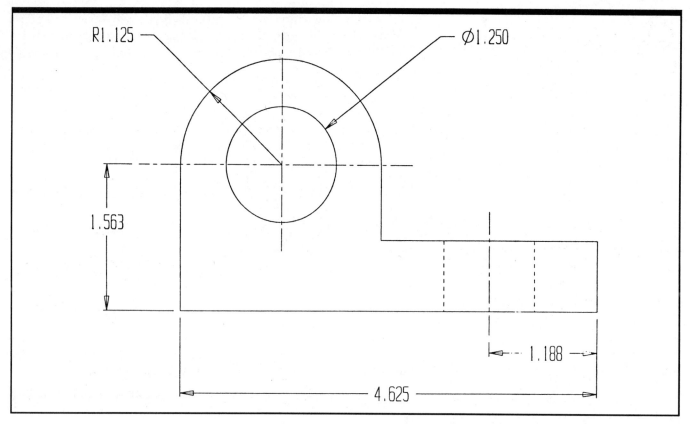

Figure 17-28 The actual size of the geometry is changed in the data base as shown by the dimensions.

to **MOVE, COPY,** or **JOIN** entities using the Selection Menu. The projection plane is defined using one of five options:

> **3 PTS**—defines a plane with three points in space using the Position Menu.
>
> **PT/LINE**—defines a plane by selecting a point in space and the endpoints of a line using the Position Menu.
>
> **2 LINES**—defines a plane by selecting two intersecting lines.
>
> **VW/DPTH**—defines a plane by specifying a view and a depth.
>
> **ENTITY**—defines a plane by selecting a line, arc, conic, or 2D spline.

PROJECT OPTION: NORMAL

The **NORMAL** option projects the selected entities perpendicular and normal to the projection plane. The Escape Code for this option is **ESC-F4-F6-F1 (X-FORM-PROJECT-NORMAL)**.

1. The **MOVE, COPY, JOIN** menu appears so you can select the appropriate option.

2. The Selection Menu will appear so you can identify the entities to project to a new plane. The circle in Figure 17-29 is selected.

3. The Plane Selection menu appears so you can select the plane onto which the entities will be projected. The selected entities will be projected perpendicular to the projection plane selected (Figure 17-30).

PROJECT OPTION: SKEWED

The **SKEWED** option projects the selected entities along a direction axis defined by the user. The Escape Code for this option is **ESC-F4-F6-F2 (X-FORM-PROJECT-SKEWED)**.

1. The **MOVE, COPY, JOIN** menu appears so you can select the appropriate option.
2. The Selection Menu will appear so you can identify the entities to project to a new plane.
3. The Plane Selection menu appears so you can select the plane onto which the entities will be projected.
4. You are prompted to choose a method for defining a projection direction. The options are:

   2 PTS—defines a projection direction or axis between two locations selected with the aid of the Position Menu.

   KEY-IN—defines a projection direction or axis between two X, Y, Z locations.

   VIEW—defines a projection direction or axis perpendicular or normal to a view.

   2D ENT—defines the projection direction or axis as perpendicular or normal to a 2D entity such as an arc, circle conic, or 2D spline.

5. The selected entities will be projected to the plane selected (Figure 17-31).

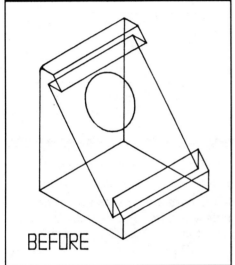

Figure 17-29 Circle on inclined surface to be projected to back plane

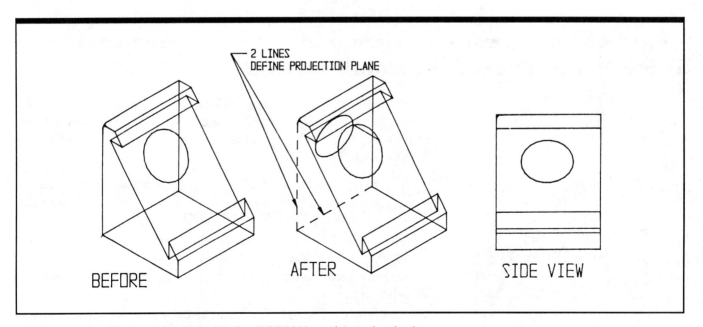

Figure 17-30 Circle projected with the NORMAL option to back plane

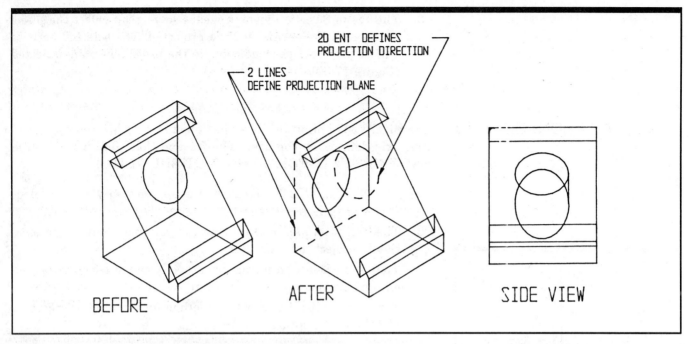

Figure 17-31   Circle projected using the SKEWED option

 POSSIBLE USE

In many situations a hole is drilled perpendicular to an inclined surface. When the drill intersects any plane that is not parallel to the inclined plane, the drill creates an elliptical-shaped hole. The **SKEWED** option creates the ellipse that results from this intersection (Figure 17-31).

THE HLX-ROT (Helix Rotation) FUNCTION

The **HLX-ROT** option rotates selected entities about a user-defined axis by a specified number of degrees, then translates them parallel to the axis. The Escape Code for this option is **ESC-F4-F7 (X-FORM-HLX-ROT)**.

1. The **MOVE, COPY, JOIN** menu appears so you can select the appropriate option. If **JOIN** is selected, you have the option of using **LINES** or **SPLINES** as your join entities.

2. The Selection Menu will appear so you can identify the entities to rotate about the axis.

3. You are prompted to enter the number of additional copies you wish to make.

4. The Position Menu is displayed to aid you in selecting the first point, and then the second point on the rotational axis. This

axis will serve as the center of the helix. If the axis is perpendicular or normal to the view or construction plane in which you are creating helix rotation, press **RETURN** when prompted for the second point.

5. Next, you must enter the rotation angle, which is the number of degrees between copies. It would be 18 for the example in Figure 17-32.

6. Last, the pitch value or number of turns per unit length is entered. The larger the value, the tighter the helix.

 POSSIBLE USE

The **HELIX ROTATION** can be used for any application that requires entities to be rotated about an axis in a helical manner. Screw threads and gears are among the most common applications.

THE C-ARRAY (Circular Array) FUNCTION

The **C-ARRAY** option rotates selected entities about a user-defined axis maintaining the orientation of the entities. This differs from **ROTATE** in that the entities are not rotated about their base position. The Escape Code for this option is **ESC-F4-F8 (X-FORM-C-ARRAY)**.

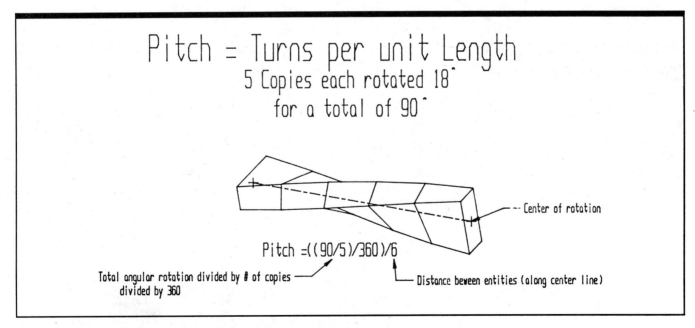

Figure 17-32 The HELIX ROTATION option

1. The **MOVE, COPY, JOIN** menu appears so you can select the appropriate option.

2. The Selection Menu will appear so you can identify the entities to rotate about the axis.

3. You are prompted to enter the number of additional copies you wish to make.

4. Next, you are prompted for the first point, and then the second point on the rotational axis. This axis will serve as the center of the circular array. If you are creating a 2D array, press **RETURN** when prompted for the second point.

5. The Position Menu is displayed to aid you in selecting the base position on the selected entities. The distance between the base position and the axis will be the radius of the circular array.

6. Last, you must enter the rotation angle, which is the number of degrees between copies. It would be 60 for the examples in Figures 17-33 and 17-34.

POSSIBLE USE

The **C-ARRAY** option can be used for any application that requires entities to be rotated about an axis without rotating the orientation of the entities.

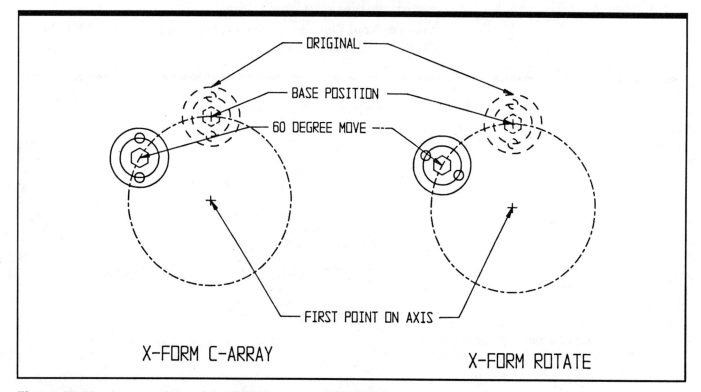

Figure 17-33 A comparison of C-ARRAY move and ROTATE move

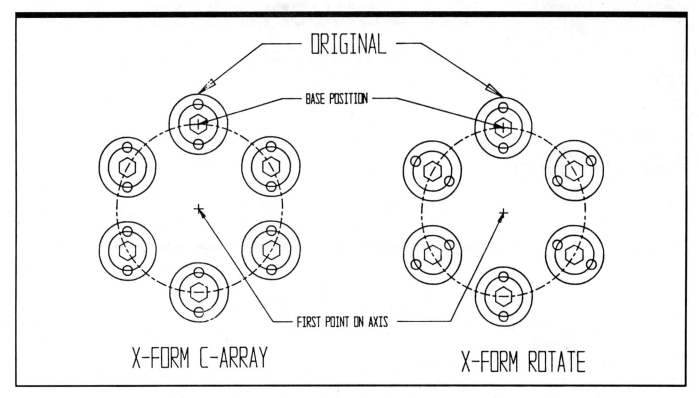

Figure 17-34   A comparison of C-ARRAY copy and ROTATE copy

REVIEW QUESTIONS

1. Describe the difference between the **MOVE** and **COPY X-FORM** options.
2. Describe the difference between the **DELTA** and **OLD-NEW** options.
3. What is the difference between the **MOVE** and **PAN** options?
4. What is the difference between the **COPY** and **JOIN X-FORM** options?
5. About which axis will entities be rotated using the **ROTATE** function?
6. Describe the difference between the **SCALE** and **DISPLAY** options.
7. Which transform option is used to create perspective drawings?
8. List the two **PROJECT** options available.

CHAPTER 18

3D Construction Techniques

OBJECTIVES

After completing this chapter, you will be able to:

- Define a 3D model using extrusion.
- Define a 3D model using extrusion with scale.
- Define a 3D model using profile revolution.
- Define a 3D model using surface revolution.
- Define new construction planes.
- Add part geometry using a defined construction plane.

3D CONCEPTUALIZATION

This is one of the most important chapters of the whole book! 3D is the area where the greatest changes in CAD are taking place. One of the most significant aids in 3D construction is viewports. Viewports give the user the ability to see more than one view of an object on the screen at the same time. This enhancement, plus the addition of construction planes, make 3D construction in CADKEY much easier. Now that you have had an opportunity to become familiar with the functions in CADKEY, let's apply them to 3D construction. It is the three-dimensional aspect of CAD that is changing the way we think, design, view, and manufacture parts in the world today. Until the development of the computer, we had been limited to designing parts in two-dimensional views or quasi-3D views. The world that we live in is three-dimensional, so it only makes sense that we should design parts as they are found in the real world. Now, with CADKEY you can design and create a 3D part in true 3 space (three-dimensional space) just as it is found in the real world.

The conceptualization process of designing a part is the area where the greatest change occurs. For individuals who have learned to think, draw, and design in 2D, the thought process will change.

Most people have been trained to create the front, top, and right side views of a part during the design process. When it comes to 3D CAD, some people want to rely on this knowledge and create those views first and then have the computer create the 3D model, but that is not how it works. Now, instead of creating the two-dimensional orthographic views first, a 3D wireframe model is constructed first. From the 3D wireframe model all of the orthographic views are pre-defined and are automatically created. The orthographic views can be seen simply by looking at the various viewports. If you wish to change a view in a viewport, invoke the Immediate Mode command **ALT-V** and enter the number of the view that you wish to see. If the view does not appear on the screen, press the Immediate Mode **ALT-A** to **AUTOSCALE**, which will bring the view onto the screen. The resulting view is an accurate orthographic view with almost all of the necessary lines in their correct places. You will have to change only the line types of some of the hidden lines and place a few additional lines to mark the center and the edge on a drill hole passing through the object. A complete layout drawing can be quickly created by saving the different views as patterns. The patterns are then retrieved and placed in one drawing, which results in the layout drawing demonstrated in the 3D dovetail tutorial.

This chapter provides a foundation in 3D construction techniques on which you can build. Once you are able to create 3D models you will be able to make the transition from a 2D thought process to a 3D thought process. Soon you will be creating 3D models first and then extracting the orthographic views from this model.

This change in the design process becomes a tremendous time saver once you become proficient with it. Once you have created a 3D wireframe model you can define an unlimited number of views with a few easy steps. Each view that you define results in a new view. There are eight pre-defined views, which are:

> **View 1 - Top**
> **View 2 - Front**
> **View 3 - Back**
> **View 4 - Bottom**
> **View 5 - Right Side**
> **View 6 - Left Side**
> **View 7 - Isometric**
> **View 8 - Axonometric** (an inverted left side out view)

Conceptually, new views are the result of the change in the viewer's position rather than a change in the position of the part. In other words, you rotate around the part to new viewing positions rather than the object being rotated in 3 space.

DEFINING A NEW VIEW

To define a new view of a part you can use the Escape Code **ESC-F6-F4-F3-F2-F3   (DISPLAY-VIEW-NEW-3   PTS-ENDENT)**. This will create an auxiliary view projecting the true size and shape of an inclined or oblique surface so that it is parallel with the screen. Refer to Chapter 14 for a more in-depth explanation.

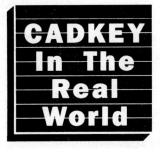

High-Tech *Speed* Ski Helmet!

Skiing is moving to a new height as a sport: pure speed. The goal in speed skiing is to go straight down the hill, as fast as possible, without getting seriously injured. Speed skiing is becoming so popular that it will be introduced as a demonstration sport in the 1992 Winter Olympic Games at Albertville, France. Speeds of well over 100 mph are now commonplace. The current men's world record is 139.02 mph!

What can you do to protect an individual who will be traveling downhill at speeds approaching 140 mph, and whose very own body will be creating the wind resistance? Particularly, how do you help to prevent head injuries?

Speed Skiing Performance Engineering Program

SS PEP, the Speed Skiing Performance Engineering Program, decided to design a new type of helmet to meet the exacting needs of this new sport.

SS PEP is a non-profit organization dedicated to implementing new technology to benefit speed skiers.

Cadkey became involved with SS PEP's helmet project on December 4, 1989. Steve Gubelmann explained to Gary Magoon, Director of Manufacturing Systems, that SS PEP's project would require genuine innovation, but at the same time it must be economically feasible. Taking advantage of existing

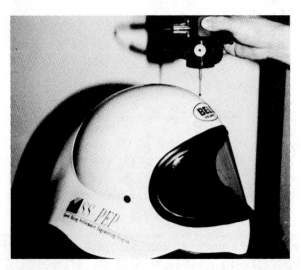

Physical prototype of aerodynamic safety helmet for speed skiing.

technology in safety-certified helmets would help to make the project economically feasible. SS PEP had considered using the SR2 alpine-skiing helmet, manufactured by Bell Helmets, as its core helmet because it has already been certified to the most demanding safety standard ever put forward for ski helmets. However, an alpine helmet does not have the aerodynamic design that speed skiing requires. Steve Gubemann, Braxton Carter, Gary Magoon and Ujjwell Trivedi, also of Cadkey, took the safety-certified SR2 helmet and an aerodynamically sound but non-certified helmet currently used in speed skiing as models and set to work.

CopyCAD, CADDInspector, and CADKEY 3

Ujjwell used CopyCAD™ and the Brown & Sharpe MicroVal® Coordinate Measuring Machine to digitize the aerodynamic outer helmet. Gary also used CopyCAD and the MicroVal to digitize the Bell SR2 as the inner safety helmet.

Gary then used a personally modified version of this reverse-engineering technology to collect accurate 3-D data of a human head, his own. Using elements of CADKEY's CADDInspector™ software, which is part of the CopyCAD system, and a Polhemus Isotrak®, a hand-held, 3-D digitizing stylus that uses a magnetic field to obtain x, y, and z coordinate data, Gary and his wife Suzanne digitized his head.

The three sets of 3-D data collected were transferred into CADKEY 3™ as individual part files. Gary created a fourth part file in which the head data, the inner helmet data, and the outer helmet data were superimposed one upon the other. One critical reason for producing this file of superimposed data was to verify that a visor on a safe aerodynamic helmet would provide an adequate field of vision for a speed skier traveling downhill in the aerodynamic tuck position.

This new part file suggested that the Bell SR2 helmet could be modified into a safe aerodynamic helmet by adding a clear visor and a fairing. The addition of the visor did not pose any particular problem. The fairing was another matter. The fairing's shape in the CADKEY part file is composed primarily of 3-D splines. The fairing starts from both ears and the back of the neck, and flares out over the shoulders and upper back. If SS PEP could get the fairing manufactured so that it would fasten securely to the Bell SR2, but in such a fashion that it could break away if necessary, they would have their innovative yet economically feasible prototype.

—Continued, next page

CADKEY In The Real World—Continued

FastSURF

The fairing would require numerical control machining. But, before it could be machined, the fairing's complex geometries would have to be transformed into a free-form surface using another CADKEY third-party product: FastSURF™. Because

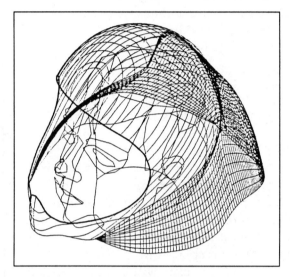

Illustration of the head data, safety-certified helmet data, and aerodynamic helmet data superimposed on each other. Notice how the data suggests the design of the fairing.

FastSURF's user interface is written in CADL™ (CADKEY Advanced Design Language), FastSURF operates as a seamless extension of CADKEY 3. With help from Fastcut personnel, Steve and Braxton created the free-form surface of the fairing themselves.

RLS Enterprises and Bell Helmets

RLS Enterprises of Burbank, California, checked the helmet's part file in CADKEY, and then used a CADL file to create the machining toolpath for the fairing on their PC-based CAM system. RLS machined the part using a polyurethane that is formulated to their proprietary specifications. The machined part served to create a mold of the fairing, that Steve and Braxton took to Bell Helmets in Norwalk, California.

Steve and Braxton arrived at Bell Helmets on February 15. Two days later, Bell Helmets had manufactured the visor and the fairing and had attached them to the helmet. SS PEP now had its physical prototype of a safe aerodynamic helmet for speed skiers. Bell Helmets shipped the helmet to Steve McKinney, coach of the U.S. Speed Skiing Team, in time for the speed skiing race in Kirkwood, California, on February 22-26, 1990.

Adapted with permission from 3-D WORLD, v. 4, no. 3, May/June, 1990.

3D MODEL CONSTRUCTION

You can construct a 3D model in several different ways. The methods discussed in this chapter are: Simple Extrusion, Extrusion with Scale, Profile Revolution, Surface Revolution, and Plane Construction.

SIMPLE EXTRUSION   The first and perhaps the most widely used method of 3D construction is the Simple Extrusion method. Extrusion means that a view is copied and placed at a new depth. When the **JOIN** option is used, lines are created along the depth axis between the original and the copy at all of the intersection points of two entities. By placing a copy of the view at a new depth, the third dimension of depth has been added to the part. Prior to this, the view only had two dimensions, width and height. This is the most basic of the

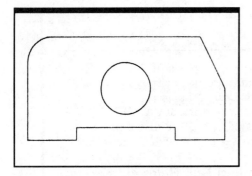

Figure 18-1 The creation of entities in the front view, view 2, to define the shape of the part

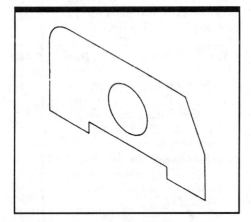

Figure 18-2 The part prior to extrusion, as viewed from view 7, the isometric view

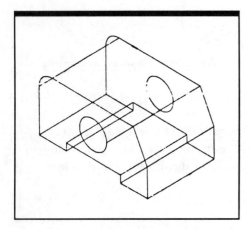

Figure 18-3 The part after extrusion to give the third dimension, depth, as seen using the world coordinate system

3D constructions but is commonly used as the building block for a complete part. Once the extrusion has taken place you can look at any of the viewports to see what your wireframe model looks like in those views.

To try this method, create a front view of a part in the view 2 viewport (Figure 18-1). The Escape Code for extruding a view is **ESC-F4-F1-F3 (X-FORM-DELTA-JOIN)**. The Selection Menu appears to aid you in selecting the entities that you wish to extrude: in many cases, you will want the **ALL DSP** (All Displayed) option. In multiple depth cases, you can singly select individual entities and extrude them to various depths.

If you extrude using the world coordinate system you can see the transformation take place in the isometric view. To change to world coordinates, press the Immediate Mode command **CTRL-V** and select **F2 WORLD** for the coordinate system. When prompted to indicate the viewport to select all entities, choose view **7**, the isometric viewport. The view should look like Figure 18-2.

You are then prompted to enter the number of copies. The default value is 1. Press **RETURN** to accept this value.

Next, you are prompted with delta world coordinates to indicate the position of the joined copy. When prompted **dx=**, press **RE-TURN** because you do not want to change the position of the copy on the X axis. When prompted **dy=**, enter a value equal to the depth because the Y axis in the world coordinate system will add depth to a part that has been created in the front view. If the value that is entered is positive, the copy will be made behind the original or into the screen. If the Y value is negative, the copy is made in front of the original, toward you. Enter a value and press **RETURN**. Last, you are prompted **dz=**. You do not want the copy moved on the Z axis so press **RETURN**. At this time the joined copy is created and you now have a 3D wireframe model (Figure 18-3). The wireframe model can be viewed from any of the other seven views. Remember, if your part does not appear when you switch to another view, press **ALT-A AUTOSCALE** and it will be displayed.

If you do the extrusion using the view coordinate system you will want to be in one of the regular orthographic viewports, preferably view 2, the front view. In the view coordinate system, the X axis (width) runs horizontally, the Y axis (height) runs vertically, and the Z axis (depth) runs in and out of the screen. The view coordinate system is the default coordinate system when you initialize CADKEY. When the copy is made you will change its position on the Z axis to give it depth.

Use the same Escape Code again, **ESC-F4-F1-F3 (X-FORM-DELTA-JOIN)**. The Selection Menu appears for you to select the entities you wish to transform and copy in the view 2 viewport. Once the entities have been selected you are prompted to enter the number of copies; 1 is the default, so press **RETURN** to accept this value.

You are then prompted with the delta view prompts to indicate the position of the copy relative to the position of the original

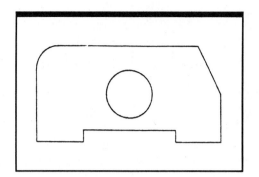

Figure 18-4   An extrusion done using the view coordinate system. There is no apparent change in the part.

entities. When prompted **dxv=**, press **RETURN** because you don't want to move on the X axis. When prompted **dyv=**, press **RETURN** because you don't want to move on the Y axis. When you are prompted **dzv=**, enter a value equal to the depth of the part. A negative value will extrude the part into the screen and a positive value will extrude the part out of the screen toward you. At first you will not see a difference (Figure 18-4), because the copies have been made parallel to each other. To see the depth dimension that has been added, look at the other viewports. The wireframe model can be seen best in the isometric view (Figure 18-5).

EXTRUSION WITH SCALE

This 3D construction technique is similar to a Simple Extrusion method except that the copy of the original entities is scaled, either reduced or enlarged. Create a 3.0 diameter circle in the top viewport, view 1 (Figure 18-6). The Escape Code for the extrusion with scale method is **ESC-F4-F4-F3 (X-FORM-SCALE-JOIN)**.

The Selection Menu is displayed to aid you in selecting the entities that you wish to copy or scale. You are then prompted to enter the number of copies. Press **RETURN** to accept the default value 1.

The next prompt requests that you indicate the scaling origin. The Position Menu appears to aid you in indicating the scaling origin. Before you indicate the scaling origin, change the working depth **CTRL-D-F1 (DEPTH-VALUE)** to the depth of the scaled copy. Enter the value **2.5**. Then, using one of the position options, indicate the scaling origin. In this case, the **CURSOR** option was used to indicate the center of the circle as the scaling origin (Figure 18-7).

Last, you are prompted to enter the scale factor. In this case the scale factor is .33. In view 1, the part looks like Figure 18-8. When viewed from the isometric view 7, it appears as Figure 18-9.

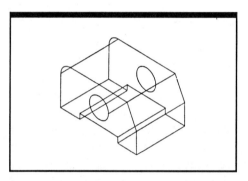

Figure 18-5   The view coordinate extrusion as viewed from view 7, the isometric view

PROFILE REVOLUTION

A profile revolution takes the profile or edge view of a part and rotates it about a center point. The profile is rotated at a degree increment specified by the user and a specified number of copies are made as the profile is rotated. The result is a 3D wireframe model that is cylindrical in shape and has the profile details of the original profile or section you created.

To get some hands-on experience with this method, create a profile in the view 2 viewport as illustrated in Figure 18-10. Use any convenient dimensions. Sometimes it may be helpful to set your construction plane (**CPLANE** in the Status Window) to the view you are presently working on. In this case, the **CPLANE** should be 2.

The Escape Code for creating a 3D model using the profile revolution method is **ESC-F4-F3-F3 (X-FORM-ROTATE-JOIN)**. Once you have invoked the Escape Code, the Selection Menu appears. Use one of the methods found on the Selection Menu to select the entities which make up the profile.

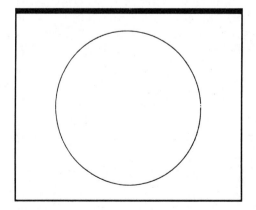

Figure 18-6   A circle created in the top view, view 1

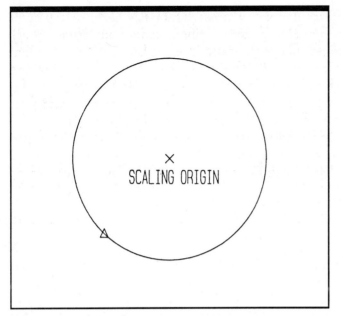

Figure 18-7 The scaling origin is indicated by selecting the circle with the CENTER option.

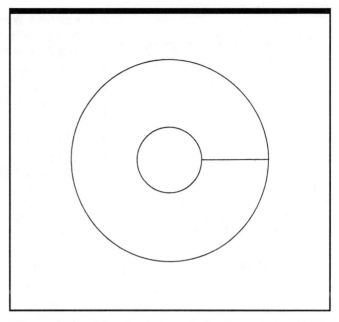

Figure 18-8 The circle extruded with a scale of .33

Next, you are prompted to enter the number of copies of the profile you wish to create as you rotate the profile. This number multiplied times the rotation angle should equal 360 degrees if you wish to create a closed cylinder. (Number of copies × rotation angle = 360 degrees). In this case create twelve copies.

When you are prompted to indicate the rotation origin, you must indicate a point in the view in which the profile is to be rotated. This is different from the view in which the profile was created. If the profile were created in the front view 2, you would rotate it in the top view. You will need to switch the **CPLANE** to view 1. Enter

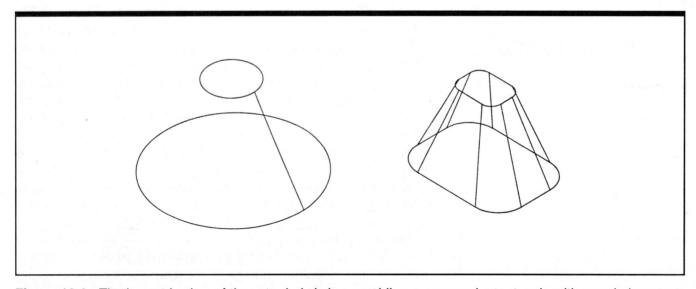

Figure 18-9 The isometric view of the extruded circle resembling a cone and a rectangle with rounded corners extruded with scale

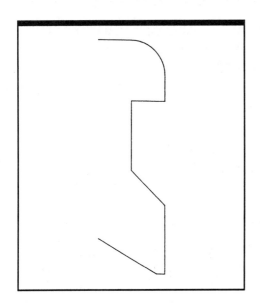

Figure 18-10 A part profile created in view 2

SURFACE REVOLUTION

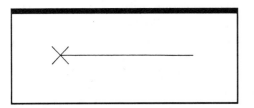

Figure 18-11 The part profile in view 1 with the rotation origin indicated (X)

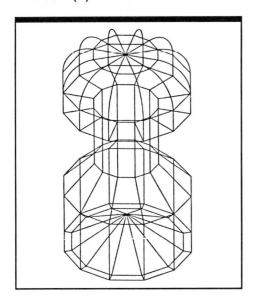

Figure 18-12 The profile rotated using twelve copies

the view number to which you wish to change, in this case view 1, and press **RETURN**. If the view is not presently visible press **ALT-A AUTOSCALE**. In this viewport the profile will appear only as a line. The Position Menu is displayed to help you indicate the rotation origin. Use the **ENDENT** option **F3** to select the endpoint of a line which serves as the point of rotation for the profile (Figure 18-11). The rotation origin could have been selected before the view was changed to clearly select the origin.

Last, you are prompted to enter the rotation angle, which should be equal to 360 divided by the number of copies. In this case a 30 degree rotation angle was used. Once the rotation angle has been entered, the profile is revolved, creating a 3D wireframe model (Figure 18-12).

In a surface revolution, a closed shape or surface (as opposed to a profile shape) is rotated about a center point. When a circle is revolved, the resulting wireframe model has a doughnut shape, known as a *torus*.

In this case, create a circle in view 1 as the surface to be revolved. Next, use the **AUTOSEG** option **F2-F6-F1** to create six arc segments on top of the circle. The start angle is 0 degrees and the end angle is 360 degrees for the **AUTOSEG** option. When prompted: **Is this correct?**, respond **F2 YES** if the circle has been divided into six equal segments (Figure 18-13).

The next thing to do is **ESCAPE** to return to the Main Menu. Then look for the edge view of the circle in the view 2 viewport. Press **ALT-A** to autoscale the edge view of the circle which currently appears as a line (Figure 18-14).

Press the Escape Code **ESC-F4-F3-F3 (X-FORM-2D ROT-JOIN)**. The Selection Menu appears and you are prompted to select the entities to rotate. Use a window to select the circle. When you are prompted to enter the number of copies, remember the formula: number of copies times rotation angle equals 360 degrees. In this case use eighteen copies.

Next, you are prompted to indicate the rotation origin. The Position Menu is displayed to aid in indicating the rotation origin. Use the **CURSOR** option and select a point to the left of the edge view of the circle that is in the same plane as the circle (Figure 18-14). The distance from the end of the edge view of the circle to the rotation origin is equal to the radius of the hole in the doughnut-shaped part that you are creating.

Last, when you are prompted to enter the rotation angle enter a value of **20** degrees and press **RETURN**. The circle will be rotated about the origin, but to see the three-dimensionality of the torus press **ALT-A** and **RETURN** to view the object in the isometric viewport (Figure 18-15).

PLANE CONSTRUCTION

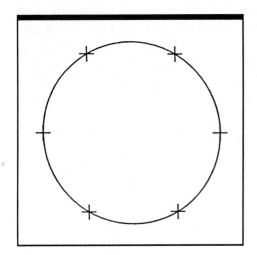

Figure 18-13 A circle in view 1 with six segments created using AUTOSEG

Figure 18-14 The edge view of the circle as seen in view 2 and the rotation origin marked (X)

In plane construction, entities are created in new planes and at new depths. Often when an inclined surface is created, entities must be placed on this surface. The following example will give you the steps to use when creating on an auxiliary surface or view other than the regular six orthographic views.

The part you will create is a 3D pyramid. Starting in view 1, create a rectangle by width and height, **F1-F1-F7-F2**. When you are prompted for the width, enter the value **4** and for the height enter the value **2.5**. You are then asked to indicate the base position; **KEYIN** the coordinate **0** when prompted for **dxv=**, **dyv=** and **dzv=**.

Next, place a point at a new depth which serves as the top of the pyramid. To create the point press **ESCAPE** and the Escape Code **ESC-F1-F4-F1** (**CREATE-POINT-POSITN**). When the Position Menu is displayed use the **KEY-IN** option to indicate a **POINT** at the center (x=2, y=1.25, z=3) of the rectangle (Figure 18-16). Change your screen to a four-view layout with **DISPLAY-VWPORTS-AUTOSET**.

Autoscale (**ALT-A**) the views and press **RETURN**. You will see the point above the rectangle in the isometric viewport (Figure 18-17). What you will do now is create lines that run between the point and the corners of the rectangle. When you are creating entities in 3D that run between planes at different depths, you must use one of the Position Menu options other than the **CURSOR** option. If the **CURSOR** option is used, the entities are created in a plane parallel to the screen and will be out of place in the other views. To begin creating the lines which run between the **POINT** and the corners press **ESCAPE** and the Escape Code **ESC-F1-F1-F1** (**CREATE-LINE-ENDPTS**). When the Position Menu is displayed choose **F2 POINT** and select the **POINT** as the start point of the line. Next, change the Position Menu option to **F3 ENDENT** and select one of the lines near the corner of the rectangle. A line is drawn from the point to the corner (Figure

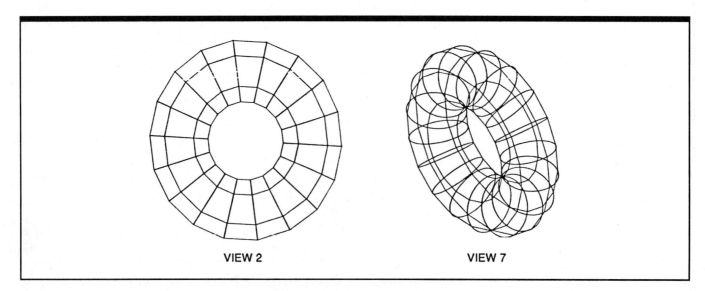

| VIEW 2 | VIEW 7 |

Figure 18-15   The torus created by the surface revolution construction

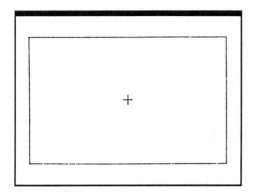

Figure 18-16   A rectangle with a point at a depth of 3

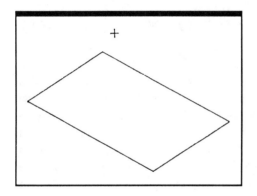

Figure 18-17   The isometric viewport of the rectangle and point

18-18). While the **ENDENT** option is still active select another corner as the start point for the line, then change the position option to **POINT** and select the point to create the second line. Repeat the previous steps to draw the last two lines and complete the pyramid.

From the Status Window select **CPLANE** (Construction Plane). A menu will appear with options for defining a construction plane. The methods of identifying a construction plane include:

3 PTS (3 POINTS)
PT/LINE (POINT & LINE)
2 LINES
VW/DEPTH (VIEW & DEPTH)
ENTITY

All of these options can be used. For this example, use **3 PTS** and endent from the Position Menu to select the three corners of one of the surfaces (Figure 18-19).

With this construction plane defined (the active depth set equal to the right side), any entities created will be on this surface. Create a circle of any size on this construction plane of the right side (Figure 18-20). Press **ESCAPE** and use the Escape Code **ESC-F1-F3-F3 (CREATE-CIRCLE-CTR+EDG)**.

Once the circle is created, look at the various viewports and you will see that the circle has been created in the plane of the right side. Now repeat this procedure and create a circle on each of the three remaining sides (Figure 18-21). Try using some of the other **CPLANE** options, such as **2 LINES**.

The 3D construction techniques discussed in this chapter give you just a starting point in the 3D world of CAD. Through your

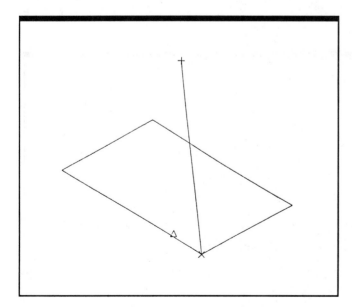

Figure 18-18   A line by end points from the point to the corner of the rectangle

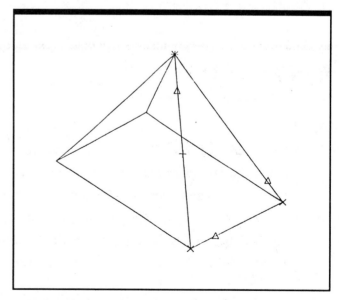

Figure 18-19   Selecting the endpoint of three lines making up the surface

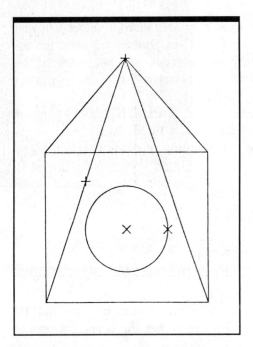

Figure 18-20 A circle created on the surface

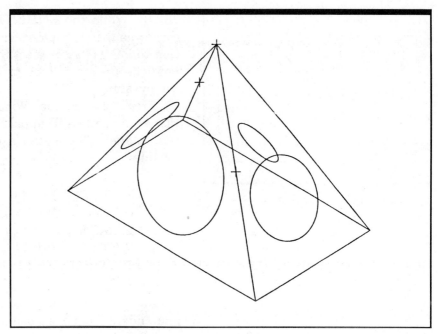

Figure 18-21 The isometric view of the pyramid with a circle created on each side

experimentation and work with the 3D dovetail tutorial you will find new methods and become more adept with 3D construction. Once you have a solid understanding of it you will never return to 2D drafting.

REVIEW QUESTIONS

1. What is one of the advantages of working in 3D?
2. Which is created first: the orthographic views of a part or a 3D wireframe model?
3. What is the difference between a simple extrusion and an extrusion with scale?
4. What is the Escape Code to define a new auxiliary view?
5. When a new view is created, has the part been rotated or the position of the viewer been rotated?
6. What is the advantage of using the world coordinate system for a simple extrusion?
7. How are the number of copies and the rotation angle determined for a profile revolution or a surface revolution?
8. What should be done when you change views and the screen is blank?
9. In plane construction, what are two ways of identifying a construction plane?

The CONTROL Options

OBJECTIVES

After completing this chapter, you will be able to:

- Determine the perimeter of a single entity or a group of entities.
- Determine the area of a closed figure.
- Determine the centroid of a closed figure.
- Determine the polar moment of a closed figure.
- Display information about an ellipse, hyperbola, and parabola.
- Determine the values assigned to an entity.
- Determine the coordinate data for a selected entity.
- Determine the distance between any two points or positions on screen.
- Determine the angle between two lines.
- Change the color, line type, line width, pen number, or outline/fill of entities.
- Display a gnomon of a 3D axis.
- Create 2D entities on a construction plane.
- Change the working depth on CADKEY.
- Determine the amount of available memory in RAM and free disk space.
- Access DOS from within the CADKEY program.

This chapter introduces the various **CONTROL** options available with CADKEY. These options are used for a wide variety of purposes ranging from geometric calculations to changing existing attributes of entities. The geometric calculations include: **PERIM** (Perimeter), **AREA/CN** (Area-Centroid), **CONIC**, and **MOMENT**. Entity attributes, which can be changed, include: **COLOR**, **L-WIDTH** (Line Width), **OUT/FIL** (Outline Fill), and **PEN #**. **CUR TRK** (Cursor Tracking), **L-LIMIT** (Line Limit), **COORD** (Coordinates—View or World), **DEPTH** (sets the working Depth), **STATUS** (Available Disk and RAM Memory), and **2D/3D** (Construction Mode) can also be toggled using the **CONTROL** function. Figure 19-1 (page 464) illustrates the **CONTROL** menu.

New Ink Technology Requires New Type of Printer

The Phaser III™ PXi Color Printer, manufactured by Tektronix, Inc. of Wilsonville, Oregon, implements a new technology in printing with ink. Based on a phase-change technology, users can print on any media, including plain paper, vellum, or cover stock, on sizes up to 12 × 18 inches. With a printing resolution of three hundred dots per inch, powered by a 24MHz RISC-based controller, and featuring Adobe PostScript Level 2, the Phaser III is well suited for printing CAD designs. To use this new ink technology required the development of a new type of printhead based on three-dimensional modeling in CADKEY.

"This printer is a mechanically intensive product that required designing at least three hundred individual parts," said Ace Van Horne, design engineer. "Some of the parts were simple; some were complex. We started with two design engineers working together with CADKEY. By the time we reached the manufacturing stage, we had fifteen engineers working as a team with CADKEY, plus people in the manufacturing-support organizations."

"We have used CADKEY for four years," Ace added. "We designed every part in the Phaser III in 3-D with CADKEY. The advantage of 3-D modeling is that one engineer can make sure that the part which he/she is designing mates properly with a corresponding part that someone else is designing."

Critical Element

The critical element in the Phaser III project was the new type of printhead needed to take advantage of the advance in ink technology: *hot-melt* or *phase-change* ink. Tektronix' previous printer models had used water-based ink. However, water-based ink can evaporate while it is in the printer's reservoir and in the printhead, plugging the holes. Water-based ink also absorbs into the paper, and this can create a fuzzy image while diluting the colors so that they are no longer brilliant.

Hot-melt or phase-change ink solves the problems associated with water-based ink. "It looks like a crayon when you insert it into the printhead," Ace Van Horne said. "As soon as it is dropped into the machine, 140-degree heat liquefies the ink. The now liquid ink shoots through a line of nozzles in the printhead onto the paper. The holes in these nozzles are two- to three-thousandths of an inch in diameter. However, because liquefied phase-change ink does not evaporate, as water-based ink does, it does not dry and clog the holes in the printhead's nozzles. The temperature of the paper going through the printer is much less than 140 degrees centigrade. The paper sucks the heat out of the ink so that the ink very rapidly goes below its melting point. The ink resolidifies on the paper as plastic, before it has a chance to bleed into the paper. After the complete image is on the page, the printer rolls the paper through two rollers to flatten the drops of plastic ink onto the paper by pressure fusing."

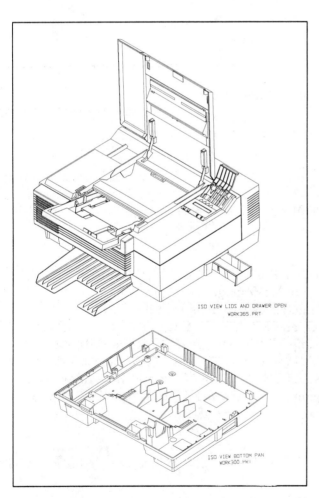

ISO VIEW LIDS AND DRAWER OPEN
WORK365.PRT

ISO VIEW BOTTOM PAN
WORK300.PRT

Isometric views of the Phaser III – PXi Color Printer

—*Continued, next page*

CADKEY In The Real World—Continued

Database Management

As the number of people in engineering and manufacturing involved in the Phaser III project increased, there was a serious need to control the database of CADKEY files shared among so many individuals. The CADKEY part files had rapidly grown to 10 megabytes. All of the personal computers of the project-team members were linked in a network with a VAX™ 8650 as a file server, running both DOS-based and UNIX-based versions of Sun Microsystems' Network File System™. "We did not share original CADKEY part files," Ace said. "We shared CADKEY pattern files over the network."

Ace purchased a CADKEY-related third-party database-management software, ACTVIEW™, produced by ALLAN CAD TOOLS of Portland, Oregon. ACTVIEW integrates with CADKEY through CADL® (CADKEY Advanced Design Language). "You can access ACTVIEW while you are working in CADKEY, and it displays an indented parts list," Ace said. "You use the ARROW keys to navigate through the parts list to find individual files. To select a particular file,

you press the space bar. You can select more than one file at a time. Pressing the ENTER key causes ACTVIEW to transfer all the tagged pattern files into your CADKEY work session. The system displays all of the geometry in your part file in its correct location and orientation, and you resume your work exactly where you had been."

"ACTVIEW has many useful features," Ace added. "One of the most useful is the ability to archive all previous revisions of a part, along with a log file describing the changes, into a single compressed file. ACTVIEW allows you to extract any one of the old revisions and to bring it back into CADKEY."

Manufacturing the Phaser III Color Printer required a considerable amount of tooling, molds for plastic parts. "Our suppliers made a number of the molds by taking IGES files from us and transferring them into their CAM systems," Ace said. "They were able to machine the molds directly from our design data. The largest and most complex molds were done without paper drawings."

Adapted with permission from **3-D World**, *v. 6, no. 3, 1992.*

THE VERIFY FUNCTION

The **VERIFY** option determines the various geometric calculations of an object. This option can determine geometric calculations of circles, polygons and general closed figures.

VERIFY OPTION: PERIM (Perimeter)

The **PERIM** option measures the actual and projected length of a single or selected group of entities. The projected and actual perimeter lengths differ only if the entities are not in the current display plane. There are two shapes that can be measured with the **PERIM** option: **CIRCLE** and **GENERAL**. The **CIRCLE** option measures circles and arcs. **GENERAL** measures a continuous string of line or curve entities.

CIRCLE

Figure 19-2A and the following steps show how to measure the perimeter of a circle.

1. Select **ESC-F7-F1-F5-F1 (CONTROL-VERIFY-PERIM-CIRCLE)**.

2. A prompt reads:

   Select arc/circle.

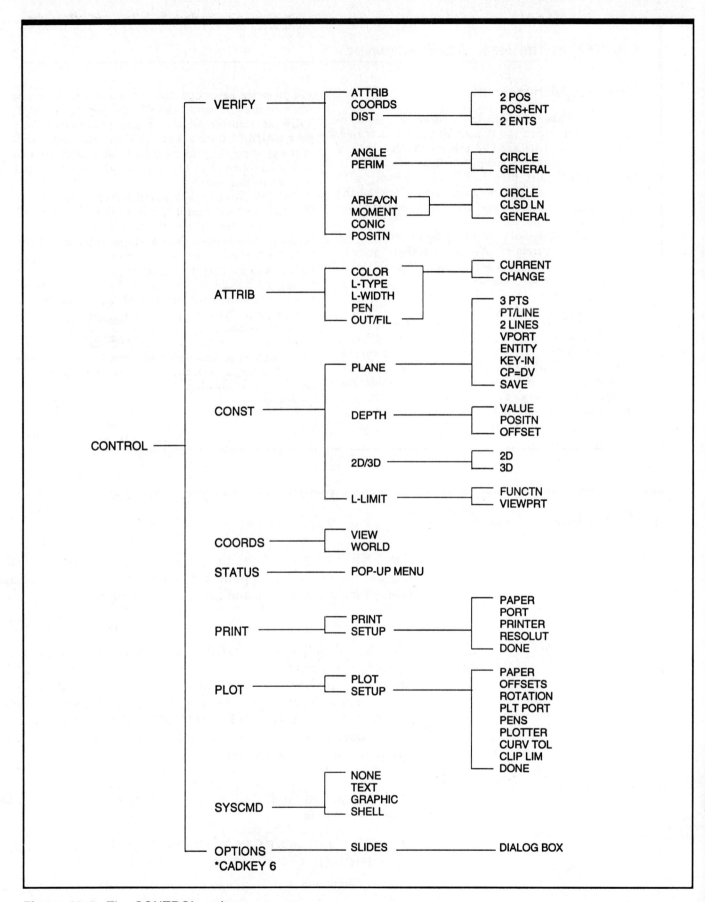

Figure 19-1 The CONTROL options

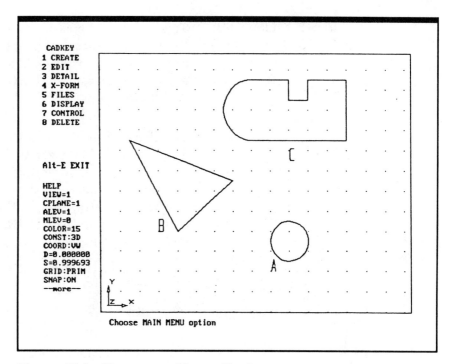

Figure 19-2 Shapes used for verification

Pick the circle to be measured by placing the cursor over the entity and pressing the mouse button.

3. A prompt lists the perimeter values:

   **Proj length=3.141593 Actual len=3.141593
   (Press RETURN).**

4. Press **ESC** to exit this option.

GENERAL When using this option, the Selection Menu is used to identify those entities to include in the calculations. Figure 19-2B and the following steps show how to use the **GENERAL** option.

1. Select **ESC-F7-F1-F5-F2 (CONTROL-VERIFY-PERIM-GENERAL).**

2. The Selection Menu is displayed. Select **F3 WINDOW** and pick the lower left and upper right corners of a rectangle that surrounds the object shown in Figure 19-2B.

3. A prompt lists the perimeter values:

   **Proj length=9.05531 Actual len=9.05531
   (press RETURN).**

4. Press **ESC** to exit this option.

This option is used for oddly shaped objects like the one in Figure 19-2C. The **WINDOW** option is used to select all the entities making up the part. For the object displayed in Figure 19-2C, the projected and actual lengths are equal to 8.856194.

VERIFY OPTION: AREA/CN (Centroid)

This option will measure the area of a closed figure and will locate the X, Y coordinate value of the centroid. The centroid is marked on screen with an asterisk-like (*) symbol. Three options define the area to be measured: **CIRCLE, CLSD LN** (Closed Line), and **GENERAL**.

CIRCLE

The **CIRCLE** option identifies an arc or circle and finds its area and centroid. If an arc is selected, a temporary dashed line is drawn between the endpoints of the arc to form a closed area. The area is then displayed in the prompt. The asterisk-like marker for the centroid is then displayed on screen. This marker is removed using the **REDRAW** option or **CTRL-R**. Figure 19-3A and the following steps show how the **CIRCLE** option is used.

1. Select **ESC-F7-F1-F6-F1 (CONTROL-VERIFY-AREA/CN-CIRCLE)**.

2. A prompt reads:

   Select arc/circle.

   Pick the circle to be measured by placing the cursor over the entity and pressing the mouse button.

3. A prompt lists the area:

   Area=0.785398 (press RETURN).

   Press **RETURN** to locate the centroid.

4. An asterisk displays the location of the centroid on screen (Figure 19-3A), and the coordinate values are listed in the Prompt Line:

   XV=6.0 YV=1.5 (press RETURN).

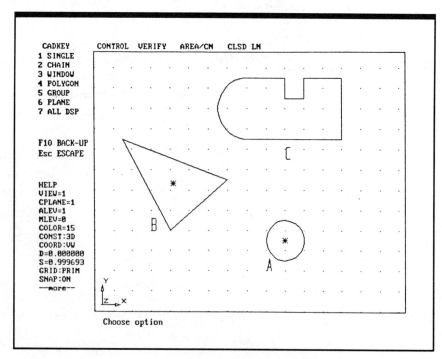

Figure 19-3 Area and centroid located using AREA/CN

5. To exit this option press **RETURN** and **ESC**.

CLSD LN   The **CLSD LN** (Closed Line) option is used to determine the area of a set of lines that define a closed figure. The line's entities for the defined area are identified using the Selection Menu. If the line entities do not make a closed area, a prompt reads: **Invalid data...press RETURN to continue**. If a closed area is selected, the area is displayed in a prompt and an asterisk-like mark is displayed on screen along with the coordinate values for the centroid. The following steps and Figure 19-3B illustrate the **CLSD LN** option.

1. Select **ESC-F7-F1-F6-F2 (CONTROL-VERIFY-AREA/CN-CLSD LN)**.

2. The Selection Menu is displayed. Select **F3 WINDOW** and pick the lower left and upper right corners of a rectangle that surrounds the object.

3. A prompt lists the area value:

    Area=3.625.

4. A second prompt is used to locate the centroid:

    Centroid: XV=3.166667 YV=2.833333.

    The centroid is marked on screen with a temporary asterisk.

5. Press **ESC** to exit this option and **CTRL-R** to clear the temporary markers.

You can use **CLSD LN** to determine the center of closed polygons for construction of new entities.

GENERAL   This option defines a set of lines and arcs that form a closed figure to determine the area and centroid. The Selection Menu is used to select the entities for the calculations. If the figure does not form a closed area, an invalid data prompt appears. CADKEY approximates the area for shapes that are not square or rectangular by filling the area with rectangular shapes called panels (Figure 19-4C). These rectangular panels are used for calculating the area. The width of the panels will determine the accuracy of the calculations and CADKEY allows you to control the panel width. The default setting is .25 and can be changed when using the **GENERAL** option, which is described in the following steps. The area and centroid for Figure 19-3C are determined using the following steps.

1. Select **ESC-F7-F1-F6-F3 (CONTROL-VERIFY-AREA/CN-GENERAL)**.

2. The Selection Menu is displayed. Select **F3 WINDOW** and pick the lower left and upper right corners of a rectangle that surrounds the object.

3. A prompt reads:

    Enter panel width interval (0.250000) =.

Use the default setting by pressing **RETURN**.

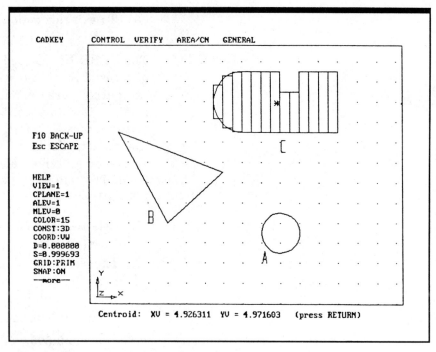

Figure 19-4 Verifying the area and centroid of a GENERAL figure

4. The panel pattern is displayed and a prompt reads:

   Area=3.555598 (press RETURN).

   Press **RETURN** to determine the location of the centroid.

5. A prompt reads:

   Centroid: XV=4.926311 YV=4.971603 (press RETURN).

   A temporary marker is displayed for the centroid location (Figure 19-4).

6. Press **RETURN** and **ESC** to exit this option and **REDRAW** to clear the markers.

   The following steps show how to change the panel width and how that affects the area calculation and the centroid location.

1. Select **ESC-F7-F1-F6-F3 (CONTROL-VERIFY-AREA/CN-GENERAL).**

2. The Selection Menu is displayed. Select **F3 WINDOW** and pick the lower left and upper right corners of a rectangle that surrounds the object.

3. A prompt reads:

   Enter panel width interval (0.250000) =.

   Enter **.1** and press **RETURN**.

4. A prompt will display the area as:

   Area=3.642605 (press RETURN).

   Notice the slight difference in areas between the two examples. Press **RETURN** to determine the location of the centroid.

5. A prompt reads:

> **Centroid: XV=4.93123 YV=4.97197 (press RETURN)**.

A temporary marker is displayed for the centroid location (Figure 19-5).

6. Press **RETURN** and **ESC** to exit this option.

VERIFY OPTION: MOMENT

The **MOMENT** option determines the polar moment of an enclosed region relative to the axis perpendicular to the screen. The user defines this axis with the Selection Menu. **MOMENT** employs three options: **CIRCLE, CLSD LN,** and **GENERAL.** The same restrictions and use of panels described for the **AREA/CN** option apply to the **MOMENT** option.

CIRCLE

Figure 19-3A and the following steps show how to use the **MOMENT-CIRCLE** option.

1. Select   **ESC-F7-F1-F7-F1   (CONTROL-VERIFY-MOMENT-CIRCLE).**

2. A prompt reads:

> **Select circle**.

Pick the circle with the cursor.

3. A second prompt reads:

> **Indicate position for axis**.

The Position Menu is displayed. For this example, pick the center of the circle using **F4 CENTER.**

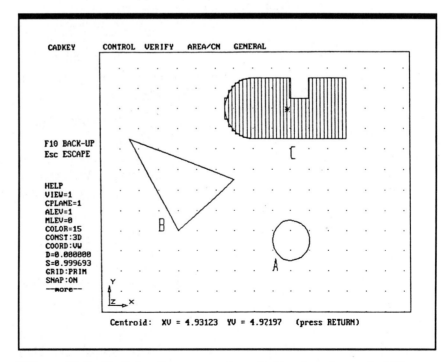

Figure 19-5 Verifying the area and centroid using .1 panel width

4. A prompt reads:

**Moment of inertia: Polar (Ix + Iy)=0.098175
(press RETURN).**

5. Press **RETURN** and **ESC** to exit this option.

CLSD LN This option determines the moment of a closed area formed by lines using a selected axis normal to the screen. Figure 19-3B and the following steps show how to use this option.

1. Select **ESC-F7-F1-F7-F2 (CONTROL-VERIFY-MOMENT-CLSD LN).**

2. The Selection Menu is used to identify the lines. For this example use **F3 WINDOW** and pick the lower left and upper right corners of a rectangle that includes all of Figure 19-3B.

3. A prompt reads:

Indicate position for axis.

The Position Menu is displayed. For this example, pick the lower corner of the triangle in Figure 19-3B using **F3 ENDENT.**

4. A prompt reads:

**Moment of inertia: Polar (Ix + Iy)=9.364583
(press RETURN).**

5. Press **RETURN** and **ESC** to exit this option.

GENERAL The **GENERAL** option determines the moment for a closed area formed by lines and arcs. **GENERAL** calculates the moment using orthogonal axes through the selected polar position. CADKEY uses panels to approximate the area for calculating the moment. Figure 19-3C and the following steps show how to use the **GENERAL** option.

1. Select **ESC-F7-F1-F7-F3 (CONTROL-VERIFY-MOMENT-GENERAL).**

2. The Selection Menu is used to identify the lines. For this example, use **F3 WINDOW** and pick the lower left and upper right corners of a rectangle that includes all of Figure 19-3B.

3. A prompt reads:

Enter panel width interval (0.25).

For this example, press **RETURN.**

4. A prompt reads:

Indicate position for axis.

The Position Menu is displayed. For this example pick the center of the arc in Figure 19-3C.

5. The panel pattern is displayed and a prompt reads:

Moment of inertia: Polar (Ix + Iy)=4.528041 (press RETURN).

Press **RETURN** to display other calculations.

6. Another prompt is displayed:

Moment of inertia: Ix(YV=5.5)=0.594064 (press RETURN).

Press **RETURN** to display more calculations.

7. A prompt reads:

Moment of inertia: Iy(XV=4.25)=3.93397 (press RETURN).

8. Press **RETURN** and **ESC** to exit this option.

VERIFY OPTION: CONIC

The **CONIC** function displays detailed information about a selected conic which CADKEY defines as an ellipse, hyperbola, or parabola. Verification data are displayed in the Prompt Line. Use **CTRL-R** to remove the temporary markers used to display the data. Twenty-three pieces of data are displayed for an ellipse, 25 for a hyperbola, and 13 for the parabola. Typical data extracted from the entity include: quadratic coefficients, start and end points, axes start and end points, and foci and centers. To use this option enter **ESC-F7-F1-F8**. Select the conic and the data are displayed in the prompt line and pressing **RETURN** displays each new line of information.

VERIFY OPTION: ATTRIB (Attributes)

The **ATTRIB** function determines various values assigned to entities, such as level number, view number, color, width, and so forth. After the entity is identified by picking, the values are displayed in the prompt area of the screen. The following steps show how this option is used.

1. Select **ESC-F7-F1-F1 (CONTROL-VERIFY-ATTRIB)**.
2. The prompt reads:

Select entity to verify.

Pick the entity. For this example a horizontal line is picked.

3. The values are listed in the prompt as:

Type=2 Form=0 Level=1 Def view=1 (press RETURN).

Press **RETURN** to view the other values.

4. A second prompt reads:

Color=1 Style=1 Width=1 Pen=1 (press RETURN).

Press **RETURN** to view other values.

5. A third prompt reads:

Unique ID = ID (press RETURN).

6. Press **RETURN** and **ESC** to exit the option.

TYPE is a reference number assigned to the entity from the data base.

FORM is a reference number assigned to the entity from the data base.

LEVEL is the level that the entity is assigned.

DEF VIEW is the number of the view in which the entity was created.

COLOR is the color number assigned to the entity.

STYLE is the line type or font value assigned to the entity.

WIDTH is the line width assigned to the entity.

PEN is the number of the pen assigned to the entity.

If the entity belongs to a group, a fourth prompt line reveals the entity's group name, group number, and subgroup number.

VERIFY OPTION: COORDS (Coordinates)

The **COORDS** option displays the coordinate data for a selected entity. After the entity is picked, coordinate values are displayed in the prompt line. X, Y, Z coordinate values for view and world coordinates are listed along with the entity's actual length (true length or normal) and projected length (length in the present view). For arcs, the coordinate values for the center of the arc, sweep angle, radius, starting angle, ending angle, start and end-points are displayed.

The **COORDS** option is selected by entering **ESC-F7-F1-F2 (CONTROL-VERIFY-COORDS)**. Use the cursor to pick the entity to verify. The coordinate information is displayed in the prompt line. Pressing **RETURN** displays more information and **F10 BACKUP** re-displays values.

VERIFY OPTION: DIST (Distance)

This option determines the distance between any two points or positions on screen (**2 POS**), between a position and an entity (**POS+ENT**), and between two entities (**2 ENTS**). The **VERIFY DIST** option is comparable to using a scale with traditional tools to measure a distance before or after an entity is placed on a drawing. The two positions are picked and the distance is displayed in the prompt line.

Figure 19-6 shows how the **POS+ENT** option is used to find the minimum distance between a shaft and a plane.

1. Select **ESC-F7-F1-F3-F3 (CONTROL-VERIFY-DIST-2 ENTS)**.

2. A prompt reads:

    Select 1st entity.

3. Pick a line on the plane closest to the shaft.

4. A prompt reads:

    Select 2nd entity.

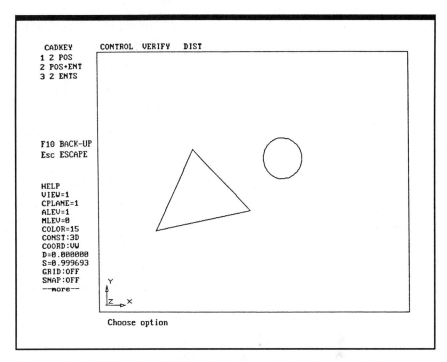

Figure 19-6   Using the DIST option to determine the shortest distance

Pick the circle on the end of the shaft. A perpendicular dashed line is drawn between the selected line and the circle and the distance is displayed in the prompt area:

Minimum projected distance=1.006149 (press RETURN).

5.   A second prompt reads:

Projected pt 1: xv=3.784455 yv=2.715545 (press RETURN).

6.   Another prompt reads:

Projected pt 2: xv=4.49591 yv=3.427 (press RETURN).

7.   The next prompt reads:

Minimum actual distance =1.006149 (press RETURN).

8.   Four more prompts display the actual X, Y, and Z for points 1 and 2.

The markers and the dashed line are removed with **REDRAW**. Figure 19-7 shows how the **2 ENT** option is used and shows the Prompt Line with the calculated distance.

VERIFY OPTION: ANGLE   This option is similar to the **VERIFY DIST** option except it is used to measure an angle. This is done by selecting the two lines that form the angle. The angle is then displayed in the prompt line. Select **ESC-F7-F1-F4 (CONTROL-VERIFY-ANGLE)**. A prompt reads: **Select first line**. Pick the first line to define the angle. A second prompt reads: **Select 2nd line**. Pick the second

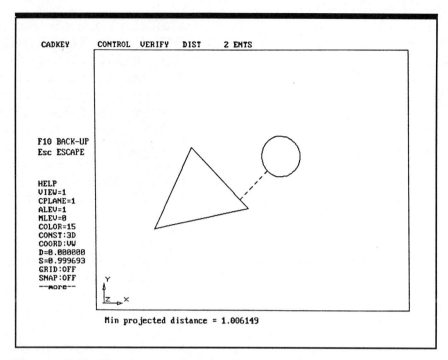

Figure 19-7 The shortest distance between a line and circle

line that defines the angle. A prompt displays the measured angle as: **Angle in display plane – 45 (press RETURN)**. Press **RETURN** to display a second prompt that reads: **Angle in plane of lines = 45 (press RETURN)**. This second prompt is measuring the true angle of the lines in three-dimensional space. The selected lines are drawn in the display plane so both angles are the same.

VERIFY OPTION: POSITN (Position)

The **VERIFY-POSITN** option determines the location of a point, such as the endpoint of a line or center of a circle. This option will show in the Prompt Line the world coordinate position of the selected point, the view coordinates, and the construction coordinates. Select **ESC-F7-F1-F9 (CONTROL-VERIFY-POSITN)**. A prompt reads: **Indicate position to verify** and the Position Menu is displayed in the menu area. To determine the position of a line endpoint select **F3 ENDENT** and pick the point. The coordinate values are shown on screen in a series of three prompts that are displayed by pressing **RETURN**.

THE ATTRIB (Attribute) FUNCTION

This **ATTRIB** option should not be confused with the **VERIFY-ATTRIB** option just described. **CONTROL-ATTRIB** is used to change the **CURRENT** modal attributes. In other words, the

CURRENT option is used to change the current **COLOR, LINE TYPE, LINE WIDTH, PEN #** and **OUT/FIL** (Outline/Fill). It is also possible to change the attribute values of existing entities. For example, you can change the line type from solid to hidden without deleting and redrawing the entity.

The **CONTROL-ATTRIB** option for CADKEY 6 has been rewritten so that when the user selects **CHANGE** or **CURRENT** a dialog box (Figure 19-8) is then displayed. All of the attribute options are found in this one dialog box. Now the user can change any one or all four attributes (line style, line width, color, and pen number) at one time. After identifying the attributes to be changed, the user then selects the entities to be changed.

ATTRIB OPTION: COLOR

This option changes the color of entities within the limits of the color palette offered with your display device. This option is selected using **ESC-F7-F2-F1 (CONTROL-ATTRIB-COLOR)** or the Immediate Mode command **ALT-X** or the Status Window command **COLOR.** There are two options: **CURRENT** and **CHANGE.** The **CURRENT** option selects the setting by keyboard input or picking from a menu (Figure 19-9). All entities drawn after the **CURRENT** option is selected will take on the new attribute. The **CHANGE** option assigns a new color to existing entities.

ATTRIB OPTION: L-TYPE (Line Type)

This option changes the line type of lines, arcs, circles, splines and witness lines. This option is assigned the Immediate Mode command **ALT-T.** When the **L-TYPE** option is selected, a menu of the four line types is displayed in the menu area of the screen (Figure

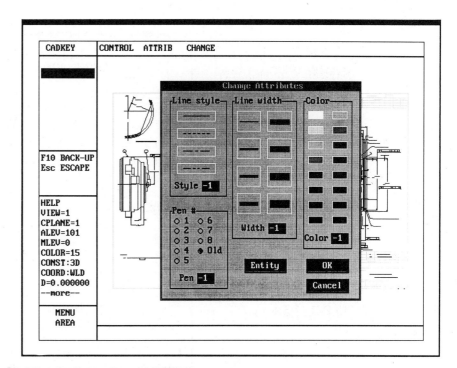

Figure 19-8 The CONTROL-ATTRIB-CHANGE dialog box for CADKEY 6

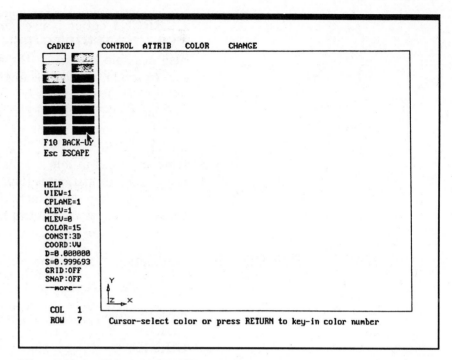

Figure 19-9 The COLOR-CHANGE menu

19-10). The new line type can be cursor-selected from the menu or **RETURN** can be pressed; a new menu is displayed in the menu area on screen and the line type can be selected from that menu. The current modal value can be changed using the **CURRENT** option, or existing entities can be changed using the **CHANGE** option or Status Window **L-TYPE**.

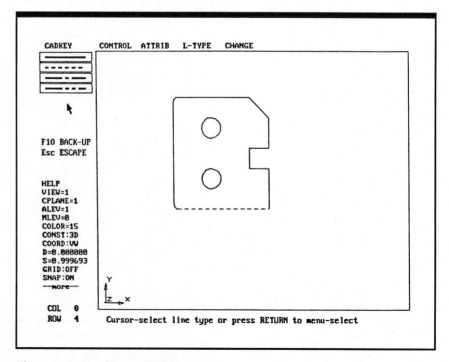

Figure 19-10 The L-TYPE menu

POSSIBLE USE

The **L-TYPE CHANGE** option is used to edit or make changes in existing entities. For example, if a line is drawn using the wrong line style or type this option is used to change the line style without having to redraw the entity. Figure 19-10 shows a drawing of a part with the base line drawn as a dashed line instead of a solid line. Figures 19-10 and 19-11 and the following steps demonstrate how to use this option to change an existing line to a different style.

1. Select **ESC-F7-F2-F2 (CONTROL-ATTRIB-L TYPE)**.
2. Select **F2 CHANGE** from the menu.
3. A menu of available line types is displayed in the menu area of the screen. Cursor-select the new line type from the menu by moving the cursor over the **SOLID** line and pressing the mouse button or space bar.
4. The Selection Menu is displayed. Select **F1 SINGLE** from the menu.
5. A prompt reads:

   Select entity 1 (press RETURN when done).

   For this example, pick the hidden line on the bottom of the part displayed in Figure 19-10.
6. Press **RETURN**. The line will immediately be changed to a solid line (Figure 19-11).
7. Press **ESC** to exit this option.

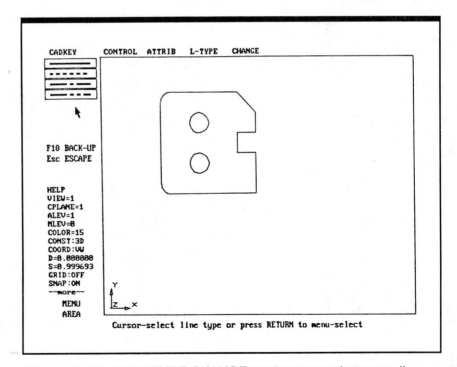

Figure 19-11   The L-TYPE-CHANGE pop-up menu changes a line

ATTRIB OPTION: L-WIDTH
(Line Width)

This option changes the width of lines, arcs, splines, witness lines, and arrows. The available thicknesses are the odd numbers from 1 (the default value) to 15 pixels (Figure 19-12). Two options are available: **CURRENT** and **CHANGE**. The Immediate Mode command **ALT-Y** can also be used to change the line width. The Status Window will display the current line width value in the **L-WIDTH** prompt and can also be used to change the line width.

POSSIBLE USE

In this example, all the visible lines in Figure 19-11 are changed from 1 to 3 line width using the **L-WIDTH** option. The following steps show how this is accomplished.

1. Select **ESC-F7-F2-F3 (CONTROL-ATTRIB-L WIDTH)**.
2. Select **F2 CHANGE**.
3. A line width menu is displayed (Figure 19-12). Pick the second line with the cursor.
4. The Selection Menu is displayed. Select **F1 SINGLE** and pick the solid lines; then press **RETURN**. All of the solid lines are drawn as line width number 3 (Figure 19-13).

ATTRIB OPTION: PEN

The **PEN #** option changes the current modal value of the pen or changes the pen number assigned to an existing entity. Entities

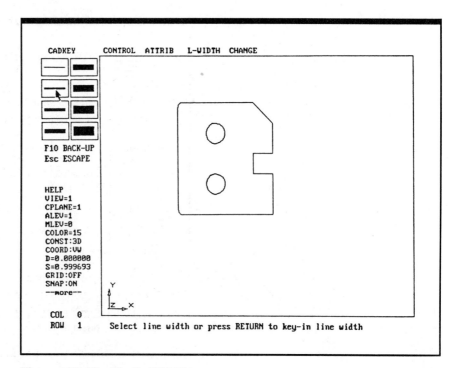

Figure 19-12 The L-WIDTH menu

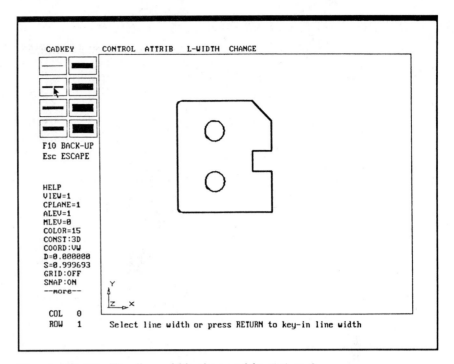

Figure 19-13   The line width changed from 1 to 3

can be assigned pen numbers 1 through 8. To change the current modal value use the **CURRENT** option or the Immediate Mode command **ALT-Z**. To define a new pen number to existing entities use the **CHANGE** option. The Status Window displays the current pen number assignment in the **PEN** prompt and can be used to change the pen number assignment. When changing the pen number of existing entities, the Selection Menu is used to select those entities to be changed.

ATTRIB OPTION: OUT/FIL (Outline/Fill)

OUT/FIL (Outline/Fill) changes polygons or polylines to outline or filled form. Filled polygons or polyline fill patterns are changed from filled to outline or vice versa. The following steps and Figure 19-14 describe how to use **OUT/FIL**.

1. Enter **ESC-F7-F2-F5 (CONTROL-ATTRIB-OUT/FIL)**.
2. **OUTLINE** and **FILLED** are displayed in the menu area. Select **F2 FILL**.
3. A prompt reads:

    Cursor select color or press RETURN to key in color number.

    A pop-up menu displaying the available colors is displayed across the top of the drawing area (Figure 19-13). (The color palette is dependent on the colors supported by the system's graphics card.) Pick a color from the menu with the cursor or enter **RETURN**. Entering **RETURN** displays the prompt:

    Enter color number from 1-255=.

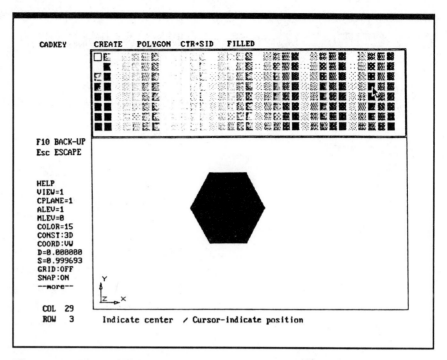

Figure 19-14 A filled polygon and the pop-up menu

Enter the color number from the keyboard.

4. The Selection Menu appears on screen. Select F3 WINDOW.
    Pick two points that surround the filled polygon or polyline.
    The color of the polygon or polyline fill is changed (Figure
    19-15). The steps are the same to change from outline to filled.

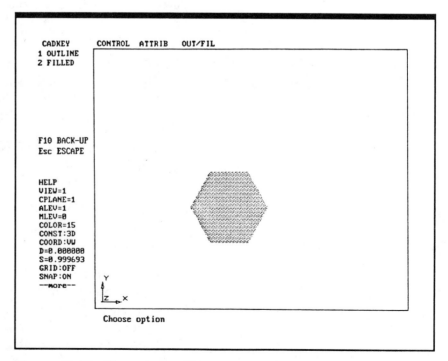

Figure 19-15 The polygon fill pattern changed

THE CONST (Construction) FUNCTION

The construction function controls working depth, alternate construction frames, and construction lines.

CONST OPTION: L-LIMIT
(Line Limit)

This option creates lines to the current window limits, instead of limiting construction to the designated endpoints. Two options are available: **FUNCTN** (function) and **VIEWPORT**. For example, in the default setting of **FUNCTN**, drawing lines using the **ENDPTS** option will draw a line between the two points picked. However, if the line limit is set for **VIEWPORT**, using the **LINE-ENDPTS** option causes a line to be drawn the full extent of the viewport and not between picked points. **ALT-L** is the immediate mode command for **L-LIMIT**.

This is represented in Figure 19-16. The top line was drawn using the **FUNCTN** line limit option where the line is drawn between the two picked points using the **ENDPTS** option. The bottom line was drawn using the **VIEWPORT** line limit option where the line is drawn to the extents of the viewport and not between the picked points. Using the **SCREEN** limits options is very useful for transferring points from one view to another in multiview drawings. **LINES** are used as construction lines on CADKEY in much the same way as they are used with traditional hand tools.

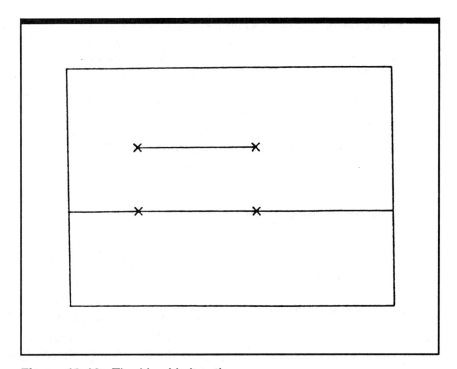

Figure 19-16 The Line Limit option

To change the current line limit setting press **ESC-F7-F3-F4** (**CONTROL-CONST-L LIMIT**) or the Immediate Mode command **ALT-L**. A prompt reads:

Choose LINE LIMITS mode (current=FUNCTN).

Select **F2 VIEWPORT** to change the setting or **F1 FUNCTN** or **RETURN** to remain the same. Press **ESC** to exit this option. After the line limit is changed it will remain in that setting until it is changed again.

CONST OPTION: 2D/3D

CTRL-W is the immediate mode for **2D/3D**. The **2D/3D** option controls the depth placement of the geometric entities **LINE**, **ARC/CIRCLE**, **RECTANGLE**, **POLYGON**, **SPLINE**, **POINT**, **POSITN**, and **CADL** point, and line coordinate output. The Status Window displays the current setting for this option as: **CONST: 2D** or **CONST: 3D**. The switch is set at 3D at system start-up. To change the current setting press **ESC-F7-F3-F3** (**CONTROL-CONST-2D/3D**) or Immediate Mode command **CTRL-W**. A prompt is displayed:

Choose mode for construction (current=3D).

If 2D construction mode is in effect, a line, for example, is created at the current working depth. The **DEPTH** option is explained later in this chapter. If the 3D mode is in effect, the same line is created at the depth of the reference position.

CONST OPTION: PLANE

ALT-K is the immediate mode for **PLANE**. The **CONST-PLANE** (construction plane) option creates new construction planes, saves construction planes, and turns existing construction planes on or off. A construction plane allows the creation of entities in a plane other than the current display plane. The Immediate Mode command **ALT-K** activates the function.

Any new construction plane created must be saved if you wish to use it again. The eight default views are the eight default construction planes. To add construction planes to the list they must be created and saved.

There are seven methods that can be used to create a construction plane:

3 PTS—Locate three points using the Position Menu. The positive direction for the X-axis is from the first position chosen through the second. The Y-axis direction is located by selecting a third point.

PT/LINE—The construction plane is created by locating a point and two endpoints of a selected line.

2 LINES—The endpoints of two lines and the intersection of the two lines form a construction plane.

VPORT—Automatically assigns the viewport's view as a construction plane.

ENTITY—An existing planar arc, circle, conic or 2D spline can be selected to create a construction plane.

KEYIN—A new construction plane is created by entering the rotation angle and rotation order using the options **TOP IN, TOP OUT, RGT IN, RGT OUT, CW,** and **CCW.**

CP=DV—Sets the construction plane to the display view.

SAVE—Saves the current construction plane.

 POSSIBLE USE

Figure 19-17 shows an object with an oblique plane. One way to create a circle on the surface of the plane with CADKEY would be to create a new view that is perpendicular to the oblique surface using **DISPLAY-VIEW-NEW-3PTS**. The depth would then be set equal to the oblique surface using **CONTROL-CONST-DEPTH-POSITN**, as explained later in this chapter. The circle could then be located on the oblique surface.

A faster method of creating the circle on the oblique surface would be to set the construction plane parallel to the surface, as explained below.

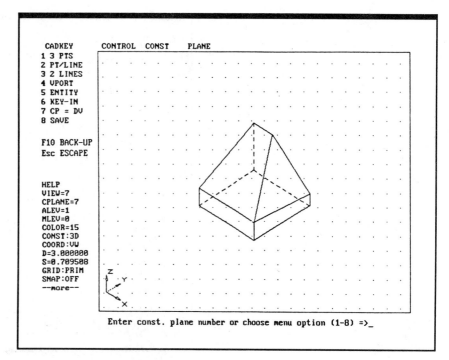

Figure 19-17   Creating a construction plane

1. Select **ESC-F7-F3-F1-F1 (CONTROL-CONST-PLANE-3 PTS)**.
2. From the Position Menu select **F3 ENDENT**. Select one of the lines in the oblique plane of Figure 19-17.
3. Select the other end of the line selected in Step 2.
4. Select another line in the oblique view. The new construction plane is created and the construction plane axis is displayed in the upper right corner of the viewport (Figure 19-18).

If a grid is turned on, it will be displayed parallel to the construction plane (Figure 19-19). Any geometry created will be drawn in the construction plane, as is the circle shown in Figure 19-19.

CONST OPTION: DEPTH

CTRL-D is the immediate mode for **DEPTH**. The **DEPTH** function assigns the current working depth for a drawing. For example, the working depth at system start-up is 0. However, if you decide to change the depth assignment for those entities being added to the drawing, it can be changed from 0 to 1, using the **DEPTH** function. There are three options used with the **DEPTH** function: **OFFSET**, **POSITN** (Position) and **VALUE**. The **DEPTH** function can be assigned using the Immediate Mode command **CTRL-D**. The Status Window can be used to view the current **DEPTH** and to change the depth option. It is displayed under the **D** heading.

The **POSITN** option defines a depth position in space. The Position Menu defines the position in space for the new working

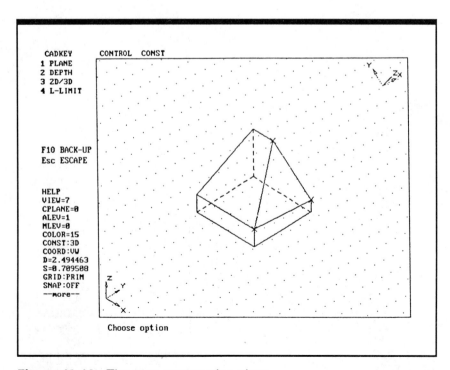

Figure 19-18 The new construction plane

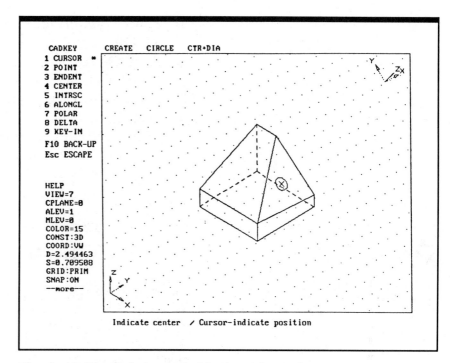

Figure 19-19 A circle created in the oblique plane

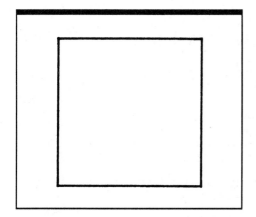

Figure 19-20 The rectangle used for the DEPTH option

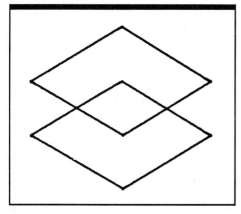

Figure 19-21 The second rectangle at a depth of 1

depth. However, option number 1 in the Position Menu is disabled because using the **CURSOR** position would be the same as indicating the current working depth. Before selecting any of the remaining options from the Position Menu, the construction switch must be set at 3D or the position will always be returned to the current working depth.

The **OFFSET** option changes the working depth by adding a specified distance to the current working depth. For example, if the current depth is set at 2, the **OFFSET** option can change the depth to 5 by entering 3. The 3 is added to the 2, equaling 5.

The **VALUE** option sets a new working depth using values entered from the keyboard. Figures 19-20 and 19-21 and the following steps show how to change the working depth and how it affects entities added to a drawing. In this example, a rectangle has been drawn at the default depth of 0. The depth is changed to 1 and another rectangle is drawn over the top of the one currently displayed in Figure 19-21. The isometric view is displayed so that you can see the effects of changing the current working depth.

1. Select **ESC-F7-F3-F2-F1 (CONTROL-CONST-DEPTH-VALUE)**.

2. A prompt reads:

   Enter new depth (current=0) =.

   Enter the number **1** and press **RETURN**.

3. The Status Window prompt **D** will change from 0 to 1. Press **ESC** to exit this function.

4. Draw a rectangle over the top of the rectangle already displayed in Figure 19-18 in view 1 using **CREATE-LINE-RECTNGL**.

5. Change the view to number **7** using the Immediate Mode command **ALT-V**. The drawing is changed (Figure 19-21). Notice that there are two rectangles displayed, one drawn at a depth of 0 and another drawn at a depth of 1.

THE COORDS (Coordinate) FUNCTION

The **COORDS** option changes the mode for view or world coordinate entry. There are two options with this command: **VIEW**, **WORLD**. The **COORDS** option has also been assigned the Immediate Mode command **CTRL-V** and Status Window under **VIEW:**. **VIEW** assigns the coordinates entered for X as horizontal, Y as vertical, and +Z as pointing out of the screen. The **WORLD** option assigns the coordinates entered relative to view 1 (top view). This means that the X, Y, Z coordinate axis directions are independent of the current working view. The setting of this switch affects **DELTA** and **KEYIN** options from the Position Menu and the **DELTA** option from the **X-FORM** menu. Prompts requesting view coordinate values read: **XV= YV= ZV=**. Prompts requesting world coordinate values read: **X= Y= Z=**, eliminating the V from the prompt. To use this command press **ESC-F7-F4** (**CONTROL-COORDS**). A prompt reads:

Choose mode for coordinate entry (current=VIEW).

Press **F2 WORLD** to change to world coordinate values, or **RETURN** to remain with view coordinate values.

THE STATUS FUNCTION

The **STATUS** option determines the amount of memory in bytes left for data storage. This option lists the amount of internal RAM memory available in the computer. It will also list the amount of free disk space left for data storage. Both of these pieces of information are important in determining the amount of storage left while working on a drawing.

Selecting this option will display a pop-up window (Figure 19-22). The pop-up window shows the RAM and disk storage total, percent used, percent available, and percent capacity.

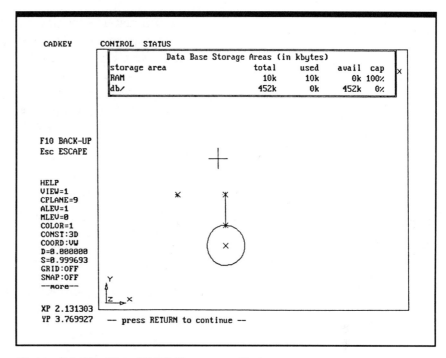

Figure 19-22   The STATUS pop-up window

THE SYS CMD (System Command) FUNCTION

The **SYSCMD** function performs DOS (Disk Operating System) operations without leaving the CADKEY program. It can also execute external programs, assuming enough RAM memory is available, without leaving CADKEY. For example, it is possible to execute a DOS command like DIR for a directory listing on a disk to determine the files and amount of space left on the disk without leaving CADKEY.

This command has some restrictions. There must be enough memory available for the DOS commands or programs. The amount of memory set aside for this is user specified, as assigned in the **CONFIG** program. This memory must be returned to the system after the **SYSCMD** option is complete. This means that no memory resident programs can be loaded by the **SYSCMD** option. Another restriction is that the **SYSCMD** option does not save the current part displayed. If the DOS program crashes the system, the part data will also be lost. So it is important that the part is saved before executing the **SYSCMD** option.

There are four options used with the **SYSCMD** function: **GRAPHIC, NONE, SHELL,** and **TEXT.** These options control the output of the **SYSCMD** functions or programs.

The **GRAPHIC** option stores the image before the program is run and is restored after the system is returned to CADKEY using the **ESC** key.

The **NONE** option suppresses the text output of the DOS command or program. The current graphic display is not changed or removed from the screen. This option does not display prompts from the DOS commands or programs.

The **SHELL** option stores the graphic image before the DOS shell is entered. After the command or program is completed, the DOS prompt is displayed. Another program can be run by typing in the name, or type **EXIT** to return to your place in the system.

The **TEXT** option functions according to your system set-up. If two screens are used, the graphic image is undisturbed and the alphanumeric terminal displays the text output. If one screen is used, the graphic image is stored and the text is displayed on the screen. Press **ESC** to return to the Main Menu and to view the graphic image. Figure 19-23 and the following steps show how to use the **TEXT** option.

1. Select **ESC-F7-F8-F2 (CONTROL-SYSCMD-TEXT)**.
2. The prompt line reads:

   Command.

   At the prompt enter DOS command **DIR A:** to list a directory of files on the disk located in the A drive.
3. The graphics image is stored and removed from the screen and replaced with a listing of the files in disk drive A (Figure 19-23).
4. To return to CADKEY press any key and **ESC** to return to the Main Menu.

```
Volume in drive A has no label
Directory of  A:\

19-1      PRT      9708     8-24-89    10:14p
19-5      PRT      8737     8-24-89    11:01p
19-12     PRT      8527     8-25-89     8:38a
19-13     PRT      8527     8-25-89     8:40a
3DBLOCK   PRT      9617     8-25-89     9:22a
19-15     PRT      9617     8-25-89     9:32a
19-16     PRT      9874     8-25-89     9:35a
19-15     HSG     23124     9-02-89     1:24p
19-16     HSG     21346     9-02-89     1:26p
19-17     HSG     24679     9-02-89     1:29p
        10 File(s)      223232 bytes free

-- press RETURN to continue --
```

Figure 19-23 Listing of files using the SYSCMD option TEXT

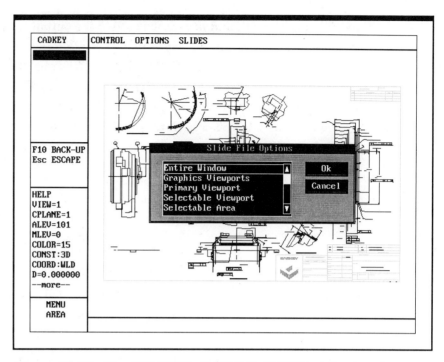

Figure 19-24   The CONTROL-OPTIONS-SLIDES dialog box for
CADKEY 6

Chapter 19 describes the use of the **CONTROL** function. However, two options, **PRINT** and **PLOT**, were not covered in this chapter because they are explained in detail in Chapter 20.

THE OPTIONS FUNCTION

In CADKEY 6, the command sequence **CONTROL-OPTIONS-SLIDES** defines portions of the display screen to be included in slide files created with the Immediate Mode command **ALT-F**. The slide files are saved in GIF format, which is portable to other software. The options include **Entire Window, Graphics Viewports, Primary Viewport, Selectable Viewport**, and **Selectable Area** (Figure 19-24).

REVIEW QUESTIONS

1.   Name the four geometric calculations that can be made using the **VERIFY** function.
2.   Describe how the centroid position is displayed on a drawing using the **AREA/CN** option.

3. What is a panel width and how does it affect the output of some geometric calculations?

4. List those attribute values displayed using the **VERIFY-ATTRIB** option.

5. How does the **CONTROL-ATTRIB** option differ from the **VERIFY-ATTRIB** option?

6. What is the difference between the **FUNCTN** and **L-LIMIT-VIEWPORT** options?

7. Describe the **STATUS** function and name the two outputs that result from executing this option.

8. Describe how the **CONST-PLANE** option is used to draw geometry on oblique surfaces.

9. List the four **SYSCMD** options.

Plotting and Printing Your Design

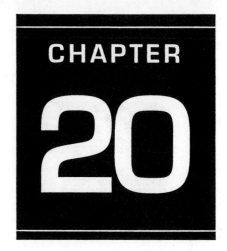

OBJECTIVES

After completing this chapter, you will be able to:

- Plot a CADKEY drawing.
- Print a CADKEY drawing.

The generation of hard copy output in CADKEY can be done in a number of different ways. CADKEY can output to a plotter or a printer. CADKEY can also plot and print a part from within the program or off-line with PLOTFAST, independent of the program. Depending upon your situation you may have a preference for one method over the other.

The on-line option allows you to plot or print the part that you currently have displayed on the screen. Through the use of the **CONTROL** function you can send the file that you are currently working on to an output device or to any other configurable device you have specified in the configuration program.

The option to print or plot outside of CADKEY is done through the off-line plotting program PLOTFAST. This program uses plot files created in CADKEY to generate a hard copy. The files can be scaled and X and Y offsets adjusted before they are sent to the output device.

THE PLOT FUNCTION

PLOT OPTION: PLOT

Auto   The plot option **AUTO** automatically sets the plotting scale factor. It sends the entities displayed on the screen to a designated plotting device. The entities are then plotted using an automatic scaling factor and X and Y offsets. This option can be executed quickly using the Escape Code **ESC-F7-F7-F1-F1 (CONTROL-PLOT-PLOT-AUTO)**.

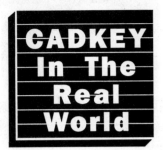

CT Scan, CADKEY and Stereolithography Team-Up for Hip-Replacement Surgery!

An accurate physical model of the femur, created from CAD data based on computer tomography scans (CT scans), offers new opportunities to surgeons preparing for hip-replacement surgery. Recent collaboration among three companies—3D Systems, Inc. of Valencia, California, developer of the StereoLithography Apparatus™; Dimensional Medicine, Inc. of Minnetonka, Minnesota, pioneers in Medical Imaging; and Cadkey, Inc.,—produced a physical model of a section of the human femur for a mutual client who manufactures hip prostheses.

A prosthesis is an artificial device used to replace a missing or damaged part of the body. An accurate model of the femur helps surgeons in their pre-surgical planning to choose and prepare ahead of time the prosthesis to be implanted during the hip-replacement operation. If successful, such a model would fill a need for improved results and patient comfort. Up to and including the present time, a hip-replacement operation must be interrupted while the surgeon selects a prosthesis and fits it to the patient.

CT Scan Data to CADKEY

The process of creating this model of the human femur began with data obtained from a real patient's CT Scan. The data consisted of 75 CT *slices,* or images, of the patient's femur taken 1.5 millimeters apart. An operator input the data into Dimensional Medicine's Maxview™ workstation. The Maxview generated contours representing the anatomy at each slice position. The contour points were then translated into a two-dimensional CADKEY® spline for each slice. The output was a CADL™ (CADKEY Advanced Design Language) file consisting of closed splines with a constant Z value, which was transferred to a CADKEY system.

Wire-Frame to Solid Model

The next step converted this wire-frame model to a solid model using CADKEY® SOLIDS. The only modification needed in the wire-frame model was the insertion of line segments between the starting points of each closed spline. Proceeding from top to bottom, closed splines with a specific Z value were joined to closed splines with the same Z value below. To run CADKEY SOLIDS, the user defines the type of solid rendering desired, and creates an output geometry file. Because 3D Systems' stereolithography software accepts only triangular polygons, it was necessary to edit the output geometry file to include this constraint as its first command. The file ran successfully through CADKEY SOLIDS.

CAD Model to Physical Prototype

The solid model was translated by the CADKEY/3D Systems Translator. This converted the solid-model file from CADKEY's .CDL format into 3D Systems' .STL format.

Manufacturing a physical model by stereolithography requires that the .STL file be processed by a *slicing* program that creates contours of the file's triangular polygons on constant Z planes. The slicing program requires that the user define the thickness of the planes or slices that it will produce in its .SLI output file. For this model, the thickness was set at .010 inch. It is the .SLI file that actually controls the laser beam which solidifies the liquid polymer in the StereoLithography Apparatus to manufacture the physical model. Six physical models were produced in 23 hours' time.

The client who requested this experiment was very impressed with the physical model's high degree of accuracy. Traditional manufacturing techniques had never been successful in producing such an accurate model of a bone, especially with repect to the interior bone-marrow canal.

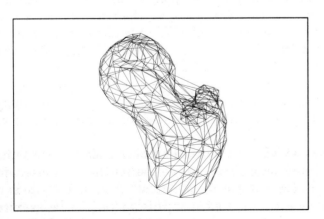

CADKEY 3-D model of hip bone.

—Continued, next page

CADKEY In The Real World—Continued

The successful merging of CT Scan data of a patient's femur with CADKEY and stereolithography to manufacture the physical model of the bone left the participants in this experiment encouraged about where pioneering applications such as this will lead.

Adapted with permission from 3-D World, v. 6, no. 1, 1992.

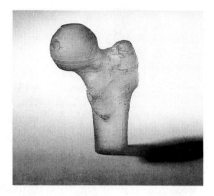

Stereolithographic model of hip bone.

When this option is selected you are prompted as to whether to display the scale factor and the X and Y offset. The **NO/YES** menu is displayed for your selection.

If **F2 YES** is selected you are then given the option to change the scale factor and the X,Y offset values. To accept the default values in parentheses press **RETURN**. To change the values enter the desired values and press **RETURN**. If **NO** is selected the default values are automatically accepted.

The automatic default scale is calculated using the current scale factor of the displayed part and the plot paper size that is specified in the **CONFIG** program. The X and Y offset values are those that were assigned in the config program. With the offset values properly assigned for your plotter, the part should automatically be centered on the plotting media.

The **PEN#/COLOR** menu is then displayed for your selection. Choose the option by which you wish to plot.

PEN# allows you to plot according to the pen number that you have assigned to each entity during the creation of the part. The pen number attribute in CADKEY only allows you to assign eight different pens. If you have a plotter that has more than eight pens, you should use the **COLOR** option.

The **PEN#** option is useful when you are working with a monochrome display and have a multi-color plotter to use. By assigning different pen numbers to the entities as they are created, you can plot your part in color.

COLOR allows you to plot colors using different pens that have been assigned according to the color that the entity is in the part. With this option it is possible to control fifteen different pens on the plotter, one for each color that is offered by CADKEY. A pen number has been designated for each of the fifteen available colors using the **CONFIG** program option to map graphics colors to plotter pen number. The default map will assign colors 1-15 to pen numbers 1-15, respectively. If you have an eight-pen plotter, you may wish to change this map to assign colors 9-15 to pens 1-8. Refer to the "Getting Started" section of the *CADKEY Users Reference Manual* for information on the **CONFIG** program.

Next, the universal Selection Menu is displayed, allowing you to plot only those entities that you have selected or All Displayed entities.

Once all of the values have been entered you are then prompted to press the **RETURN** to begin plotting. If you wish to interrupt the plotting process press the **ESC** key. This terminates the flow of data to the plotter, but the plotter will not stop until its buffer is free of data.

The **NORMAL** option plots your part files with all of the overlapping lines. It plots every entity in the database. It may be quicker, but if you have a very complicated part the end result may be many lines on top of lines and arcs or circles on edge lines with inconsistent line weight.

The **OPTIMAL** option removes all of the lines on top of lines and arcs or circles on edge lines. It may take longer to plot this way but the end result is a clearer plot which is easier to read.

 POSSIBLE USE

When you need a plot quickly of all the entities that are displayed on the screen regardless of the scale, the **AUTO** option is the best method. The entire part will be scaled to fill the maximum limits of the paper.

KEY-IN   The **KEY-IN** option sends the displayed entities to the specified plotting device. The plotting scale value is input by the user in this option. This function can be accessed quickly by using the Escape Code **ESC-F7-F7-F1-F2 (CONTROL-PLOT-PLOT-KEY-IN)**.

The prompt requests whether you wish to display the scale factor and X,Y offsets. The **NO/YES** menu is displayed for your selection. Choose **F2 YES** to display each of these values because no scale factor has been calculated for you in this option. You must enter a scale factor value and press **RETURN**.

The next prompt displays the offset values as they are assigned in the **CONFIG** program. If you wish to change the offset values enter the new values and press **RETURN**.

The **PEN #/COLOR** menu is displayed for you to select the method by which you wish to plot. Refer to the previous option for a review of this menu. If the **NO** option is selected you are prompted by the **PEN #/COLOR** menu.

Next, the universal Selection Menu is displayed, allowing you to plot only those entities that you have selected or All Displayed entities.

Finally, with all of the values assigned you are prompted to ready the plotter and press **RETURN** to begin plotting. To terminate the plotting process press **ESC**.

 POSSIBLE USE

When you are plotting a part that is displayed on the screen and you want it plotted at a specific scale, the **KEY-IN** option allows you to specify the scale at which you wish to plot. Sometimes, when plotting on a certain size paper, you will have to scale your drawing so that it fits using the **KEY-IN** option. In Figure 20-1 the part was plotted using the **KEY-IN** option and the scale **.5** was used.

PLOT OPTION: SETUP

The **SETUP** option allows the user to temporarily configure the plotter without exiting CADKEY and running the **CONFIG** program. The setting will be valid only for the current working session and will not change the existing CONFIG.DAT file. The Escape Code for this option is **ESC-F7-F7-F2 (CONTROL-PLOT-SETUP)**.

The options that can be configured are listed below.

> **PAPER**—allows the user to select paper size for plotting.
> **OFFSETS**—defines the X and Y offsets for the image.

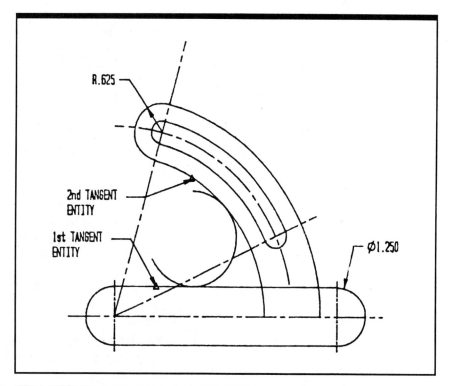

Figure 20-1   A part plotted with KEY-IN scale of .5

ROTATION—allows the image to be rotated on the paper.

PLT PORT—defines the port to which the plotter is connected.

PENS—defines the sort order, speed, and width of the pens.

PLOTTER—selects the type of plotter and its parameters.

CURV TOL—the number 1 generates the smoothest curve but requires more time to plot.

CLIP LIM—accepts the defaults or refer to Reference Manual.

DONE—returns you to the plot menu.

POSSIBLE USE

Most of the plotter options should already be set in the configuration file. The one option that changes most frequently is the paper size. You can configure the plotter for the most commonly used paper size, but, if you wish to use another size paper, you can simply make the required changes while you are plotting on-line.

CONTROL-PLOT FOR CADKEY 6

In CADKEY 6 after **CONTROL-PLOT** the selection menu appears for the user to select the entities to plot. This is followed by a dialog box (Figure 20-2) with several options which in previous

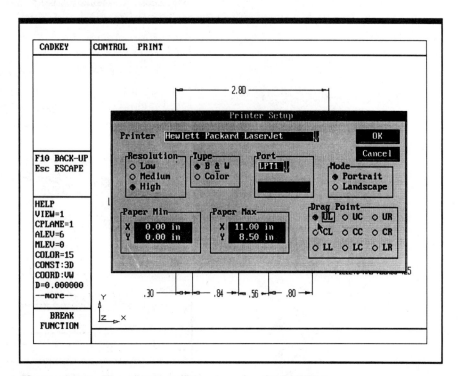

Figure 20-2 The plotting dialog box for CADKEY 6

versions were separate menus. From this dialog box you can select your **Plotter, Port, Paper Size, Scale, Paper Drag Point, Curve Tolerance** and **Optimize**. To change any of these options just click on it. The only option which is unique to CADKEY 6 is the **Drag Point**. The drag point options include six locations on the piece of paper which will be dragged into position over the drawing to be plotted. Figures 20-3 and 20-4 illustrate an A-size sheet of paper being dragged by the Center Center location into position around the drawing. The options include **UL** (upper left), **UC** (upper center), **UR** (upper right), **CL** (center left), **CC** (center center), **CR** (center right), **LL** (lower left), **LC** (lower center), and **LR** (lower right). The best option is to simply press the **Enter** key and the drawing will be automatically centered.

THE PRINT FUNCTION

PRINT OPTION: PRINT

The **PRINT** option allows you to send the entities displayed on the screen to a specified printer for a hard copy printout on paper. Hard copy can be generated off-line using a printer also. The

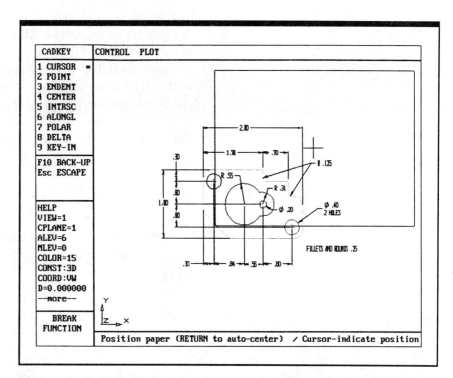

Figure 20-3 The plotting drag box equal to the paper size for CADKEY 6

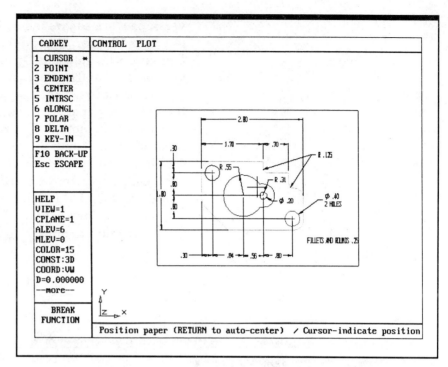

Figure 20-4 The plotting drag box with the drawing centered for CADKEY 6

program to do this is called **RPLOT**. To access the on-line option quickly, use the Escape Code **ESC-F7-F6-F1 (CONTROL-PRINT-PRINT)**.

Once this option is selected you are prompted to align the printer paper with the printer head. When the paper is in place and the printer is on, press **RETURN** to begin printing. The part or drawing that is displayed on the screen is dumped and the data is rasterized. The printer receives this data and the part is automatically scaled to fit on the paper.

To terminate the printing process turn the printer off. After a short period of time, the **ABORT/RETRY** menu will appear. Select **ABORT** to cancel printing.

 POSSIBLE USE

In many cases, only a printer is available for hard copy output. Previously, the printer output was poor because it was limited to the resolution of the screen. Now, with the EGA and VGA standard of resolution the output is greatly improved. Figure 20-5 is an example of hard copy from an EPSON dot matrix printer. A laser printer will provide an even clearer print.

PRINT OPTION: SETUP

The **SETUP** option allows the user to temporarily configure the printer without exiting CADKEY and running the CONFIG program. The setting will be valid only for the current working session and will not change the existing CONFIG.DAT file. The Escape Code for this option is **ESC-F7-F6-F2 (CONTROL-PRINT-SETUP)**.

The options that can be configured are listed below.

PAPER—sets the paper X and Y minimums and maximums.
PORT—identifies port to which the printer is connected.
PRINTER—selects the type of printer in use.
RESOLUT—selects the resolution quality desired.
DONE—returns you to the print menu.

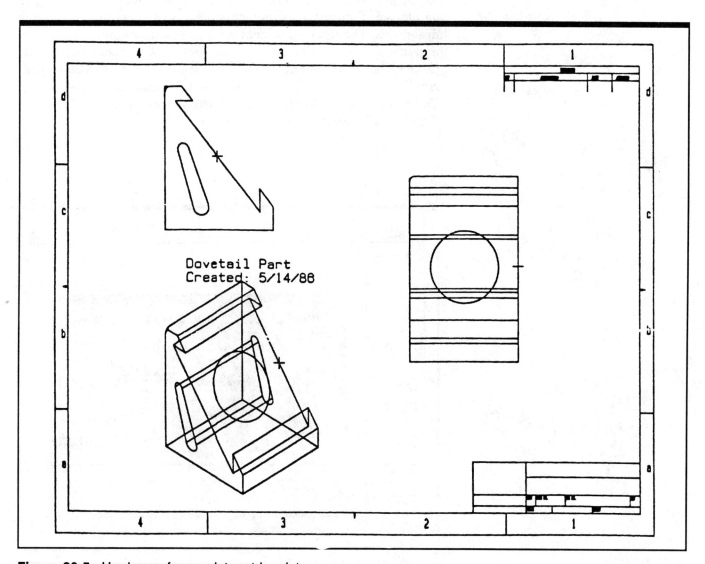

Figure 20-5 Hard copy from a dot matrix printer

POSSIBLE USE

The printer options should already be set in the configuration file. If you were trying a new printer you could connect it to an available port and simply make the required setup changes while you are on-line.

CONTROL-PRINT FOR CADKEY 6

In CADKEY 6, after **CONTROL-PRINT** you will see a dialog box (Figure 20-6) with several options which in previous versions were separate menus. From this dialog box you can select your **Printer, Port, Resolution, Type, Paper Min, Paper Max, Mode**, and **Paper Drag Point**. To change any of these options just click on it. The only option unique to CADKEY 6 is the **Drag Point**. The drag point options include six locations on the piece of paper which will be dragged into position over the drawing to be plotted. Figure 20-7 illustrates an A-size sheet of paper being dragged by the upper left location into position around the drawing. The options include **UL** (upper left), **UC** (upper center), **UR** (upper right), **CL** (center left), **CC** (center center), **CR** (center right), **LL** (lower left), **LC** (lower center), and **LR** (lower right). The best option is to simply press the **Enter** key, and the drawing will be automatically centered.

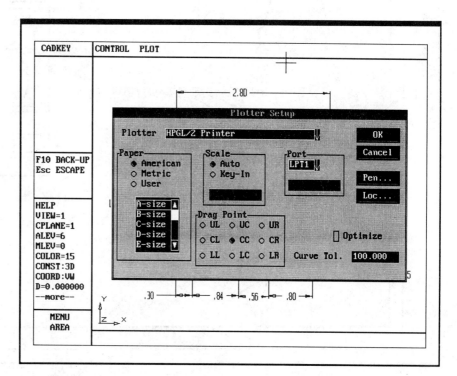

Figure 20-6 The PRINT dialog box for CADKEY 6

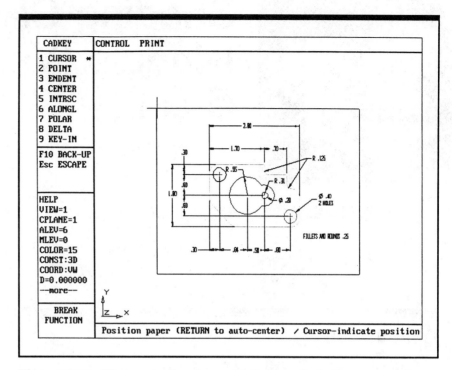

Figure 20-7   The paper drag box with the handle in the upper left corner

REVIEW QUESTIONS

1. What is the Escape Code for on-line plotting with the **KEY-IN** option?
2. When plotting in CADKEY, what are X and Y offsets used for?
3. When you plot, you have the option to plot by **PEN#** or **COLOR**. What is the difference?
4. What is the difference between plotting and printing?

CHAPTER 21

Shading CADKEY 5 Models (Version 5 Only)

OBJECTIVES

After completing this chapter, you will be able to:

- Create a wireframe drawing that can be used as input into CADKEY Solids.
- Create a shaded rendered model.
- Change the displayed colors of a rendered model.
- Change the light source of a rendered model.
- Create and view a smooth shaded rendering.
- Create and view mass property calculations for a model.

CADKEY Solids is a compatible program for CADKEY that creates a solid model from wireframe geometry. The 3D model is constructed from lines or polygons using CADKEY and stored as a CADL file, or constructed by using 3D primitives and Boolean operations. CADKEY Solids is then used to process the model. CADKEY Solids can create shaded rendered models, animation, mass property calculations, automatic hidden line removal, and perspective views. This chapter provides an overview of CADKEY Solids' rendering options.

THE SOLIDS PROCEDURE

The solids process transforms a 3D wireframe model into a collection of planar polygons that defines the bounding surfaces of the solid model. Processed models are full 3D descriptions of a solid. CADKEY Solids is capable of producing solid model renderings that are:

Wireframes with hidden lines removed (Figure 21-1a).

Wireframes with hidden lines replaced with dashed lines (Figure 21-1b).

Filled and shaded solid models with hidden lines removed (Figure 21-1c).

Perspective views of the model (Figure 21-1d).

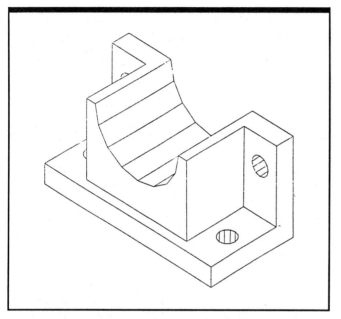

Figure 21-1a   Hidden lines removed using the HIDDEN rendering option

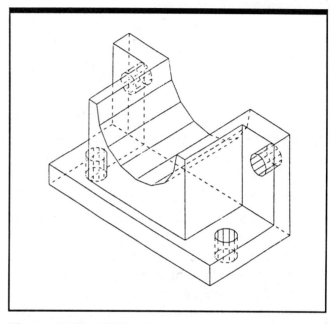

Figure 21-1b   Hidden lines as dashed lines using the DASHED option

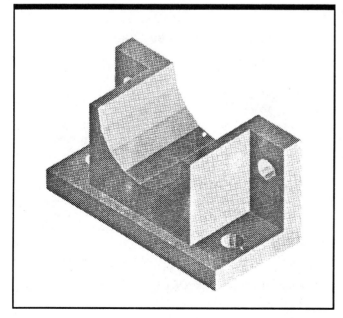

Figure 21-1c   A shaded model created using the FILL option

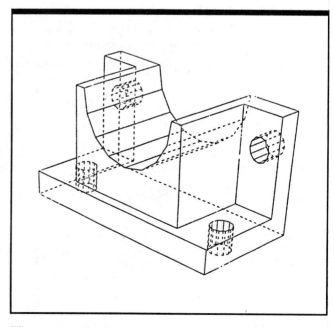

Figure 21-1d   A perspective view created using the PERSPC option

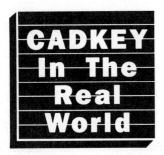

Ergo Computing Designs and Builds Bricks Using CADKEY

Tom Spalding believes that a small company with the right tools can compete with giants and accomplish great things. Spalding, president of Ergo Computing, Peabody, Massachusetts, and designer of the innovative Brick™ computer, credits the use of CADKEY software with allowing his small team of engineers to complete this highly complex design project under a very tight deadline. "If Ergo is the David among the Goliaths of computer makers, then CADKEY is our slingshot," Tom says. "It was the big reason why we were able to get the Brick to market quickly."

The process began in late 1988, when Tom Spalding and Keith Kowal, Ergo's vice president of hardware engineering, saw the need for a portable personal computer that did not trade power and performance for the convenience of portability. They defined a 386SX system that was no bigger than a standard Webster's dictionary (3 × 8 × 11 inches). To be truly useful, the sytem had to fit into half a briefcase. "Everyone who came into Ergo got the briefcase test," Tom said. "We made sure that our prototype would fit into a standard briefcase." After only eleven months of design time, in November 1989 Tom and Keith were able to demonstrate the system. In May 1990, they formally introduced it.

Incredibly Tight Design Constraints

Keith and Tom both have experience in mechanical engineering, and Keith is also an electrical engineer. To achieve the form factor with functionality that they wanted, they knew that they would be working with incredibly tight design constraints. "Many of our clearances are less than thirty thousandths of an inch," Tom said. "We had to design in 3-D, and we already knew that the best 3-D CAD software on the market was CADKEY." In March 1989, Ergo contracted with an award-winning industrial-design firm (also CADKEY users) Bleck Design Group of Chelmsford, Massachusetts, to collaborate in the mechanical design of their computer and to give it an ergonomic and aesthetic package.

The Brick was designed entirely in CADKEY, except for the electronics. Keith and his design team, Aomsak Audcharevorakul and Dave Dion, used software from LSI Logic, Inc., Milpitas, California, to design four large application-specific integrated circuits using gate-array technology for the Brick's disk drive interface, system bus manager, modem interface, and graphics interface. These four ASIC chips allowed them to miniaturize everything to fit the concept of creating an 80386-level computer, with one to eight megabytes of random access memory, and between 44 and 212 megabytes of hard disk storage, that would fit into a 3 × 8 × 11-inch form factor.

Need to Get to Market Quickly

After the Ergo team knew that their new product idea was feasible, they realized that they had a very narrow window of opportunity to get their product to market. They had to cut the time involved in the design and manufacturing cycles. Ergo contracted with LSI Logic to manufacture Keith's four custom-designed ASIC chips, and Ergo engaged Philips North America to manufacture the printed-circuit boards that Keith's team had designed. Through Bleck Design Group, Ergo made contact with several specialized manufacturers. It is significant to note that the basic reason why Ergo selected Bleck Design Group and most of its manufacturing vendors for the Brick, other than the quality of their workmanship, was the fact that they use CADKEY or products that link with CADKEY.

Ergo Computing's 8" x 11" x 3" Brick (on the left side of the monitor) displaying the original CADKEY part file from which it was built.

—*Continued, next page*

CADKEY In The Real World—Continued

Almost Paperless Design and Manufacture

"We designed and built the Brick by exchanging databases with our collaborators and vendor," Tom Spalding said. "We did not create detailed, dimensional hard-copy drawings until after the Brick was done, and we needed to produce the technical documentation for it." Jim Bleck, president of Bleck Design Group, echoed Tom's thoughts. "Controlling the design process by the database rather than by paper drawings frees us to spend more time creating innovative design and packaging," Jim said. "Models cost less to produce from a 3-D database than from detailed drawings, and they get done more quickly."

A Variety of Specialized Manufacturers

Bleck Design Group sent its database of files to another CADKEY user, Santin Engineering of Beverly, Massachusetts, to have a quick model made. "Using the CADKEY database, Santin Engineering produced the model even more quickly than they had agreed to do," Jim Bleck said.

After receiving the model, Bleck Design sent CADKEY part files or CADKEY IGES files to several specialized vendors to manufacture various parts for the Brick. Using CADKEY IGES files with a Computervision system, Bermo, Incorporated of Circle Pines, Minnesota, produced fourteen different sheetmetal parts within ten working days. PTA Corporation of Des Plaines, Illinois, used CADKEY IGES files with a Calma system to build the tooling for plastic injection molding of the Brick's front and rear bezels. Tech Prototype of Merrimack, New Hampshire, another

CADKEY user, produced twelve other injection-molded parts for the Brick.

Leominster Die Service of Leominster, Massachusetts, also received CADKEY IGES files from Bleck Design Group to design and produce injection-mold tooling for the Brick's corner pieces. Leominster Die used the CADKEY IGES files with their SmartCAM™ CNC manufacturing system. They delivered the finished injection molds to Custom Molded Plastics, Inc., also of Leominster, in eight weeks. Custom Molded Plastics did the actual injection molding of the corner pieces using Ato-Chem, Inc.'s PEBAX™, a new, nylon-based, rubberlike plastic.

Chatel Engineering Co. of Lowell, Massachusetts, made the aluminum extrusions for the Brick's exterior case. Using aluminum contributed to achieving not only the heat dissipation necessary for the Brick's compact electronics, but also the Brick's weight of 8.3 pounds. Although Chatel Engineering does not yet use computer-aided design, Lou Chatel, Sr., President of the company, furnished the extrusion-related data to John Thrailkil for entry into Bleck Design Group's Brick database.

Unexpected Bonus

When the technicians at Ergo assembled the parts into their first working prototype Brick, they were amazed to find that everything fit together perfectly the first time. "This was only possible because of the 3-D software," Tom Spalding said.

Adapted with permission from **3-D WORLD**, *v. 5, no. 1, 1991.*

Any view of the model can be rendered in shaded color or as a wireframe. The user has control of the color and shading of the model. Display colors of the model are user controlled, and background colors can be assigned. The shading of the model is controlled by repositioning the light source.

CADKEY Solids is an integrated system that uses CADKEY input for post-processing, shading, and input into other software programs (Figure 21-2).

INPUT REQUIREMENTS

CADKEY Solids uses a CADL file as the input geometry. The CADL file is produced from a CADKEY file and must meet certain specifications before it is used with CADKEY Solids. The CADKEY model must meet the following requirements:

All boundaries must be planar and closed. Each wireframe model must represent a closed, contained volume.

All features of the model must contribute to the definition of the model. (ANSI standard symbols, such as center lines, must be removed.)

Surface edges that are curved or twisted must also have the opposite edge curved or twisted, and the endpoints must be connected with lines. If the surface does not meet these requirements, the curved or twisted entity must be segmented with a four-sided

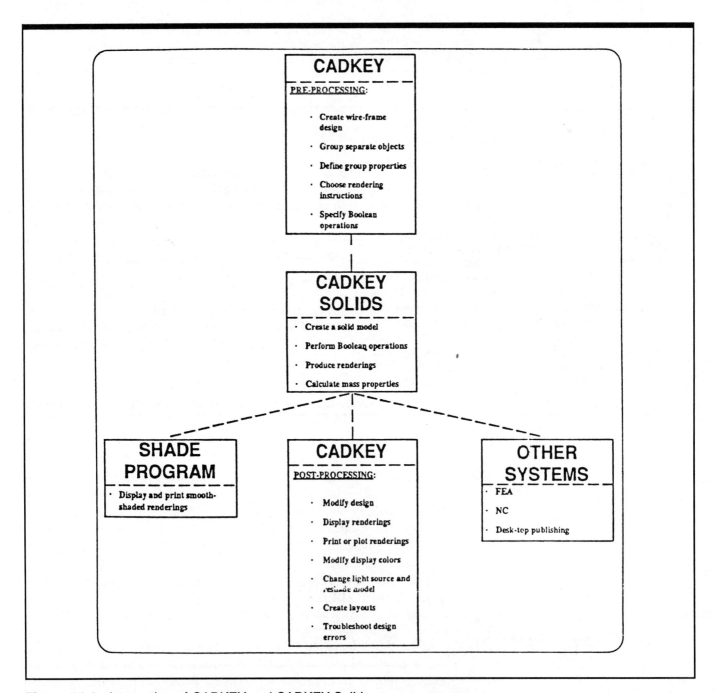

Figure 21-2   Integration of CADKEY and CADKEY Solids (Courtesy of Cadkey, Inc.)

polygon mesh. Figure 21-3 shows a valid surface on the left and an invalid surface on the right. CADKEY Solids will automatically generate a ruled polygon mesh between the curved surfaces of the object shown on the left in Figure 21-3.

Each solid object should be assigned to a separate group to increase processing speed and accuracy. Solids created with primitives automatically are assigned to different groups.

Surfaces bounded by straight lines must be coplanar.

Surface boundaries cannot cross each other, but they can have common endpoints and edges.

The volume of one wireframe model must not share any of its volume with another model. Boolean operations must be used for these types of models.

RENDERING A WIREFRAME MODEL

CADKEY Solids can use a wireframe model created from CADKEY or a solid model created with CADKEY Solids using Boolean operations. This section will describe how to create a rendered model of a 3D wireframe object created with CADKEY. The wireframe model must meet the specifications described earlier in this chapter or else errors will occur when it is being processed by CADKEY Solids. Review Chapters 5 and 18 for 3D wireframe construction techniques.

DEFINING THE CADL FILE

After a valid wireframe model is created with CADKEY, it must be converted to a CADL file before processing with CADKEY Solids. Figure 21-4 is a 3D model created with CADKEY that is made into a CADL file using the following steps.

Figure 21-3 *Left,* valid surface; *right,* invalid surface (Courtesy of Cadkey, Inc.)

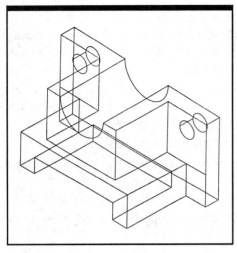

Figure 21-4   A valid CADKEY model

1. Load the 3D model and display it on screen by entering **ESC-F5-F1-F2 (FILES-PART-LOAD)**. Enter the part filename and **RETURN**.
2. Enter **ESC-F5-F4-F1 (FILES-CADL-OUTPUT)**.
3. A prompt reads:

   Enter CADL file name ():.

   Type a filename, such as **RENDER**, and press **RETURN**.
4. A prompt reads:

   Choose output option.

   Use the Select Menu to define the entities to be made into a CADL file. For this example enter **F6-GRP TBL, F3-ALL ENT, F9-DONE**.
5. As the CADL file is being created a prompt reads:

   Writing to CADL file...please wait.

   The file is created and saved in the CDL subdirectory under the filename PRACTICE.CDL.

LOADING A CADL FILE

Before creating a rendering, the CADL file must be loaded. The following steps are used to demonstrate loading CADL files.

1. Start with a blank screen and enter **ESC-F5-F4-F2 (FILES-CADL-EXECUTE)**.
2. A prompt reads:

   Enter CADL file name:

   Type in the filename **RENDER** and press **RETURN**.
3. The model appears on screen (Figure 21-4). Use **ALT-A** to display all the model if necessary.

   Before a model is processed, create any views to be rendered by using the **DISPLAY, VIEW, NEW** options.

LOADING CADKEY SOLIDS

After the wireframe model is saved as a CADL file CADKEY Solids is loaded by selecting **F9 APPLIC** and choosing **CADKEY Solids** from the pop-up window (Figure 21-5). After Solids is loaded the menu shown in Figure 21-6 is displayed with six options: **FRAMES, PROCESS, RESULTS, CONFIG, FILES,** and **EXIT**.

THE CONFIG MENU

The **CONFIG** menu displayed in Figure 21-7 defines the system environment, light source, colors and shading, and materials.

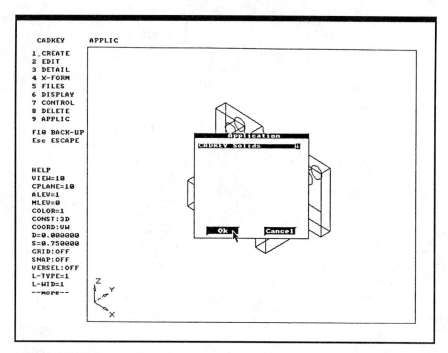

Figure 21-5   The Application menu

Normally you would want to make these settings before processing a model.

THE SYSTEM OPTION

Selecting **F1 SYSTEM** displays the dialog box shown in Figure 21-8. This option defines graphic tolerances and identifies the locations for your text editor and CADL files. The text **EDITOR**

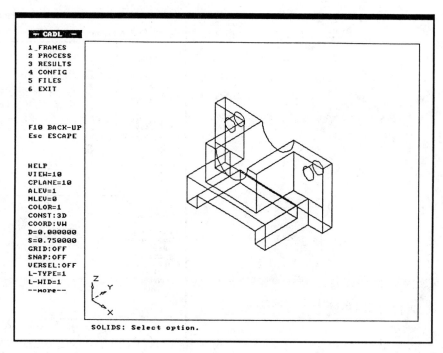

Figure 21-6   SOLIDS main menu

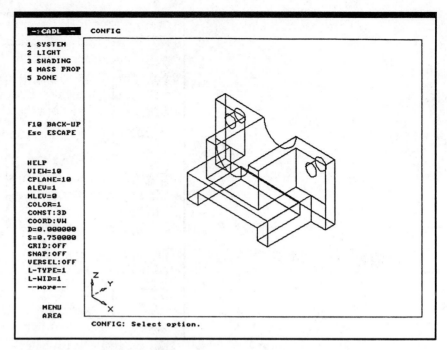

Figure 21-7 The CONFIG menu

can be the DOS command EDLIN or some word processor.
SMOOTHNESS controls the smoothing factor used in shaded
output. The default setting is .75. The closer the value is to one
the smoother the results and the longer it takes to process the
model. **SLA SUPPORT** is for stereo lithography support. **SM
BACKGROUND** sets the background color for the **SMOOTH-
NESS** option. The default setting is black. The options listed in
the dialog box are changed by picking the button.

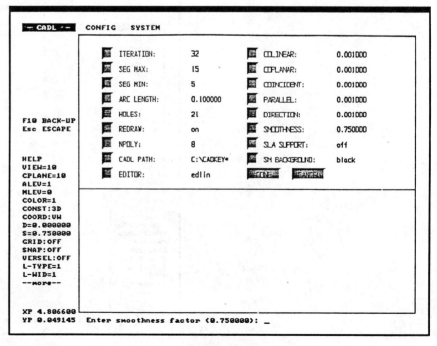

Figure 21-8 The system pop-up menu

THE LIGHT OPTION

Selecting **F2 LIGHT** from the Config Menu displays a 3D coordinate system and the direction for the current light source (Figure 21-9). The direction of the light is defined from the red point to the origin of the world coordinate system. Select one of the menu items to change the current light source. In this example the Z option (F3) is selected from the Light Menu. A prompt reads:

Enter the Z component [-2 to 2] (1.00000):.

F1 **CURSOR** is also displayed in the menu area. The cursor option is used to locate a new Z location by cursor pick. For this example enter **2** and press **RETURN**. Figure 21-10 shows how the light source location is moved higher on the screen.

THE SHADING OPTION

Selecting **F3 SHADING** displays the color scroll bars shown in Figure 21-11. The solids palette consists of 256 colors, which are created from CADKEY's sixteen base colors. To change the six base colors, select one of the Base Color buttons and pick a new base color from the sixteen displayed in the menu area of the CADKEY screen (Figure 21-12). Each of the six base colors has a pre-defined shading range which can be adjusted to add realism to your rendered model. The range of colors is controlled by changing the start and end colors. Clicking on the start or end buttons displays red, green, and blue slider bars to adjust the shading value for that color (Figure 21-13). Select **DONE** when you finish adjusting the shading. The **OUTLINE** button toggles the wireframe outline on or off. When the outline option is on a double border line surrounds the button (Figure 21-13). Outlining displays a high intensity outline of all polygon edges for filled renderings so that each polygon is distinctly represented on the

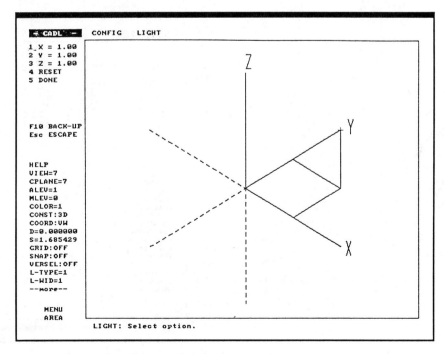

Figure 21-9   The LIGHT 3D coordinate reference

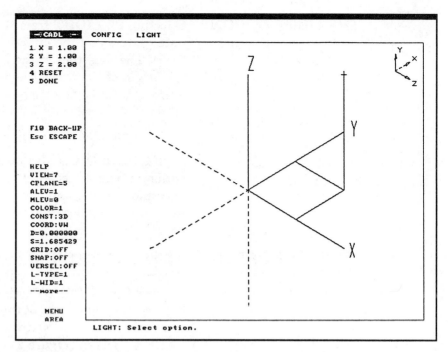

Figure 21-10 The LIGHT 3D coordinate reference coordinate after change

shaded model. The **RESET** button returns the shading values for all colors back to their default settings. After color adjustments are made, select **DONE** to return to the **CONFIG** menu.

MASS PROPERTY OPTIONS

Selecting **F4 MASS PROP** from the **CONFIG** menu displays three new menu options: **GROUP, FILE,** and **DONE**. Mass

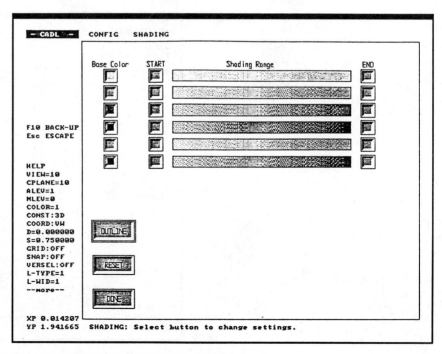

Figure 21-11 The SHADING menu

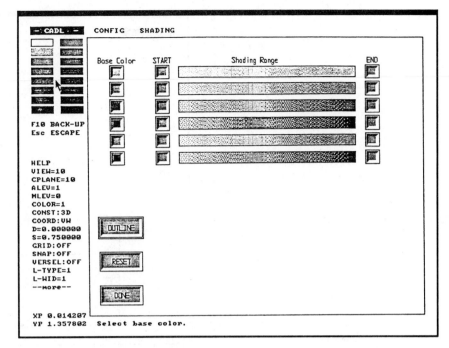

Figure 21-12 Changing the base color

properties define the material for each group within your solid model to create a realistic smooth rendering. Material properties are also used to calculate mass property values. Mass property settings can be changed by selecting **F2 FILE** from the **CONFIG** menu; this displays the pop-up window shown in Figure 21-14. Click on the **MATERIAL FILE** button to load or create a new material file. If it is a preexisting material file click on **LOAD**. If

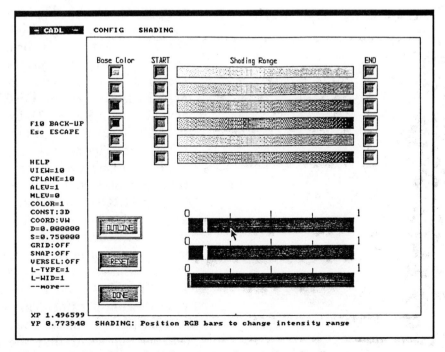

Figure 21-13 Changing the start color on the shading range

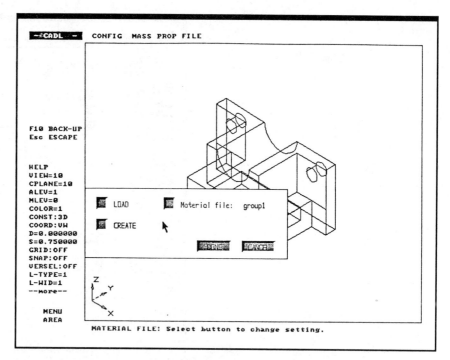

Figure 21-14 The mass property FILE menu

the material file is new, click on **CREATE**. For this example we will use the default material file called GROUP 1. Click on **DONE** to return to the **MASS PROP** menu.

Assemblies or Boolean models having more than one group can be assigned more than one mass property value. For example, one part could be assigned iron and another part brass. The mass property **GROUP-SELECT** option assigns properties to each group in a model. Selecting **F1-F1 (GROUP-SELECT)** and clicking on any entity of the displayed object group displays the pop-up menu shown in Figure 21-15. The values currently displayed are the default settings for brass. These settings can be changed by clicking on the appropriate button or a new material could be assigned from the materials list loaded with **FILE**. Selecting **MATERIAL** from the **GROUP** menu displays the pop-up window shown in Figure 21-16. The materials shown in the list are from CADKEY's default materials group. For this example **IRON** is selected. The density, specular reflection, diffuse settings, and material name are changed on the group menu to reflect the change in materials from brass to iron.

The **GROUP** menu assigns values for angular velocity and area weighted densities for analysis. The **SPECULAR** button assigns an intensity of specularly reflected light. Specular reflection is a measure of mirror qualities of a surface. Values are between 0 and 1. Shiny surfaces have higher coefficient values. **REFLEC-TIVITY** defines the intensity of specularly reflected light and your viewing position. This reflective light creates a highlight or hotspot on the object. The values are between 0 and 200. High settings represent that the intensity of the specular reflection rapidly falls off as your view point shifts away from the direction

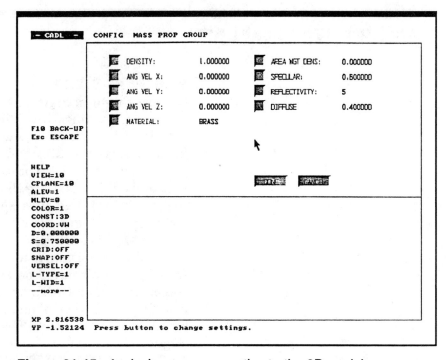

Figure 21-15   Assigning mass properties to the 3D model

of the reflection. **DIFFUSE** controls the diffuse reflection of surface light. Values are assigned between 0 and 1. Dull surfaces are assigned values closer to 1.

After all the variables have been assigned to the model choose the **DONE** button, then **DONE** from the mass properties menu.

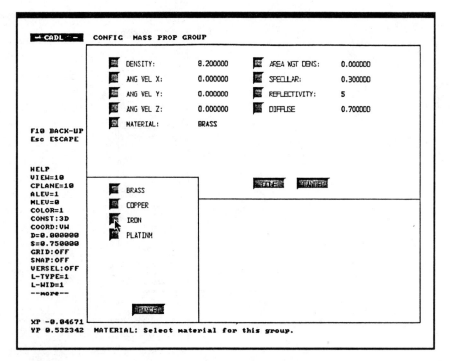

Figure 21-16   Assigning iron as the material property

THE FILES COMMAND

Selecting **F5 FILES** from the Solids Menu displays the pop-up window shown in Figure 21-17. Select a button to change any of the default settings listed in the dialog box. The **FILES** option determines which files to use and the type of models to create. Select the type of files and names to use on your model before processing the model. Located on the left of the menu is the file type and on the right are the file names. For example, after clicking on **SMOOTH OUT**, a **SMOOTH FILE** option appears in the right column (Figure 21-18).

The **MODEL IN** option names the CADL file to be used as input when processing a model. This file name can be different from the one displayed on screen. **POLYGON OUT** gets a polygon output file in CADL format after processing. **RENDER OUT** determines whether a hidden line output file is generated and saved in CADL format. The default name of the file is hidden (Figure 21-17).

The **SMOOTH OUT** option generates a smooth shaded output file. **MASS PROP** generates a mass properties text file. **GROUP INFO** determines which group input file to use for processing. This information determines mass property descriptions for each group in the model. After clicking on **GROUP INFO**, a **MATERIAL FILE** button appears to the right. Clicking on **MATERIAL FILE** displays a pop-up window (Figure 21-18) that specifies the name of the material file to be used for processing the model. **INPUT TYPE** specifies either a wireframe or polygon model as input for processing. A polygon model is faster

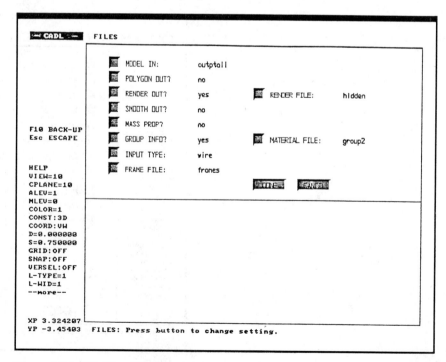

Figure 21-17 The FILES menu

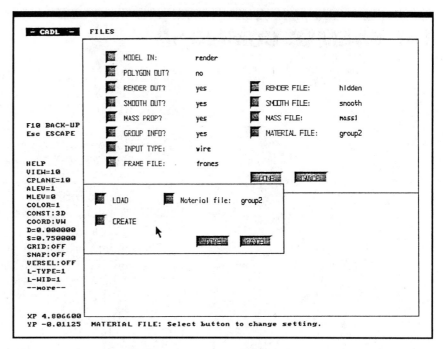

Figure 21-18 MATERIAL FILE pop-up menu

to process, but only surface property calculations are possible and the model must be made entirely of polygon entities.

The **FRAME FILE** button sets the name of the frame file to be set up before processing, loads a previous frame file, or changes the name of the frame file. **DONE** is used after all the changes have been made to the **FILES** menu (Figure 21-19).

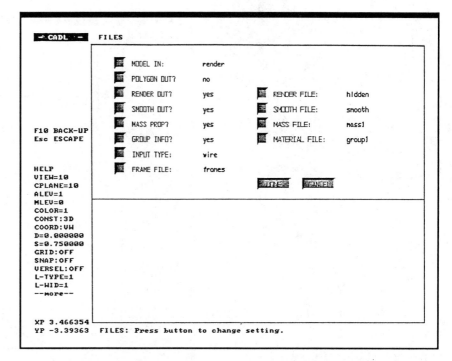

Figure 21-19 FILES menu after changes have been made

THE FRAMES COMMAND

FRAMES lets you define the type of view used to visualize the solid model. The type of view is created using a pop-up window (Figure 21-20). The options in the **FRAME** dialog box define the format, appearance, and orientation for each solid rendering, as well as Boolean operations to be performed when processing the model. The **FRAMES** option will create frames that can be shown in sequence, like a slide show, after processing your model. You are limited to twenty frame definitions per frame file. **FRAMES** lets you define the format, appearance, and orientation for each solid rendering.

The column on the left of the **FRAMES** pop-up window is used to add, delete, and view rendering descriptions. The options on the right define the current frame. The total number of frames appears at the bottom left of the pop-up window.

The **ADD** option inserts a new frame after the current frame. **NEXT** displays the next frame settings. **PREVIOUS** will display the previous frame definitions. **DELETE** deletes the current frame. **FILE** displays a pop-up window (Figure 21-21). This option will name a frame file, load an existing file, or create a new one.

Selecting **WINDOW** from the **FRAMES** dialog box displays a pop-up menu (Figure 21-22) that sets the rendered frame to single or 4 view. Select **4 VIEW** from the pop-up window to see another pop-up window (Figure 21-23). Picking the button in the **4 VIEW** pop-up window displays a prompt to enter a new view number for one of the quadrants. For this example, CADKEY's standard views will be used: View 1 in the upper-left, view 2 in

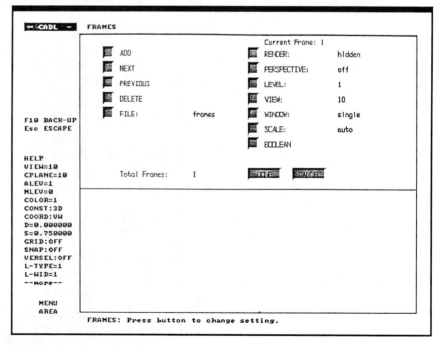

Figure 21-20 The FRAMES menu

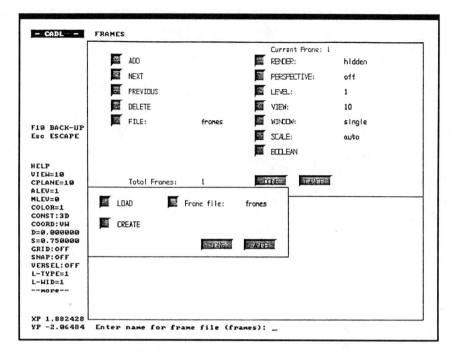

Figure 21-21 The FRAMES-FILE menu

the lower-left, view 5 in the lower-right, and view 7 in the upper-right. After making the frame definition select **DONE** to return to the Solids main menu.

PROCESSING A MODEL

To process a solid model select **F2 PROCESS** from the Solids Menu. Selecting Process displays a menu with two choices: **QUIET RUN** and **STAT RUN**. The **STAT RUN** option displays messages during processing, then prompts you to press **ENTER** when the processing is complete. For large models **STAT RUN** takes longer to process, but it gives you the option to read messages to determine the progress of the process. **QUIET RUN** processes a model faster, but it does not display messages. This option is more useful when rendering a model that has already been successfully processed.

DISPLAYING THE RESULTS

Selecting **F3 RESULTS** from the Solids menu displays a menu with six options: **RENDERED, SMOOTH, MASS PROP, GRAPH ERR, ERROR LOG** and **COLOR**. To view the results

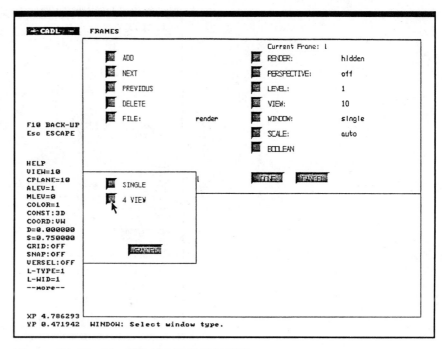

Figure 21-22 The FRAMES-WINDOW menu

of the rendered model after processing, select **RENDERED** from the Results Menu. Figure 21-24 shows what is displayed after selecting **RENDERED** using the **4 VIEW** hidden line option.

Selecting **SMOOTH** displays a new menu, letting you view either a Gouraud or a Phong shaded model. **MASS PROPERTIES** displays the mass properties of a rendered model. **ERROR LOG** displays error messages for an invalid solid model that was

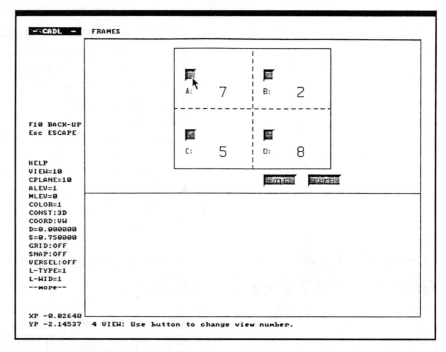

Figure 21-23 Changing the views to be displayed in each window

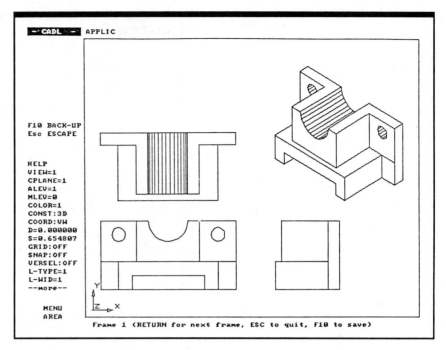

Figure 21-24 The 4 VIEW hidden line removed rendering

processed. The **COLOR** option makes changes to the rendered model using slider bars.

CREATING MULTIPLE FRAMES

Selecting the **RENDER** option from the **FRAME** dialog box displays a pop-up window (Figure 21-25) that selects the type of view used to display the processed solid model. Select **DASHED**

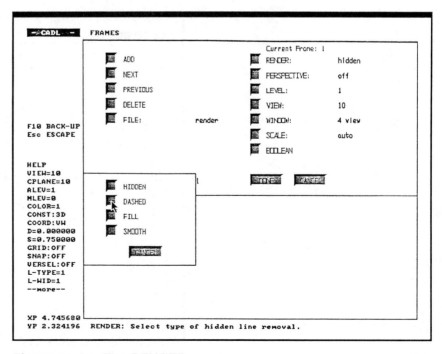

Figure 21-25 The RENDER menu

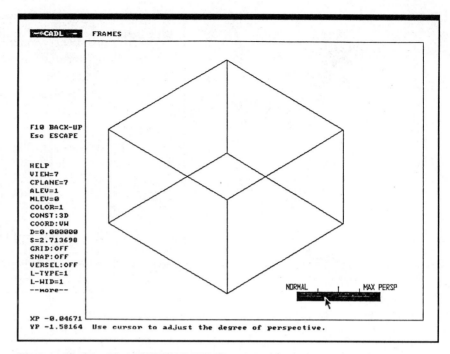

Figure 21-26 The PERSPECTIVE reference box

to create a model with hidden lines drawn. **PERSPECTIVE** displays a reference box (Figure 21-26). Figure 21-26 is the default setting of no perspective. Pointing at the scroll bar and moving it from its normal position to a new position to the right will cause the reference box to change. Figure 21-27 shows the effect on the reference box when the scroll bar is adjusted to near 40% perspective. This visual cue helps you determine what your model

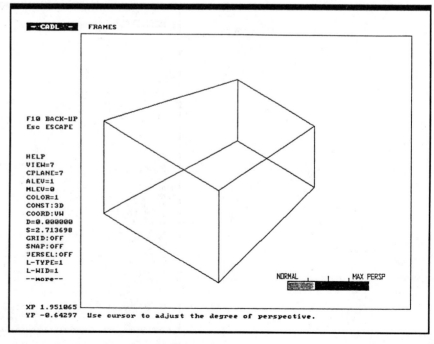

Figure 21-27 The PERSPECTIVE reference box after changing

will look like at a specific perspective setting. Frame 1 will be a hidden line perspective model.

To create another frame with different settings click on the **ADD** button. The Total Frame number located near the lower-left portion of the frame dialog box will change to 2. Click on **RENDER** button and select **FILL**.

Create frame 3 by clicking on **ADD**. Click on **RENDER** and select **SMOOTH**. Select **DONE** to exit and save the frame file for processing. From the Solids Main Menu select **F2-F1** (**PROCESS-QUIET RUN**). You can use **QUIET RUN** because the model was already processed when the **4 VIEW** rendering was created.

To view the three frame rendering, select **F3 RESULTS** from the Solids main menu. To view the dashed line perspective model choose **F1 RENDERED** from the results menu. Figure 21-28 shows what is displayed on screen. To save a rendered image select **F10** and enter the name of the file. By pressing **RETURN**, the next frame (the filled model) is automatically displayed (Figure 21-29). After this model is displayed you will see the color menu. If you want to change to change the displayed color of the rendering, select a color. The RGB (red, green, blue) sliders appear (Figure 21-30).

To view frame three (the smooth mode), select **F2 SMOOTH** from the results menu. There are two smooth options: Gouraud and Phong. A Gouraud shaded model calculates light intensity polygon by polygon. Phong shading calculates light intensity pixel by pixel. Because there are many more pixels than there are polygons, Phong shading takes longer to calculate but results in a much better shaded model. Pressing **P** will print the shaded

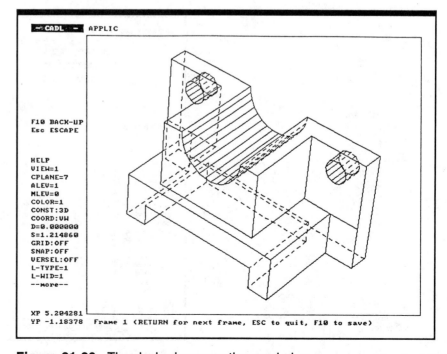

Figure 21-28 The dashed perspective rendering

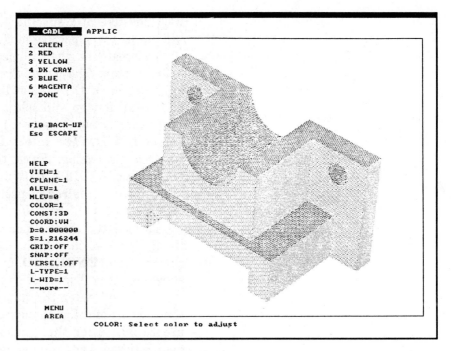

Figure 21-29 The filled rendering

model. Pressing **Q** will cancel the display and return you to the results menu. Figure 21-31 shows a Phong shaded model.

VIEWING MASS PROPERTY CALCULATIONS

The mass properties of a model can be created when using the CAD-KEY Solids program. When Figure 21-24 was processed, the mass properties were also calculated and saved in a file called MASS1.

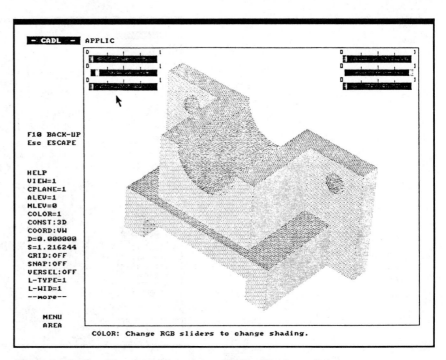

Figure 21-30 Changing the color of the filled model

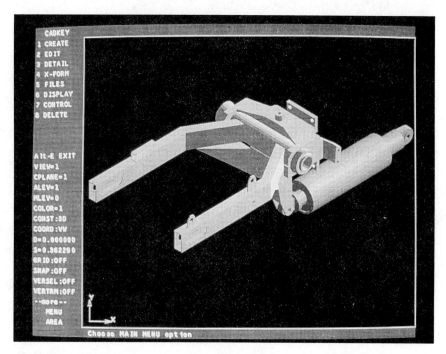

Figure 21-31   A smooth shaded 3D model

The mass properties output is stored as an ASCII text file and is viewed by using a text editor or word processor. Figure 21-32 shows the mass property output for Figure 21-24.

There are two parts in the mass properties report. The first part includes information for each grouped solid object in the rendering. The second part of the report lists the total mass properties of the solid model, such as volume, mass, area, centroid, and many other

```
    2:
    3: STATUS: --------- Mass Properties of Group render ---------
--  4:
    5: STATUS:    Volume    :        14.679
    6: STATUS:    Area      :        59.241
    7: STATUS:    Material  : IRON
    8: STATUS:    Density   :          7
    9: STATUS:    Mass      :        102.75
   10: STATUS:    Area weighted density :             0
   11: STATUS:    Surface mass :         6087.2
   12: STATUS:    Centroid : x =         3.6678  y =         3.1436  z =      -
1.3608
   13:
   14: STATUS:    Moment of inertia, axis through centroid :
   15:               Ixx =        92.505   Iyy =        190.22  Izz =        2
02.45
   16:               Ixy = -8.5759e-006   Iyz =        2.4664  Ixz = 6.8443
e-006
   17:
   18: STATUS:    Moment of inertia, axis through origin :
   19:               Ixx =        1298.2   Iyy =        1762.8  Izz =        2
608.2
   20:               Ixy =        1184.8   Iyz =        -437.1  Ixz =       -5
12.86
Continue (Y/N)?
```

Figure 21-32   The mass properties displayed on screen

properties. The ASCII file can be added to a CADKEY part file by using the disk based **NOTE** option.

REVIEW QUESTIONS

1. List the input requirements for a solid model.
2. List the types of rendered models possible with CADKEY Solids.
3. Explain the difference between Gouraud and Phong shading.
4. How many frames can be created in a single frame file?
5. Explain the difference between the two processing options.

CADKEY Solid Modeling (Version 5 Only)

CHAPTER 22

OBJECTIVES

After completing this chapter, you will be able to:

- Create a box, cone, cylinder, sphere, torus, or wedge primitive shape.
- Use the Boolean operations *union, interference, plane,* and *difference* to create models.
- Assign properties to a model.
- View an error file for an invalid model.

When creating 3D geometric models of a design, three types can be created with CAD: wireframe, surface, and solid. *Wireframe* models describe the edges or boundaries of a part. Points between the boundaries of a part cannot be defined with wireframe models. CADKEY creates wireframe models. *Surface* modeling is a technique used to define and describe the surfaces of a 3D model. They not only define the edges of a model, but also the surfaces between the edges. Surface models can be shaded with various colors and light sources.

A *solid* model looks like a surface model but it is very different. A solid model is a mathematically complete representation of a part. This type of model includes information about the interior of the part. As the model is created, different properties, such as material and mass, can be assigned. There are two methods of creating solid models: *constructive solid geometry (CSG)* and *boundary representation (B-rep)*.

Boundary representations are a collection of surfaces, points, and curves. These entities are then combined to create a solid image. B-rep models are used for complex and unusual shapes, such as aircraft wings and fuselages.

CSG modeling is a process used to design complex shapes using such simple solid geometric primitives as blocks, cylinders, cones, spheres, and other shapes. These shapes are called *primitives* and are combined through the use of Boolean operations called union, intersection, and difference. CSG works well for regular geometric shapes that can be easily created using primitive solids.

CADKEY and CADKEY Solids can be used to create CSG and B-rep solid models. To create a B-rep model, create the 3D wireframe model with CADKEY, then process the wireframe model with CADKEY Solids. To create a CSG model, use the CADKEY Solids CADL program SP (Solid Primitives), then process it with CADKEY Solids.

Composite Engineering Builds Winning Aquatic Racing Gear

Composite Engineering of Concord, Massachusetts, fuses CADKEY and CADKEY SURFACES with hand craftsmanship to produce championship-level aquatic racing equipment.

"If you have ever paddled a canoe or a kayak," said James Lawrence, Director of Engineering at Composite Engineering, "you know that when you insert a regular paddle into the water and pull it back, it propels the craft forward, but there are lots of little whirlpools around the paddle blade. Some of these vortices represent energy that has been lost, energy expended that does not move the craft forward." Composite Engineering's newest design for kayak paddles is a modification of the wing-shaped paddles developed since 1981 by Swedish and Norwegian canoeists Stefan Lindeberg, Lief Hakensson, and Einar Rasmussen. These Scandinavian paddles gave canoeists and kayakers a two-to-three second improvement in competitive performance in 500-meter events. This two-to-three second decrease in time, from the start of a race to its finish, represents an overall performance improvement of approximately 2% to 3%.

"Our design attempts to reduce the amount of human energy wasted with each paddle stroke," James continued. "The paddle is double ended, and each paddle blade is gull-wing shaped with an airfoil twist like an airplane propeller. Water passes across the blade in the same way that air flows over and under the wing of an airplane. The paddle slices the water.

It gives forward motion to the kayak or canoe without disturbing the water unnecessarily.

First, we entered the full 3-D path of a paddle into CADKEY from video camera measurements of U.S. Olympic Team kayakers. After analyzing this motion, we designed the paddle as a three-dimensional model in CADKEY, and we used CADKEY SURFACES to create the surface of the blade."

New Types of Paddles

"An ideal paddle design is a compromise among four elements," James said. "A paddle can have (1) a perfect shape for insertion into the water, (2) a perfect shape for the path that it makes in the water, (3) a perfect shape for coming out of the water, and (4) a perfect shape for the time that the blade is in the air. Remember, a paddle blade spends half of its time in the air."

James illustrated how Composite Engineering's kayak paddles for the 1992 Summer Olympics went from design to production. "We used Einar Rasmussen's paddle as a model. Then we modified it for our initial paddle design. We found that the right-hand blade and the left-hand blade of Rasmussen's paddle did not work exactly the same way in the water. We used a computer-based measuring device created by John Lawson of Petersham, Massachusetts, for measuring boats to measure the right and left blades of Rasmussen's paddle in detail. John Lawson's measuring device pinpointed minor, yet apparently significant, differences in the right and left paddle blades. We fed the data from John Lawson's measuring device into CADKEY; then we made our modifications to Rasmussen's design, creating blades that are exact mirror images of each other."

"Since the boats are extremely difficut to balance," Ted added, "it is an advantage to make the strokes perfectly symmetrical. The mathematical design of a paddle is extremely difficult because the flow changes dramatically through a stroke. We design for an improvement in efficiency through the main power phase of the stroke, but we must modify the design based on the paddler's evaluation of its handling. After we had created the basic design in CADKEY, it was much faster and more accurate to incorporate modifications to make new designs. This permitted us to try more ideas and to improve the design more quickly."

Two-person kayak and paddles, designed with CADKEY and CADKEY SURFACES, in competition.

—Continued, next page

CADKEY In The Real World—Continued

Enter CADKEY SURFACES

Composite Engineering focused its further design effort on 11 out of 24 cross sections or airfoil stations, approximately five centimeters apart, in the paddle blade. "Near the tip of the blade these stations are closer than five centimeters," James added. These cross sections/stations are parallel sets of 2-D cubic splines. "After we had defined the stations, we created a four-sided complex surface using the paddle blade's leading edge, its trailing edge, and all the intermediate airfoil stations. This surface intersects all of the splines at each station. We then created an offset surface, that is, a surface reduced from the original complex surface by the thickness of the laminate in the paddle blade. This feature of CADKEY SURFACES is truly amazing. We used the intersection function to generate quickly all of the intermediate stations needed to construct a male mold called a *plug*."

Composite Engineering plotted the top view of the paddle at full scale, and also the front views of all 24

airfoil stations. The plots of the front views of the stations served as templates for cutting the stations out of plywood. The plot of the top view served as a map for placing the airfoil stations, now cut from plywood, in their correct locations. "After they were correctly in place," James said, "we filled the spaces in between with plywood specifically cut to make sure that the surface of the paddle fairs evenly in all directions, from one station to the next, throughout the length and width of the paddle blade. We then glued all of these pieces of plywood together to make the plug."

Handcrafted Manufacturing

Composite Engineering used the plug to produce a female mold from which they make the laminated kayak paddle blades with careful hand craftsmanship. A technician lays a layer of carbon fiber preimpregnated with epoxy, by hand, in the mold. This material goes by the nickname *prepreg*. Next, the technician places a layer of foaming epoxy, one-tenth-inch thick, on top of the carbon fiber. Then, the technician adds a second layer of prepreg carbon fiber to complete the laminate sandwich, which now goes into the vacuum-bag process for heat curing. Vacuum bagging is a process that takes advantage of atmospheric pressure to create a uniform distribution of an adhesive, such as epoxy, strongly bonding two or more materials firmly together. Composite Engineering uses heat in conjunction with the vacuum-bag process because heat assists the molecules of the epoxy resin and of the hardener to flow together evenly to form direct molecule-to-molecule bonds. After the epoxy has cured for approximately six hours at 250 degrees Fahrenheit, the technician removes the paddle blade and deburrs the edges.

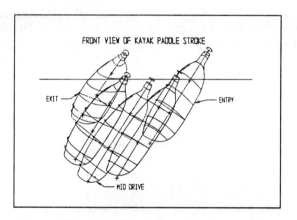

FRONT VIEW OF KAYAK PADDLE STROKE

EXIT

ENTRY

MID DRIVE

3-D model in CADKEY of the motion in a paddle stroke.

Adapted with permission from **3-D World**, *v. 6, no. 2, 1992.*

SOLID PRIMITIVES

The solid primitive program SP can be run any time you are working in CADKEY (Figure 22-1). It will allow the creation, placement, and Boolean operation for blocks, wedges, cones, cylinders, spheres, or tori (Figure 22-2). The type of primitive to create is selected from a menu (Figure 22-3). A prompt requests that the parameters for defining the size and orientation of the primitive be input. For example, to define a block (rectangular prism), the length of each face must be input. A prompt then requests that a position be chosen to place the block in 3D space using the Position Menu.

The following examples demonstrate how to use CADKEY to create and place solid primitives. The primitive program is loaded into CADKEY by selecting **ESC-F5-F4-F2 (FILES-CADL-EXECUTE)**. A prompt reads: **Enter CADL filename ( ) =.** Enter **SP** and **RETURN.** The file is loaded and you can begin creating primitive shapes.

CONFIGURING SOLID PRIMITIVES

Option 7, **CONFIG**, is used to change the number of transverse sections for cones, cylinders, and spheres. Transverse sections are the number of lines created to give the 3D model definition. Add more transverse sections to increase the accuracy of the model. The default setting is 5 for cones and cylinders, and 7 for spheres; it should be changed before the model is created. **CONFIG** is also used to change the coplanarity tolerance from its default setting of .000010. This is the setting used to determine if two solid primitives occupy the same space.

CREATING BLOCKS

A *block* is a prism that can be used to create cubes, and rectangular prisms. A block is defined by the lengths of three faces and its position in 3D space. **VIEW, WORLD**, and **DEFINE** are the three options used to define the plane that will contain the block primitive shape.

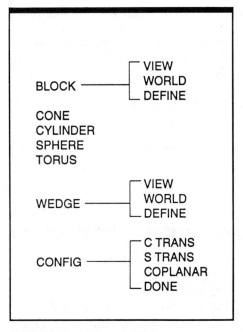

Figure 22-1 The SOLID PRIMITIVE menu

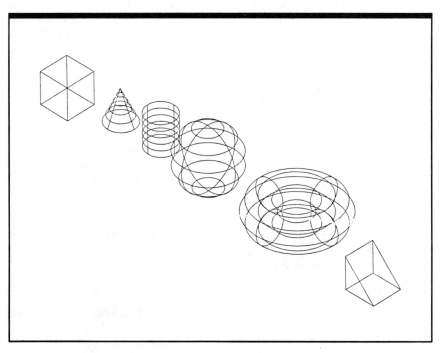

Figure 22-2 Primitive shapes created with CADKEY Solids

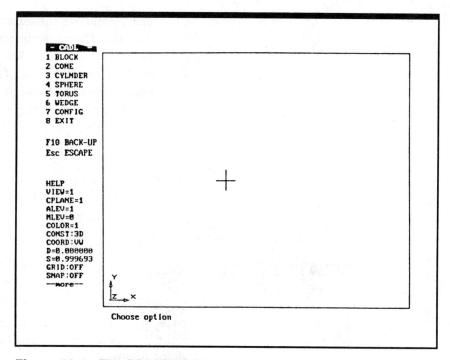

Figure 22-3 The PRIMITIVES main menu

1. To create a block enter **F1 BLOCK**.

2. A prompt reads:

   Choose X-Y plane option.

   This prompt is requesting that the plane that contains the base of the block be defined. There are three options: **VIEW**, **WORLD**, and **DEFINE**. **VIEW** defines a plane that is parallel to the current display plane. **WORLD** creates the base of the block in a plane parallel to the XY plane in World Coordinates. **DEFINE** creates a plane for the base of the block by using **3 PTS**, **PT/LINE**, **2 LINES**, **VW/DPTH**, or **ENTITY**. For this example, select **F2 WORLD**.

3. A prompt reads:

   Displacement along X axis (1.000000) =.

   Enter **4** and **RETURN**. Select **F10 BACKUP** if you want to enter a different value.

4. A prompt reads:

   Length along Y axis (1.000000) =.

   Enter **2** and **RETURN**.

5. A prompt reads:

   Length along Z axis (1.000000) =.

   Enter **3** and **RETURN**.

6. A prompt reads:

   Indicate corner point.

The corner of the block must be located in 3D space. The Position Menu is displayed to locate the corner. Select **F9 KEY-IN.**

7. Enter **0,0,0** for the X,Y,Z coordinate points. **A prompt reads: Is this O.K. (YES)?** Enter **RETURN.** The block is drawn on screen. Select **ALT-V** then enter **7** and **ALT-A** to display the isometric view of the box automatically scaled (Figure 22-4).

    To erase the block select **CTRL-Q** and pick one of the entities. A prompt will say that the entity belongs to a group. To erase the whole block select **F1 GROUP** then **RETURN. ESC** returns you to the Primitive Solids main menu.

CREATING CONES

A cone is a basic geometric solid that is defined by entering its height, base radius, and top radius, then its position in space. Entering a value greater than zero for the top radius creates a truncated cone (a cone with its point cut off). The number of transverse sections is changed by selecting **F7 CONFIG** from the Primitive main menu.

1. Select **F2 CONE** from the Solids Primitive main menu.

2. A prompt reads:

    Length of the cone (1.000000) =.

    This prompt is requesting a value for the height along the axis of the cone. Enter a value of **4** and **RETURN.**

3. A prompt reads:

    Radius of the larger face (0.500000) =.

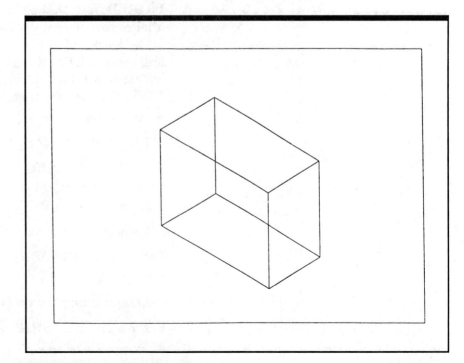

Figure 22-4 A primitive box

This prompt is requesting the radius for the base of the cone. Enter a value of **2** and **RETURN**.

4. A prompt reads:

   Radius of the smaller face (0.000000) =.

   This prompt is requesting the radius for the other end of the cone. Enter **0** and **RETURN** to create a full cone.

5. A prompt reads:

   Choose method for defining cone axis.

   TWO PTS and **PLANE** appear in the menu. The **TWO PTS** option defines the cone's axis by selecting two points in 3D space. **PLANE** defines the cone's axis by identifying a plane that contains the base of the cone using **3 PTS**, **PT/LINE**, **2 LINES**, **VW/DPTH**, and **ENTITY** options. Select **F1 TWO PTS** from the menu. A prompt reads:

   Indicate 1st pt. along the axis.

6. Select **F1 CURSOR** from the Position Menu. Turn on the snap (**CTRL-X**), tracking (**CTRL-T**), world coordinates (**F3 WORLD**), grid (**CTRL-G**), and **F3 ALL DSP**.

7. In view 2, move the cursor to 3,0,2 using the cursor read-out in the lower left corner of the screen. Pick that point.

8. A prompt reads:

   Indicate 2nd pt. along the axis.

   Move the cursor to 3,0,3 and pick that point.

9. A temporary arrow is drawn on screen showing the defined axis direction (Figure 22-5). A prompt reads:

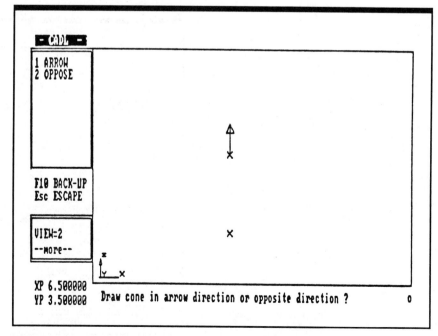

Figure 22-5   A temporary arrow to indicate the direction of the cone's axis

Draw cone in arrow direction or opposite direction?

Selecting the **ARROW** direction draws the cone so the small radius or top of the cone is in the direction of the arrow. Selecting **OPPOSE** draws the cone so the small radius of the cone is in the opposite direction of the arrow. Select **F1 ARROW** for this example.

10. A prompt reads:

Indicate position for base face center.

The center of the base of the cone is located by using the Position Menu. Select **F9 KEY-IN** and enter **3,0,2** for the X,Y,Z coordinates for the center of the base.

11. The cone is drawn on screen. Select **F2 YES** to accept the drawn cone. Enter **ALT-V** then **7** and **ALT-A** to create the view shown in Figure 22-6.

By entering a value greater than zero in Step 4, a truncated cone is created. Figure 22-7 shows a cone created by using the same steps but substituting a value of .5 in Step 4.

CREATING A CYLINDER

Cylinders are basic geometric solids that are created by defining the height and radius, rather than position in space. CADKEY creates a cylinder by entering the length (height), radius, direction of the axis in 3D space, and the center point for the base of the cylinder in 3D space.

1. Enter **F3 CYLNDER** from the Solids Primitive main menu.

2. A prompt reads:

Length of the cylinder (1.000000) =.

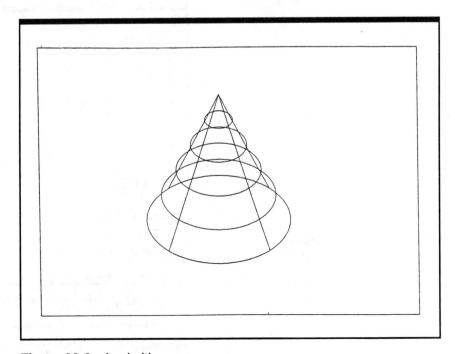

Figure 22-6 A primitive cone

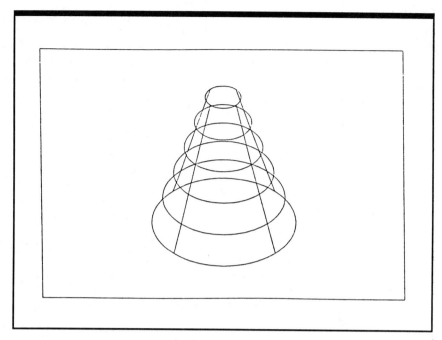

Figure 22-7   A truncated primitive cone

Enter the height of **4** and **RETURN**.

3. A prompt reads:

Radius of the cylinder (0.500000) =.

Enter a radius of **1.5** and **RETURN**.

4. A prompt reads:

Choose method for defining cylinder axis.

Use the **TWO PTS** option by entering **F1**.

5. A prompt reads:

Indicate 1st point along the axis.

Select **KEYIN** from the Position Menu and enter **0,0,0**.

6. A prompt reads:

Indicate 2nd point along the axis.

Select **KEYIN** from the Position Menu and enter **0,4,0**.

7. An arrow is drawn on screen indicating the axis position. A prompt reads:

Draw cylinder in arrow direction or opposite direction?

Select **F1 ARROW**.

8. A prompt reads:

Indicate position for base face center.

Locate the center of the base of the cylinder by selecting **F9 KEYIN** from the Position Menu. Enter **0,0,0**. Enter **F2 YES** to place the cylinder. Enter **ALT-V** then **7** and **ALT-A** to create the view of the cylinder shown in Figure 22-8.

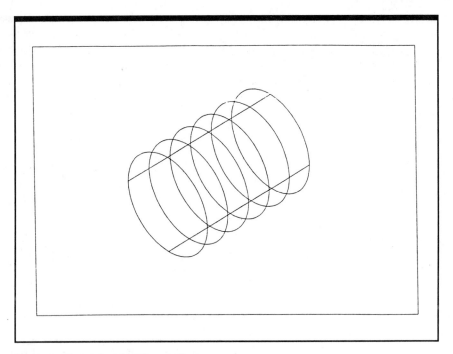

Figure 22-8   A primitive cylinder

CREATING A SPHERE

A sphere is a basic geometric solid that is defined by entering a radius and its position in 3D space.

1. Enter **F4 SPHERE** from the Solids Primitive main menu.
2. A prompt reads:

    Enter radius (1.000000) =.

    Enter a value of **4** and **RETURN**.
3. A prompt reads:

    Indicate the center.

    Select **F9 KEY-IN** from the Position Menu and enter **0,0,0**.
4. Enter **F2 YES** to place the sphere. Enter **ALT-V** then **7** and **ALT-A** to display the sphere (Figure 22-9).

CREATING A TORUS

A torus is a solid geometric shape that is defined by the outside radius, the radius of the cross section, and its position in 3D space.

1. Enter **F5 TORUS** from the Solids Primitive main menu.
2. A prompt reads:

    Radius of the torus (1.000000) =.

    Enter a value of **4** and **RETURN** to define the outside radius of the torus.
3. A prompt reads:

    Radius of cross section (1.000000) =.

    Enter a value of **.75** and **RETURN** to define the cross section radius of the torus.

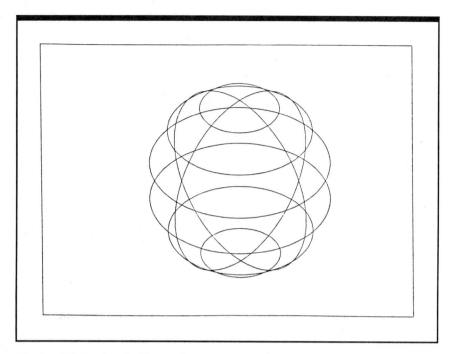

Figure 22-9   A primitive sphere

4. A prompt reads:

    Choose method for defining torus plane.

    There are two options: **NORMAL** and **PLANE**. **NORMAL** is used to draw the torus normal (perpendicular) to a vector defined using two points in 3D space. **PLANE** is used to draw the torus on a plane defined by using the following options: **3 PTS, PT/LINE, 2 LINES, VW/DPTH**, and **ENTITY**. For this example, select **F1 NORMAL**.

5. Select **F1 CURSOR** from the Position Menu and define the first point in the plane by moving the cursor to 0,0,0 and picking.

6. Select the second point by moving the cursor to 0,0,1 and picking.

7. A prompt reads:

    Indicate position for torus center.

    Select **F9 KEY-IN** from the Position Menu.

8. Key in the values **0,0,0**. The torus is drawn on screen. Select **F2 YES** to accept the torus position.

9. Enter **ALT-V** then **7** and **ALT-A** to view the torus (Figure 22-10).

CREATING A WEDGE   The wedge option is used to create a triangular prism. A wedge is created by defining its length along the X, Y, and Z axes, and its position in 3D space (Figure 22-11). A long and short length are defined along the X axis to create the wedge.

1. Select **F6 WEDGE** from the Solids Primitive Main Menu.

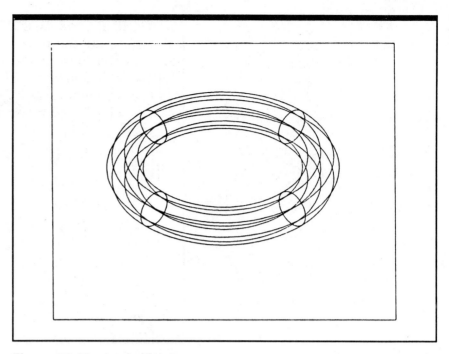

Figure 22-10   A primitive torus

2.   A prompt reads:

Choose X-Y plane option.

This is the method used to define the plane that will contain the base of the wedge. There are three options: **VIEW,**

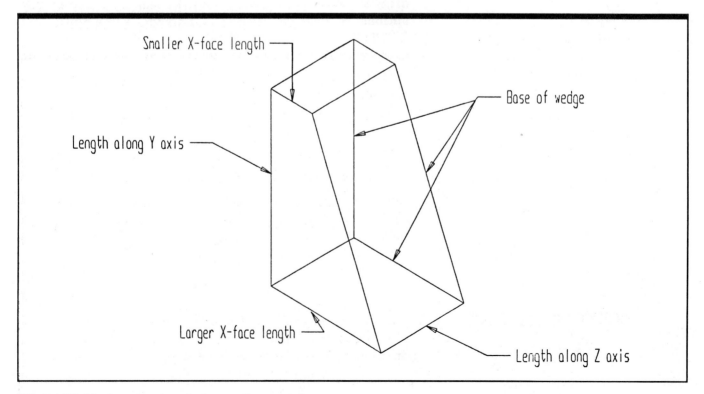

Figure 22-11   Important parts in creating a wedge

WORLD, and **DEFINE**. **VIEW** creates the base of the wedge parallel to the current display plane. **WORLD** creates the base of the wedge parallel to the XY plane in World Coordinates. **DEFINE** is used to create the base of the wedge using **3 PTS, PT/LINE, 2 LINES, VW/DPTH,** and **ENTITY**. For this example, select **F1 VIEW**.

3. A prompt reads:

   Displacement along X axis for larger face (1.000000) =.

   Enter a value of **4** and **RETURN** (Figure 22-12).

4. A prompt reads:

   Displacement along X axis for the smaller face (1.000000) =.

   Enter a value of **1.5** and **RETURN**.

5. A prompt reads:

   Length along Y axis (1.000000) =.

   Enter **6** and **RETURN**.

6. A prompt reads:

   Length along Z axis (1.000000) =.

   Enter **3** and **RETURN**.

7. A prompt reads:

   Indicate corner point.

   Select **F9 KEY-IN** from the Position Menu and enter **0,0,0** to position the lower left corner of the wedge. Select **F2 YES** to accept the placement of the wedge. Enter **ALT-V** then **7** and **ALT-A** to view the wedge (Figure 22-12).

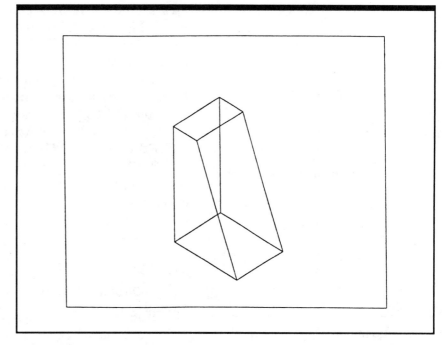

Figure 22-12 A primitive wedge

To exit the Solids Primitive Menu and return to CADKEY enter **F8 EXIT**. A prompt reads:

> **Exit SP?**

and a **NO/YES** menu is displayed. Select **F2 YES** to return to CADKEY.

COMBINING SOLID PRIMITIVES

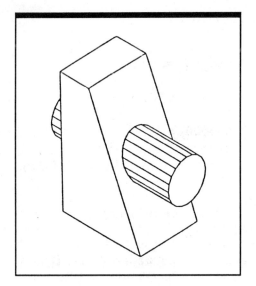

Figure 22-13   The union of a wedge and a cylinder

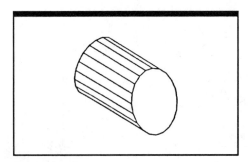

Figure 22-14   The intersection between the wedge and the cylinder

By combining solid primitive shapes, you can create complicated solid parts. CADKEY Solids uses Boolean operations on primitive shapes to create solid objects. Union, interference, difference, and plane are the Boolean operators used with CADKEY Solids.

Union creates a new object by calculating the union or sum of two selected parts. Figure 22-13 shows the union of a wedge and a cylinder. *Interference* creates a new object by calculating the intersection of two parts. Figure 22-14 shows the intersection created by the cylinder and wedge. *Difference* creates a new object by subtracting one object from another. Figure 22-15 shows the wedge with a hole drilled in that was created by subtracting the hole from the wedge. This example shows the advantage of using Boolean operations to create part files. The hole as it enters the inclined surface of the wedge is elliptical, not round, in shape. Trying to draw an accurate ellipse on the surface of the inclined plane is a difficult process. CADKEY Solids automatically calculates the elliptical hole when the model is processed. *Plane* creates a new object by sectioning a selected object with a cutting plane. Figure 22-16 shows a section view of the difference of the wedge and cylinder.

The following steps demonstrate how to use Boolean operations to create different designs of the same primitive shapes. Before starting to create the object enter **ALT-V** and change the view to **2**. Enter **CTRL-V** to change the coordinate system to **WORLD**. Load the Solids Primitive program by entering **F5-F4-F2** (**FILES-CADL-EXEC**), then enter **SP** and **RETURN**.

The **WEDGE** option is used to create a triangular prism. A wedge is created by defining its length along the X, Y, and Z axes, and its position in 3D space (Figure 22-11). A long and short length is defined along the X axis to create the wedge.

1. Select **F6 WEDGE** from the Solids Primitive Main Menu.
2. A prompt reads:

> **Choose X-Y plane option.**

This is the method used to define the plane that will contain the base of the wedge. There are three options: **VIEW**, **WORLD**, and **DEFINE**. For this example, select **F1 VIEW**.

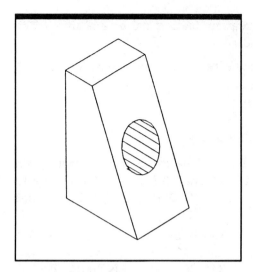

Figure 22-15   The model after the cylinder is subtracted from the wedge

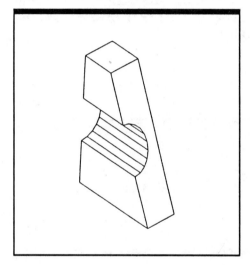

Figure 22-16   A section view of the model

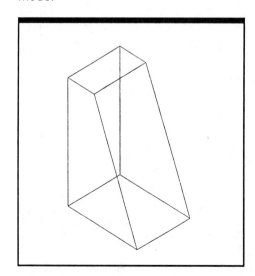

Figure 22-17   The wedge used for Boolean operations

3. A prompt reads:

Displacement along X axis for larger face (1.000000) =.

Enter a value of **4** and **RETURN.**

4. A prompt reads:

Displacement along X axis for the smaller face (1.000000) =.

Enter a value of **1.5** and **RETURN.**

5. A prompt reads:

Length along Y axis (1.000000) =.

Enter **6** and **RETURN.**

6. A prompt reads:

Length along Z axis (1.000000) =.

Enter **3** and **RETURN.**

7. A prompt reads:

Indicate corner point.

Select **F9 KEY-IN** from the Position Menu and enter **0,0,0** to position the lower-left corner of the wedge. Enter **F2 YES** to place the wedge. Enter **ALT-V** then **7** and **ALT-A** to view the wedge (Figure 22-17).

The next steps describe how to create the cylinder and position it in 3D space so that it intersects the wedge.

1. Enter **ESC-F3 CYLNDER** from the Solids Primitive Main Menu.

2. A prompt reads:

Length of the cylinder (1.000000) =.

Enter the height of **6** and **RETURN.**

3. A prompt reads:

Radius of the cylinder (0.500000) =.

Enter a radius of **1** and **RETURN.**

4. A prompt reads:

Choose method for defining cylinder axis direction.

Use the **TWO PTS** option by entering **F1.**

5. A prompt reads:

Indicate 1st point along the axis.

Select **F9 KEY-IN** from the Position Menu and enter world coordinates **-1,-1.5,3.**

6. A prompt reads:

Indicate 2nd point along the axis.

Enter **5,-1.5,3.**

7. An arrow is drawn on screen indicating the axis position (Figure 22-18). A prompt reads:

   Draw cylinder in arrow direction or opposite direction?

   Select **F1 ARROW**.

8. A prompt reads:

   Indicate position for base face center.

   Locate the center of the base of the cylinder by selecting **F9 KEY-IN** from the Position Menu. Enter **-1,-1.5,3** to create the model shown in Figure 22-19. Enter **F2 YES** to accept the cylinder position.

9. Save the geometry as a CADL file by entering **ESC-ESC-F1 (OUTPUT)**. Enter a filename of **WEDGECYL** and **RETURN**. Enter **F6-F3-F9 (GRP TBL-ALL ENT-DONE)**.

10. Erase the drawing from the screen by entering **ESC-F8-F1-F7-F1 (DELETE-SELECT-ALL DSP-ALL)**. Load the CADL file by entering **ESC-F5-F4-F2 (FILES-CADL-EXECUTE)**, and enter **WEDGECYL**.

BOOLEAN OPERATIONS

The following steps demonstrate how to use Boolean operations to create different objects.

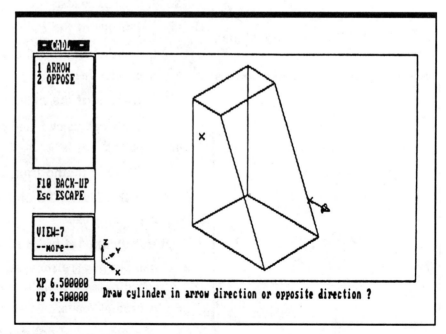

Figure 22-18 A temporary arrow to indicate the direction of the cylinder's axis

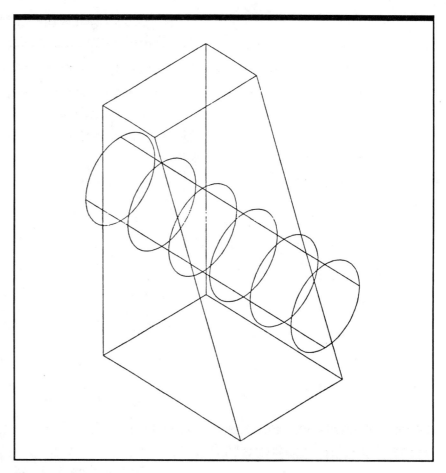

Figure 22-19   The completed wedge and cylinder ready for Boolean operations

The Boolean command is located on the Frames menu located on the Solids main menu. The Solids program is loaded by entering **ESC-F9 (APPLIC)**. After entering the Solids menu select **F4 CONFIG** to preset any system, light, shading, or mass property variables. For this example, all system variables are left in their default settings because only Boolean operations are to be executed then displayed.

Select **F5 FILES** from the Solids main menu. The Files dialogue box is displayed (Figure 22-20). Pick the **MODEL IN** button and enter **WEDGECYL** as the CDL file name. Set Render Out to **YES**, then select **NO** for all the other options. Name the frame file **BOOLS**. Select **DONE** to save the settings.

1. To begin the Boolean procedures select **F1 FRAMES** from the Solids main menu.

2. Select **Boolean** from the Frames dialogue box.

3. The Boolean Menu is displayed (Figure 22-21). There are three basic Boolean operations: union, intersection and difference. The **UNION** option creates a new object by adding two objects together. The **INTER** option creates a new object by determining the common volume of two selected entities.

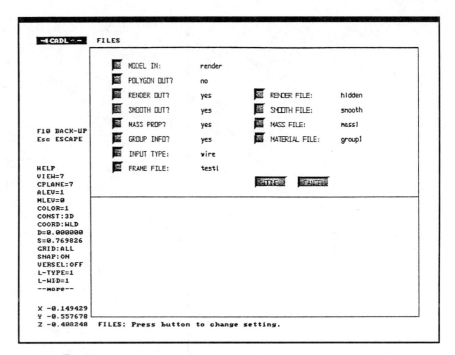

Figure 22-20 The files dialogue box

The **DIFF** option creates a new model by subtracting the volume of the second group from the first. The **PLANE** option on the Boolean Menu is used to create a section view of a solid object by defining a plane of intersection. The **LIST** option is used to display a table showing defined Boolean operations, their associated groups, and the frames where they are used. The **LINEFILE** option creates a CADL file of the resulting Boolean operations so that

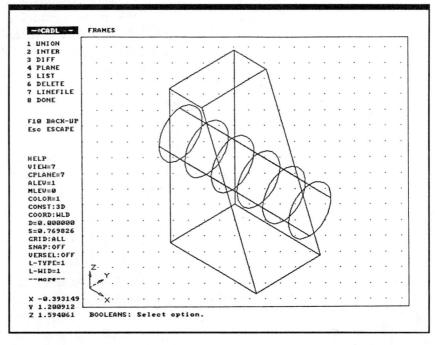

Figure 22-21 The Boolean menu

the model can be edited in CADKEY. To create the first Boolean operation select **F1 UNION** from the Boolean Menu.

4. There are two options: **SELECT** and **KEYIN**. Use **KEYIN** to enter the group number of the objects to be unioned.

5. Use **SELECT** to cursor pick the objects. For this example, choose **F1 SELECT**.

6. A prompt reads:

    Select group.

    Use the cursor to pick the wedge.

7. A prompt reads:

    Choose method for selecting second group of union.

    Enter **F1 SELECT**, then cursor pick the cylinder.

8. Select **F5 LIST** to display the Boolean operations, groups of objects, and frame number for the first Boolean process (Figure 22-22).

9. Each Boolean operation must be displayed using a different frame. Select **F8 DONE** to return to the Frames dialogue box.

10. Select **ADD** to create the second frame.

11. Pick **BOOLEAN** to display the Boolean menu.

12. Select **F2 INTER** to begin the process to create an intersection operation.

13. Use **SELECT** to cursor pick first the wedge, then the cylinder.

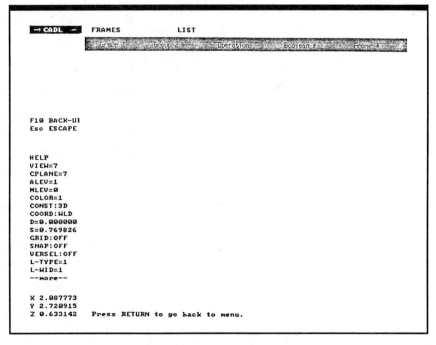

Figure 22-22   Using the LIST option to view the Boolean settings

14. Use **F5 LIST** to check the Boolean operation and groups assigned to frame 2.

15. Select **F8 DONE** to return to the Frames dialogue box.

16. Select **ADD** to create the third frame.

17. Pick **BOOLEAN** to display the Boolean menu.

18. Select **F3 DIFF** to begin the process to create a difference operation.

19. Use **SELECT** to cursor pick first the wedge, then the cylinder.

20. Use **F5 LIST** to check the Boolean operation and groups assigned to frame 3.

21. Select **F8 DONE** to return to the Frames dialogue box.

22. Select **ADD** to create the fourth frame.

23. Pick **BOOLEAN** to display the Boolean menu.

24. Select **F4 PLANE** to begin the process to create a plane that will section the objects.

25. Use **KEYIN** to define the position of the cutting plane in 3D space.

26. A prompt reads:

> **Enter group number (5):.**

Select the default group number by entering **RETURN**.

27. A menu is displayed with five selection options to define the cutting plane. A prompt reads:

> **Identify plane.**

Select **F4 VW/DPTH** to define the cutting plane by specifying a view and a depth.

28. A prompt reads:

> **Enter view number from 1 –9 =>:.**

Enter view number **2** and **RETURN**.

29. A prompt reads:

> **Choose option for setting depth:.**

Two options are available: **VALUE** and **POSITION**. Select **F2 POSITN**.

30. The Position menu is displayed. Select **F4 CENTER**.

31. Pick the base circle of the cylinder to define the depth of the cutting plane as the center of the cylinder and parallel to the selected view. A temporary square with diagonal dashed lines is drawn indicating the position of the cutting plane just defined (Figure 22-23).

32. A prompt reads:

> **Accept plane (YES)?**

Select **F2 YES** to accept the displayed position for the plane.

33. A prompt reads:

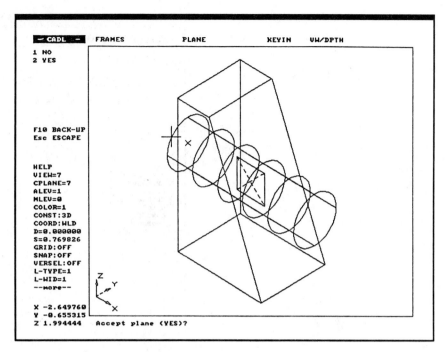

Figure 22-23 Defining the cutting plane

Select side of plane to keep.

Using **ENDENT** from the Position menu, pick an end point of one of the lines at the rear of the wedge.

34. Select **F5 LIST** to display the Boolean operations assigned to each frame.

35. Select **F8 DONE** to complete the Boolean operations and return to the Frames dialogue box.

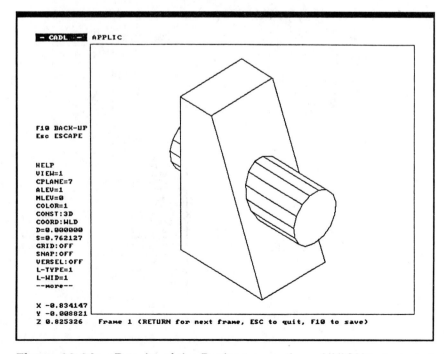

Figure 22-24a Results of the Boolean operations: UNION

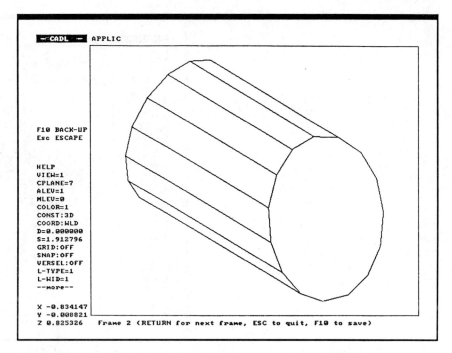

Figure 22-24b Results of the Boolean operations: INTER

36. Click on **DONE** to return to the Solids main menu.

37. Process the frames by selecting **F2 PROCESS**, then **F2 STAT RUN**.

38. After the processing is completed, enter **F3 RESULTS**, then **F1 RENDERED** to view the Boolean operations (Figure 22-24a–d, starting on page 547).

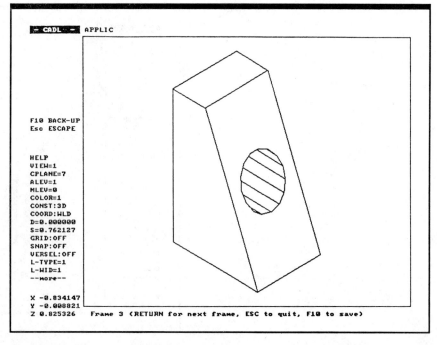

Figure 22-24c Results of the Boolean operations: DIFF

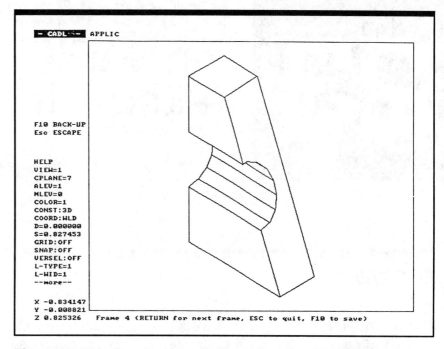

```
- CADL -- APPLIC




F10 BACK-UP
Esc ESCAPE


HELP
VIEW=1
CPLANE=7
ALEV=1
MLEV=0
COLOR=1
CONST:3D
COORD:WLD
D=0.000000
S=0.827453
GRID:OFF
SNAP:OFF
VERSEL:OFF
L-TYPE=1
L-WID=1
--more--

X -0.834147
Y -0.008821
Z 0.825326    Frame 4 (RETURN for next frame, ESC to quit, F10 to save)
```

Figure 22-24d   Results of the Boolean operations: PLANE

REVIEW QUESTIONS

1. List the solid primitives that can be created with CADKEY Solids.
2. Name the CADL program that is executed to create primitive shapes.
3. List the VALUES that must be entered to create a

> **BLOCK**
> **CONE**
> **CYLINDER**
> **SPHERE**
> **TORUS**
> **WEDGE**

4. List the Boolean operations that can be done with CADKEY Solids.
5. List the properties that can be assigned to a model.

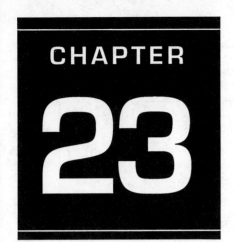

CHAPTER

23

Solids Modeling for CADKEY's "PICTURE IT"

OVERVIEW

PICTURE IT is not just CADKEY SOLIDS, fully integrated. SOLIDS worked on the principle of tessellating (breaking down) an entire model into linear segments to which it applied polygons and then combined those segments in a logical manner.

PICTURE IT is a shape recognition and rendering system. Most of the modeling techniques used for solids modeling are the same for PICTURE IT. However, there are a few exceptions. Consult your PICTURE IT documentation for a complete list. Many of the same results that were available to you with SOLIDS are also available with PICTURE IT. Smooth-shaded images are now projected inside of CADKEY to your viewport. See Figure 23-1.

PICTURE IT renders your 3-D wire-frame design as a shaded solid model. When you load and execute PICTURE IT, the program analyzes the displayed geometry in order to recognize

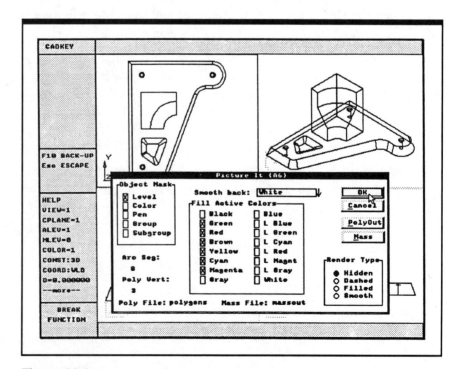

Figure 23-1

the solid objects implicit in your design. PICTURE IT then renders the entire model, hiding those portions of any surface that appear behind another.

Previously, CADKEY SOLIDS used geometry associated by the GROUP function to distinguish the individual solid components of your model. With PICTURE IT, you don't need to group geometry to render your model. However, PICTURE IT does depend on you to resolve any ambiguities in your design. Ambiguities can occur where the geometry of one solid component of your part is adjacent to another component, or in cases where an explicit hole exists in the part. You resolve ambiguities by assigning special attributes, such as color, to the distinguishing features of your wire-frame design. Such topics will be covered later in this chapter.

If you have already made groups to separate your model into distinct solid components, don't worry. PICTURE IT can still distinguish solid objects by group or subgroup. However, you'll probably find that PICTURE IT's enhanced ability to distinguish between solid objects within your wire-frame model by attribute type makes rendering fast and easy.

GEOMETRY CONSTRUCTION

CADKEY PICTURE IT works with the currently loaded part file and displayed levels. You must have the PICTURE IT CDE loaded (open) before you can execute it. If it is not loaded, choose **FILES, CDE, OPEN** and type "Picture." This will open or load the CDE. The next step is to actually execute the CDE. This can be done in one of two ways. The first is to choose it off of the **APPLIC** menu. The second method is to use the **FILES, CDE, EXECUTE** command and type in "picture_it."

Step 1:   Create a model. The construction techniques used for SOLIDS will also work for PICTURE IT, with two exceptions. In PICTURE IT, four curve intersections (where four curves meet), are automatically recognized, no meshing is required. Two curve "surfaces" that meet at the same position or are joined with a single line will not be recognized properly. It is necessary to add additional geometry for correct processing. (See Construction section following for detailed instructions.)

Step 2:   Choose **PICTURE IT** from the **APPLIC** menu. This brings up the PICTURE IT dialog box. There are eight sections and four buttons in the dialog box. See Figure 23-2.

Step 3:   Choose **OK**. This starts the PICTURE IT CDE with the default settings.

GEOMETRIC CONSTRUCTION TIPS AND TECHNIQUES

Extrusions   All extrusions (profiles that project normal or skewed) must have all corresponding endpoints joined. When you use the XFORM options for profiling, use the JOIN option. See Figure 23-3.

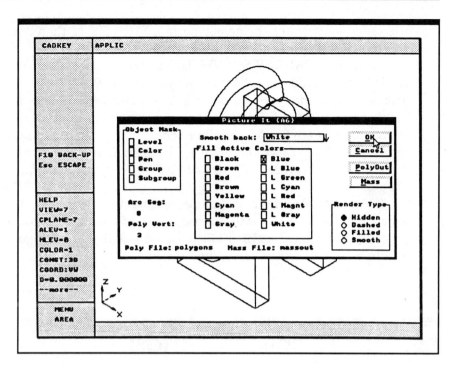

Figure 23-2

Inserting Fillets

You must join fillets with join lines as well. See Figure 23-4.

Through Holes and Cylinders

All through hole entities (entities that go from one plane to another) must have lines that join corresponding endpoints. Cylinders are represented by two circles that have the same definition view, joined with a line. PICTURE IT, a shape recognition system, sees this combination and breaks it into four 90 degree arcs. This is why circles seem to have more segments than a standard arc or fillet. See Figure 23-5.

Two-Curve Sections (Two-Sided "Surfaces")

An area bound by two curves that form just two corners between them does not define a unique surface to PICTURE IT. You need to give

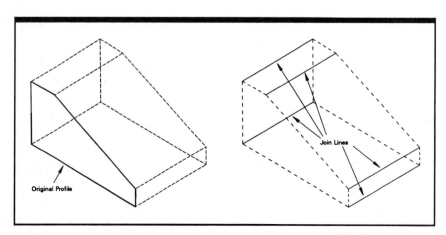

Figure 23-3

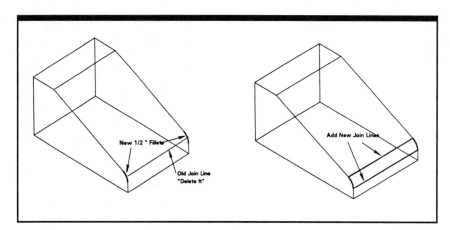

Figure 23-4

PICTURE IT more information, either by creating a line that segments the arc into two pieces as shown in Figure 23-6, or by meshing the area with polygons. After creating the line in the first option, you must then break both arcs to that line.

Three-Curve Intersections (Three-Curve Mesh Generation)

When three sides of an object are filleted or curved where they meet, it usually results in a three curve intersection. PICTURE IT does not know how to interpret this situation. If you place a polygon mesh over the three curves, PICTURE IT will then shade the area for you. It is not necessary to mesh the other corners after you successfully mesh one corner. Simply use the **XFORM, MIRROR** command and transform the first mesh to the opposite corner and then mirror those two meshes from one side to the other so that all corners will be meshed. See Figure 23-7.

Building Your Model Modularly

The best technique for successfully processing your entire model through PICTURE IT is to break the model down into smaller pieces. If you can't figure out the pieces, PICTURE IT never will.

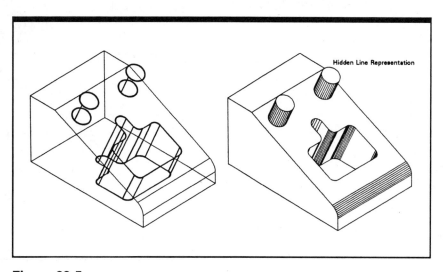

Figure 23-5

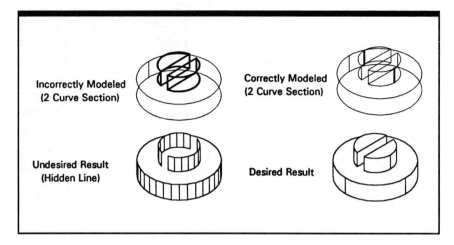

Figure 23-6

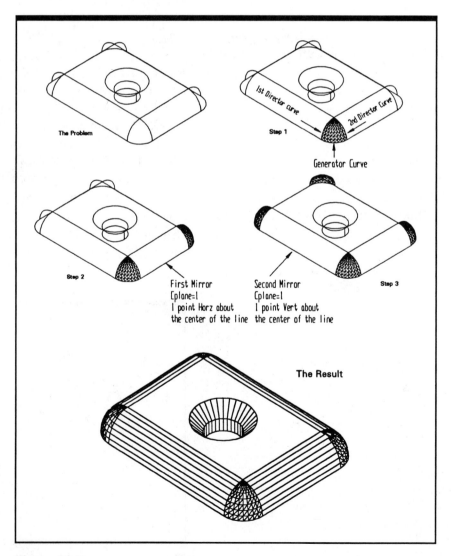

Figure 23-7

Placing the different pieces of your model on different levels will help you to keep organized. Make sure you process each piece successfully before moving on to the rest of your model. See Figure 23-8.

Explicit Holes

A more complicated situation exists for the case where lines (or other curves) join the geometry defining a hole to the external surface boundaries of the object. In such a case, it can be ambiguous which closed volumes describe a hole and which describe solid portions of the object. Figure 23-9 shows an example where the explicit hole geometry is inseparable from the geometry defining the external edges.

To resolve the ambiguity, assign the phantom line style - Line Style 4, to a closed set of curves that together describe a cross-sectional cut through the solid. This phantom cut gives PICTURE IT the information it needs to correctly render objects with explicit holes.

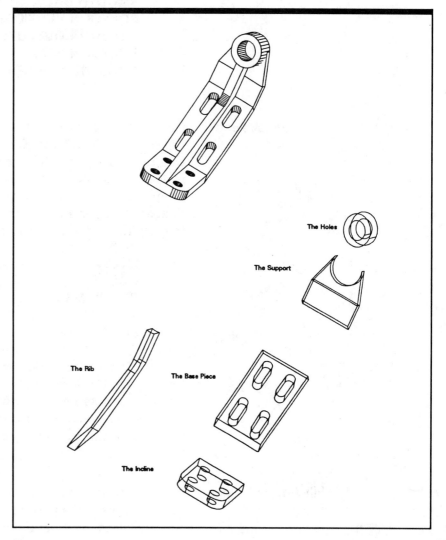

Figure 23-8

OPTIONS

ASSIGNING ATTRIBUTES

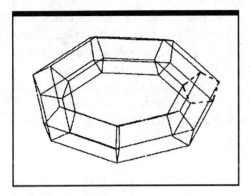

Figure 23-9

PICTURE IT can render a simple, single component part without assigning special attributes. However, when your wire-frame design describes an assembly of component parts, PICTURE IT requires more information to distinguish one component, or object, from another. You provide this information by assigning different attributes to distinguish one object from another. PICTURE IT can distinguish between objects on the basis of any combination of the attribute types listed below.

The sections following describe the PICTURE IT options available through the PICTURE IT and Mass Properties dialog boxes:

> **OBJECT MASK**
> **FILL ACTIVE COLORS**
> **RENDER TYPE**
> **ARC SEGMENTATION**
> **SMOOTH BACK**
> **POLYGON VERTICES**
> **POLYGON OUTPUT**
> **POLYGON FILE**
> **MASS PROPERTIES AND MASS PROPERTY OUTPUT**

Object Mask

Choosing one or more of the attribute types listed below defines the Object Recognition Mask (Object Mask) that PICTURE IT uses to distinguish one object from another. Before analyzing your wire-frame design, the program separates the entities according to each unique combination of attributes that you choose. PICTURE IT then analyzes entities sharing a common set of attributes in order to recognize individual objects and their features.

> **LEVEL**
> **COLOR**
> **PEN NUMBER**
> **GROUP**
> **SUBGROUP**

PICTURE IT considers any objects that share the same Object Recognition Mask attribute set as a single object when calculating mass properties for that attribute set. If you choose "Level" only, PICTURE IT treats objects on the same level as if they are connected by rigid, massless rods. See Figure 23-10.

Fill Active Colors

The PICTURE IT palette contains 256 colors derived from CADKEY's 16 base colors. You can choose from 1 to 16 of these as "Fill Active Colors" and define a shading range for each. PICTURE IT renders any objects drawn in a chosen "Fill Active Color" in shades

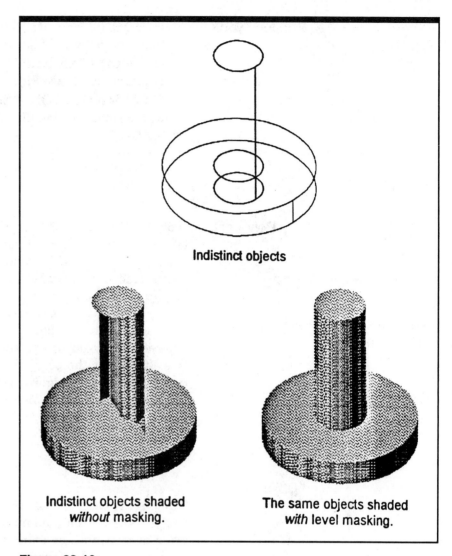

Indistinct objects

Indistinct objects shaded
without masking.

The same objects shaded
with level masking.

Figure 23-10

of that color. PICTURE IT shades objects drawn in any other color to the shading range you define for the first "Fill Active Color" you choose.

Suppose your wire-frame model uses all 16 base colors, and you turn on only six "Fill Active Colors" in the PICTURE IT dialog box. The ten colors that aren't chosen default to the first color having an (x) in its check box. PICTURE IT renders objects drawn in these colors in shades of the first "Fill Active Colors" check box you choose. For example, if you choose green, red, yellow, magenta, and blue as "Fill Active Colors," and portions of your wire-frame model are brown, those portions appear as shades of green in the shaded image. This happens because the green check box appears first out of the colors you chose as "Fill Active Colors."

NOTE: If your graphics configuration is for a 16-color card, PICTURE IT approximates the shading ranges you assign by rendering your model in dithered shades of the chosen "Fill Active Colors."

Render Types

HIDDEN LINE REPRESENTATION—Displays the wire-frame model with hidden lines removed. See Figure 23-11.

DASHED LINE REPRESENTATION—Changes hidden lines to dashed lines. See Figure 23-12.

FILLED POLYGON REPRESENTATION—Shows the model with faceted shading. See Figure 23-13.

SMOOTH—Lets you create a smoothly shaded image of your solid model so it looks like a real object. Surfaces appear smooth rather than faceted or plated.

Poly Vert

"Poly Vert" is CADKEY shorthand for polygon vertices. This option lets you specify the maximum number of vertices in the polygons PICTURE IT uses when rendering a solid model. PICTURE IT creates surface-fill polygons with the largest number of vertices possible, up to the number you assign.

For example, suppose PICTURE IT creates a polygon surface-fill for a square surface boundary. PICTURE IT can either create one four-sided polygon or two three-sided polygons to render the surface. If you set the polygon vertices to three, PICTURE IT creates two three-sided polygons. If you choose a value of 4 or greater, PICTURE IT creates a single four-sided polygon. The more polygons PICTURE IT creates, the longer the processing takes. For PICTURE IT, you can choose any value from 3 through 8. The default is 3.

Arc Seg

"Arc Seg" is CADKEY shorthand for arc segmentation. This option lets you specify the minimum number of segments PICTURE IT divides arcs, conics, or splines into when creating the polygons to render your model. The higher the value you enter, the smoother your model looks. You can choose any value from 4 through 1000.

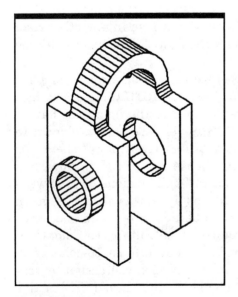

Figure 23-11

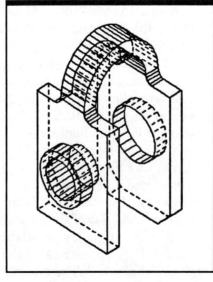

Figure 23-12

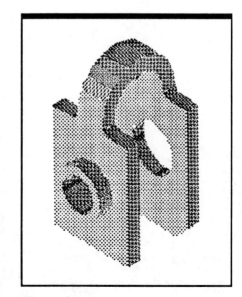

Figure 23-13

The default is 4. PICTURE IT may subdivide an arc into fewer segments than the "Arc Seg" value if that value results in segments shorter than the minimum size CADKEY can process.

Smooth Back   The "Smooth Back" option lets you specify a background color for your smooth-shaded model. You can choose from any of CADKEY's 16 base colors. When you use the Smooth option, your model appears in the background color that you choose with this option.

PolyOut   "PolyOut" saves the polygons PICTURE IT uses to display your filled or smoothly-shaded model to both a CADL file (.cdl), and a stereo lithography (.sla) file in the default cdl directory.

The CADL file saves polygon outlines (no fill).

You can use the polygon .cdl file as a solid model for the CAD-KEY Advanced Modeler, and the .sla file for third party stereo lithography packages.

Click on the "PolyOut" option button. PICTURE IT writes the polygon model to two files with the same default filename. One file has a .cdl filename extension and the other a .sla extension.

Poly File   "Poly File" assigns a name for both the default CADL file and .sla file storing the polygon output that the PolyOut option generates from your model.

Mass   "Mass" lets you perform an analysis of the mass (inertial) properties of one or more objects making up your design. PICTURE IT individually analyzes the mass properties of object(s) within each Object Recognition Mask attribute set. Define the density and angular velocity for each object, or attribute set, to compute its results.

The units of the mass properties results agree with the units of the Object Properties you enter for each object or attribute set. For example, if you configure CADKEY units of length as centimeters, then the corresponding object data units are as follows:

STANDARD UNITS BASED ON CENTIMETERS, GRAMS, AND SECONDS

Density	g/cm^3
angular velocity	radians per second
volume	cm^3
area	cm^2
mass	g
moment of inertia	$g(cm)^2$
kinetic energy	$g(cm^2/sec^2)$
angular momentum	$g(cm^2/sec)$
radius of gyration	cm

Object Properties

Object properties are values you assign to the density and angular velocity of objects within your model. Enter values of the following parameters:

> **DENSITY**—Mass per unit volume. The units of measure must agree with the wire-frame length units you specify when configuring CADKEY. See "Set Program Options" in the "Configuration" chapter of Getting Started for more information.
>
> **AWD**—Area-weighted density is mass per unit area of the faceted surface enveloping a solid object. Think about this as the density of a material coating the solid. This value depends on the Density you assign here and the current Arc Seg and Poly Vert assignments in the PICTURE IT dialog box. PICTURE IT uses AWD, not Density, to calculate mass property results.
>
> **ANG VEL**—Angular velocity is the rate of rotation or revolution about a point. The value you enter can refer to a revolution about the origin or a rotation about the object's centroid. PICTURE IT calculates results for both types of angular motion. Angular velocity is expressed in terms of three vectors, one component for each of the angular velocities about the x, y, and z axes of the display view. A typical unit for angular velocity is radians per second.

MASS PROPERTY RESULTS

PICTURE IT calculates the following results for each attribute set within your model:

> **VOLUME**—Total space occupied by the faceted solid.
>
> **AREA**—Total area of the faceted solid.
>
> **MASS**—The product of Volume times Density.
>
> **SURF MASS**—The product of Area times AWD.

The Mass Properties dialog box lists results for each Object Recognition Mask attribute set. Each attribute set consists of one or more objects having common attributes from chosen attribute types. See "Object Mask" in this section for more information.

You can enter a value signifying an angular velocity about either the centroid or the origin. PICTURE IT uses angular velocity to calculate the following results:

> **MOMENT OF INERTIA**—The angular equivalent of mass, moment of inertia expresses the resistance of an object to rotational change, or to stopping rotation, about a given axis.
>
> **Ixx**—Moment of inertia about the centroid x axis or an axis parallel to the origin x axis.

Iyy—Moment of inertia about the centroid y axis or an axis parallel to the origin y axis.

Izz—Moment of inertia about the centroid z axis or an axis parallel to the origin z axis.

Ixy—Moment of inertia about the diagonal axis in the centroid xy plane or an axis parallel to the origin xy plane.

Iyz—Moment of inertia about the diagonal axis in the centroid yz plane or an axis parallel to the origin yz plane.

Ixz—Moment of inertia about the diagonal axis in the centroid xz plane or an axis parallel to the origin xz plane.

KINETIC ENERGY—The energy of a rotating or revolving object: K.E.$=I\omega^2$, where I and ω represent the sum of the moment of inertia about the vector sum of the component angular velocities. PICTURE IT calculates this value from average weighted density and the value you entered for angular velocity. This calculated value refers only to energy due to the object's angular motion.

ANGULAR MOMENTUM—The resistance to change in angular velocity. L = Iω PICTURE IT calculates this from the values for moment of inertia, angular velocity, and average weighted density.

RADIUS OF GYRATION—The ratio of moment of inertia to volume. This is the same as the radius of a thin hoop of the same mass that has the same moment of inertia as the object.

APPENDIX A

The CADKEY Menus

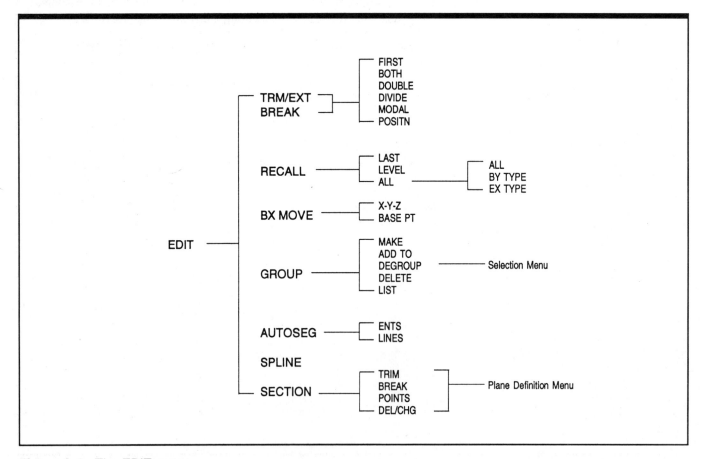

Figure A-1 The EDIT menu

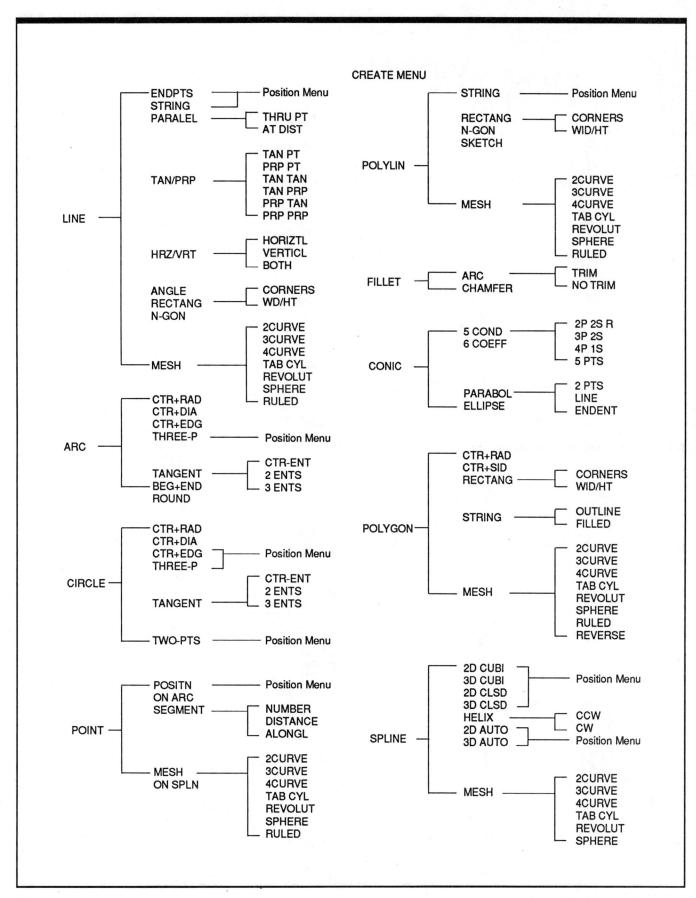

Figure A-2   The CREATE menu

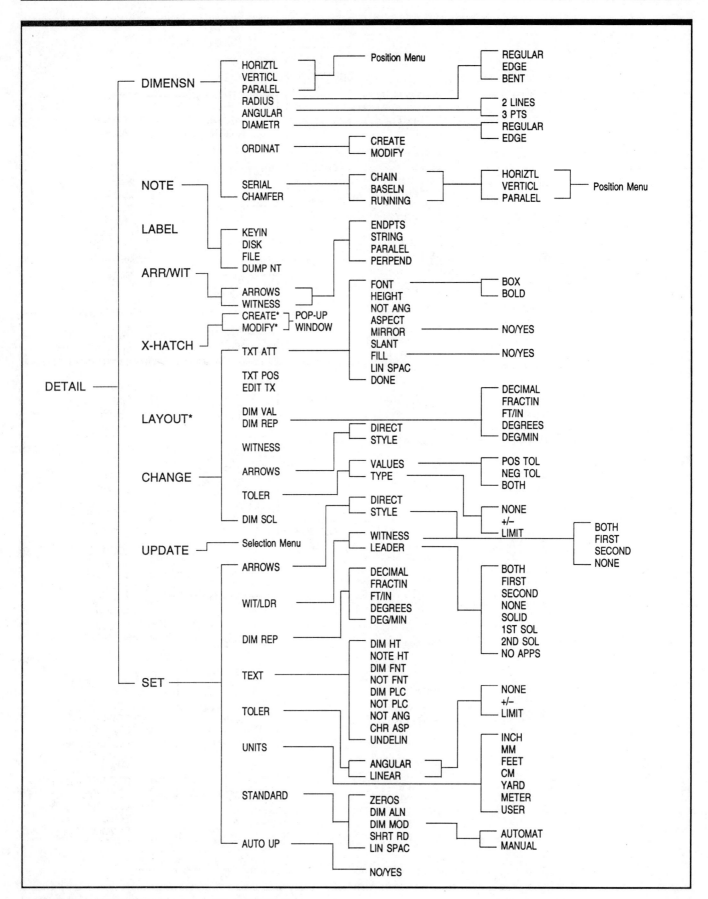

Figure A-3 The DETAIL menu (*CADKEY 6)

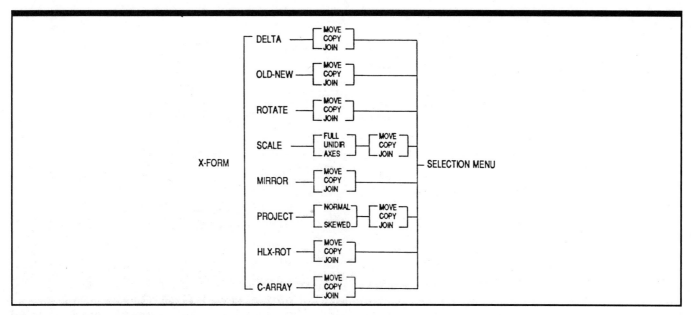

Figure A-4   The X-FORM menu

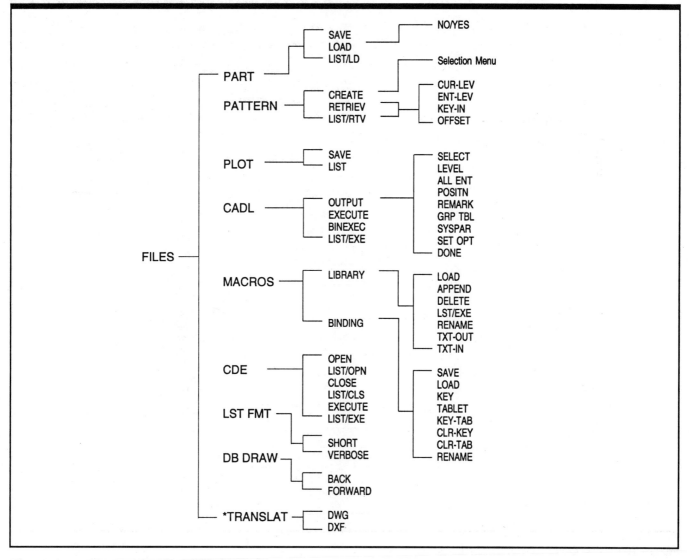

Figure A-5   The FILES menu (*CADKEY 6)

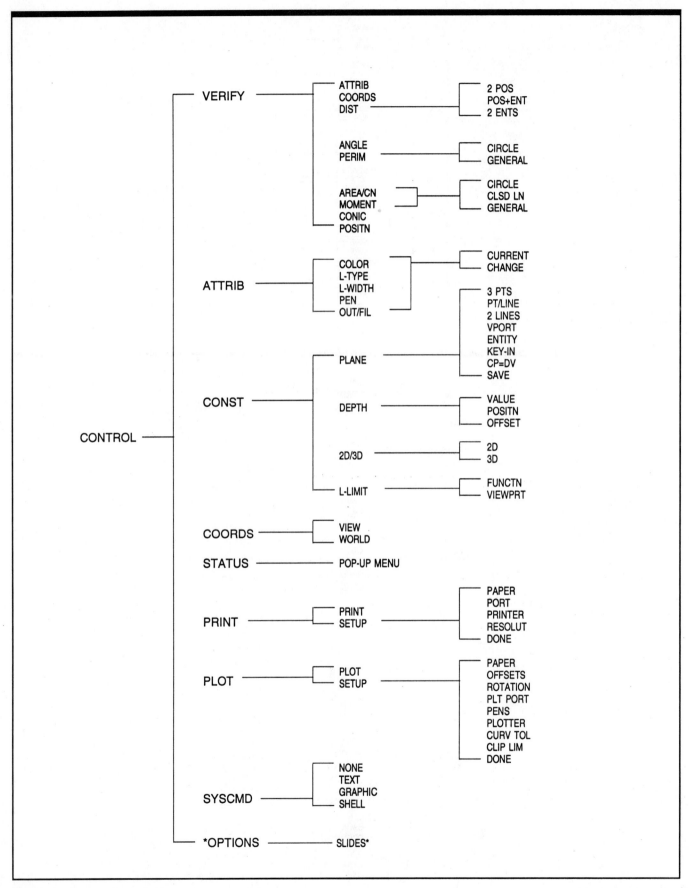

Figure A-6 The CONTROL menu (*CADKEY 6)

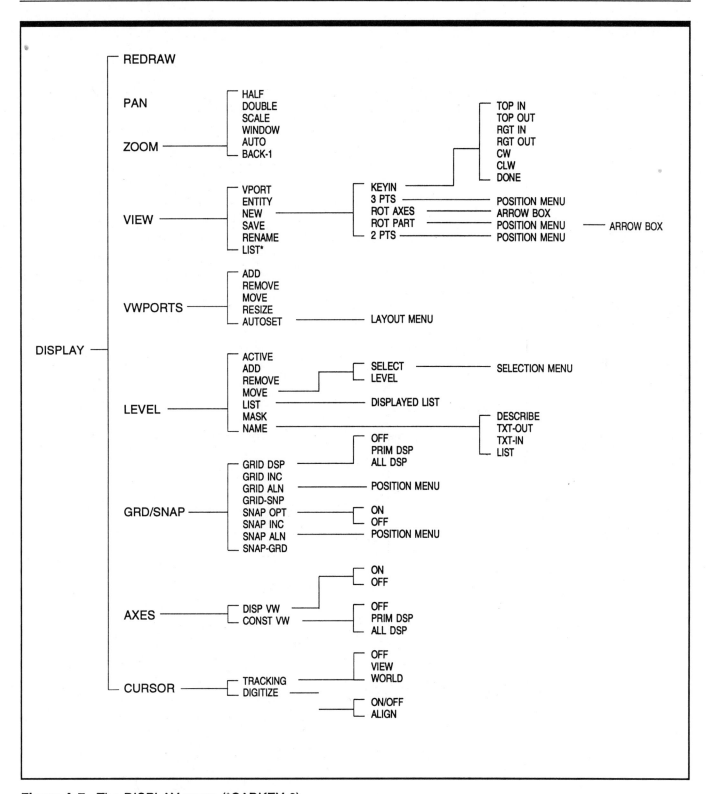

Figure A-7   The DISPLAY menu (*CADKEY 6)

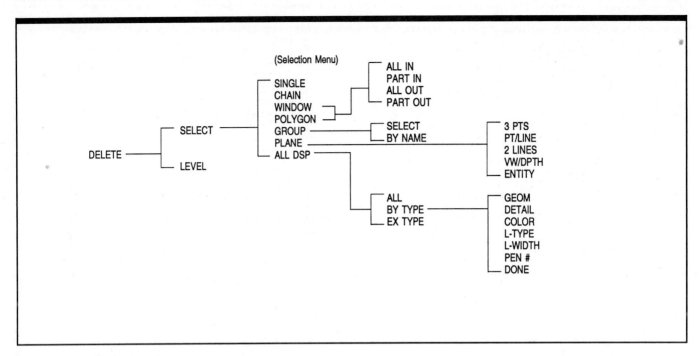

Figure A-8 The DELETE menu

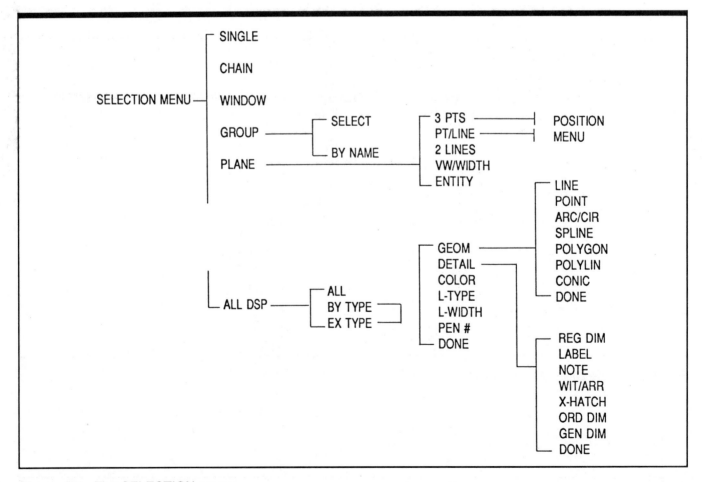

Figure A-9 The SELECTION menu

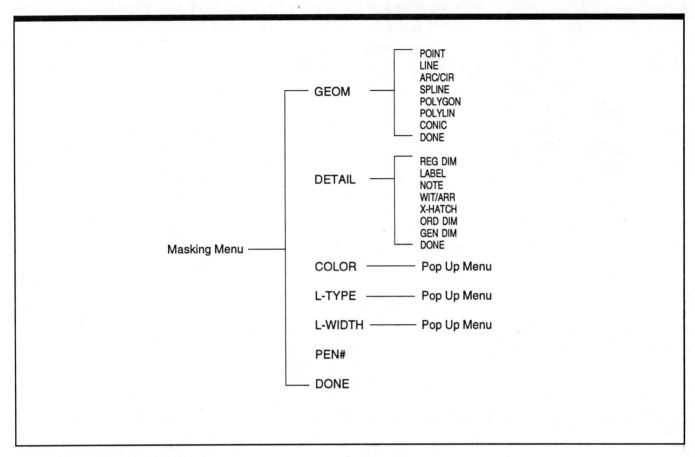

Figure A-10 The MASKING menu

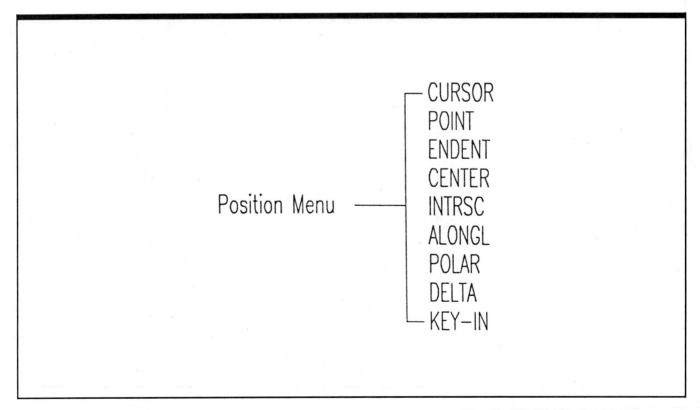

Figure A-11 The POSITION menu

CADKEY Problems and Solutions

PROBLEM 1

PROBLEM STATEMENT Draw the two given views and add the missing view.

CADKEY Setup Create a **SETUP** file.
Change to view 2, **ALT-V** and enter **2**.
Turn on snap, **CTRL-X**.
Cut the snap increment in half (press **PgDn** key).
Turn on cursor tracking coordinates, **CTRL-T**, and select **VIEW**.
Turn on grid, **CTRL-G**, and select **ALL DSP**.
Enlarge the drawing window, **ALT-H**.
Save **SETUP** file, **CTRL-F** and enter filename **SETUP**.

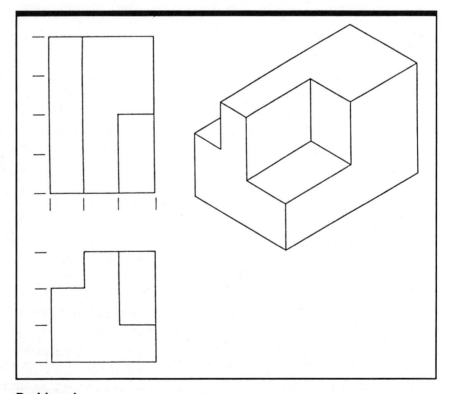

Problem 1

PROCEDURE

Approach 1: Draw Mode

1. Use the **CREATE-LINE-STRING** or **RECTANG** command to define the outside boundary of the views.

2. Use the **CREATE-LINE-ENDPTS** command to place individual lines for interior details.

3. **ALT-T** to change line type to hidden (2).

Approach 2: Orthographic Projection

1. Use the **CREATE-LINE-STRING** and **ENDPTS** commands to create the front view.

2. Create a 45 degree mitre line off the front view.

3. Use **CREATE-LINE-HRZ/VRT-VERTICL-ENDENT** to project the width of the front view to the top view.

4. Use **CREATE-LINE-HRZ/VRT-HORIZTL** to project the height of the front view to the right side view.

5. Use the **CREATE-LINE-HRZ/VRT** command to transfer depth to mitre line and between top and right side view.

6. Use **EDIT-TRM/EXT-DOUBLE** or **FIRST** to trim projection lines between the appropriate edges.

Approach 3: Three-Dimensional Modeling

1. Use the **CREATE-LINE-STRING** and **ENDPTS** commands to create the front view.

2. Once the front view has been created, change to view 7 (**ALT-V** and enter **7**) and use **CTRL-V** to change to world coordinates.

3. At this time you are ready to add the depth dimension to the height and width of the front view. Use **X-FORM-DELTA-JOIN** and whichever selection option is appropriate.

4. Enter **1** when prompted to enter the number of copies. The copy does not move on the X axis (width); it should move at a negative distance equal to the depth for the Y axis and does not move on the Z axis (height). Therefore, enter **0** for the X and Z values and a negative value equal to the depth for the Y axis.

5. When adding additional lines, arcs, or circles in the isometric view *do not use the* **CURSOR** option from the Position Menu.

6. Use **ALT-V** to change the front, top or right side view and use **FILES-PATTERN-CREATE** to create a pattern file for each view.

7. Save the wireframe model (**CTRL-F**) and **FILES-PART-LOAD** your **SETUP** file.

8. Use the **FILES-PATTERN-RETRIEVE** option to retrieve a border and all of the desired views.

PROBLEM 2

PROBLEM STATEMENT   Draw the two given views and add the missing view.

CADKEY Setup   Load the **SETUP** file created in Problem 1.
FILES-PART-LOAD-NO. Enter filename **SETUP**.

PROCEDURE

Approach 1: Draw Mode

1. Use the **CREATE-LINE-STRING** or **RECTANG** command to define the outside boundary of the views.
2. Use the **CREATE-LINE-ENDPTS** command to place individual lines for interior details.
3. Use the Position Menu option **ENDENT** and select opposite corners of the side view for the diagonal line.

Approach 2: Orthographic Projection

1. Use the **CREATE-LINE-STRING** and **ENDPTS** commands to create the front view.
2. Create a 45 degree mitre line off front view.
3. Use **CREATE-LINE-HRZ/VRT-VERTICL-ENDENT** to project the width of the front view to the top view.

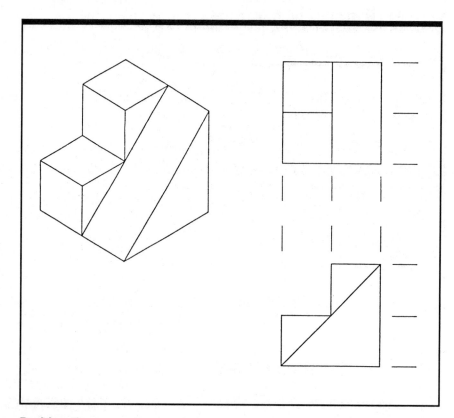

Problem 2

4. Use **CREATE-LINE-HRZ/VRT-HORIZTL** to project the height of the front view to the right side view.

5. Use the **CREATE-LINE-HRZ/VRT** command to transfer depth to mitre line and between top and right side views.

6. Use **EDIT-TRM/EXT-DOUBLE** or **FIRST** to trim projection lines between the appropriate edges.

Approach 3:
Three-Dimensional Modeling

1. Use the **CREATE-LINE-STRING** and **ENDPTS** commands to create the front view.

2. Once the front view has been created change to view 7 (**ALT-V** and enter **7**) and use **CTRL-V** to change to world coordinates.

3. At this time you are ready to add the depth dimension to the height and width of the front view. Use **X-FORM-DELTA-JOIN** and whichever selection option is appropriate.

4. Enter **1** when prompted to enter the number of copies. The copy does not move on the X axis (width); it should move at a negative distance equal to the depth for the Y axis and does not move on the Z axis (height). Therefore, enter **0** for the X and Z values and a negative value equal to the depth for the Y axis.

5. When adding additional lines, arcs, or circles in the isometric view *do not use the* **CURSOR** option from the Position Menu.

6. Use **ALT-V** to change the front, top or right side view and use **FILES-PATTERN-CREATE** to create a pattern file for each view.

7. Save the wireframe model (**CTRL-F**) and **FILES-PART-LOAD** your **SETUP** file.

8. Use the **FILES PATTERN-RETRIEVE** option to retrieve a border and all of the desired views.

PROBLEM 3

PROBLEM STATEMENT   Draw the two given views and add the missing view.

CADKEY Setup   Load the **SETUP** file created in Problem 1.
FILES-PART-LOAD-NO. Enter filename **SETUP**.

PROCEDURE

Approach 1: Draw Mode

1. Use the **CREATE-LINE-STRING** or **RECTANG** command to define the outside boundary of the views.

2. Use the **CREATE-LINE-ENDPTS** command to place individual lines for interior details.

3. Use **ALT-T** to change line type to hidden.

Approach 2: Orthographic Projection

1. Use the **CREATE-LINE-STRING** and **ENDPTS** commands to create the front view.

2. Create a 45 degree mitre line off front view.

3. Use **CREATE-LINE-HRZ/VRT-VERTICL-ENDENT** to project the width of the front view to the top view.

4. Use **CREATE-LINE-HRZ/VRT-HORIZTL** to project the height of the front view to the right side view.

5. Use the **CREATE-LINE-HRZ/VRT** command to transfer depth to mitre line and between top and right side views.

6. Use **EDIT-TRM/EXT-DOUBLE** or **FIRST** to trim projection lines between the appropriate edges.

7. Use **ALT-T** to change line type to hidden. Use the **EDIT-BREAK-FIRST** option to break the vertical line in the side view.

Approach 3: Three-Dimensional Modeling

1. Use the **CREATE-LINE-STRING** and **ENDPTS** commands to create the front view.

2. Once the front view has been created, change to view 7 (**ALT-V** and enter **7**) and use **CTRL-V** to change to world coordinates.

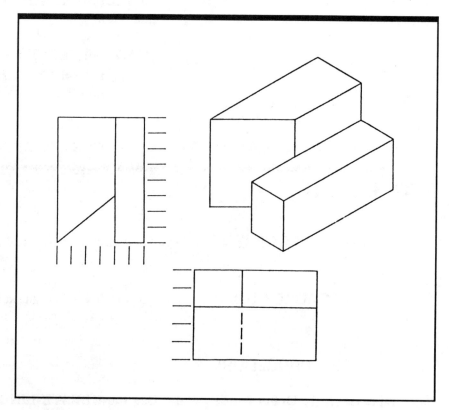

Problem 3

3. At this time you are ready to add the depth dimension to the height and width of the front view. Use **X-FORM-DELTA-JOIN** and whichever selection option is appropriate.

4. Enter **1** when prompted to enter the number of copies. The copy does not move on the X axis (width); it should move at a negative distance equal to the depth for the Y axis and does not move on the Z axis (height). Therefore, enter **0** for the X and Z values and a negative value equal to the depth for the Y axis.

5. When adding additional lines, arcs, or circles in the isometric view, *do not use the* **CURSOR** option from the Position Menu.

6. Use **ALT-V** to change the front, top or right side view and use **FILES-PATTERN-CREATE** to create a pattern file for each view.

7. Save the wireframe model (**CTRL-F**) and **FILES-PART-LOAD** your **SETUP** file.

8. Use the **FILES-PATTERN-RETRIEVE** option to retrieve a border and all of the desired views.

PROBLEM 4

PROBLEM STATEMENT   Pick the best front view and draw the top, front and right side views.

CADKEY Setup   Load the **SETUP** file created in Problem 1.
FILES-PART-LOAD-NO. Enter filename **SETUP**.

PROCEDURE

Approach 1: Draw Mode

1. Use the **CREATE-LINE-STRING** or **RECTANG** command to define the outside boundary of the views.

2. Use the **CREATE-LINE-ENDPTS** command to place individual lines for interior details.

3. Use **ALT-T** to change line type to hidden.

Approach 2: Orthographic Projection

1. Use the **CREATE-LINE-STRING** and **ENDPTS** commands to create the front view.

2. Create a 45 degree mitre line off the front view.

3. Use **CREATE-LINE-HRZ/VRT-VERTICL-ENDENT** to project the width of the front view to the top views.

4. Use **CREATE-LINE-HRZ/VRT-HORIZTL** to project the height of the front view to the right side views.

5. Use the **CREATE-LINE-HRZ/VRT** command to transfer depth to mitre line and between top and right side views.

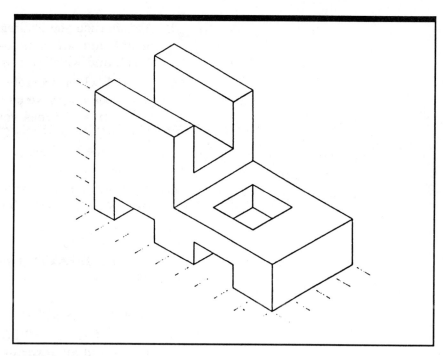

Problem 4

6. Use **EDIT-TRM/EXT-DOUBLE** or **FIRST** to trim projection lines between the appropriate edges.

7. Use **ALT-T** to change line type to hidden. Use the **EDIT-BREAK-DOUBLE** option to break the vertical line in the top view.

Approach 3: Three-Dimensional Modeling

1. Use the **CREATE-LINE-STRING** and **ENDPTS** commands to create the front view.

2. Once the front view has been created, change to view 7 (**ALT-V** and enter **7**) and use **CTRL-V** to change to world coordinates.

3. At this time you are ready to add the depth dimension to the height and width of the front view. Use the **X-FORM-DELTA-JOIN** and whichever selection option is appropriate.

4. Enter **1** when prompted to enter the number of copies. The copy does not move on the X axis (width); it should move at a negative distance equal to the depth for the Y axis and does not move on the Z axis (height). Therefore, enter **0** for the X and Z values and a negative value equal to the depth for the Y axis.

5. When adding additional lines, arcs, or circles in the isometric view *do not use the* **CURSOR** option from the Position Menu.

6. Use **ALT-V** to change the front, top or right side view and use **FILES-PATTERN-CREATE** to create a pattern file for each view.

7. Save the wireframe model (**CTRL-F**) and **FILES-PART-LOAD** your **SETUP** file.

8. Use the **FILES-PATTERN-RETRIEVE** option to retrieve a border and all of the desired views.

PROBLEM 5

PROBLEM STATEMENT

Pick the best front view and draw the top, front and right side views.

CADKEY Setup

Load the **SETUP** file created in Problem 1.
FILES-PART-LOAD-NO. Enter filename **SETUP**.

PROCEDURE

Approach 1: Draw Mode

1. Use the **CREATE-LINE-STRING** or **RECTANG** command to define the outside boundary of the views.
2. Use the **CREATE-LINE-ENDPTS** command to place individual lines for interior details.
3. Use **ALT-T** to change line type to hidden.

**Approach 2:
Orthographic Projection**

1. Use the **CREATE-LINE-STRING** and **ENDPTS** commands to create the front view.
2. Create a 45 degree mitre line off front view.

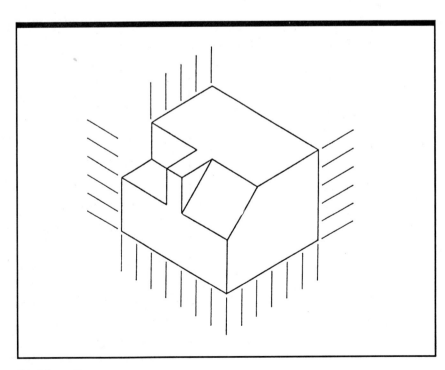

Problem 5

3. Use **CREATE-LINE-HRZ/VRT-VERTICL-ENDENT** to project the width of the front view to the top view.

4. Use **CREATE-LINE-HRZ/VRT-HORIZTL** to project the height of the front view to the right side view.

5. Use the **CREATE-LINE-HRZ/VRT** command to transfer depth to mitre line and between top and right side views.

6. Use **EDIT-TRM/EXT-DOUBLE** or **FIRST** to trim projection lines between the appropriate edges.

7. Use **ALT-T** to change line type to hidden.

PROBLEM 6

PROBLEM STATEMENT

Pick the best front view and draw the top, front and right side views.

CADKEY Setup

Load the **SETUP** file created in Problem 1.
FILES-PART-LOAD-NO. Enter filename **SETUP**.

PROCEDURE

Approach 1: Draw Mode

1. Use the **CREATE-LINE-STRING** or **RECTANG** command to define the outside boundary of the views.

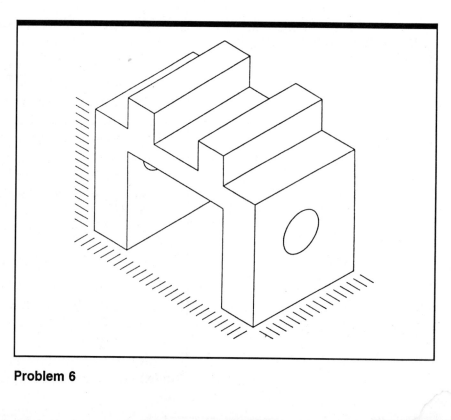

Problem 6

2. Use the **CREATE-LINE-ENDPTS** command to place individual lines for interior details.

3. Use **CREATE-CIRCLE-CTR+DIA** for drill holes.

4. Use **ALT-T** to change line type to hidden (2).

5. Use **ALT-T** to change line type to center (3).

Approach 2:
Orthographic Projection

1. Use the **CREATE-LINE-STRING** and **ENDPTS** commands to create the front view.

2. Create a 45 degree mitre line off front view.

3. Use **CREATE-LINE-HRZ/VRT-VERTICL-ENDENT** to project the width of the front view to the top view.

4. Use **CREATE-LINE-HRZ/VRT-HORIZTL** to project the height of the front view to the right side view.

5. Use the **CREATE-LINE-HRZ/VRT** command to transfer depth to mitre line and between top and right side views.

6. Use **EDIT-TRM/EXT-DOUBLE** or **FIRST** to trim projection lines between the appropriate edges.

7. Use **CREATE-CIRCLE-CTR+DIA** for drill hole.

8. Place center lines (**ALT-T-3**) using **CREATE-LINE-HRZ/VRT-BOTH** and **CENTER** from the Position Menu.

9. Project the hidden lines (**ALT-T-2**) from the **INTRSC** (intersection) of the circle and the center line.

10. Use the **EDIT-BREAK-DIVIDE** option to break the hidden lines as needed.

Approach 3:
Three-Dimensional Modeling

1. Use the **CREATE-LINE-STRING** and **ENDPTS** commands to create the front view.

2. Once the front view has been created, change to view 7 (**ALT-V** and enter **7**) and use **CTRL-V** to change to world coordinates.

3. Add the depth dimension to the height and width of the front view. Use the **X-FORM-DELTA-JOIN** and whichever selection option is appropriate.

4. Enter **1** when prompted to enter the number of copies. The copy does not move on the X axis (width); it should move at a negative distance equal to the depth for the Y axis and does not move on the Z axis (height). Therefore, enter **0** for the X and Z values and a negative value equal to the depth for the Y axis.

5. When adding additional lines, arcs, or circles in the isometric view *do not use the* **CURSOR** option from the Position Menu.

6. Use **ALT-V** to change the front, top or right side view and use **FILES-PATTERN-CREATE** to create a pattern file for each view.

7. Save the wireframe model (**CTRL-F**) and **FILES-PART-LOAD** your **SETUP** file.

8. Use the **FILES-PATTERN-RETRIEVE** option to retrieve a border and all of the desired views.

PROBLEM 7

PROBLEM STATEMENT   Draw the view shown.

CADKEY Setup   Load the **SETUP** file created in Problem 1.
FILES-PART-LOAD-NO. Enter filename **SETUP**.

PROCEDURE

1. Use **ALT-T** to change line type to center (3).
2. Use the **CREATE-LINE-ENDPTS** to lay out the center lines.
3. Use **CREATE-CIRCLE-CTR+DIA** and position the center of the circle at the **INTRSC** (intersection) of the horizontal and vertical center lines.
4. Use **ALT-T** to change line type as needed. Create the remaining circles using the **CENTER** position of the previous circle to position the center of the current circle.
5. Create the drill hole at the 3 o'clock position. Use the **INTRSC** of the center lines to position the circle.

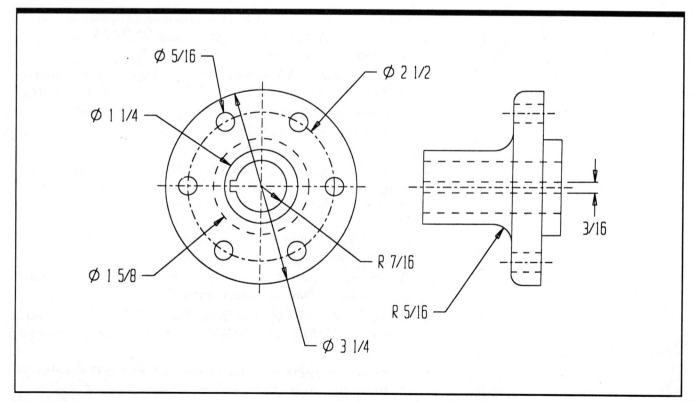

6. Use the **X-FORM-ROTATE-COPY** option to make five copies of the drill hole 60 degrees apart.
7. Use the **CREATE-LINE-PARALEL** option for creating the keyway. Use **EDIT-TRM/EXT-DOUBLE** to trim the keyway lines.
8. Place remaining center lines for drill holes.

PROBLEM 8

PROBLEM STATEMENT   Draw the view shown, including dimensions and notes.

CADKEY Setup   Load the **SETUP** file created in Problem 1. **FILES-PART-LOAD-NO**. Enter filename **SETUP**.

PROCEDURE
1. Use **ALT-T** to change line type to center (3).
2. Use **CREATE-LINE-ENDPTS** to lay out the center lines.
3. Use **CREATE-CIRCLE-CTR+DIA** and position the center of the circles at the **INTRSC** (intersection) of the horizontal and vertical center lines.

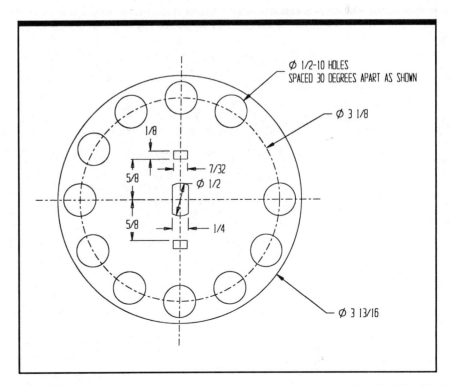

Problem 8

4. Use **ALT-T** to change line type as needed. Create the remaining circles using the **CENTER** position of the previous circle to position the center of the current circle.

5. Create a **.5 DIA** drill hole in the **CENTER** and at the 12 o'clock position. Use the **INTRSC** option of the center lines to position the circle.

6. Use the **X-FORM-ROTATE-COPY** option to make twelve copies of the drill hole 30 degrees apart.

7. Use the **CREATE-LINE-PARALEL-AT DIST** option for creating the slot.

8. Use **EDIT-TRM/EXT-DOUBLE** to trim the parallel lines.

9. Use **CTRL-A** to set the direction of the arrows for the dimensions to **IN** or **OUT**.

10. Use the **DETAIL-DIMENSION** commands **DIAMETER**, **HORIZTL**, and **VERTICL** as needed.

11. Use the **DETAIL-NOTE-KEYIN** command and type the note followed by **SAVE** to place the note.

PROBLEM 9

PROBLEM STATEMENT   Draw the view shown, including dimensions and notes.

CADKEY Setup   Load the **SETUP** file created in Problem 1.
FILES-PART-LOAD-NO. Enter filename **SETUP**.

PROCEDURE

1. Use **ALT-T** to change line type to **CENTER** (**3**).

2. Use **CREATE-LINE-ENDPTS** to lay out the center lines.

3. Use **CREATE-CIRCLE-CTR+DIA** and position the center of the circles at the **INTRSC** (intersection) of the horizontal and vertical center lines.

4. Use **ALT-T** to change line type as needed and create the remaining circles using the **CENTER** position of the previous circle to position the center of the current circle.

5. Create a **.3125 DIA** drill hole in the center.

6. Use the **EDIT-BREAK-BOTH** option to break the center lines into four separate lines. Make sure line type is set to **CENTER** (**3**).

7. Use **CREATE-LINE-ANGLE** to locate the remaining center lines to position the remaining circles.

8. Use the **CREATE-LINE-PARALEL-AT DIST** option for creating the slots and keyway in the center. Use **EDIT-**

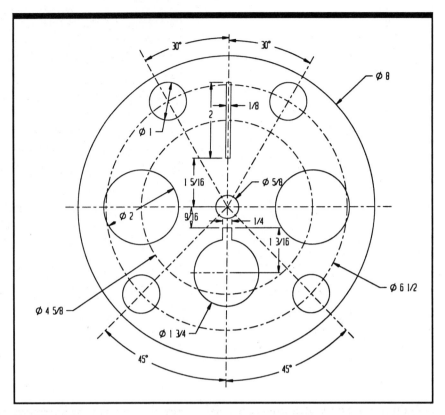

Problem 9

TRM/EXT-DOUBLE or DIVIDE to trim the circle and lines used to construct the slots and keyway.

9. Place remaining center lines for drill holes.

10. Use CTRL-A to set the direction of the arrows for the dimensions.

11. Use the DETAIL-DIMENSION commands DIAMETR, HORIZTL, VERTICL, and ANGLE as needed.

12. Use the DETAIL-NOTE-KEYIN command to type the required note, followed by SAVE to place the note.

PROBLEM 10

PROBLEM STATEMENT   Draw the view shown, including dimensions and notes.

CADKEY Setup   Load the SETUP file created in Problem 1.
FILES-PART-LOAD-NO. Enter filename SETUP.

PROCEDURE   1. Make one of each of the detail features and use the X-FORM-ROTATE-COPY command to make three copies 90 degrees apart rotated about the center.

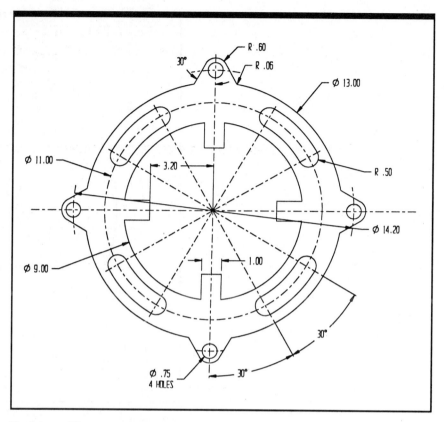

Problem 10

2. Use **ALT-T** to change line type to **CENTER (3)** and use the **CREATE-LINE-ENDPTS** to lay out the center lines.

3. Use **CREATE-CIRCLE-CTR+DIA** and position the center of the circle at the **INTRSC** (intersection) of the horizontal and vertical center lines.

4. Use **ALT-T** to change line type as needed and create the remaining circles using the **CENTER** position of the previous circle to position the center of the current circle.

5. Use the **EDIT-BREAK-BOTH** option to break the center lines into four separate lines. Make sure line type is set to **CENTER (3)**.

6. Use **CREATE-LINE-ANGLE** to locate the center lines to position the round end slot.

7. Use **CREATE-CIRCLE-CTR+RAD** to locate the round end at 30 and 60 degrees.

8. Use the **CREATE-FILLET-ARC-NO TRIM** option to locate the 5.25 radius arc and the 5.75 radius arc tangent to the round ends.

9. Use the **CREATE-LINE-PARALEL-AT DIST** option for creating the slot on the center circle. Use **EDIT-TRM/EXT-DOUBLE** for the first time to trim the circle, then use **DIVIDE** once the circle becomes an arc.

10. Use **CTRL-A** to set the direction of the arrows for the dimensions.

11. Use the **DETAIL-DIMENSN** commands **DIAMETR**, **HORIZTL**, **VERTICL**, and **ANGLE** as needed.

12. Use the **DETAIL-NOTE-KEYIN** command to type the required note, followed by **SAVE** to place the note.

PROBLEM 11

PROBLEM STATEMENT

Draw the views shown, including dimensions and notes.

CADKEY Setup

Load the **SETUP** file created in Problem 1.
FILES-PART-LOAD-NO. Enter filename **SETUP**.

PROCEDURE

1. Start with the top view.

2. Use **ALT-T** to change line type to **CENTER** (3) and use **CREATE-LINE-ENDPTS** to lay out the center lines.

3. Use **CREATE-CIRCLE-CTR+DIA** and position the centers of the circles at the **INTRSC** (intersection) of the horizontal and vertical center lines.

4. Use **CREATE-LINE-TAN TAN** to create the four tangent lines.

5. Then use **EDIT-TRM/EXT-DOUBLE** to trim the circles into arcs. Use the **DIVIDE** option to trim the 6" diameter circle a second time.

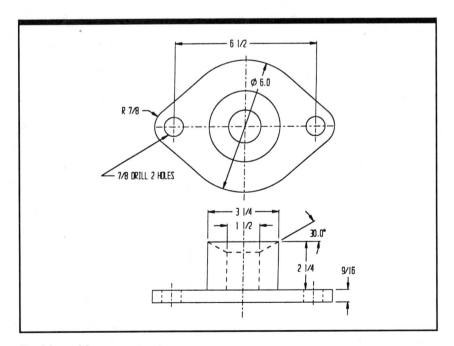

Problem 11

6. Use **CREATE-LINE-HRZ/VRT-VERTICL-INTRSC** using the intersection of the arc and circles with the center lines to project the width of the top view to the front view. Use **ALT-T** to change line type to hidden as needed.

7. Use **EDIT-TRM/EXT-DOUBLE** or **FIRST** to trim projection lines between the appropriate edges.

8. Use **CTRL-A** to set the direction of the arrows for the dimensions to **IN** or **OUT**.

9. Use the **DETAIL-DIMENSION** commands **DIAMETR**, **HORIZTL**, and **VERTICL** as needed.

10. Use the **DETAIL-NOTE-KEYIN** command and type the note, followed by **SAVE** to place the note.

PROBLEM 12

PROBLEM STATEMENT   Draw the view shown, including dimensions and notes.

CADKEY Setup   Load the **SETUP** file created in Problem 1.
FILES-PART-LOAD-NO. Enter filename **SETUP**.

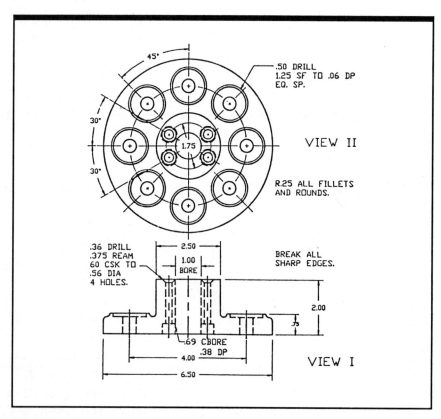

Problem 12

PROCEDURE

1. Use **ALT-T** to change line type to **CENTER** (3).
2. Use **CREATE-LINE-ENDPTS** to lay out the center lines.
3. Use **CREATE-CIRCLE-CTR+DIA** and position the center of the circle at the **INTRSC** (intersection) of the horizontal and vertical center lines.
4. Use **ALT-T** to change line type as needed and create the remaining circles using the **CENTER** position of the previous circle to position the center of the current circle.
5. Create a **.5 DIA** drill, **1.25 SF**, and 1.5 concentric circle at the 3 o'clock position.
6. Use the **X-FORM-ROTATE-COPY** option to make seven copies of the drill hole 45 degrees apart.
7. Create the countersunk hole and use the **X-FORM-MIRROR-COPY** command to properly place the four holes.
8. Use the **CREATE-LINE-HRZ/VRT-VERTICL-INTRSC** using the intersection of the arc and circles with the center lines to project the width of the top view to the front view. Use **ALT-T** to change line type to hidden as needed.
9. Use **EDIT-TRM/EXT-DOUBLE** or **FIRST** to trim projection lines between the appropriate edges.
10. Use **CTRL-A** to set the direction of the arrows for the dimensions to **IN** or **OUT**.
11. Use the **DETAIL-DIMENSN** commands **DIAMETR**, **HORIZTL**, and **VERTICL** as needed.
12. Use the **DETAIL-NOTE-KEYIN** command and type the note, followed by **SAVE** to place the note.

PROBLEM 13

PROBLEM STATEMENT

Draw an isometric projection. Do not include dimensions.

CADKEY Setup

Load the **SETUP** file created in Problem 1.
FILES-PART-LOAD-NO. Enter filename **SETUP**.

PROCEDURE

1. Use the **CREATE-LINE-STRING** commands to create the front view.
2. Once the front view has been created, change to view 7 (**ALT-V** and enter **7**) and use **CTRL-V** to change to world coordinates.
3. At this time, you are ready to add the depth dimension to the height and width of the front view. Use **X-FORM-DELTA-JOIN** and whichever selection option is appropriate.

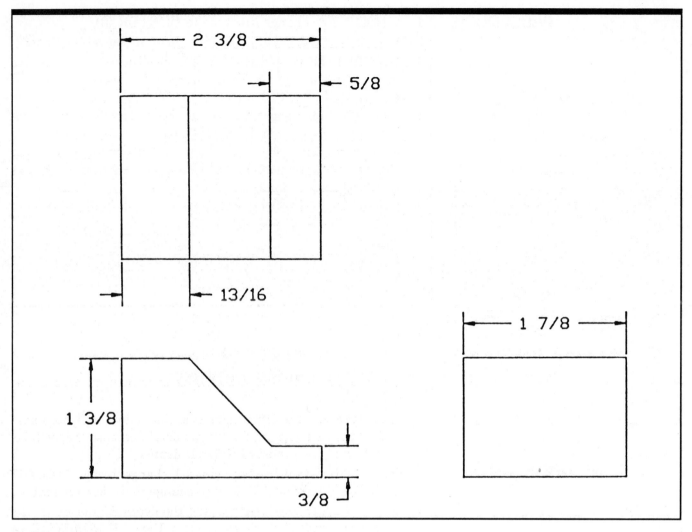

Problem 13

4. Enter **1** when prompted to enter the number of copies. The copy does not move on the X axis (width); it should move at a negative distance equal to the depth for the Y axis and does not move on the Z axis (height). Therefore, enter **0** for the X and Z values and a negative value equal to the depth for the Y axis.

PROBLEM 14

PROBLEM STATEMENT   Draw an isometric projection. Do not include dimensions.

CADKEY Setup   Load the **SETUP** file created in Problem 1.
FILES-PART-LOAD-NO. Enter filename **SETUP**.

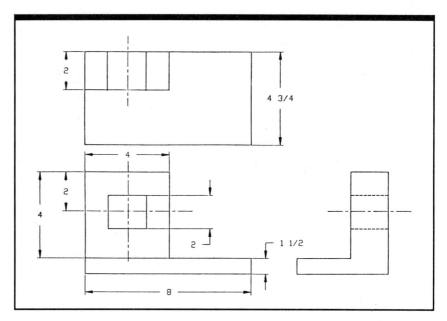

Problem 14

PROCEDURE

1. Use the **CREATE-LINE-STRING** command to create the front view.

2. Make sure that the leftmost vertical line is two separate lines, one for the base and one for the top detail, because they will be transformed and copied to different depths.

3. Once the front view has been created, change to view 7 (**ALT-V** and enter **7**) and use **CTRL-V** to change to world coordinates.

4. At this time, you are ready to add the depth dimension to the height and width of the front view. Use **X-FORM-DELTA-JOIN** and the **SINGLE** selection option to select first the entities which make up the top detail and transform them to the proper depth on the negative Y axis.

5. Enter **1** when prompted to enter the number of copies. The copy does not move on the X axis (width); it should move at a negative distance equal to the depth for the Y axis and does not move on the Z axis (height). Therefore, enter **0** for the X and Z values and a negative value equal to the depth for the Y axis.

6. Repeat steps 4 and 5 for the base entities.

PROBLEM 15

PROBLEM STATEMENT

Draw an isometric projection. Do not include dimensions.

CADKEY Setup

Load the **SETUP** file created in Problem 1.
FILES-PART-LOAD-NO. Enter filename **SETUP**.

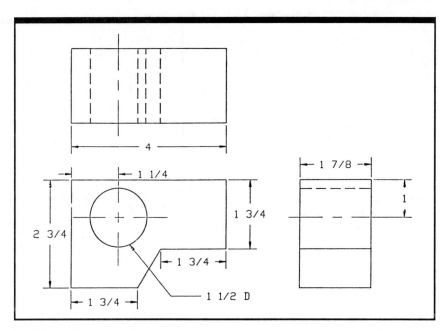

Problem 15

PROCEDURE
1. Use the **CREATE-LINE-STRING** command to create the front view.
2. Once the front view has been created, change to view 7 (**ALT-V** and enter **7**) and use **CTRL-V** to change to world coordinates.
3. At this time you are ready to add the depth dimension to the height and width of the front view. Use **X-FORM-DELTA-JOIN** and whichever selection option is appropriate.
4. Enter **1** when prompted to enter the number of copies. The copy does not move on the X axis (width); it should move at a negative distance equal to the depth for the Y axis and does not move on the Z axis (height). Therefore, enter **0** for the X and Z values and a negative value equal to the depth for the Y axis.

PROBLEM 16

PROBLEM STATEMENT
Draw an isometric projection. Do not include dimensions.

CADKEY Setup
Load the **SETUP** file created in Problem 1.
FILES-PART-LOAD-NO. Enter filename **SETUP**.

PROCEDURE
1. Use **ALT-V** to change to view 5 and use the **CREATE-LINE-STRING** and **PARALEL** commands to create the right side view.

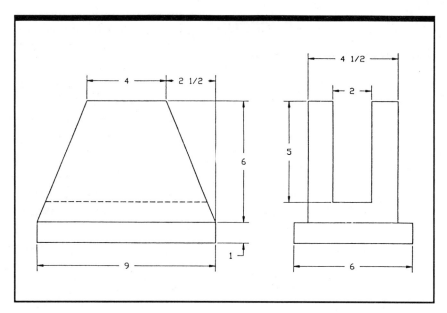

Problem 16

2. Once the front view has been created, change to view 7 (**ALT-V** and enter **7**) and use **CTRL-V** to change to world coordinates.

3. At this time you are ready to add the width dimension to the height and depth of the right side view. Use **X-FORM-DELTA-JOIN** and whichever selection option is appropriate (**ALL DSP**).

4. Enter **1** when prompted to enter the number of copies. The copy should move at a distance equal to the width on the X axis. It does not move on the Y axis (depth) and does not move on the Z axis (height). Therefore, enter a value equal to the width for the X axis (**5-3/4**) and **0** for the Y and Z values.

5. Use **CREATE-LINE-ENDPTS** (**ENDENT** for the start point and **ALONGL** for the endpoint) to draw the angled lines. Use **EDIT-TRM/EXT-FIRST** and **CTRL-Q** to trim and delete unnecessary lines.

PROBLEM 17

PROBLEM STATEMENT   Draw an isometric projection. Do not include dimensions.

CADKEY Setup   Load the **SETUP** file created in Problem 1.
FILES-PART-LOAD-NO. Enter filename **SETUP**.

PROCEDURE

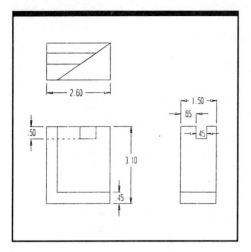

Problem 17

1. Use **ALT-V** to change to view 1 and use the **CREATE-LINE-STRING** and **PARALEL** commands to create the right side view.

2. Once the top view has been created, change to view 7 (**ALT-V** and enter **7**) and use **CTRL-V** to change to world coordinates.

3. At this time you are ready to add the height dimension to the width and depth of the right side view. Use **X-FORM-DELTA-JOIN** and whichever selection option is appropriate (single) to select the middle two lines to be moved −.5 on the Z axis.

4. Enter **1** when prompted to enter the number of copies. The copy should move at a distance equal to the height on the Z axis.

5. Repeat steps 3 and 4 to move the three sides of the triangle to a depth of 2.65 and to move the four outside edge lines to the depth of 3.10.

PROBLEM 18

PROBLEM STATEMENT

Draw an isometric projection. Do not include dimensions.

CADKEY Setup

Load the **SETUP** file created in Problem 1.
FILES-PART-LOAD-NO. Enter filename **SETUP**.

PROCEDURE

1. Use **ALT-V** to change to view 1, the top view. Use the **CRE-ATE-LINE-STRING** and **CREATE-CIRCLE-CTR+DIA** commands to create the top view.

2. Once the top view has been created, change to view 7 (**ALT-V** and enter **7**) and use **CTRL-V** to change to world coordinates.

3. At this time you are ready to add the height dimension to the width and depth of the top view. Use **X-FORM-DELTA-JOIN** and whichever selection option is appropriate (**ALL DSP**).

4. Enter **1** when prompted to enter the number of copies. The copy does not move on the X axis (width); it should not move on the Y axis (depth) and it does move on the Z axis (height). Therefore, enter **0** for the X and Y values and a value equal to the first level of height (**.625**) for the Z axis.

5. Repeat steps 3 and 4. This time use **SINGLE** and select only the two circles from the base position. **COPY** them to a Z value of **1.25**.

6. Use **EDIT-TRM/EXT-FIRST** and **CTRL-Q** to trim and delete unnecessary lines. The circles must be trimmed using the **DOUBLE** option.

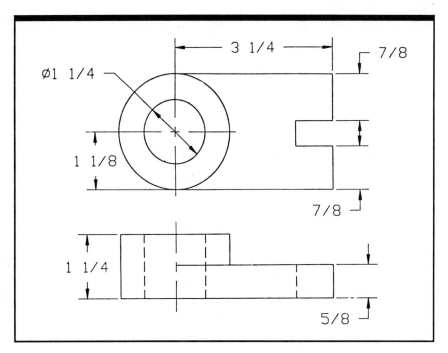

Problem 18

7. Use **CTRL-W** to change the construction method to 2D. Use **CREATE-LINE-TANGENT** to create the line tangent to the two arcs.

PROBLEM 19

PROBLEM STATEMENT   Determine the best front view and draw the top, front and right side views, including dimensions and notes.

CADKEY Setup   Load the **SETUP** file created in Problem 1. **FILES-PART-LOAD-NO**. Enter filename **SETUP**.

PROCEDURE

Approach 1: Draw Mode

1. Use the **CREATE-LINE-STRING** or **RECTANG** command to define the outside boundary of the views.
2. Use the **CREATE-LINE-ENDPTS** command to place individual lines for interior details.
3. Use **ALT-T** to change line type to hidden (2) or center (3).
4. Use **CREATE-CIRCLE-CTR+DIA** and **CREATE-FILLET** as needed.
5. Use **CREATE-ARC-CTR+RAD** with a start angle of **270** and an end angle of **90**.

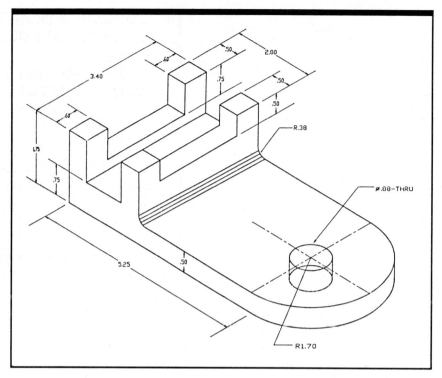

Problem 19

Approach 2:
Orthographic Projection

1. Use the **CREATE-LINE-STRING** and **FILLET** commands to create the front view.
2. Create a 45 degree mitre line off front view.
3. Use **CREATE-LINE-HRZ/VRT-VERTICL-ENDENT** to project the width of the front view to the top view.
4. Use **CREATE-LINE-HRZ/VRT-HORIZTL** to project the height of the front view to the right side view.
5. Use the **CREATE-LINE-HRZ/VRT** command to transfer depth to mitre line and between top and right side views.
6. Use **EDIT-TRM/EXT-DOUBLE** or **FIRST** to trim projection lines between the appropriate edges.

Approach 3:
Three-Dimensional Modeling

1. Use the **CREATE-LINE-STRING** and **FILLET** commands to create the front view.
2. Once the front view has been created, change to view 7 (**ALT-V** and enter **7**) and use **CTRL-V** to change to world coordinates.
3. Add the depth dimension to the height and width of the front view. Use **X-FORM-DELTA-JOIN** and whichever selection option is appropriate.
4. Enter **1** when prompted to enter the number of copies. Enter **0** for the X and Z values and a negative value equal to the depth for the Y axis. Edit model as needed and add circles and arcs.

5. Use **ALT-V** to change the front, top or right side view and use **FILES-PATTERN-CREATE** to create a pattern file for each view.

6. Save the wireframe model (**CTRL-F**) and **FILES-PART-LOAD** your **SETUP** file.

7. Use the **FILES-PATTERN-RETRIEV** option to retrieve a border and all of the desired views.

Dimensioning

1. Use **DETAIL-DIMENSN-HORIZTL** or **VERTICL** (**END-ENT**). Select the origin of the first witness (extension) line and then the second witness line. Then select the position of the dimension text.

2. Use **CTRL-A** to set Arrows **OUT**. Use **DETAIL-DIMENSN-DIAMETR** or **RADIUS**. Select the circle or arc you wish to dimension, indicate the text position and whether the leader shoulder should be to the left or right of the text.

3. **DETAIL-NOTE-KEYIN** to type in note; press **RETURN** for additional lines. When finished with the note, press **RETURN** twice (CADKEY versions 1 through 3.0) or **SAVE TXT** (version 3.1 or higher). Then indicate text position.

PROBLEM 20

PROBLEM STATEMENT

Determine the best front view and draw the top, front and right side views, including dimensions and notes.

CADKEY Setup

Load the **SETUP** file created in Problem 1.
FILES-PART-LOAD-NO. Enter filename **SETUP**.

PROCEDURE

Approach 1: Draw Mode

1. Use the **CREATE-LINE-STRING** or **RECTANG** command to define the outside boundary of the views.

2. Use the **CREATE-LINE-ENDPTS** command to place individual lines for interior details.

3. Use **ALT-T** to change line type to hidden (2) or center (3).

4. Use **CREATE-CIRCLE-CTR+DIA** and **CREATE-FILLET** as needed.

5. Use **CREATE-ARC-CTR+RAD** with a start angle of **270** and an end angle of **90** for the right arc and the reverse for the left arc.

Approach 2: Orthographic Projection

1. Use the **CREATE-LINE-STRING** and **FILLET** commands to create the front view.

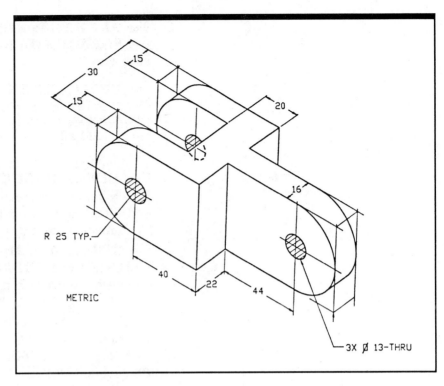

Problem 20

2. Create a 45 degree mitre line off front view.

3. Use **CREATE-LINE-HRZ/VRT-VERTICL-ENDENT** to project the width of the front view to the top view.

4. Use **CREATE-LINE-HRZ/VRT-HORIZTL** to project the height of the front view to the right side view.

5. Use the **CREATE-LINE-HRZ/VRT** command to transfer depth to mitre line and between top and right side views.

6. Use **EDIT-TRM/EXT-DOUBLE** or **FIRST** to trim projection lines between the appropriate edges.

Approach 3: Three-Dimensional Modeling

1. Use the **CREATE-LINE-STRING ARC** and **CIRCLE** commands to create the front view, as described above.

2. Once the front view has been created, change to view 7 (**ALT-V** and enter **7**) and use **CTRL-V** to change to world coordinates.

3. Add the depth dimension to the height and width of the front view. Use **X-FORM-DELTA-JOIN** and whichever selection option is appropriate.

4. Enter **1** when prompted to enter the number of copies. Enter **0** for the X and Z values and a negative value equal to the depth for the Y axis. Edit model as needed and add circles and arcs.

5. Use **ALT-V** to change the front, top or right side view and use **FILES-PATTERN-CREATE** to create a pattern file for each view.

6. Save the wireframe model (**CTRL-F**) and **FILES-PART-LOAD** your **SETUP** file.

7. Use the **FILES-PATTERN-RETRIEV** option to retrieve a border and all of the desired views.

Dimensioning

1. Use **DETAIL-DIMENSN-HORIZTL** or **VERTICL (END-ENT)**. Select the origin of the first witness (extension) line and the second witness line; then select the position of the dimension text.

2. Use **CTRL-A** to set arrows **OUT**. Use **DETAIL-DIMENSN-DIAMETR or RADIUS**. Select the circle or arc you wish to dimension, and indicate the text position and whether the leader shoulder should be to the left or right of the text.

3. Use **DETAIL-NOTE-KEYIN** to type in note; press **RETURN** for additional lines. When finished with note, press **RETURN** twice (CADKEY versions 1 through 3.0) or **SAVE TXT** (version 3.1 or higher). Then indicate text position.

PROBLEM 21

PROBLEM STATEMENT

Determine the best front view and draw the top, front and right side views, including dimensions and notes.

CADKEY Setup

Load the **SETUP** file created in Problem 1. **FILES-PART-LOAD-NO**. Enter filename **SETUP**.

PROCEDURE

Approach 1: Draw Mode

1. Use the **CREATE-LINE-STRING** or **RECTANG** command to define the outside boundary of the views.

2. Use the **CREATE-LINE-ENDPTS** or **PARALEL** command to place individual lines for interior details.

3. Use **ALT-T** to change line type to hidden (2) or center (3).

4. Locate the center of the circle(s) and or arc. Use **CREATE-CIRCLE-CTR+DIA**.

5. Use **EDIT-TRM/EXT-FIRST** or **BOTH** as needed to trim lines and **DOUBLE** for the circle.

Approach 2: Orthographic Projection

1. Use the **CREATE-LINE-STRING** and **ENDPTS** commands to create the front view.

2. Create a 45 degree mitre line off front view.

3. Use **CREATE-LINE-HRZ/VRT-VERTICL-ENDENT** to project the width of the front view to the top view.

4. Use **CREATE-LINE-HRZ/VRT-HORIZTL** to project the height of the front view to the right side view.

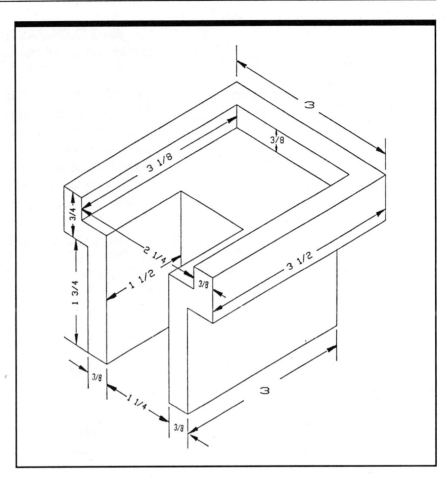

Problem 21

5. Use the **CREATE-LINE-HRZ/VRT** command to transfer depth to mitre line and between top and right side views.

6. Use **EDIT-TRM/EXT-DOUBLE** or **FIRST** to trim projection lines between the appropriate edges.

Approach 3: Three-Dimensional Modeling

1. Use the **CREATE-LINE-STRING** and **ENDPTS** commands to create the front view, as described above.

2. Once the front view has been created, change to view 7 (**ALT-V** and enter **7**) and use **CTRL-V** to change to world coordinates.

3. Add the depth dimension to the height and width of the front view. Use **X-FORM-DELTA-JOIN** and whichever selection option is appropriate for the detail you are selecting to copy.

4. Enter **1** when prompted to enter the number of copies. Enter **0** for the X and Z values and a negative value equal to the depth for the Y axis of the front detail. Use a positive value for the depth of the back plate. Edit model as needed.

5. Use **ALT-V** to change the front, top or right side view and use **FILES-PATTERN-CREATE** to create a pattern file for each view.

6. Save the wireframe model (**CTRL-F**). Use **FILES-PART-LOAD** on your **SETUP** file.

7. Use the **FILES-PATTERN-RETRIEV** option to retrieve a border and all of the desired views.

Dimensioning

1. Use **DETAIL-DIMENSN-HORIZTL** or **VERTICL** (**END-ENT**). Select the origin of the first witness (extension) line and the second witness line. Then select the position of the dimension text.
2. Use **CTRL-A** to set arrows **OUT**. Use **DETAIL-DIMENSN-DIAMETR** or **RADIUS**. Select the circle or arc you wish to dimension, and indicate the text position and whether the leader shoulder should be to the left or right of the text.
3. Use **DETAIL-NOTE-KEYIN** to type in the note; press **RE-TURN** for additional lines. When you are finished with the note, press **RETURN** twice (CADKEY versions 1 through 3.0) or **SAVE TXT** (version 3.1 or higher). Then indicate text position.

PROBLEM 22

PROBLEM STATEMENT

Determine the best front view and draw the top, front and right side views, including dimensions and notes.

CADKEY Setup

Load the **SETUP** file created in Problem 1.
FILES-PART-LOAD-NO. Enter filename **SETUP**.

PROCEDURE

Approach 1: Draw Mode

1. Use the **CREATE-LINE-STRING** or **RECTANG** command to define the outside boundary of the views.
2. Use the **CREATE-LINE-ENDPTS** or **PARALEL** command to place individual lines for interior details.
3. Create vertical construction lines **PARALEL** at 1.5 from either edge and at 3.25 in the middle.
4. Use **EDIT-BREAK-FIRST** to break the middle horizontal line into four segments. Use **CTRL-Q** to delete the construction lines.
5. Use the **CREATE-LINE-ANGLE** command. Use an angle of 30 for the second segment at the first break point and –30 for the third segment at the second break point.
6. Use **ALT-T** to change line type to hidden (2) or center (3).
7. Use approach 2 below for the top and right side views.
8. Use **EDIT-TRM/EXT-FIRST** or **BOTH** as needed to trim lines.

Approach 2:
Orthographic Projection

1. Use the **CREATE-LINE-STRING** and **ENDPTS** commands to create the front view, as described above.
2. Create a 45 degree mitre line off front view.
3. Use **CREATE-LINE-HRZ/VRT-VERTICL-ENDENT** to project the width of the front view to the top view.
4. Use **CREATE-LINE-HRZ/VRT-HORIZTL** to project the height of the front view to the right side view.
5. Use the **CREATE-LINE-HRZ/VRT** command to transfer depth to mitre line and between top and right side views.
6. Use **EDIT-TRM/EXT-DOUBLE** or **FIRST** to trim projection lines between the appropriate edges.

Approach 3:
Three-Dimensional Modeling

1. Use the **CREATE-LINE-STRING** and **ENDPTS** commands to create the front view, as described above.
2. Once the front view has been created, change to view 7 (**ALT-V** and enter **7**) and use **CTRL-V** to change to world coordinates.
3. Add the depth dimension to the height and width of the front view. Use **X-FORM-DELTA-JOIN** and whichever selection option is appropriate for the detail you are selecting to copy.
4. Enter **1** when prompted to enter the number of copies. Enter **0** for the X and Z values and a negative value equal to the depth for the Y axis. Edit model as needed and add circles and arcs.

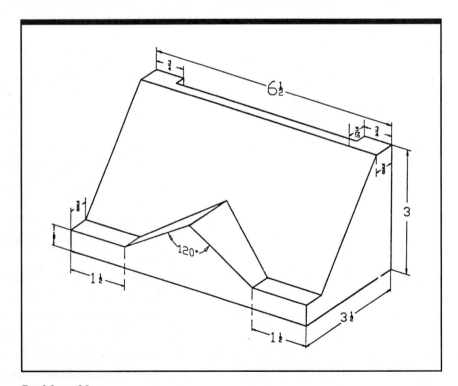

Problem 22

5. Use **ALT-V** to change the front, top or right side view and use **FILES-PATTERN-CREATE** to create a pattern file for each view.

6. Save the wireframe model (**CTRL-F**) and use **FILES-PART-LOAD** on your **SETUP** file.

7. Use the **FILES-PATTERN-RETRIEV** option to retrieve a border and all of the desired views.

Dimensioning

1. Use **DETAIL-DIMENSN-HORIZTL** or **VERTICL** (**END-ENT**). Select the origin of the first witness (extension) line and the second witness line. Then select the position of the dimension text.

2. Use **CTRL-A** to set arrows **OUT**. Use **DETAIL-DIMENSN-DIAMETR** or **RADIUS**. Select the circle or arc you wish to dimension; indicate the text position and whether the leader shoulder should be to the left or right of the text.

3. Use **DETAIL-NOTE-KEYIN** to type in note; press **RE-TURN** for additional lines. When finished with note, press **RETURN** twice (CADKEY versions 1 through 3.0) or **SAVE TXT** (version 3.1 or higher). Then indicate text position.

PROBLEM 23

PROBLEM STATEMENT

Determine the best front view and draw the top, front and right side views, including dimensions and notes.

CADKEY Setup

Load the **SETUP** file created in Problem 1.
FILES-PART-LOAD-NO. Enter filename **SETUP**.

PROCEDURE

Approach 1: Draw Mode

1. Use the **CREATE-LINE-STRING** or **RECTANG** command to define the outside boundary of the views.

2. Use the **CREATE-LINE-ENDPTS** command to place individual lines for interior details.

3. Use **ALT-T** to change line type to hidden (2) or center (3).

4. Use **CREATE-CIRCLE-CTR+RAD** and then **TRM/EXT-DOUBLE** for the arc in the front view.

5. Use Approach 2 below and **CREATE-SPLINE-2D CUBI** to create the top view.

Approach 2: Orthographic Projection

1. Use the **CREATE-LINE-STRING** and **CIRCLE** commands to create the front view.

2. Create a 45 degree mitre line off front view.

3. Use **CREATE-POINT-SEGMENT** to place 9 points on the 180 degree arc in the front view.

4. Use **CREATE-LINE-HRZ/VRT-VERTICL-POINT** to project the location of the points in the front view to the top view.

5. Use **CREATE-LINE-HRZ/VRT-HORIZTL** to project the height of the points in the front view to the right side view.

6. Use the **CREATE-LINE-HRZ/VRT** command to transfer depth location to mitre line and between top and right side views.

7. Use **EDIT-TRM/EXT-DOUBLE** or **FIRST** to trim projection lines between the appropriate edges.

Approach 3:
Three-Dimensional Modeling

1. Use the **CREATE-LINE-STRING, ARC,** and **CIRCLE** commands to create the front view as described above.

2. Once the front view has been created, change to view 7 (**ALT-V** and enter **7**) and use **CTRL-V** to change to world coordinates.

3. Add the depth dimension to the height and width of the front view. Use **X-FORM-DELTA-JOIN** and whichever selection option is appropriate.

4. Enter **1** when prompted to enter the number of copies. Enter **0** for the X and Z values and a negative value equal to the depth for the Y axis. Edit model as needed and add circles and arcs. Create the arc in the right side, add a **POINT-SEGMENT** and transform it along the Y axis to find the intersection of the arc with the inclined surface. Edit as needed.

5. Use **ALT-V** to change the front, top or right side view and use **FILES-PATTERN-CREATE** to create a pattern file for each view.

6. Save the wireframe model (**CTRL-F**) and use **FILES-PART-LOAD** on your **SETUP** file.

7. Use the **FILES-PATTERN-RETRIEV** option to retrieve a border and all of the desired views.

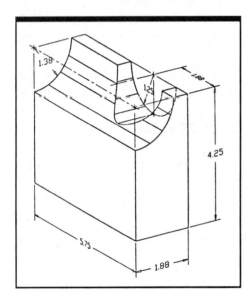

Problem 23

Dimensioning

1. Use **DETAIL-DIMENSN-HORIZTL** or **VERTICL** (**END-ENT**). Select the origin of the first witness (extension) line and the second witness line. Then select the position of the dimension text.

2. Use **CTRL-A** to set arrows **OUT**. Use **DETAIL-DIMENSN-DIAMTER** or **RADIUS**. Select the circle or arc you wish to dimension, and indicate the text position and whether the leader shoulder should be to the left or right of the text.

3. Use **DETAIL-NOTE-KEYIN** to type in note; press **RETURN** for additional lines. When finished with note, press **RETURN** twice (CADKEY versions 1 through 3.0) or **SAVE TXT** (version 3.1 or higher). Then indicate text position.

PROBLEM 24

PROBLEM STATEMENT

Determine the best front view and draw the top, front and left side views, including dimensions and notes.

CADKEY Setup

Load the **SETUP** file created in Problem 1.
FILES-PART-LOAD-NO. Enter filename **SETUP**.

PROCEDURE

Approach 1: Draw Mode

1. Use the **CREATE-LINE-STRING** or **RECTANG** command to define the outside boundary of the views.
2. Use the **CREATE-LINE-PARALEL** command to place individual lines for interior details and center lines.
3. Use **ALT-T** to change line type to hidden (2) or center (3).
4. Use **CREATE-ARC-CTR+RAD** with a start angle of 90 and an end angle of 180 for the left arc in the front view. Use a start angle of 0 and an end angle of 180 for the second arc in the front view.
5. Use the **CREATE-LINE-TANGENT** command to create the line tangent to the two arcs. Trim the arc as necessary using the first option.
6. Use the **CREATE-LINE-ENDPTS** option. Use **ALONGL** to locate the first point on the dovetail angle and **POLAR** to locate the endpoint. Use 120 degrees for the left angle and 60 degrees for the right angle.
7. Use **CREATE-FILLET-ARC** for the .15 and .5 radii.

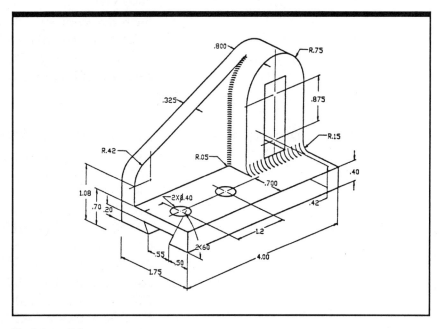

Problem 24

Approach 2:
Orthographic Projection

1. Use the **CREATE-LINE-STRING** and **ARC** commands to create the front view.

2. Create a 45 degree mitre line off front view.

3. Use **CREATE-LINE-HRZ/VRT-VERTICL** to project the width in the front view to the top view.

4. Use **CREATE-LINE-HRZ/VRT-HORIZTL** to project the height of the points in the front view to the right side view.

5. Use the **CREATE-LINE-HRZ/VRT** command to transfer depth location to mitre line and between top and right side views.

6. Use **EDIT-TRM/EXT-DOUBLE** or **FIRST** to trim projection lines between the appropriate edges.

Approach 3:
Three-Dimensional Modeling

1. Use the **CREATE-LINE-STRING, ARC,** and **CIRCLE** commands to create the front view, as described above.

2. Once the front view has been created, change to view 7 (**ALT-V** and enter **7**) and use **CTRL-V** to change to world coordinates.

3. Add the depth dimension to the height and width of the front view. Use **X-FORM-DELTA-JOIN** and whichever selection option is appropriate.

4. Enter **1** when prompted to enter the number of copies. Enter **0** for the X and Z values and a negative value equal to the depth for the Y axis. Edit model as needed and add circles and arcs. Create the arc in the right side, add a **POINT-SEGMENT** and transform it along the X axis. Edit as needed.

5. Use **ALT-V** to change the front, top or right side view and use **FILES-PATTERN-CREATE** to create a pattern file for each view.

6. Save the wireframe model (**CTRL-F**) and use **FILES-PART-LOAD** on your **SETUP** file.

7. Use the **FILES-PATTERN-RETRIEV** option to retrieve a border and all of the desired views.

Dimensioning

1. Use **DETAIL-DIMENSN-HORIZTL** or **VERTICL** (**END-ENT**). Select the origin of the first witness (extension) line and then the second witness line. Then select the position of the dimension text.

2. Use **CTRL-A** to set arrows **OUT.** Use **DETAIL-DIMENSN-DIAMETR** or **RADIUS**. Select the circle or arc you wish to dimension, and indicate the text position and whether the leader shoulder should be to the left or right of the text.

3. Use **DETAIL-NOTE-KEYIN** to type in note; press **RE-TURN** for additional lines. When finished with note, press **RETURN** twice (CADKEY versions 1 through 3.0) or **SAVE TXT** (version 3.1 or higher). Then indicate text position.

PROBLEM 25

PROBLEM STATEMENT   Draw the view shown, including dimensions and notes.

CADKEY Setup   NOTE: CADKEY must be configured for metric using the **CON-FIG** program option for Program Options.
Load the **SETUP** file created in Problem 1.
FILES-PART-LOAD-NO. Enter filename **SETUP**.

PROCEDURE
1. Use **ALT-T** to change line type to center (3).
2. Use **CREATE-LINE-ENDPTS** to lay out the center lines.
3. Use **ALT-T** to change line type as needed. Use **CREATE-CIRCLE-CTR+DIA** and position the center of the circle at the intersection of the horizontal and vertical center lines.
4. Create the remaining circles using the center position of the previous circle to position the center of the current circle.
5. Use the **CREATE-ARC-TANGENT-2 ENTS** option to place the R 25 arc (select inside the tangent points) and the R 94 arc (select outside the tangent points).
6. Use **EDIT-TRM/EXT-DOUBLE** to trim circles between the tangent points.
7. Use **CTRL-A** to set the direction of the arrows for the dimensions to **IN** or **OUT**.

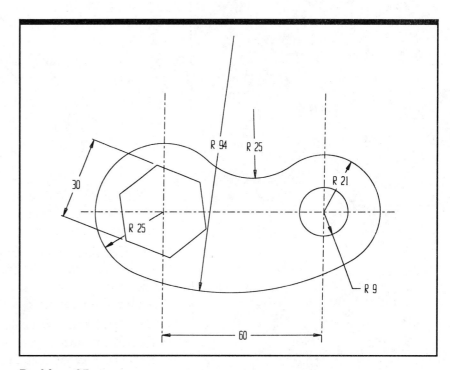

Problem 25

8. Use the **DETAIL-DIMENSN** commands **DIAMETR**, **HORIZTL**, and **PARALEL** as needed.

PROBLEM 26

PROBLEM STATEMENT   Draw the view shown, including dimensions and notes.

CADKEY Setup   Load the **SETUP** file created in Problem 1.
FILES-PART-LOAD-NO. Enter filename **SETUP**.

PROCEDURE
1. Use **ALT-T** to change line type to center (3).
2. Use **CREATE-LINE-ENDPTS** and **ANGLE** to lay out all the center lines.
3. Use **ALT-T** to change line type as needed. For all arcs use the **CREATE-CIRCLE-CTR+DIA** and position the center of the circle at the intersection of the horizontal and vertical center lines and the angled and curved center lines. The circles will be trimmed later to ensure an accurate intersection.

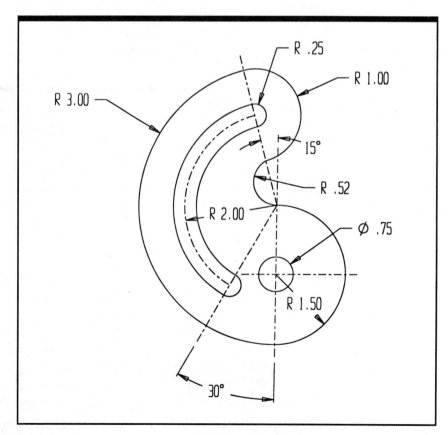

Problem 26

4. Create the remaining circles using the center position of the previous circle to position the center of the current circle.

5. Use the **CREATE-ARC-ROUND** option to place the R.63 arc. Select the tangent point and the endpoint using the **ENDENT** option.

6. Use the **CREATE-ARC-TANGENT-2 ENTS** option to place the large outside arc (select outside the tangent points).

7. Use **EDIT-TRM/EXT-DOUBLE** to trim the circles between the tangent points.

8. Use **CTRL-A** to set the direction of the arrows for the dimensions to **IN** or **OUT**.

9. Use the **DETAIL-DIMENSN** commands **RADIUS** and **ANGLE** as needed.

PROBLEM 27

PROBLEM STATEMENT

Draw the view shown, including dimensions and notes.

CADKEY Setup

NOTE: CADKEY must be configured for metric using the **CONFIG** program option for Program Options.
Load the **SETUP** file created in Problem 1.
FILES-PART-LOAD-NO. Enter filename **SETUP**.

PROCEDURE

1. Use **ALT-T** to change line type to center (3).

2. Use **CREATE-LINE-ENDPTS** to lay out the center lines.

3. Use **ALT-T** to change line type as needed. Use the **CREATE-CONIC-ELLIPSE** and position the center of the circle at the intersection of the horizontal and vertical center lines. Subtract 24mm from the major and minor diameters and create a second ellipse.

4. Create concentric circles using **CREATE-CIRCLE-CTR+DIA** with diameters of 44, 50, and 76.

5. Use **CREATE-LINE-PARALEL-AT DIST** to create lines parallel to the horizontal center line at distances of 5 and 6 millimeters.

6. Use **EDIT-TRM/EXT-DOUBLE** to trim 5mm lines between the 50mm circle, and the 6mm lines between the 76mm circles and the inside ellipse. Then trim the 76mm circle between the inside ellipse.

7. Use **X-FORM-ROTATE-COPY** to make 3 copies of the 5mm lines with a rotation angle of 45 degrees about the center of the circles.

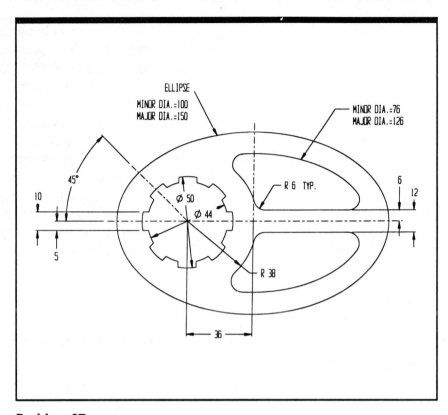

Problem 27

8.   Use the **DOUBLE** option to trim the first segment out of the
     44 and the 50mm circles. Then use the **DIVIDE** method to
     trim the remaining segments.

9.   Use the **FILLET-ARC-NO TRIM** option to place the R6
     fillets.

10. Use trim **FIRST** and **DIVIDE** options to trim the lines and
     arcs to their intersection points.

11. Use **CTRL-A** to set the direction of the arrows for the di-
     mensions to **IN** or **OUT**.

12. Use the **DETAIL-DIMENSN** commands **DIAMETR, HORI-
     ZONTL,** and **PARALEL** as needed.

13. Use the **DETAIL-LABEL** option for R6 (TYP.).

PROBLEM 28

PROBLEM STATEMENT   Draw the view shown, including dimensions and notes.

CADKEY Setup   Load the **SETUP** file created in Problem 1.
FILES-PART-LOAD-NO. Enter filename **SETUP**.

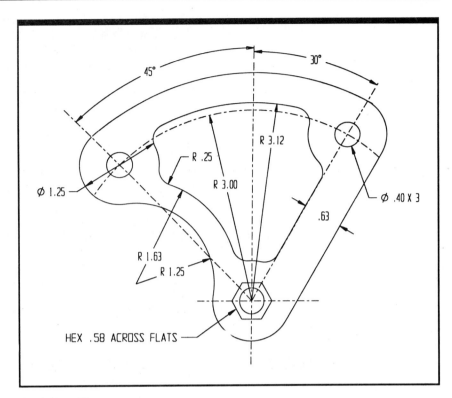

Problem 28

PROCEDURE

1. Use **ALT-T** to change line type to center (3).
2. Use **CREATE-LINE-ENDPTS** and **ANGLE** to lay out all the center lines.
3. Use **ALT-T** to change line type as needed. For 1.25 diameter arcs use **CREATE-CIRCLE-CTR+DIA** and position the center of the circle at the intersection of the horizontal and vertical center lines and the angled and curved center lines. The circles will be trimmed later to ensure an accurate intersection.
4. Create the remaining circles using the center position of the previous circle to position the center of the current circle.
5. Use **CREATE-LINE-N-GON** with **6** sides; distance across flat is a radius of .28.
6. Use the **CREATE-ARC-TANGENT-2 ENTS** option to place the R1.25 arc (select inside the tangent points).
7. Use **CREATE-ARC-CTR+RAD** for the R1.63 arc with a start angle of 0 and an end angle of 90 at the center of the R1.25 arc.
8. Use the **LINE-TANGENT (TAN TAN** or **2 ENTS)** option and **CREATE-LINE-PARALEL-AT DIST** of .63.
9. Use the **CREATE-FILLET-ARC-NO TRIM** option to place the R.25 interior arcs.
10. Use **EDIT-TRM/EXT-DOUBLE** to trim the circles between the tangent points for the first segment to be removed and the **DIVIDE** option for the second segment to be removed.

11. Use **CTRL-A** to set the direction of the arrows for the dimensions to **IN** or **OUT**.

12. Use the **DETAIL-DIMENSN** commands **RADIUS** and **ANGLE** as needed.

PROBLEM 29

PROBLEM STATEMENT   Draw the view shown, including dimensions and notes.

CADKEY Setup   Load the **SETUP** file created in Problem 1.
FILES-PART-LOAD-NO. Enter filename **SETUP**.

PROCEDURE
1. Use **ALT-T** to change line type to center (3).
2. Use **CREATE-LINE-ENDPTS** or **ANGLE** and **CREATE-ARC-CTR+RAD** with start angles of 10 degrees and end angles of 160 to lay out all the center lines.
3. Use **ALT-T** to change line type as needed. For R1 and R5/8 arcs use the **CREATE-CIRCLE-CTR+RAD** and position the center of the circle at the intersection of the horizontal and vertical center lines and the angled and curved center

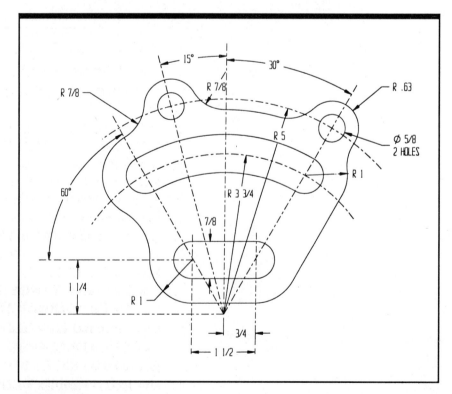

Problem 29

lines. The circles will be trimmed later to ensure an accurate intersection.

4. Create the remaining circles using the center position of the previous circle to position the center of the current circle.

5. Use the **CREATE-ARC-TANGENT-2 ENTS** option to place the R1 3/16 arc (select inside the tangent points).

6. Use the **LINE-TANGENT (TAN TAN** or **2 ENTS)** option and **CREATE-LINE-PARALEL-AT DIST** of 1.

7. Use the **CREATE-FILLET-ARC-NO TRIM** option to place the R7/8 arcs.

8. Use **EDIT-TRM/EXT-DOUBLE** to trim the circles between the tangent points for the first segment to be removed and the **DIVIDE** option for the second segment to be removed.

9. Use **CTRL-A** to set the direction of the arrows for the dimensions to **IN** or **OUT**.

10. Use the **DETAIL-DIMENSN** commands **RADIUS** and **ANGLE** as needed.

PROBLEM 30

PROBLEM STATEMENT Draw the view shown, including dimensions and notes.

CADKEY Setup Load the **SETUP** file created in Problem 1.
FILES-PART-LOAD-NO. Enter filename **SETUP**.

PROCEDURE

1. Use **ALT-T** to change line type to center (3).

2. Use **CREATE-LINE-ENDPTS** to lay out all the center lines.

3. Use **ALT-T** to change line type as needed. For R1.25 and R.75 arcs use **CREATE-CIRCLE-CTR+RAD** and position the center of the circle at the intersection of the horizontal and vertical center lines. The circles will be trimmed later to ensure an accurate intersection.

4. Create the remaining .75 diameter circles using the center position of the previous circle to position the center of the current circle.

5. Use the **LINE-TANGENT (TAN TAN** or **2 ENTS)** option for the lines tangent to the .75 diameter circles.

6. Use **CREATE-LINE-PARALEL-AT DIST** of 5.5, .5, .625 for the horizontal lines in the base.

7. Use **CREATE-LINE-PARALEL** to place the vertical lines in the base.

8. Use **EDIT-TRM/EXT-BOTH** to trim the corners of the base and the **DOUBLE** option for trimming the circles.

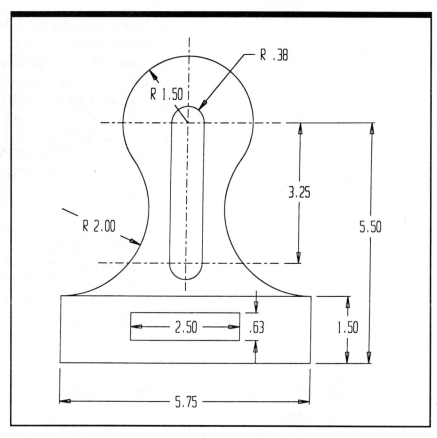

Problem 30

9. Use **CREATE-FILLET-ARC-NO TRIM** for the R2.00 ARC.
10. Use **CTRL-A** to set the direction of the arrows for the dimensions to **IN** or **OUT**.
11. Use the **DETAIL-DIMENSN** commands **HORIZTL, VERTICAL, RADIUS** and **ANGLE** as needed.

PROBLEM 31

PROBLEM STATEMENT   Pick the best front view and draw the top front and auxiliary views. Do not include dimensions.

CADKEY Setup   Load the **SETUP** file created in Problem 1.
FILES-PART-LOAD-NO. Enter filename **SETUP**.

PROCEDURE

Approach 1: Draw Mode   1. Use the **CREATE-LINE-STRING** or **RECTANG** command to define the outside boundary of the views.

2. Use the **CREATE-LINE-ENDPTS** command to place individual lines for interior details.

3. Use **ALT-T** to change line type to hidden (2).

4. **CREATE-LINE-PARALEL** to the inclined surface.

5. **CREATE-LINE-PERPEND** or **TAN/PRP** between the endpoint of the inclined surface and the auxiliary surface.

6. Use **CREATE-ARC** and **CIRCLE** as needed to complete the auxiliary view.

**Approach 2:
Orthographic Projection**

1. Use the **CREATE-LINE-STRING** and **ENDPTS** commands to create the front view.

2. Use the **CREATE-LINE-HRZ/VRT-VERTICL-ENDENT** to project the width of the front view to the top view.

3. **CREATE-LINE-PARALEL** to the inclined surface.

4. **CREATE-LINE-PERPEND** or **TAN/PRP** between the endpoint of the inclined surface and the auxiliary surface.

5. Use **EDIT-TRM/EXT-DOUBLE** or **FIRST** to trim projection lines between the appropriate edges.

**Approach 3:
Three-Dimensional Modeling**

1. Use the **CREATE-LINE-STRING** and **ENDPTS** commands to create the front view.

2. Once the front view has been created, change to view 7 (**ALT-V** and enter **7**) and use **CTRL-V** to change to world coordinates.

3. At this time you are ready to add the depth dimension to the height and width of the front view. Use **X-FORM-DELTA-JOIN** and whichever selection option is appropriate.

4. Enter **1** when prompted to enter the number of copies. The copy does not move on the X axis (width); it should move at a negative distance equal to the depth for the Y axis and does not move on the Z axis (height). Therefore, enter **0** for the X and Z values and a negative value equal to the depth for the Y axis.

5. To display the auxiliary view, use **DISPLAY-VIEW-DEFINE-3 PTS** using the **ENDENT** option to select the first two endpoints to determine the X axis and the third point to determine the Y axis.

6. Switch to the new view that has been created and place the arcs and circles at the correct depths.

7. Use **ALT-V** to change the front, top or right side view and use **FILES-PATTERN-CREATE** to create a pattern file for each view.

8. Save the wireframe model (**CTRL-F**) and use **FILES-PART-LOAD** on your **SETUP** file.

9. Use the **FILES-PATTERN-RETRIEV** option to retrieve a border and all of the desired views.

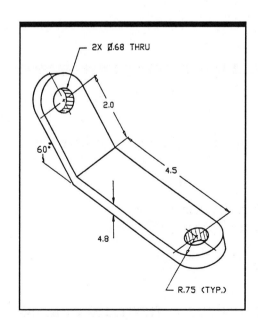

Problem 31

PROBLEM 32

PROBLEM STATEMENT   Pick the best front view and draw the top front and auxiliary views. Do not include dimensions.

CADKEY Setup   Load the **SETUP** file created in Problem 1.
FILES-PART-LOAD-NO. Enter filename **SETUP**.

PROCEDURE

Approach 1: Draw Mode
1. Use the **CREATE-LINE-STRING** or **RECTANG** command to define the outside boundary of the views.
2. Use the **CREATE-LINE-ENDPTS** command to place individual lines for interior details.
3. Use **ALT-T** to change line type to hidden (2).
4. **CREATE-LINE-PARALEL** to the inclined surface.
5. **CREATE-LINE-PERPEND** or **TAN/PRP** between the endpoint of the inclined surface and the auxiliary surface.
6. Use **CREATE-ARC** and **CIRCLE** as needed to complete the auxiliary view.

Approach 2: Orthographic Projection
1. Use the **CREATE-LINE-STRING** and **ENDPTS** commands to create the front view.
2. Use the **CREATE-LINE-HRZ/VRT-VERTICL-ENDENT** to project the width of the front view to the top view.
3. **CREATE-LINE-PARALEL** to the inclined surface.

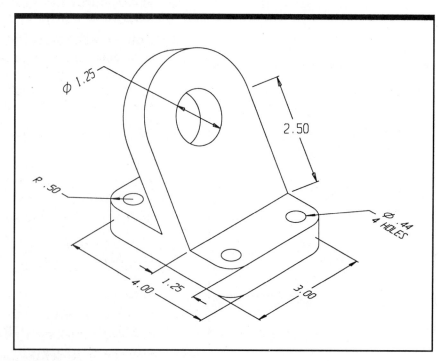

Problem 32

4. **CREATE-LINE-PERPEND** or **TAN/PRP** between the endpoint of the inclined surface and the auxiliary surface.

5. Use **EDIT-TRM/EXT-DOUBLE** or **FIRST** to trim projection lines between the appropriate edges.

Approach 3: Three-Dimensional Modeling

1. Use the **CREATE-LINE-STRING** and **ENDPTS** commands to create the front view.

2. Once the front view has been created, change to view 7 (**ALT-V** and enter **7**) and use **CTRL-V** to change to world coordinates.

3. At this time you are ready to add the depth dimension to the height and width of the front view. Use **X-FORM-DELTA-JOIN** and whichever selection option is appropriate.

4. Enter **1** when prompted to enter the number of copies. The copy does not move on the X axis (width); it should move at a negative distance equal to the depth for the Y axis and does not move on the Z axis (height). Therefore, enter **0** for the X and Z values and a negative value equal to the depth for the Y axis.

5. To display the auxiliary view, use **DISPLAY-VIEW-DEFINE-3 PTS**, using the **ENDENT** option to select the first two endpoints to determine the X axis and the third point to determine the Y axis.

6. Switch to the new view that has been created and place the arcs and circles at the correct depths.

7. Use **ALT-V** to change the front, top or right side view and use **FILES-PATTERN-CREATE** to create a pattern file for each view.

8. Save the wireframe model (**CTRL-F**) and use **FILES-PART-LOAD** on your **SETUP** file.

9. Use the **FILES-PATTERN-RETRIEV** option to retrieve a border and all of the desired views.

PROBLEM 33

PROBLEM STATEMENT

Pick the best front view and draw the top front and auxiliary views. Do not include dimensions.

CADKEY Setup

Load the **SETUP** file created in Problem 1.
FILES-PART-LOAD-NO. Enter filename **SETUP**.

PROCEDURE

Approach 1: Draw Mode

1. Use the **CREATE-LINE-STRING, ANGLE,** or **ENDPTS** commands to define the outside boundary of the front and top views.

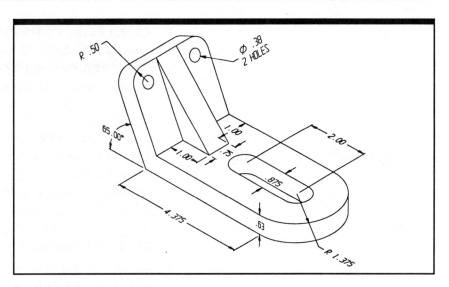

Problem 33

2. Use the **CREATE-LINE-ENDPTS** command to place individual lines for interior details.

3. Use **ALT-T** to change line type to hidden (2).

4. **CREATE-LINE** or **POLYLIN-SKETCH** to create the freehand break line in the top view.

5. **CREATE-LINE-PARALEL** to the inclined surface.

6. **CREATE-LINE-PERPEND** or **TAN/PRP** between the endpoint of the inclined surface and the auxiliary surface.

7. Lay out the center lines for the auxiliary view.

8. Use **CREATE-ARC, CIRCLE,** and **FILLET** as needed to complete the auxiliary view.

Approach 2:
Orthographic Projection

1. Use the **CREATE-LINE-STRING** and **ENDPTS** commands to create the front view.

2. Use **CREATE-LINE-HRZ/VRT-VERTICL-ENDENT** to project the width of the front view to the top view.

3. Use the **CREATE-FILLET-ARC** command for the .58 round corners.

4. **CREATE-LINE-PARALEL** to the inclined surface.

5. **CREATE-LINE-PERPEND** or **TAN/PRP** between the endpoint of the inclined surface and the auxiliary surface.

6. Use **EDIT-TRM/EXT-DOUBLE** or **FIRST** to trim projection lines between the appropriate edges.

7. Create the auxiliary view as described in steps 4 through 8 in Approach 1.

Approach 3:
Three-Dimensional Modeling

1. Use the **CREATE-LINE-STRING** and **ENDPTS** commands to create the front view.

2. Once the front view has been created, change to view 7 (**ALT-V** and enter **7**) and use **CTRL-V** to change to world coordinates.

3. At this time you are ready to add the depth dimension to the height and width of the front view. Use **X-FORM-DELTA-JOIN** and whichever selection option is appropriate.

4. Enter **1** when prompted to enter the number of copies. The copy does not move on the X axis (width); it should move at a negative distance equal to the depth for the Y axis and does not move on the Z axis (height). Therefore, enter **0** for the X and Z values and a negative value equal to the depth for the Y axis.

5. To display the auxiliary view, use **DISPLAY-VIEW-DEFINE-3 PTS**. Use the **ENDENT** option to select the first two endpoints to determine the X axis and the third point to determine the Y axis.

6. Switch to the new view that has been created and place the arcs, fillets and circles at the correct depths.

7. Use **ALT-V** to change the front, top or right side view and use **FILES-PATTERN-CREATE** to create a pattern file for each view.

8. Save the wireframe model (**CTRL-F**) and use **FILES-PART-LOAD** on your **SETUP** file.

9. Use the **FILES-PATTERN-RETRIEV** option to retrieve a border and all of the desired views.

PROBLEM 34

PROBLEM STATEMENT

Pick the best front view and draw the top front and auxiliary views, including dimensions and notes.

CADKEY Setup

Load the **SETUP** file created in Problem 1.
FILES-PART-LOAD-NO. Enter filename **SETUP**.

PROCEDURE

Approach 1: Draw Mode

1. Use the **CREATE-LINE-STRING, ANGLE,** or **ENDPTS** command to define the outside boundary of the front and top views.

2. Use the **CREATE-LINE-ENDPTS** command to place individual lines for interior details.

3. Use **ALT-T** to change line type to hidden (2).

4. **CREATE-POLYLIN-SKETCH** to create the free-hand break line in the top view.

5. **CREATE-LINE-PARALEL** to the inclined surface.

6. **CREATE-LINE-PERPEND** or **TAN/PRP** between the endpoint of the inclined surface and the auxiliary surface.

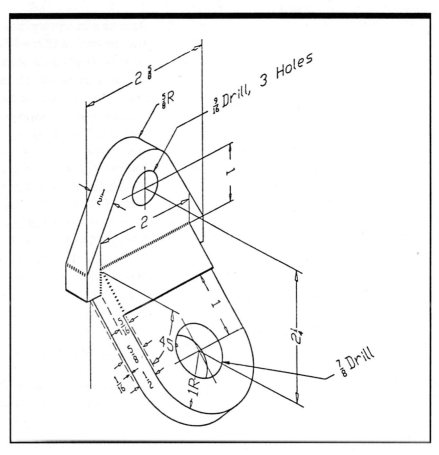

Problem 34

7. Lay out the center lines for the auxiliary view.

8. Use **CREATE-ARC, CIRCLE**, and **FILLET** as needed to complete the auxiliary view.

Approach 2: Orthographic Projection

1. Use the **CREATE-LINE-STRING** and **ENDPTS** commands to create the front view.

2. Use **CREATE-LINE-HRZ/VRT-VERTICL-ENDENT** to project the width of the front view to the top view.

3. Use the **CREATE-FILLET** command for the 1/8 round corners.

4. **CREATE-LINE-PARALEL** to the inclined surface.

5. **CREATE-LINE-PERPEND** or **TAN/PRP** between the endpoint of the inclined surface and the auxiliary surface.

6. Use **EDIT-TRM/EXT-DOUBLE** or **FIRST** to trim projection lines between the appropriate edges.

7. Create the auxiliary view as described in steps 4 through 8 in Approach 1.

Approach 3: Three-Dimensional Modeling

1. Use the **CREATE-LINE-STRING** and **ENDPTS** commands to create the front view.

2. Once the front view has been created, change to view 7 (**ALT-V** and enter **7**) and use **CTRL-V** to change to world coordinates.

3. At this time you are ready to add the depth dimension to the height and width of the front view. Use **X-FORM-DELTA-JOIN** and whichever selection option is appropriate.

4. Enter **1** when prompted to enter the number of copies. The copy does not move on the X axis (width); it should move at a negative distance equal to the depth for the Y axis and does not move on the Z axis (height). Therefore, enter **0** for the X and Z values and a negative value equal to the depth for the Y axis.

5. To display the auxiliary view, use **DISPLAY-VIEW-DEFINE-3 PTS**. Use the **ENDENT** option to select the first two endpoints to determine the X axis and the third point to determine the Y axis.

6. Switch to the new view that has been created and place the arcs, fillets and circles at the correct depths.

7. Use **ALT-V** to change the front, top or right side view and use **FILES-PATTERN-CREATE** to create a pattern file for each view.

8. Save the wireframe model (**CTRL-F**) and use **FILES-PART-LOAD** on your **SETUP** file.

9. Use the **FILES-PATTERN-RETRIEV** option to retrieve a border and all of the desired views.

Dimensioning

1. Use **DETAIL-DIMENSN-HORIZTL** or **VERTICL** (**ENDENT**). Select the origin of the first witness (extension) line and the second witness line. Then select the position of the dimension text.

2. Use **CTRL-A** to set arrows **OUT**. Use **DETAIL-DIMENSN-DIAMETR** or **RADIUS**. Select the circle or arc you wish to dimension, and indicate the text position and whether the leader shoulder should be to the left or right of the text.

3. Use **DETAIL-NOTE-KEYIN** to type in the note; press **RETURN** for additional lines. When finished with the note, press **RETURN** twice (CADKEY versions 1 through 3.0) or **SAVE TXT** (version 3.1 or higher). Then indicate text position.

PROBLEM 35

PROBLEM STATEMENT   Pick the best front view and draw the top front and auxiliary views, including dimensions and notes.

CADKEY Setup   Load the **SETUP** file created in Problem 1.
FILES-PART-LOAD-NO. Enter filename **SETUP**.

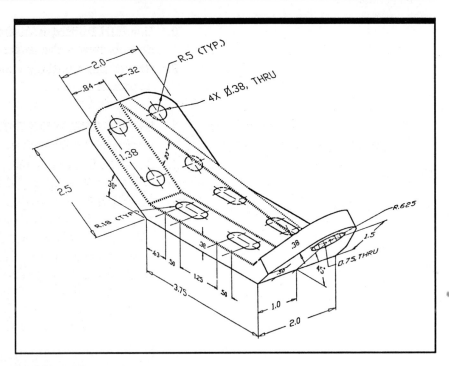

Problem 35

PROCEDURE

Approach 1: Draw Mode

1. Use the **CREATE-LINE-STRING, ANGLE,** or **ENDPTS** commands to define the outside boundary of the front and top views.
2. Use the **CREATE-LINE-ENDPTS** command to place individual lines for interior details.
3. Use **ALT-T** to change line type to hidden (2).
4. **CREATE-LINE-PARALEL** to the inclined surface.
5. **CREATE-LINE-PERPEND** or **TAN/PRP** between the endpoint of the inclined surface and the auxiliary surface.
6. Lay out the center lines for the auxiliary view.
7. Use **CREATE-ARC, CIRCLE,** and **FILLET** as needed to complete the auxiliary view.

Approach 2: Orthographic Projection

1. Use the **CREATE-LINE-STRING** and **ENDPTS** commands to create the front view.
2. Use **CREATE-LINE-HRZ/VRT-VERTICL-ENDENT** to project the width of the front view to the top view.
3. Use the **CREATE-FILLET** command for the 1/8 round corners.
4. **CREATE-LINE-PARALEL** to the inclined surface.
5. **CREATE-LINE-PERPEND** or **TAN/PRP** between the endpoint of the inclined surface and the auxiliary surface.

6. Use **EDIT-TRM/EXT-DOUBLE** or **FIRST** to trim projection lines between the appropriate edges.

7. Create the auxiliary view as described in steps 4 through 7 in Approach 1.

Approach 3:
Three-Dimensional Modeling

1. Use the **CREATE-LINE-STRING** and **ENDPTS** commands to create the front view.

2. Once the front view has been created, change to view 7 (**ALT-V** and enter **7**) and use **CTRL-V** to change to world coordinates.

3. At this time you are ready to add the depth dimension to the height and width of the front view. Use **X-FORM-DELTA-JOIN** and whichever selection option is appropriate.

4. Enter **1** when prompted to enter the number of copies. The copy does not move on the X axis (width); it should move at a negative distance equal to the depth for the Y axis and does not move on the Z axis (height). Therefore, enter **0** for the X and Z values and a negative value equal to the depth for the Y axis.

5. To display the auxiliary view, use **DISPLAY-VIEW-DEFINE-3 PTS**. Use the **ENDENT** option to select the first two endpoints to determine the X axis and the third point to determine the Y axis.

6. Switch to the new view that has been created and place the arcs, fillets and circles at the correct depths.

7. Use **ALT-V** to change the front, top or right side view. Use **FILES-PATTERN-CREATE** to create a pattern file for each view.

8. Save the wireframe model (**CTRL-F**) and use **FILES-PART-LOAD** on your **SETUP** file.

9. Use the **FILES-PATTERN-RETRIEV** option to retrieve a border and all of the desired views.

Dimensioning

1. Use **DETAIL-DIMENSN-HORIZTL** or **VERTICL** (**END-ENT**). Select the origin of the first witness (extension) line and the second witness line. Then select the position of the dimension text.

2. Use **CTRL-A** to set arrows **OUT**. Use **DETAIL-DIMENSN-DIAMETR** or **RADIUS**. Select the circle or arc you wish to dimension, and indicate the text position and whether the leader shoulder should be to the left or right of the text.

3. Use **DETAIL-NOTE-KEYIN** to type in the note; press **RE-TURN** for additional lines. When you are finished with the note, press **RETURN** twice (CADKEY versions 1 through 3.0) or **SAVE TXT** (version 3.1 or higher). Then indicate text position.

PROBLEM 36

PROBLEM STATEMENT   Draw the front view and a full section view, including dimensions.

CADKEY Setup   Load the **SETUP** file created in Problem 1.
FILES-PART-LOAD-NO. Enter filename **SETUP**.

PROCEDURE   NOTE: This problem can be approached any of the three ways
that have been described for the multiview problems.

1. Lay out the center lines using **CREATE-LINE-ENDPTS**.
2. Use **CREATE-CIRCLE-CTR+DIA** for the circles.
3. Create the right side view by whichever approach you wish to use.
4. Use **EDIT-TRM/EXT-FIRST** and **DIVIDE** to eliminate any
    unnecessary lines.
5. Use the **ALEV** option in the Status Window to change the
    active level to 5. Also change the current color.
6. Using **CREATE-LINE-STRING-ENDENT**, trace the areas
    that will be filled with the section lines.
7. Now change **ALEV** to 6. Use **DISPLAY-LEVELS-LIST** to
    turn off all levels except 5 and 6.
8. Use **DETAIL-X-HATCH-BRICK** and **WINDOW** the sec-
    tions to be filled.

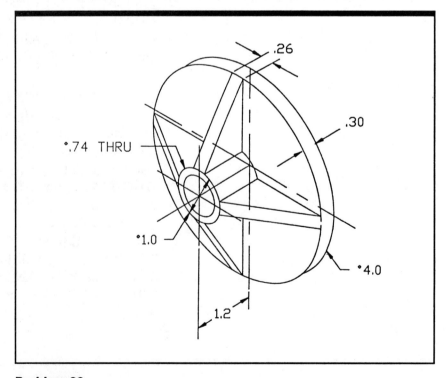

Problem 36

9. Again use **DISPLAY-LEVELS-LIST** to turn off level 5 and turn on any other levels you may have turned off.

Dimensioning

1. Use **DETAIL-DIMENSN-HORIZTL** or **VERTICL** (**END-ENT**). Select the origin of the first witness (extension) line and the second witness line. Then select the position of the dimension text.

2. Use **CTRL-A** to set arrows **OUT**. Use **DETAIL-DIMENSN-DIAMETR** or **RADIUS**. Select the circle or arc you wish to dimension, and indicate the text position and whether the leader shoulder should be to the left or right of the text.

3. Use **DETAIL-NOTE-KEY IN** to type in a note; press **RE-TURN** for additional lines. When finished with the note, press **RETURN** twice (CADKEY versions 1 through 3.0) or **SAVE TXT** (version 3.1 or higher). Then indicate text position.

PROBLEM 37

PROBLEM STATEMENT   Draw the front view and a full section view, including dimensions.

CADKEY Setup   Load the **SETUP** file created in Problem 1.
FILES-PART-LOAD-NO. Enter filename **SETUP**.

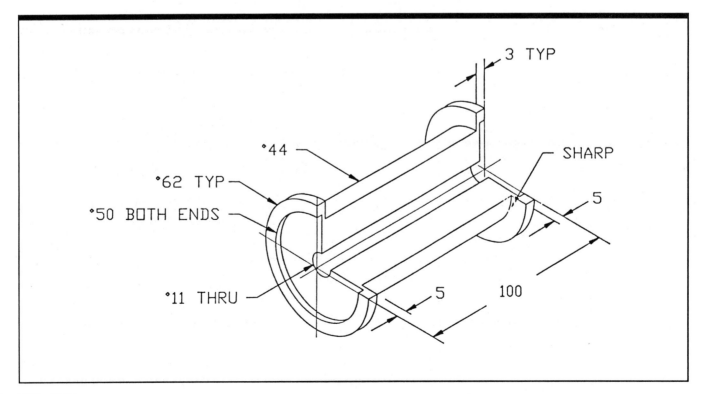

PROCEDURE

NOTE: This problem can be approached any of the three ways that have been described for the multiview problems.

1. Lay out the center lines using **CREATE-LINE-ENDPTS**.
2. Use **CREATE-CIRCLE-CTR+DIA** for the circles.
3. Create the right side view by whichever approach you wish to use.
4. Use **EDIT-TRM/EXT-FIRST** and **DIVIDE** to eliminate any unnecessary lines.
5. Use **ALEV** option in the status window to change the active level to 5. Also change the current color.
6. Using **CREATE-LINE-STRING-ENDENT**, trace the area that will be filled with the section lines.
7. Now change **ALEV** to 6. Use **DISPLAY-LEVELS-LIST** to turn off all levels except 5 and 6.
8. Use **DETAIL-X-HATCH-BRICK** and **WINDOW** the section to be filled.
9. Again use **DISPLAY-LEVELS-LIST** to turn off level 5 and turn on any other levels you may have turned off.

Dimensioning

1. Use **DETAIL-DIMENSN-HORIZTL** or **VERTICL** (**END-ENT**). Select the origin of the first witness (extension) line and the second witness line. Then select the position of the dimension text.
2. Use **CTRL-A** to set arrows **OUT**. Use **DETAIL-DIMENSN-DIAMETR** or **RADIUS**. Select the circle or arc you wish to dimension, and indicate the text position and whether the leader shoulder should be to the left or right of the text.
3. Use **DETAIL-NOTE-KEYIN** to type in the note; press **RETURN** for additional lines. When finished with the note, press **RETURN** twice (CADKEY versions 1 through 3.0) or **SAVE TXT** (version 3.1 or higher). Then indicate text position.

PROBLEM 38

PROBLEM STATEMENT

Draw a perspective view, including dimensions and notes.

CADKEY Setup

Load the **SETUP** file created in Problem 1.
FILES-PART-LOAD-NO. Enter filename **SETUP**.

PROCEDURE

1. **CREATE-LINE-RECTANG** with a width of 22 and a height of 17.
2. Use **ALT-A** to autoscale. **CREATE-LINE-PARALEL-AT DIST** of **5** for the horizon and **9** for the picture plane.
3. Draw the front and top views in their appropriate positions.

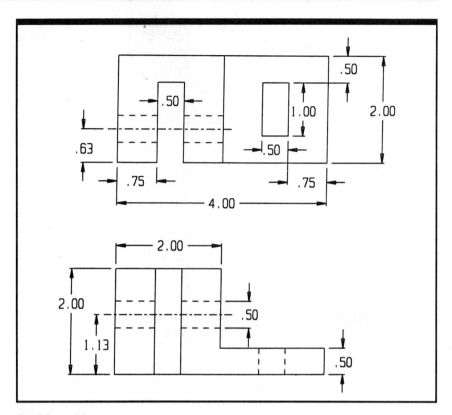

Problem 38

4. **CREATE-POINT-POSITN** for the vanishing points and the station point.

5. Use **CREATE-LINE-ENDPTS** to **ENDENT** to **POINT** on the Position Menu to find the intersecting point of the edges with the picture plane.

6. Use **CREATE-LINE-HRZ/VRT-VERTICL** through the **INTRSC** of the **PP** and the line to the station point.

7. Continue to use these methods to draw the perspective view. Delete construction lines as they become unnecessary and use **EDIT-TRM/EXT** as needed.

PROBLEM 39

PROBLEM STATEMENT   Pick the best front view and draw the top front and auxiliary views, including dimensions and notes.

CADKEY Setup   Load the **SETUP** file created in Problem 1.
FILES-PART-LOAD-NO. Enter filename **SETUP**.

PROCEDURE   1. **CREATE-LINE-RECTANG** with a width of **22** and a height of **17**.

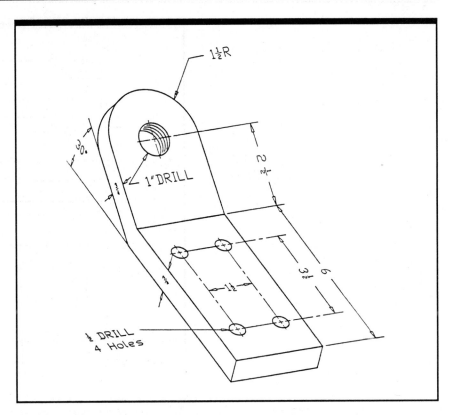

Problem 39

2. Use **ALT-A** to autoscale. **CREATE-LINE-PARALEL-AT DIST** of **5** for the horizon and **9** for the picture plane.

3. Draw the front and top views in their appropriate positions.

4. **CREATE-POINT-POSITN** for the vanishing points and the station point.

5. Use **CREATE-LINE-ENDPTS** to **ENDENT** to **POINT** on the Position Menu to find the intersecting point of the edges with the picture plane.

6. Use **CREATE-LINE-HRZ/VRT-VERTICL** through the **INTRSC** of the **PP** and the line to the station point.

7. Continue to use these methods to draw the perspective view. Delete construction lines as they become unnecessary and use **EDIT-TRM/EXT** as needed.

PROBLEM 40

PROBLEM STATEMENT   Draw two perspectives; omit dimensions and notes.

CADKEY Setup   Load the **SETUP** file created in Problem 1.
FILES-PART-LOAD-NO. Enter filename **SETUP**.

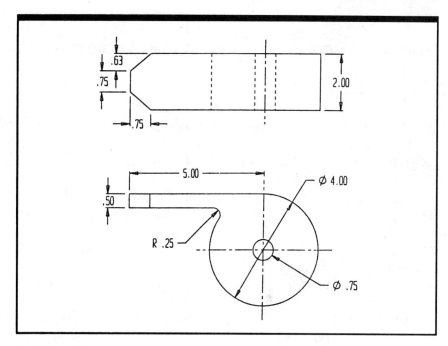

Problem 40

PROCEDURE

1. **CREATE-LINE-RECTANG** with a width of **22** and a height of **17**.

2. Use **ALT-A** to autoscale. **CREATE-LINE-PARALEL-AT DIST 2.5** for the horizon and **10** for the picture plane.

3. Draw the front and top views in their appropriate positions.

4. **CREATE-POINT-POSITN** for the vanishing points and the station point.

5. Use **CREATE-LINE-ENDPTS** to **ENDENT** (edge) to **POINT** (station point) on the Position Menu to find the instersecting point of the edges with the picture plane.

6. Use **CREATE-LINE-HRZ/VRT-VERTICL** through the **INTRSC** of the **PP** and the line to the station point.

7. Continue to use these methods to draw the perspective view. Delete construction lines as they become unnecessary and use **EDIT-TRM/EXT** as needed.

ABORT (text editor) Exits the text editor without changing the text.

ACTIVE Chooses the level to which newly created entities are assigned.

ADD (LEVEL) Enters a level number for display.

ADD (VWPORT) Adds a new viewport to the display.

ADD TO Adds entities to existing ordinate dimensions or groups.

ALIGN (DIGITIZE) Orients your drawing to the active tablet area when in digitizing mode.

ALIGN (ORDINAT) Aligns ordinate dimensions to the base ordinate.

ALIGNED Creates and arranges all dimension ordinate text corresponding to text positioning of the base ordinate.

ALL Selects all entities displayed for interaction or recall.

ALL (RECALL) Recalls for display all entities deleted since the last redraw.

ALL DSP Chooses all entities displayed on the screen for selection.

ALL IN (POLYGON, WINDOW) Selects the entities lying totally inside the polygon or window.

ALL OUT (POLYGON, WINDOW) Selects the entities lying totally outside the polygon or window.

ALONG C Creates a point on a curve at a specified distance from a designated endpoint.

ALONGL Locates a position on a line that is a certain distance from a designated endpoint.

ANGLE (CONIC) Defines the slope of a conic at a selected point with a keyed-in value for the tangent vector at that point (2D construction mode only).

ANGLE (LINE) Creates a line entity at a given angle to a reference line.

ANGLE (2D CUBI, 3D CUBI) Determines the direction of a tangent vector for a spline by setting the angle between the tangent vector and the positive X axis.

ANGLE (VERIFY) Calculates the angle formed by two intersecting lines.

ANGULAR (DETAIL) Dimensions the angle between two intersecting lines.

ANGULAR (TOLER) Assigns tolerancing values to angular dimensions.

APPEND Merges the contents of a macro library with the contents of the currently active library.

APPEND (text editor) Joins the current line with the following line.

ARC Creates arc entities (portions of circles).

ARC (FILLET) Creates a tangent arc between two intersecting curve entities.

ARCLEN Defines the arc length of the helix curve.

ARCLNTH Divides the edge curve of a mesh into segments of a specific length.

AREA/CN Calculates the area and X,Y coordinates of the centroid in selected closed figures.

ARR/WIT Creates arrows or witness lines as drafting entities.

ARROWS Changes or sets the direction or style of arrows relative to the witness lines.

ASPECT Assigns the character width to height ratio of text.

AT DIST Creates a parallel line at a designated distance from a specified entity.

ATTRIB (CONTROL) Verifies or changes the attributes and form assigned to selected entities (i.e., color, line type, line width, pen number, outlined, filled).

ATTRIB (VERIFY) Displays attribute specifications of a selected entity.

AUTO Adjusts the size of your drawing so it is fully displayed on the screen or fits on the paper for plotting.

AUTOMAT Enables automatic calculation of dimension values.

AUTOSEG Creates a specified number of entity or line segments of equal lengths on a selected entity.

AUTO SET Assigns the number of viewports to the display.

AUTO UP Controls automatic updating of the displayed values of dimensions that are selected during pattern file retrieval and X-FORM operations.

AXES Defines and displays two icons that identify the orientation of the world coordinate axes and the construction view axes relative to the currently displayed view.

AXIS   Defines the distance measured along the major axis of the helix curve.

BACK-1   Returns your drawing to the previous display.

BASE PT   Moves or modifies selected entities by designating old and new base point positions.

BASE LN (DIMENSN, SERIAL)   Creates a series of linear dimensions that have one reference point in common—the baseline point.

BEG+END   Creates an arc by specifying start and end points and the included angle.

BENT RAD   Creates bent radius dimensions. The radius leader is bent, somewhat like a bolt of lightning, to indicate that the endpoint of the leader is not at the center of an arc or circle.

BINDING   Assigns existing macros to keys or tablet positions, and provides means for storing, activating, and modifying these assignments.

BINEXEC   Reads in a previously compiled CADL file and stores data primitives as entities.

BOTH (BREAK)   Divides two intersecting curve entities into two or three entities at their common intersections.

BOTH (DETAIL)   Displays both witness lines or leaders.

BOTH (TOLER)   Assigns negative and positive values to tolerancing.

BOTH (TRM/EXT)   Trims or extends intersecting entities to each other.

BREAK   Divides curve entities into separate entities at their intersections.

BREAK (SECTION)   Breaks selected entities at their intersections with a user-defined plane.

BX MOVE   Modifies or moves entities that have been selected using a rubberbox window by base point or by assigning X, Y, Z delta view coordinates.

BY TYPE   Selects entities by type.

CADL   CADKEY Advanced Design Language.

CENTER   Locates the center of a selected entity as a specified position.

CHAIN   Selects a series of connected curves by defining the start and end entities in the chain and the direction in which you want the chain to be selected.

CHAIN (DIMENSN, SERIAL)   Creates chain dimensioning by selecting the points to be dimensioned.

CHAMFER   Creates an angled line that intersects two selected curve entities.

CHANGE (ATTRIB)   Defines a new color, line type, line width, or pen number to existing entities.

CHANGE (DETAIL)   Defines a new detail assignment to selected entities.

CHANGE (VIEW)   Modifies the current display view.

CHG END   Modifies the end condition of an existing spline.

CHR ASP   Assigns the character width to height ratio of the text in dimensions, notes, and labels.

CIRCLE   Creates two-dimensional circles of any size.

CIRCLE (VERIFY)   Calculates an arc's or circle's area/centroid, moment of inertia, or perimeter calculations.

CLR-KEY   Cancels the assignment of a macro to a key.

CLR-TAB   Cancels the assignment of a macro to a tablet position.

CLSD LN   Selects a set of line entities that define a closed region for area/centroid and moment of inertia calculations.

COLOR   Assigns color to entities (monochrome screens are disabled for this function).

COLOR (masking)   Allows you to mask by color.

CONIC (CREATE)   Constructs a conic section (i.e., ellipse, hyperbola, parabola).

CONIC (VERFY)   Specifies a selected conic and calculates and displays detailed information about that conic type.

CONST (AXES)   Defines and displays the construction icon, which identifies the orientation of the construction view coordinate axes relative to the current display plane.

CONST (DEPTH)   Allows you to assign the current working depth.

CONST VW (AXES)   Turns the construction plane icon on or off.

CONTROL   Alters various modal parameters, displays detailed information about selected entities, allows you to print and plot, and accesses DOS from within the program.

COORD   Displays coordinate data for a selected entity.

COORDS   Sets view or world coordinate entry.

COPY   Produces copies of the selected entities without changing the original entities.

CORNERS   Creates a rectangle by defining its first and last corner.

CREATE   Allows you to construct geometric entities on the screen.

CREATE (FILES)   Creates a pattern file given the requested information.

CREATE (ORDINAT)   Creates a string of ordinate dimensions on the horizontal, vertical, or parallel axis.

CTR+DIA   Creates arcs and circles by specifying a center position and diameter.

CTR+EDG   Creates arcs and circles by specifying the start and end angles, center, and edge points.

CTR-ENT   Creates an arc or a circle with a specified center position and tangent to a selected entity.

CTR+RAD   Creates arcs, circles, or polygons by specifying a center position and numerical radius.

CTR+SID   Creates polygons by specifying a center position and the number of sides.

CUR ATT   Changes the attributes of entities that intersect, or lie on a selected side of, a user-defined plane to those attributes currently in effect.

CUR-LEV   Assigns pattern file entities to the current active level during pattern file retrieval.

CUR TRK   Displays the current coordinate positions (X,Y) of the screen cursor.

CURRENT   Sets the existing modal attribute in effect.

CURSOR   Locates a position using the cursor and the current working depth.

CURSOR (DIGITIZE)   Activates the tablet as a coordinate recording device.

DB DRAW   Controls the order in which entities are drawn in the data base.

DBL UP   Assigns both the positive (+) and negative (–) symbols to the same line for tolerance values that are the same.

DECIMAL   Changes or sets the type of decimal units used for dimensioning.

DEFINE (VIEW)   Defines a new display view.

DEGROUP   Ungroups selected entities from a group.

DEL/CHG   Deletes or changes the existing attributes of entities that lie on a user-selectable side of a defined plane.

DEL END (text editor)   Deletes all characters from the current cursor position to the end of the line.

DELETE   Erases levels, displayed entities, groups, or macros.

DELETE (SECTION)   Deletes entities that intersect a user-defined plane.

DEL LIN (text editor)   Deletes the current line and displays the preceding line.

DELTA   Indicates a screen position by entering coordinates relative to a reference point.

DEPTH   Changes the working depth.

DETAIL   Creates drafting entities that add dimensions and text to a drawing.

DETAIL (masking)   Allows you to select drafting entities for masking.

DIAMETR   Dimensions the diameter of a circle in degrees or radians.

DIGITIZE   Turns digitizing mode on and off, and sets up your active digitizing area.

DIM ALN   Aligns the dimension text using horizontal or vertical positioning.

DIM FNT   Changes or defines the type of font used for displayed dimensions.

DIM HT   Changes or defines the height of displayed dimensions.

DIM MOD   Permits you to toggle between automatic calculation of dimension values and manual entry of dimension text.

DIM PLC   Sets the placement position of text in dimensions.

DIM REP   Changes the existing linear, radius, diameter, or ordinate dimension's representation to decimal feet, inches, or fractions.

DIM SCL (dimension scale)   Changes the value of a dimension by a designated scale factor.

DIM VAL (dimension value)   Changes dimension values.

DIMENSN   Calculates different radius, diameter, distance, and ordinate dimensions.

DIRECT   Alters or sets the direction of the arrows relative to witness lines.

DISK   Reads in an existing ASCII file of no more than 1024 characters.

DISP VW   Turns the display view icon on or off.

DISPLAY   Affects how your drawing looks on the screen, but does not physically alter or change coordinates of the drawing itself.

DISPLAY (AXES)   Displays the desired icon (world or construction).

DIST   Calculates the actual and projected distance between two positions, a position and an entity, or two lines.

DISTNCE   Segments a selected curve by creating a series of evenly spaced points starting with a designated endpoint on the curve.

DIVIDE (BREAK)   Breaks a selected curve into three separate entities defined by two intersections, with the inner portion assigned the current attributes.

DIVIDE (TRM/EXT)   Separates a curve entity into two entities by selecting the portion of the entity to be trimmed or extended.

DOUBLE (BREAK)   Breaks a selected curve into three separate entities defined by two intersections, with the outer portions taking on current attributes.

DOUBLE (DISPLAY)   Scales the current drawing to twice its size.

DOUBLE (TRM/EXT)   Trims or extends a curve to two selected curve entities.

DUMP NT   Outputs the text from an existing note to a disk-based ASCII text file.

EDIT   Changes or revises a drawing.

EDIT TX   Calls the on-line text editor, which is used to modify text in existing dimensions, labels, disk-based notes, and keyed-in notes, and to create text for manually entered dimensions, labels, and keyed-in notes.

ELLIPSE   Creates an elliptical arc that has start, end, and center points at indicated positions, and major and minor axes with specified lengths and orientations.

ENDENT   Locates the endpoint of an entity as a position.

ENDENT (CONIC:5 COND)   Defines the slope of a conic at a selected point with the tangent vector that starts at the end of a selected entity.

ENDENT (PARABOL)   Defines the directrix of a parabola as the line that is tangent to a selected curve entity at the end of the curve that is closest to the selection point.

ENDENT (PROJECT)   Defines a direction parallel to the tangent vector at the end of a selected entity.

ENDENT (2D CUBI, 3D CUBI)   Specifies a tangent vector for a spline by using the tangency at the end of an existing entity.

ENDPTS   Creates a line entity, arrow, or witness line by specifying the start and end points.

ENTITY   Defines a selection plane using a selected line, arc, or 2D spline.

ENTITY (PROJECT)   Defines a projection plane with a selected line, arc, conic, or 2D spline.

ENTITY (VIEW)   Defines view using a planar entity.

ENT-LEV   Assigns retrieved pattern file entities to the levels that were in effect when the pattern was created.

ENTS   Creates a specified number of segments at equal lengths, on selected line or arc entities.

EX TYPE   Selects all entities except those specified.

EXECUTE   Reads in an ASCII-format (uncompiled) CADL file, stores data primitives as entities, and executes CADL commands.

EXIT   Leaves the program and returns you to the current DOS prompt.

FILE (NOTE)   Reads in an existing ASCII file with no size limitation.

FILES   Creates, retrieves, or lists plot, part, and pattern files. Also accesses macros, CADL files, and DXF files.

FILL   Allows you to select a font and fill its character area.

FILLED   Fills in a polygon or polyline with a color from the color palette.

FILLET   Connects two nonparallel curve entities with an arc or an angled line.

1ST SOL   Displays a leader line underneath the dimension, and extends to the first dimension point.

FIRST (BREAK)   Divides the first curve selected into two or three separate entities with current attributes assigned.

FIRST (DETAIL)   Displays the first of two witness lines or leaders.

FIRST (TRM/EXT)   Trims or extends the first entity selected.

5 COND   Creates a conic section (ellipse, hyperbola, parabola) based on five specified conditions. Specified conditions include indicated positions that the conic passes through, the slope of the conic at one or two indicated points, or a rho value.

5 POINT   Creates a conic section based on indicated positions for five points on the entity.

FONT   Assigns type of text font displayed.

4 CURVE   Creates a mesh defined by two generator curves and two director curves. The mesh approximates the blending of the four two-curve surfaces.

4PT 1SL   Creates a conic section based on indicated positions for four points on the entity, and the slope of the conic at one of the indicated positions.

FRACTIN   Changes or assigns the current dimension to a fractional value.

FT/IN   Changes the current dimension value to feet and inches mode.

FUNCTN   Sets the length of lines to the definition of the function.

GENERAL   Calculates the area/centroid, moment of inertia, or perimeter in a set of entities (excluding points) whose projections define a closed region in the current view.

GEOM   Allows you to select geometric entities for masking.

GRAPHIC   Executes system commands or programs without exiting the program environment.

GRID   Displays a two-dimensional set of dots on the screen, which are spaced according to the X and Y coordinates.

GRID ALN   Aligns a point on the drawing to a point on the displayed grid, or to the snap resolution grid, or to the snap resolution assigned.

GRID DISP   Turns grid on and off.

GRID INC   Sets the X and Y coordinate of the grid.

GRID=SNAP   Sets the current grid points to the snap resolution.

GROUP   Links entities together so they can be selected individually or as a complete unit.

HALF   Scales the current drawing to one-half its size.

HEIGHT   Changes text height.

HELIX   Creates a 3D cubic spline, which is an approximation to a helix curve of arbitrary orientation.

HORIZTL (DIMENSN)   Displays the horizontal dimension between two designated points.

HORIZTL (SET)   Positions the horizontal alignment for text.

HRZ/VRT   Draws horizontal and vertical line entities.

IN   Redisplays dimension arrows pointing inward.

INS LIN (text editor)   Inserts a blank line above the current line.

INTRSC   Locates a position at the intersection of two designated entities.

JOIN   Copies selected entities and connects their endpoints.

JOIN LN (text editor)   Joins the current line with the following line.

KEY   Assigns (binds) a macro to a key.

KEYIN   Defines a direction by entering delta values for Cartesian coordinates from the keyboard.

KEYIN (CONIC:5 COND)   Defines the slope of a conic at a selected point with keyed-in delta-X and delta-Y values. In 3D construction mode, a delta-Z value is also required.

KEY-IN   Enters a Cartesian coordinate position from the keyboard.

KEY-IN (NOTE)   Enters a note via the keyboard.

KEY-IN (RETRIEV, LST/RTV)   Enters a level assignment for retrieved pattern file entities via the keyboard.

KEY IN (PLOT)   Inputs a plotting scale value.

KEY-TAB   Assigns (binds) a specified key to tablet position.

L-LIMIT   Allows certain line creation methods to produce lines to the length of the current viewport limits.

L-TYPE   Defines the type of lines drawn: solid, dashed, center-line, and phantom.

L-TYPE (masking)   Allows you to mask by line type.

L-WIDTH   Defines the width of the line according to the number of pixels assigned.

L-WIDTH (masking)   Allows you to mask by line width.

LABEL   Creates a label (with leaders and arrows) for a part or parts of a drawing.

LAST   Recalls the last entity deleted.

LEAD   Assigns zeros before the decimal point of any dimension or tolerance less than 1.

LEADER   Controls the display of leader lines in dimensions.

LEVELS   Adds, changes, deletes, or recalls layers of a drawing.

LIBRARY   Stores, loads, or modifies the contents of a macro library.

LIMIT   Displays the high limit of tolerancing above the low limit.

LINE   Constructs line entities on the screen.

LINE (PARABOL)   Defines the directrix of a parabola with an existing line.

LINEAR   Assigns tolerance values to linear dimensions.

LINES   Creates a specified number of line segments of equal length on reference line and curve entities.

LIST (FILES)   Lists the names of all plot files in a specified directory.

LIST (GROUP)   Lists group name and contents.

LIST (LEVELS)   Generates list of levels on the screen to turn on or off.

LIST/EXE   Lists and executes existing CADL files.

LOAD   Retrieves a designated DXF file, macro library, binding file, or part file from a specified (or default) directory.

LST/EXE   Lists the names of all macros in a library and allows you to cursor-select a macro for execution.

LST/LD   Lists the contents of the current part file directory and allows you to activate a desired file via cursor selection.

LST/RTV   Lists the contents of the current pattern directory and allows you to activate a desired file via cursor selection.

LST FMT   Assigns type of filename listing desired, short to long.

MACROS   A sequence of commands that are named, recorded, and stored in a macro library to be recalled and executed during any idle state of system operation.

MAKE   Groups selected entities.

MANUAL   Allows manual entry of dimension text.

MASK (LEVELS)   Screens out desired levels in a part.

MESH (LINE)   Allows you to generate a line mesh approximation of a two-, three-, or four-curve swept surface, a tabulated cylinder, surface of revolution, sphere, or ruled surface.

MESH (POINT)   Allows you to generate a point mesh approximation of a two-, three-, or four-curve swept surface, a tabulated cylinder, surface of revolution, sphere, or ruled surface.

MESH (POLYGON)   Allows you to generate a polygon mesh that is an approximation of a two-, three-, or four-curve swept surface, a tabulated cylinder, ruled surface, surface of revolution, or sphere. It is also possible to change the direction (normals) of a grouped polygon's mesh.

MESH (POLYLIN)   Allows you to generate a polyline mesh approximation of a two-, three-, or four-curve swept surface, a tabulated cylinder, surface of revolution, sphere, or ruled surface.

MESH (SPLINE)   Allows you to generate a spline mesh approximation of a two-, three-, or four-curve swept surface, a tabulated cylinder, surface of revolution, or sphere.

MIRROR (TXT ATT)   Changes a selected note by rotating the entity about its mirroring axis within its current text position.

MIRROR (X-FORM)   Duplicates a selected entity about the mirroring axis in the plane of the screen.

MODAL (BREAK)   Divides multiple curve entities, based on intersections with a common entity, into two separate entities and assigns current attribute values.

MODAL (TRM/EXT)   Continuously trims or extends selected curve entities to selected line or curve entities.

MODIFY   Alters ordinate dimensions in four different ways: adding, moving, aligning, or removing.

MOMENT   Calculates the moment of inertia of a selected closed region.

MOVE (LEVELS)   Moves entities to specified levels.

MOVE (ORDINAT)   Moves a subordinate to a new position.

MOVE (VWPORTS)   Moves viewport to a new position.

MOVE (X-FORM)   Operates on selected entities by changing the position and/or scale of the selected entities.

MOV NODE   Moves an existing node on a spline.

N-GON   Creates a line or polyline with equal length sides.

NATURAL   Assigns a curvature of zero to the start or end condition of the spline.

NEG TOL   Assigns negative tolerancing values.

NEW   Defines a new view using rotation techniques.

NO ARRS   Displays a solid leader line (without arrows).

NO TRIM   Allows selected entities to remain in their original state when arc fillets are added.

NON-ALN   Creates ordinate dimensions and text according to positions specified by the user.

NONE (SYSCMD)   Executes a system command or program without exiting the program environment and suppresses text output.

NONE (TOLER)   Displays no tolerancing with a dimension.

NONE (WITNESS/ LEADER)   Displays no leader or witness lines in a dimension.

NORMAL   Defines a direction perpendicular to a selected plane.

NOT ANG   Changes or sets the angle of text.

NOT FNT   Changes or sets the type of font used for notes or labels displayed on the screen.

NOT PLC   Sets the placement position of text in notes and labels.

NOTE   Adds general note text and comments to your part.

NOTE HT   Changes or sets the height of notes or labels displayed on the screen.

NUMBER   Creates a designated number of equally spaced points on a selected curve.

NUMSEGS   Divides the edge curve of a mesh into a specific number of equal length segments.

OFFSET (DEPTH)   Allows you to change the working depth at which entities are created.

OFFSET (FILES)   Calculates level assignments for retrieved pattern entities by adding a user-specified integer to the level number that each entity had when the pattern was created.

OMIT (APPEND)   Ignores the new macro when merging two libraries that contain macros with the same name.

OMIT (RENAME)   Exits the RENAME option without renaming either macro.

ON ARC   Creates point entities on a selected arc.

ON SPLN   Recreates node points in the form of point entities on selected splines.

ORDINAT   Creates and edits ordinate dimensions.

OUT   Displays dimension arrows pointing outward.

OUT/FIL   Changes the current status of polygons or closed polylines to outline or filled form.

OUTLINE   Assigns the active entity color to a selected polygon/polyline outline.

OUTPUT (CADL)   Writes entities and remarks to a CADL file.

PAN   Views any part of a drawing that extends beyond the limits of the screen by moving a view window over a desired part.

PARABOL   Creates a parabola with a specified directrix, focus, and start and end points.

PARALEL (ARR/WIT)   Creates witness lines or arrows parallel to a designated line.

PARALEL (DIMENSN)   Dimensions the distance between two parallel points.

PARALEL (LINE)   Creates a line parallel to an existing line.

PART   Contains basic information about the current working part or drawing in file form.

PART IN (POLYGON, WINDOW)   Selects the entities lying at least partly inside the polygon or window.

PART OUT (POLYGON, WINDOW)   Selects the entities lying at least partly outside the polygon or window.

PATTERN   Creates a part-independent file containing entities that can be added to an existing part file.

PEN #   Assigns a pen number to the current drawing for plotting purposes.

PEN # (masking)   Allows you to mask by pen number.

PERIM   Measures the actual and projected length of a single or selected group of entities.

PERPEND   Creates an arrow or witness line perpendicular to a line.

PLANE   Defines a plane using orientation and depth for selection.

PLANE (CONST)   Defines the construction plane.

PLOT (FILES)   Codes the current part or pattern for plotting.

PLOT (CONTROL)   Begins the plotting procedure.

± (Positive or Negative)   Displays dimensions followed by a plus-minus expression.

POINT   Indicates a point entity as the location for a position.

POINT (CREATE)   Creates point entities (+) on the screen.

POINTS   Creates point entities on selected entities at the point of intersection with a user-defined plane.

POLAR   Indicates a reference point using the X,Y view coordinates, along with the current depth for the X,Y,Z world coordinate position.

POLYGON   Creates a polygon entity with equal length sides.

POLYLINE   Creates a polyline entity.

POS+ENT   Calculates the distance between a position and a selected geometric entity.

POS TOL   Assigns positive tolerancing values.

POSITN (BREAK)   Divides a curve entity into two separate entities near a given position.

POSITN (DEPTH)   Indicates a depth position in 3D space.

POSITN (POINT)   Creates a point entity at an indicated position.

POSITN (TRM/EXT)   Allows you to trim or extend a curve entity to a position.

POSITN (VERIFY)   Displays the world, display view, and construction view coordinates for a position you wish to verify.

PREFIX   Allows you to select one of the standard prefix characters, input a character string, or input no characters when dimensioning.

PRINT   Sends a screen dump to printer.

PROJECT   Maps selected entities onto a specified plane, along a designated direction.

PRP PRP   Creates a line perpendicular to two selected entities.

PRP PT   Creates a line perpendicular to a selected entity and through a specified point.

PRP TAN   Creates a line perpendicular to a selected entity and tangent to another entity.

PT/LINE (PLANE)   Defines a selection plane using a point in space and two endpoints of a selected line entity.

PT/LINE   Defines a projection plane with a point in space and the endpoints of a selected line.

RADIUS   Dimensions the radius of a circle or arc in degrees or radians.

RECALL   Returns to the screen entities erased since the last REDRAW option was selected.

RECTANG   Creates a rectangle using four separate lines or one continuous polygon or polyline entity.

REDRAW   Refreshes the screen, removing existing screen symbols.

REMOVE (LEVEL)   Removes designated levels from the display.

REMOVE (ORDINAT)   Deletes unwanted ordinate dimensions.

REMOVE (VWPORTS)   Deletes existing viewport.

REN-NEW   When a macro is renamed or read in to the current library and its name matches an existing macro in the current library, this option can be used to enter a new name for the macro.

REN-OLD   When a macro is renamed or read in to the current library and its name matches an existing macro in the current library, this option can be used to rename the existing macro in the current library.

REPLACE   When a macro is renamed or read in to the current library and its name matches an existing macro in the current library, this option can be used to replace the existing macro in the current library.

RESIZE   Enlarges or reduces a selected viewport.

RETRIEV   Recalls a pattern file and adds it to the currently displayed part.

REVERSE   Changes the direction (normals) of a grouped polygon's mesh.

REVOLUT   Creates a mesh approximating the surface formed by rotating a generator curve about an axis by a specific angular distance.

ROTATE   Rotates entities about the axis pointing into the screen, at the X,Y view location specified.

ROT AXES   Rotates the view axes to desired view.

ROT PART   Rotates the actual part on the display.

ROTS   Defines the number of rotations in a helix curve.

ROUND   Draws an arc tangent to the endpoint of a previously created line or arc.

RULED   Creates a mesh that is an approximation to the surface formed by simultaneously moving a line (generator) along two director curves.

RUNNING (DIMENSN, SERIAL)   Creates a superimposed running dimension: a chainlike series of dimensions that have one reference point in common—the baseline point.

SAVE   Stores macro binding, part, and plot files under the specified filename.

SAVE (VIEW)   Stores current view.

SAVE TX (text editor)   Saves the edited text, and either redraws the edited drafting entity or prompts for a text position.

SCALE   Increases and decreases the size of your drawing by real coordinates.

SCREEN   Extends the length of lines to the limits of the screen.

2ND SOL   Displays the leader line underneath the dimension and extends to the second dimension point.

SECOND   Displays only the second of two witness or leader lines.

SECTION   Divides selected entities in a drawing at a defined plane for manipulating.

SEGMENT   Creates point entities based on segmenting a selected curve.

SELECT   Moves selected entities.

SERIAL   Creates linear dimensions (i.e., baseline, chain, superimposed running dimensions) on a series of specified points.

SET   Reassigns the modal parameters that are used to automatically dimension and label a drawing.

SET X,Y   Sets the spacing between grid points or the snap resolution along the X and Y axes.

SHELL   Enters the system environment without leaving the system environment.

SHRT R/D   Allows you to create radius or diameter dimensions with shortened leaders.

SINGLE   Single selects entities.

6 COEFF   Creates a conic section (i.e., ellipse, hyperbola, parabola) with specified values for the quadratic coefficients, and indicated positions for the start and end points.

SKETCH   Generates a continuous polyline.

SKEWED   Defines a direction that is not perpendicular to the specified projection plane.

SLANT   Allows you to slant text at an angle. The angle must be between −31° and +31°.

SNAP   Sets a resolution for cursor movement.

SNAP ALN   Aligns the snap resolution to the current grid position.

SNAP INC   Allows you to enter the X and Y coordinates of the snap resolution.

SNAP OPT   Turns snapping on and off.

SNAP=GRD   Sets the snap resolution to the current grid display.

SOLID   Displays one solid leader line (ISO/DIN).

SPHERE   Creates a mesh approximating a sphere created in one of the following ways: a) by specifying the center and radius of the sphere, as well as the number of latitude and longitude lines; or b) by specifying the center and diameter of the sphere, as well as the number of latitude and longitude lines.

SPLINE   Creates a smooth curve passing through a defined set of nodes, which has continuity of slope and curvature at all points.

SPLINE (EDIT)   Alters the shape of an existing spline.

STANDRD   Assigns whether trailing or leading zeros are displayed with a dimension, controls whether dimension text is aligned with leaders, and toggles between automatic dimensioning and manual entry of dimension text.

STATUS   Displays the amount of memory left for program usage (in bytes).

STRING   Draws continuous lines, polygons, polylines, arrows, or witness lines by requesting continuous endpoints.

STYLE   Alters or assigns user-defined arrowhead styles.

SYSCMD   Allows external or system processes to be run from inside the CADKEY environment.

TABLET   Assigns a macro to a specified column or row position on the tablet.

TAB CYL   Creates a mesh approximating the surface defined by sweeping a generator curve along a vector axis (director) of a specified length.

TANGENT (ARC, CIRCLE)   Creates an arc or a circle tangent to one, two, or three selected entities.

TANGENT (2D CUBI, 3D CUBI)   Calculates the X and Y vector components (determining the slope Y/X) and magnitude of the desired tangent vector for a spline.

TAN/PRP   Creates a line tangent or perpendicular to two selected entities or to an entity and a position.

TAN PRP   Creates a line tangent to a selected entity and perpendicular to another entity.

TAN PT   Creates a line tangent to a selected entity and through a specified point.

TAN TAN   Creates a line tangent to two selected entities.

TEXT   Alters the size, font, or alignment of displayed text.

TEXT (SYSCMD)   Executes a system command or program without exiting the program environment through the use of the system's graphics setup.

3 CURVE   Creates a mesh defined by either two generator curves and a director curve, or by one generator curve and two director curves. The mesh approximates the blending of the two two-curve surfaces.

3D AUTO   Creates a 3D spline through a path of minimum distance point entities.

3D CLSD   Creates a 3D cubic parametric spline that begins and ends at the first point selected, appearing as an unbroken loop.

3D CUBI   Creates a 3D cubic parametric spline.

3 ENTS   Creates an arc or a circle tangent to three selected entities.

THREE-P   Creates an arc or circle by specifying the start, middle, and endpoints.

3 PTS (PLANE)   Defines a selection plane using three positions in space.

3 PTS (PROJECT)   Defines a projection plane with three noncollinear points in space.

3 PTS (VIEW)   Defines a new display plane with three indicated points.

3PT 2SL   Creates a conic section based on indicated positions for three points on the entity, and the slope of the conic at two of the indicated points.

THRU PT   Creates a parallel line through a designated position on the screen.

TOLER   Specifies or changes the total amount by which an angular or linear dimension is permitted to vary.

TRAIL   Displays zeros in a dimension after the decimal point.

TRACKING   Displays current location of the cursor's coordinates.

TRANS-A   Translates or shifts entities from one location to another by referencing an old and new base position.

TRM/EXT (EDIT)   Trims or extends curve entities to projected intersections with other line and curve entities.

TRIM (FILLET)   Trims two selected entities to the intersections of the entities with a newly created arc fillet.

TRIM (SECTION)   Trims selected entities to their intersection with a user-defined plane.

2 CURVE   Creates a mesh defined by a selected generator and director curve. The mesh approximates a swept surface formed by moving (sweeping) the generator along the director curve.

2D AUTO   Creates a 2D cubic parametric spline through a path of minimum distance point entities.

2D CLSD   Creates a 2D cubic parametric spline that starts and ends at the first point selected, appearing as an unbroken loop.

2D CUBI   Creates a 2D cubic parametric spline.

2D/3D   Sets the construction mode to 2D or 3D.

2D/3D (EDIT)   Automatically converts a 2D spline to 3D form.

2 ENTS   Creates an arc or a circle tangent to two entities with a specified radius.

2 ENTS (DIST)   Calculates the distance between two geometric entities.

2 LINES (PLANE)   Defines a selection plane using the endpoints of two lines and their 3D intersection.

2 LINES (PROJECT)   Defines a projection plane with the endpoints of two selected lines and the 3D intersection (or projected 3D intersection) of the lines.

TWO-PTS   Creates a circle by specifying the diameter.

2 PTS   Positions a tangent vector at a spline's endpoint by indicating two points.

2 PTS (CONIC:5 COND)   Defines the slope of a conic at a selected point with the tangent vector that passes through two indicated positions (or projections of the positions).

2 PTS (PARABOL)   Defines the directrix of a parabola as a line with endpoints at two indicated positions.

2 PTS (PROJECT)   Defines a direction parallel to a line that joins two selected points in space.

2P 2S R   Creates a conic section with indicated start and end points, defined slopes at the start and end points, and a specified rho value.

2 POS   Calculates the distance between two positions.

TXT ATT (text attributes)   Changes font, text height, note angle, character aspect ratio of dimensions, notes, and labels. Mirrors selected notes.

TXT-IN   Reads in an ASCII text file and converts it to a macro.

TXT-OUT   Outputs a macro to an ASCII text file.

TXT POS   Changes text position.

TYPE   Assigns the type of tolerance used.

UNDERLN   Underlines each line of text, in dimensions, notes, and labels, from the first character to the end character.

UNITS   Specifies units for dimensioning other than the units used when creating the part.

UPDATE   Updates certain attributes to the current modal values.

USER   Assigns a user-defined scale factor for the current part units.

VALUE   Assigns a value for the current working depth.

VALUES   Changes tolerance values on a dimension which already has a tolerance assigned to it.

VER DIM   Turns the YES/NO menu displayed after each dimension on or off.

VERIFY   Displays the area, attributes, moment of inertia, coordinate data, angles, perimeter, and distance information about selected entities.

VERSEL   Sets the status of the selection verification switch.

VERIFY (CONIC)   Specifies the type of conic selected and displays detailed information relevant to that conic type.

VERTICL (DIMENSN)   Displays the vertical dimension between two points.

VERTICL (SET)   Aligns position for text.

VIEW (COORDS)   Allows you to construct a part in view coordinates.

VIEW (DISPLAY)   Defines or changes the display view.

VIEWPORT   Limits the length of line entities to viewport boundaries.

VWPORTS   Allows you to add more than one viewport to your display.

VW/DPTH   Defines a selection plane using a specified view and depth.

VW/DPTH (PROJECT)   Defines a projection plane with a specified view and depth.

WID/HT   Constructs a rectangle using given values for width and height.

WINDOW   Produces rubberbox that can be shrunk or enlarged to fit around a specific entity or area of your drawing for selecting, editing, or transforming purposes.

WIT/LDR   Specifies or changes the witness lines or leader lines for display.

WITNESS   Creates two lines that extend from the points being dimensioned; used as a drafting entity.

WITNESS (CHANGE/ SET)   Specifies or changes the witness lines for display (i.e., both, none, first, last).

WORLD (COORDS)   Displays the world icon, which identifies the orientation of the world coordinate axes relative to the currently displayed view.

X-FORM (Transform)   Performs coordinate changes on a drawing or its selected parts.

X-HATCH   Fills in a specified, closed-in area with selected line patterns.

X-Y-Z   Moves or modifies selected entities using the X, Y, and Z delta view coordinates assigned.

0 REP   Controls the display of zero tolerancing.

ZEROS   Controls the display of zeros in dimensions.

ZOOM   Displays a close-up view of your drawing for examination and to add finer detail.

Active area   The block on a tablet that returns coordinate data.

Active instance   The instance that is currently active in a drawing layout. You know an instance is active when CADKEY surrounds it with a dashed, highlighted box. Only one instance can be active at a time.

Address   A memory location where data can be stored.

Addressable memory   Memory usable by an application.

Alphanumeric   A name made up of letters and numbers.

ANSI   The acronym for American National Standards Institute.

Arc   An open curve.

Arrowhead   Arrowheads show the beginning and/or end of the dimension line. You can select closed- or open-style arrowheads.

ASCII   An acronym for American Standard Code for Information Interchange. A seven-bit standard code adopted to facilitate the interchange of data among various types of data processing and data communications equipment.

Associativity   The relationship between a drawing instance in Drawing Layout Mode and the original part file in Model Mode, and also the relationship between an instance and its dimensions in DLM.

Attribute   The line or color characteristics assigned to an entity or group of entities.

AUTOEXEC.BAT   A DOS file that you can create to perform special start-up procedures each time you start your computer.

Axonometric view   A projection that appears inclined with three sides showing and horizontal and vertical distances drawn to scale.

Base plane   The plane normal to the scaling direction, which contains the base position.

Baud rate   A unit for measuring data transmission speed. One baud is one bit per second.

Bind   When you bind a macro to your keyboard, you assign it to a specific set of keys. Whenever you press these keys in the correct combination, CADKEY starts the macro.

Binding   Holds information on key and tablet areas assigned to a macro.

BIOS   The acronym for Basic Input/Output System. A small program that resides on a ROM chip and that acts as the basic interface between your applications and hardware (printers, plotters, and so forth).

Blanking   To "hide" an instance-specific entity in DLM so that it disappears from view.

Booting   Starting the system.

Buffer   A temporary storage area.

Cabling diagram   A diagram showing connections and physical locations of system or unit cables, and used to facilitate field installation and repair wiring systems.

CAD   The acronym for Computer-Aided Drafting or Computer-Aided Design. Using CAD, you design a graphic image on the screen, and store information about the image in a database. Working with CAD software, a computer or workstation, an input device, such as a mouse or tablet, and a plotter or printer, you translate you design into a drawing. Different applications can share the information in the CAD database.

CAD/CAM   The acronym for Computer-Aided Design/Computer-Aided Manufacturing. CAD/CAM combines the design and manufacturing process into one process. Combining CAD and CAM is accomplished by computers and software sharing information from the CAD system's database.

CADD   Computer-Aided Design and Drafting. *See* CAD.

CADL   The acronym for CADKEY Advanced Design Language. Use CADL to take CADKEY part file information and import it to other tools such as SOLIDS or as output from other tools into a CADKEY part file.

CADUTIL   A utility program that lets you convert part and pattern files to be compatible with the most recent version of CADKEY.

CAE   The acronym for Computer-Aided Engineering.

CAM   The acronym for Computer-Aided Manufacturing. It combines CAD with manufacturing processes. CAM uses computer-controlled machines and robotics for assembly, material handling, measuring, and inspection. Some of the information for the manufacturing processes can be gleaned from the CAD database.

Cartesian coordinates   x, y, z absolute coordinates.

CDE   The acronym for CADKEY Dynamic Extensions.

Centroid   The center of mass in an object.

Chain dimension   A series of dimension text that you place by selecting points to dimension.

Chamfer   A cut on a corner or edge of an object.

CIM   The acronym for Computer-Integrated Manufacturing. This is total automation of the manufacturing process—from concept to finished product—ready for shipment.

Circle   A closed curve.

Click on   The process of pressing the mouse button.

Colinear points   Points that lie on the same line.

Color palette   A selection of colors provided by the program.

Compression   A way of compacting data in order to decrease the disk space needed to store it. Compressed data cannot be used until it is decompressed. Your CADKEY 386 disks are shipped to you compressed. They are automatically decompressed during installation.

CONFIG.SYS   A system file created in DOS.

Configuration   The assigned hardware and computer system setup for a particular period of operation.

Conic   The intersection of a plane with a pair of right circular cones that have the same axis and intersect only at the tips. Depending on how the plane intersects the cones, an ellipse, a hyperbola, a parabola, or a degenerate conic is formed. The following equation is the quadratic representation of a conic: $A*x2 + B*x*y + C*y2 + D*x + E*y + F = 0$

Conventional memory   The first 640K of memory in your computer. The memory that can be directly accessed by DOS (and by applications that run under DOS). All DOS-based personal computers have conventional memory.

Cursor   An on-screen indicator that matches your mouse movement. Wherever you move your mouse or digitizer, the cursor matches the moves on screen. CADKEY uses a cross-hair cursor, +. Also, a symbol (+) used to indicate the active position in 3D space or to select entities or their parts.

Cursor button   A button assigned on the digitizer stylus or mouse which denotes a selection or position when pressed.

Cursor control   Part of an input device (digitizer or mouse) used for selection and position.

Cursor indicate   To define a position or location with the cursor.

Cursor select   To select an option, setting, entity, or position using the cursor.

Cursor tracking   Allows you to continuously display the cursor's x, y coordinates.

Data primitive   An entity that a graphics program can draw, store, and manipulate.

Database management   Defines the structure of data for accessing, entering, and deleting information.

Decompression   The method of expanding compressed data and programs so that they can be read. Your CADKEY 386 files are automatically decompressed during installation.

Default   A preset value that the system uses whenever you do not specify some other value. As an example, during installation the default source and target drives are A> and C>, respectively.

Delimiter   A special character that separates parts in a set of data.

Detailing   Adding dimensions, cross-hatching, notes, labels, or pointers to the drawing. Detailing adds to the impact of the drawing—taking it from a series of lines to an integrated whole that gives you the big picture at a glance.

Dialog box   In a graphical user interface, a special window displayed by the system or application to solicit a response to various options available from the user.

Diameter   The distance from one side of a circle through the center to the other side.

Digitizer   A tablet or pad which converts graphic information into digital values.

Digitizing area   Block where the graphics cursor returns coordinate data or menu/status selections.

Digitizing mode   To convert a desired part to digital form.

Director curve   In meshing, a director curve determines the direction of the sweep.

Disassociation   What happens when CADKEY no longer recognizes the relationship or link between a part file in Model Mode and instance in Drawing Layout Mode, or between an instance and its dimensions in DLM.

Disk drive specifier   The drive where you store files, or from where you retrieve files.

DOS   The acronym for Disk Operating system.

Drafting entity   Type of entity used in dimensioning that appears in the view of creation; for example, witness lines, arrows, labels, notes.

Drawing instances   Different views of the part file you created in Model Mode that you position in a drawing layout. You can place up to 200 drawing instances in a single drawing layout. Drawing instances are also known as instances.

Drawing layout   An area on your screen comparable to a flat piece of paper. You define the paper boundaries and can then position different views, or drawing instances, of a part file for plotting and printing as a layout. Drawing layouts are also known as layouts.

Drawing Layout Mode   The CADKEY system mode that you enter when you've finished creating a part file in Model Mode and want to detail and position different views of the part file in a layout format. Drawing Layout Mode is also known as DLM.

Ellipse   A set of points within a plane whose sum of the distance between one fixed point to any two other fixed points is constant. The fixed points are called *foci*. A conic is not an ellipse unless the constant is greater than the distance between the fixed points. If the fixed points are the same, the ellipse is a circle.

Entity   Basic individual drawing component. CADKEY entities include circles, lines, arcs, points, conics, polylines, polygons, and fillets.

Extended memory   On 386 systems, memory above one megabyte.

Extension   An addition to the filename; in effect, the file's last name. CADKEY uses a three-letter extension that identifies the file type. Filename extensions start after the separator period. For example, *prt* is the extension for part files.

File   A collection of data, in any form, that is stored.

Filename   The name you give to a file. For example, in DOS, filenames can be up to eight characters long. A period separates the filename from its three-character extension. Different operating systems use different filenaming conventions, so check your system's operation manual for information on filenames.

File locking   For network users, a way to lock a file by a user name. File locking prevents multiple users on a network from accessing the same part file simultaneously and making changes to it. File locking does not affect single-user systems.

Fillet   A rounded interior corner.

Fixed instance   A drawing instance that remains stationary and acts as a reference when aligning instances in DLM.

Foci   Fixed points.

Font   The style of text.

Function keys   A set of twelve keys located on the keyboard. These keys allow you to choose menu options displayed by pressing the assigned function key.

Generator curve   In meshing, the contour that CADKEY sweeps along the director curve.

Geometric entity   Components of the drawing itself; for example lines, points, and arcs.

Ghost box   A highlighted box that surrounds an active instance when you use the INSTANCE, CREATE or LOCATE options in DLM. The ghost box automatically scales to the minimum and maximum boundaries for each instance.

Gnomon   x, y, z, world coordinate direction icon.

Grid   A set of points that crisscross at 90 degree angles.

Grouping   A collection of entities.

Hatch lines   Each of the lines in a cross-hatch pattern.

Hexadecimal   A number associated with the base 16 number system. The base 16 number system uses 0–9 and letters a–f. A–f is equivalent to the base 10 system's 10–15. Hexadecimal numbers grow less quickly than decimal numbers because they are based on powers of 16 and not powers of 10.

Hidden line removal   Removes lines from an object that you could not or would not see if the object were solid.

Hyperbola   A set of points in a plane such that the difference of the distances from any point in the set to two fixed points—the *foci*—is a constant. A conic is not a hyperbola unless the constant is less than the distance between the foci.

Icon   A small on-screen symbol that simplifies access to a program, command or data file.

Immediate mode command Specified functions that may be invoked at any place in the menu structure.

Initialize To start for the first time.

Input device A device that transfers data or signals into a processor system, for example, a tablet or a mouse.

Instance-specific entities Any entity in a layout that represents an entity that belongs to the original part file in Model Mode. Drawing instances consist of instance-specific entities until you modelize them.

Island An area inside another area.

ISO An acronym for International Standards Organization.

Layout-specific entities Entities in Drawing Layout Mode that are not associated to the original part file in Model Mode. When you return to Model Mode, you do not see any layout-specific entities that you create in DLM. When you modelize instance-specific entities, they become layout-specific entities.

Leader lines Thin solid lines that show the direction or extent of dimensions. A leader line is usually broken for the placement of a dimension.

Line An entity that you create when you connect two endpoints.

Line-spacing factor The ratio between the current text height and the line spacing.

Magnitude As used in a vector, the length of the line segment. The higher the magnitude value, the more the slope vector affects the spline's shape.

Masking Allows you to screen or designate certain entity or attribute types when making a selection.

Math coprocessor A chip that takes over most mathematical calculations, freeing the main processor for other tasks.

Memory resident A program that stays permanently in your system's main memory.

Model Mode CADKEY's default system mode and regular construction environment. Model Mode is also known as MM.

Modelize To change entities, instances, or layouts from instance-specific to layout-specific geometry using the MODELIZE option in DLM. When you change something from instance-specific to layout-specific geometry, it loses its link, or association, to the original part file in Model Mode.

NC The acronym for Numerical Control, which is basically computer-controlled milling.

N-gon A polygon with three or more sides.

Nodes Also known as *knot points,* a set of points through which you create a spline.

Numeric keypad A set of keys found to the right of the keyboard which are used to direct the cursor on the screen (up, down arrows, and so on).

Numeric keys The number keys 1-0 located on the top row of the keyboard.

Off-line program A program which is initiated in a DOS environment rather than from within another program.

On-line calculator Assigns variables and evaluates algebraic expressions according to given syntax rules.

Ordinate dimension A dimension that uses a surface from which to measure dimensions.

Origin The intersecting point of coordinate axes.

Orthogonal Intersecting or lying at right angles.

Parabola A set of points in a plane, whose distance from any point in the set to another given point—the focus—is equal to the distance from that point to a given line—the directrix.

Parity An error checking procedure used for checking data transferred within a computer or between computers.

Part A complete file that contains all the information about a drawing's views, entity attributes, scale, and so on; sometimes called a drawing.

Part file What CADKEY uses to store all the information about your part or drawing. CADKEY creates it with a .prt extension.

Pathname The location of a directory or file within the system. Specify a pathname by typing the drive letter, followed by a directory name, one or more subdirectories, if needed, and a filename. A backslash separates each name. The following example shows a typical DOS pathname command sequence: C:\CADKEY\PRT\TRUCKS.PRT.

Pattern A component file that contains only entity information, independent of a part file.

Pattern file What CADKEY uses to store entities from a part file that can be merged into other part files.

Pen velocity The speed at which the pen moves; usually associated with a plotter.

Peripherals An input or output device connected to a computer.

Perpendicular Being at a right angle to a given plane or line.

PLOTFAST A program which allows you to acquire pen plots from existing plot files.

Plotter A device used to draw charts, diagrams, and other line-based graphics. Pen plotters draw on a paper or transparencies with one or more colored pens. Electrostatic plotters draw a pattern of electrostatically charted dots on the paper and then apply toner and fuse it in place.

Point A location without any dimension. CADKEY represents a point as a small plus sign (+).

Polar reference axis A vector pointing out of the screen and passing through the projection of the selected point to the current view.

Polygon Any closed figure with sides consisting of straight lines.

Printer A computer peripheral that puts text or a computer-generated image on paper or on another medium such as a transparency.

Projected intersection The intersection at which two lines meet as if they were extended.

Prompt Instructions or information displayed across the bottom of the screen.

Protected mode The mode built into the 80386 or 40386 chip that allows it to run multiple applications and sessions without conflict.

Pull-down menu A menu that appears when you cursor select certain options. Pull-down menus remain open until you exit them.

Radius The distance from the center of an arc or circle to any point on the circumference.

RAM disk A part of memory that the user defines as a logical disk drive (an area of RAM that you can use just like a hard disk or a floppy disk). When using a RAM drive, remember to store your data to a physical drive (the hard disk or a floppy disk) before turning off your computer.

Real mode The default memory mode used by DOS and its applications. Real mode limits the system to running one application at a time, and memory addressable to less than one megabyte.

Reference point A designated point to which a function is applied.

Register manager Stores and retrieves variables for use in the system's on-line calculator.

Rubberband An "elastic" line attached to the cursor, which stretches when the cursor is moved.

Rubberband box A flexible box that attaches to your cursor in different functions. You can stretch and shrink the box and pull it around different entities. By completely surrounding an object with a rubberband box, you can often move the object by dragging it to a new place.

Scale factor A value by which you divide or multiply the displayed part for scaling purposes.

Section A profile of a part cut through by an intersecting plane.

Segment Defines the section of a spline between two nodes. CADKEY stores the segments in parametric cubic form—that is, x, y, and z are functions of the same parameter value—with parameter values ranging from 0 to 1 inclusive, where t = 0 defines the first node, and t = 1, the second.

Snap increment Controls the horizontal and vertical movement of the cursor.

Soft clip limits Setting soft clip limits lets you maximize the plotting area. Soft clip limits involve the area of the paper on which the plotter cannot plot because of hardware constraints such as pinch rollers. Check your plotter manual for more information on clip limits.

Spline looping A spline that intersects itself.

Stylus A pointing device used with a digitizer to specify position and location.

Subgroup A subdivision of a group.

Subordinates Ordinates you create after the base ordinate.

Tangent vector A directed line segment that CADKEY can translate in 3-D coordinates to intersect with exactly one point on a given segment of a spline.

Toggle Turns a function on or off by the press of a key.

Tolerancing The total amount which a dimension is allowed to vary.

Transformation Changes the form, appearance, or placement of an entity.

TSR The acronym for Terminate and Stay Resident. A program, driver, utility, or software tool, which—when you exit it—stops running, terminates, but stays in memory, resident, so that you can call it up by selecting an option, pressing a hot key, or typing a command.

Unblank To redisplay an entity that you hid from view.

VCPI   The acronym for Virtual Control Program Interface; an interface designed by Phar Lap Software and Quarterdeck Office Systems. This interface lets 80386 control programs coexist on a single system.

Vector   A directed line segment.

View coordinates   The x, y, z axis is relative to the screen, where x is horizontal, y is vertical, and z is pointing out from the screen.

View dependent   The current part is not visible in any other view or perspective.

Viewport   The space on your screen that displays your drawing. It can vary in size and shape. Your part and all of its pieces cannot extend past the boundaries of the viewport.

Witness lines   Thin solid lines that project from a dimensioned object to indicate the extent of the leader lines.

World coordinates   The standard Cartesian coordinate system consisting of an x, y, z axis relative to view 1 in the system, a top view. The 0,0 coordinates are located in the bottom left corner of the screen.

Order Form

Sharpen your CADKEY skills with
The CADKEY Electronic Drafting Workbook Disk

Now available for your CADKEY workstation, *The CADKEY Electronic Drafting Workbook Disk* is full of CADKEY part files and additional drafting practice problems. Great for use with *Using CADKEY 5 & 6*, or as a stand-alone practice disk at your workstation, *The CADKEY Electronic Drafting Workbook Disk* can help you acquire the skills you need in today's high-tech CAD world.

To Order: Simply fill in the following information and mail this form today.
Or, call 1-800-347-7707 and press "1" for customer service.

Order #	Item	Qty.	List Price*	Subtotal
0-8273-6525-X	CADKEY Electronic Drafting Workbook Disk		$12.95 (net)	

Prices subject to change without notice.
Prices may differ outside the continental United States.

Sub-total	
Add State, Local Taxes	
Total Order	

Check one:

❏ Charge my (circle one): VISA or MasterCard
Card # _____ Exp. Date _____ Signature _____

❏ A check is enclosed for $ _____ (I have included the appropriate sales tax or attached a tax exempt certificate.)

❏ Enclosed is my purchase order number: _____

Ship To:

Name _____

Street Address _____

_____

City/State/Zip _____

Phone (home) _____ (office) _____

Signature _____

Bill To: (If different than "Ship To")

Name _____

Street Address _____

_____

City/State/Zip _____

Phone (home) _____ (office) _____

❏ Please have a Delmar representative contact me.

DO 819

Return To:

United States
Delmar Publishers Inc.
3 Columbia Circle
P.O. Box 15015
Albany NY 12212-5015
1-800-347-7707 • 518-464-3500
Fax 518-464-0301

Europe, Middle East, Africa
International Thomson Publishing
Berkshire House
168-173 High Holborn
London WC1V 7AA, UK
44-71-497-1422 • Fax 44-71-497-1426

Canada
Nelson Canada
1120 Birchmount Road
Scarborough, Ontario M1K 5G4, Canada
416-752-9100 • Fax 416-752-9646

Asia Pacific & Hawaii
International Thomson Publishing Asia
38 Kim Tian Road, #01-05
Kim Tian Plaza
Singapore 0316
2-272-6497 • Fax 2-272-6498

Latin America, Puerto Rico
Thomson International Publishing
20 Park Plaza
Boston, MA 02116
617-423-4210 • Fax 617-423-4325

Australia/New Zealand
Thomas Nelson Australia
102 Dodds Street
South Melbourne 3205
Victoria, Australia
61 3 685-4111 • Fax 61 3 685-4199

Japan
International Thomson Publishing
Kyowa Building, 3rd Floor
2-2-1 Kirakawacho Chiyoda-Ku
Tokyo 102 Japan
81-33-221-1385 • Fax 81-33-237-1459